T0091863

METHODS IN MOLECULAR BIOLOGY™

Series Editor
John M. Walker
School of Life Sciences
University of Hertfordshire
Hatfield, Hertfordshire, AL10 9AB, UK

For further volumes:
http://www.springer.com/series/7651

Chromatin Remodeling

Methods and Protocols

Edited by

Randall H. Morse

Wadsworth Center, New York State Department of Health, Albany, NY, USA

Editor
Randall H. Morse
Wadsworth Center
New York State Department of Health
Albany, NY, USA
morse@wadsworth.org

ISSN 1064-3745 e-ISSN 1940-6029
ISBN 978-1-61779-476-6 e-ISBN 978-1-61779-477-3
DOI 10.1007/978-1-61779-477-3
Springer New York Dordrecht Heidelberg London

Library of Congress Control Number: 2011943306

Printed on acid-free paper

Humana Press is part of Springer Science+Business Media (www.springer.com)

Preface

No modern molecular biologist needs convincing of the central importance of chromatin to gene regulation in eukaryotes. From large-scale domains to individual gene promoters, gene expression is regulated by histone modifications and histone variants, nucleosome positioning, and nucleosome stability. A panoply of approaches has evolved in the laboratory over the past three decades to study chromatin structure and its alterations, and methods of investigating chromatin remodeling – changes in nucleosome structure or position with respect to the incorporated DNA, or in histone modifications – have progressed rapidly over the past 10 years. Here are presented a wide array of protocols for studying chromatin remodeling. We include methods for investigating chromatin remodeling in vitro and in vivo, in yeast, plants, and mammalian cells, and at local and global levels. Both gene-specific and genome-wide approaches are covered, and in recognition of the increasing prevalence of the latter type of study, the final two chapters focus on bioinformatic/computational approaches to analyzing genome-wide data on chromatin structure.

We hope that this volume will be used by readers in more than way. The obvious utility is as a direct guide to a multitude of techniques, served up as recipes with introductions and special "Notes" sections. These latter sections, which provide extra tips and background for the user, have been an especially appreciated feature of these volumes over the years. We also hope the chapters may be read comparatively and have allowed some overlap among approaches to this end, to permit the reader to benefit from delving into differences as to how different authors approach similar problems. Finally, simply perusing the titles may refresh or illuminate readers as to the variety of methods available to study chromatin remodeling and perhaps invigorate their next grant application or project.

This volume has benefited from a group of authors with tremendous collective experience and authority, who have generously given time and effort to provide these detailed explications of protocols. I am grateful for their terrific cooperation in this project.

Albany, NY, USA *Randall H. Morse*

Contents

Contributors

BUNGO AKIYOSHI • *Basic Sciences Division, Fred Hutchinson Cancer Research Center, Seattle, WA, USA*
TREVOR K. ARCHER • *Laboratory of Molecular Carcinogenesis, National Institute of Environmental Health Sciences, National Institutes of Health, Research Triangle Park, NC, USA*
SONGJOON BAEK • *Laboratory of Receptor Biology and Gene Expression, National Cancer Institute, National Institutes of Health, Bethesda, MD, USA*
BLAINE BARTHOLOMEW • *Department of Biochemistry and Molecular Biology, Southern Illinois School of Medicine, Carbondale, IL, USA*
PETER B. BECKER • *Adolf-Butenandt Institute, Ludwig-Maximilians-Universität, Munich, Germany*
SUE BIGGINS • *Basic Sciences Division, Fred Hutchinson Cancer Research Center, Seattle, WA, USA*
JEF D. BOEKE • *High Throughput Biology Center, Johns Hopkins University School of Medicine, Baltimore, MD, USA*
ANDREW BOWMAN • *The Wellcome Trust Biocentre, University of Dundee, Dundee, Scotland, UK*
GENE O. BRYANT • *Molecular Biology Program, Memorial Sloan Kettering Cancer Center, New York, NY, USA*
STEPHANIE BYRUM • *Department of Biochemistry and Molecular Biology, University of Arkansas for Medical Sciences, Little Rock, AR, USA*
KAIRONG CUI • *Laboratory of Molecular Immunology, National Heart, Lung and Blood Institute, National Institutes of Health, Bethesda, MD, USA*
JUNBIAO DAI • *School of Life Sciences, Tsinghua University, Beijing, China*
RUSSELL P. DARST • *Department of Biochemistry and Molecular Biology, University of Florida and Shands Cancer Center, University of Florida College of Medicine, Gainesville, FL, USA*
ANN DEAN • *Laboratory of Cellular and Developmental Biology, National Institutes of Health, Bethesda, MD, USA*
TANJA DURBIC • *The Donnelly Centre for Biomolecular Research, University of Toronto, Toronto, ON, Canada*
VAMSI K. GANGARAJU • *Yale Stem Cell Institute, New Haven, CT, USA*
MARC R. GARTENBERG • *Department of Pharmacology, Robert Wood Johnson Medical School, University of Medicine and Dentistry of New Jersey, Piscataway, NJ, USA; The Cancer Institute of New Jersey, New Brunswick, NJ, USA*
MARINELLA GEBBIA • *The Donnelly Centre for Biomolecular Research, University of Toronto, Toronto, ON, Canada*
DANIEL GINSBURG • *Department of Biomedical Sciences, Long Island University, Brookville, NY, USA*

CHHABI K. GOVIND • *Department of Biological Sciences, Oakland University, Rochester, MI, USA*
GORDON L. HAGER • *Laboratory of Receptor Biology and Gene Expression, National Cancer Institute, National Institutes of Health, Bethesda, MD, USA*
JEFFREY C. HANSEN • *Department of Biochemistry and Molecular Biology, Colorado State University, Fort Collins, CO, USA*
JEFFREY J. HAYES • *Department of Biochemistry and Biophysics, University of Rochester Medical Center, Rochester, NY, USA*
STEVEN HENIKOFF • *Division of Basic Sciences, Fred Hutchinson Cancer Research Center and Howard Hughes Medical Institute, Seattle, WA, USA*
ALAN G. HINNEBUSCH • *Laboratory of Gene Regulation and Development, Eunice Kennedy Shriver National Institute of Child Health and Human Development, National Institutes of Health, Bethesda, MD, USA*
JAE-WAN HUH • *Department of Biochemistry and Molecular Biology, University of Ulsan College of Medicine, Seoul, South Korea*
JUAN JOSE INFANTE • *Bionaturis, Seville, Spain*
TATIANA S. KARPOVA • *Laboratory of Receptor Biology and Gene Expression, National Cancer Institute, National Institutes of Health, Bethesda, MD, USA*
CHRISTINE M. KIEFER • *Laboratory of Cellular and Developmental Biology, National Institutes of Health, Bethesda, MD, USA*
MICHAEL P. KLADDE • *Department of Biochemistry and Molecular Biology, University of Florida and Shands Cancer Center, University of Florida College of Medicine, Gainesville, FL, USA*
PHILIPP KORBER • *Molecular Biology Unit, Adolf-Butenandt-Institut, University of Munich, Munich, Germany*
VERANDRA KUMAR • *National Botanical Research Institute, Council of Scientific and Industrial Research, Rana Pratap Marg, Lucknow, UP, India*
SIGNE LARSON • *Department of Biochemistry and Molecular Biology, University of Arkansas for Medical Sciences, Little Rock, AR, USA*
G. LYNN LAW • *Department of Microbiology, University of Washington, Seattle, WA, USA*
BING LI • *Department of Molecular Biology, UT Southwestern Medical Center, Dallas, TX, USA*
NING LIU • *Department of Biochemistry and Biophysics, University of Rochester Medical Center, Rochester, NY, USA*
NIRAJ LODHI • *National Botanical Research Institute, Council of Scientific and Industrial Research, Rana Pratap Marg, Lucknow, UP, India*
VERENA K. MAIER • *Department of Molecular Biology, Massachusetts General Hospital, Department of Genetics, Harvard Medical School, Boston, USA*
DAVIDE MAZZA • *Laboratory of Receptor Biology and Gene Expression, National Cancer Institute, National Institutes of Health, Bethesda, MD, USA*
JAMES G. MCNALLY • *Laboratory of Receptor Biology and Gene Expression, National Cancer Institute, National Institutes of Health, Bethesda, MD, USA*
FLORIAN MUELLER • *Groupe Imagerie et Modélisation, Institut Pasteur, CNRS, URA 2582, Paris, France*

NANCY H. NABILSI • *Department of Biochemistry and Molecular Biology, University of Florida and Shands Cancer Center, University of Florida College of Medicine, Gainesville, FL, USA*
GEETA J. NARLIKAR • *Department of Biochemistry and Biophysics, University of California, San Francisco, CA, USA*
COREY NISLOW • *The Donnelly Centre for Biomolecular Research, University of Toronto, Toronto, ON, Canada*
TOM OWEN-HUGHES • *The Wellcome Trust Biocentre, University of Dundee, Dundee, Scotland, UK*
CAROLINA E. PARDO • *Department of Biochemistry and Molecular Biology, University of Florida and Shands Cancer Center, University of Florida College of Medicine, Gainesville, FL, USA*
CHUONG D. PHAM • *AstraZeneca R&D Boston, Waltham, MA, USA*
SANTHI PONDUGULA • *Department of Biochemistry and Molecular Biology, University of Florida and Shands Cancer Center, University of Florida College of Medicine, Gainesville, FL, USA*
KRISHAN MOHAN RAI • *National Botanical Research Institute, Council of Scientific and Industrial Research, Rana Pratap Marg, Lucknow, UP, India*
AMOL RANJAN • *National Botanical Research Institute, Council of Scientific and Industrial Research, Rana Pratap Marg, Lucknow, UP, India*
CHUN RUAN • *Department of Molecular Biology, UT Southwestern Medical Center, Dallas, TX, USA*
SAMIR V. SAWANT • *National Botanical Research Institute, Council of Scientific and Industrial Research, Rana Pratap Marg, Lucknow, UP, India*
GAVIN R. SCHNITZLER • *Molecular Cardiology Research Institute, Tufts Medical Center, Boston, MA, USA*
TINA SHAHIAN • *Department of Biochemistry and Biophysics, University of California, San Francisco, CA, USA*
HILLEL I. SIMS • *Department of Biology, Rosenstiel Basic Medical Sciences Research Center, Brandeis University, Waltham, MA, USA*
MALA SINGH • *National Botanical Research Institute, Council of Scientific and Industrial Research, Rana Pratap Marg, Lucknow, UP, India*
SUNIL KUMAR SINGH • *National Botanical Research Institute, Council of Scientific and Industrial Research, Rana Pratap Marg, Lucknow, UP, India*
SHERRI K. SMART • *Department of Biochemistry and Molecular Biology, University of Arkansas for Medical Sciences, Little Rock, AR, USA*
TIMOTHY J. STASEVICH • *Graduate School of Frontier Biosciences, Osaka University, Suita, Osaka, Japan*
MYONG-HEE SUNG • *Laboratory of Receptor Biology and Gene Expression, National Cancer Institute, National Institutes of Health, Bethesda, MD, USA*
HEATHER J. SZERLONG • *Department of Biochemistry and Molecular Biology, Colorado State University, Fort Collins, CO, USA*
ALAN J. TACKETT • *Department of Biochemistry and Molecular Biology, University of Arkansas for Medical Sciences, Little Rock, AR, USA*
SHEILA S. TEVES • *Division of Basic Sciences, Fred Hutchinson Cancer Research Center and Molecular and Cellular Biology Program, University of Washington, Seattle, WA, USA*

ITAY TIROSH • *Department of Molecular Genetics, Weizmann Institute of Science, Rehovot, Israel*
ILA TRIVEDI • *National Botanical Research Institute, Council of Scientific and Industrial Research, Rana Pratap Marg, Lucknow, UP, India*
KEVIN W. TROTTER • *Laboratory of Molecular Carcinogenesis, National Institute of Environmental Health Sciences, National Institutes of Health, Research Triangle Park, NC, USA*
KYLE TSUI • *The Donnelly Centre for Biomolecular Research, University of Toronto, Toronto, ON, Canada*
TOSHIO TSUKIYAMA • *Basic Sciences Division, Fred Hutchinson Cancer Research Center, Seattle, WA, USA*
ASHWIN UNNIKRISHNAN • *Basic Sciences Division, Fred Hutchinson Cancer Research Center, Seattle, WA, USA*
CHRISTIAN J. WIPPO • *Molecular Biology Unit, Adolf-Butenandt-Institut, University of Munich, Munich, Germany*
ELTON T. YOUNG • *Department of Biochemistry, University of Washington, Seattle, WA, USA*
MIYONG YUN • *Department of Molecular Biology, UT Southwestern Medical Center, Dallas, TX, USA*
KEJI ZHAO • *Laboratory of Molecular Immunology, National Heart, Lung and Blood Institute, National Institutes of Health, Bethesda, MD, USA*

Chapter 1

Strain Construction and Screening Methods for a Yeast Histone H3/H4 Mutant Library

Junbiao Dai and Jef D. Boeke

Abstract

A mutant library consisting of hundreds of designed point and deletion mutants in the genes encoding *Saccharomyces cerevisiae* histones H3 and H4 is described. Incorporation of this library into a suitably engineered yeast strain (e.g., bearing a reporter of interest), and the validation of individual library members is described in detail.

Key words: Histones modification, Chromatin remodeling, High-throughput mutagenesis

1. Introduction

The eukaryotic genome consists of hundreds of thousands of its organizational building blocks, the nucleosomes, which are arrayed in a highly condensed form to allow the genome to fit into the nucleus. Each nucleosome is formed by wrapping a histone octamer with 146–147 base pairs of double-strand DNA. While packaging of DNA into chromatin is necessary to fit the genome into a tight space, it also forms a potent obstacle for transcription, DNA replication, recombination, repair, and other dynamic processes. Biological evidence has shown that DNA can however be rendered accessible by at least three basic mechanisms including posttranslational modifications (PTMs) of histone proteins, incorporation of histone variants into nucleosomes, and ATP-dependent chromatin remodeling.

Numerous covalent PTMs have been identified at the N-terminal flexible "tail" domains of histones and within the structured core regions (1, 2), which can be added or removed at specific locations

Randall H. Morse (ed.), *Chromatin Remodeling: Methods and Protocols*, Methods in Molecular Biology, vol. 833,
DOI 10.1007/978-1-61779-477-3_1, © Springer Science+Business Media, LLC 2012

in chromatin in response to biological cues. This dynamic alteration of nucleosome composition underlies the ability of histones to carry out specific roles, directly or indirectly, through the recruitment of accessory factors, highlighting the mechanistic significance of modifiable histone residues within the nucleosome. For example, methylation on histone H3 lysine 79 leads to loss of Sir3p binding to nucleosomes, leading to general transcriptional competence (3, 4). Additionally, a growing body of evidence indicates that cellular processes may be regulated redundantly or cooperatively by multiple modifications. For example, the acetylation of histone H4 tail and the trimethylation of histone H3K4 are both intimately associated with transcription initiation (5, 6) whereas the repair of double strand breaks requires both H4 S1 phosphorylation and H3 K79 methylation (7, 8) indicating that combinations of modifications on histone residues may also contribute to specific cellular processes.

In addition to PTMs of histone proteins, histone variants can also be incorporated into specific nucleosomes in chromatin to replace the canonical histones to carry out specialized functions. For example, replacement of histone H3 by its variant Cse4 marks the centromeric *Saccharomyces cerevisiae* nucleosome (9), which leads to the association of multiple proteins to form the kinetochore in budding yeast. In another case, studies from multiple organisms including yeast (10), *Drosophila* (11), *Caenorhabditis elegans* (12), and human (13) have revealed that the histone H2A variant H2A.Z-containing nucleosomes are especially enriched at the +1 position near the transcriptional start site (TSS). Recent work has shown that in *Arabidopsis*, the H2A.Z-containing nucleosomes have the ability to "sense" changes in temperature and allow DNA to unwrap progressively as temperature rises, and therefore lead to the activation and inactivation of hundreds of temperature regulated genes (14).

Another mechanism to open chromatin and facilitate transcription is through chromatin remodeling complexes, which utilize the energy from hydrolyzing ATP to enhance sliding of histone octamer along the DNA or transfer the octamer in *trans*. One good example is Gal4-mediated Gal gene induction in yeast (15). Upon galactose addition, the inhibition of the transcriptional activator Gal4 is alleviated, resulting in the recruitment of chromatin remodeling complex SWI/SNF to Gal gene promoters, where it removes the nucleosomes and facilitate the rapid activation of transcription.

PTMs, histone variants and chromatin-remodeling are not mutually exclusive but often interdependent. For instance, at least in vitro studies suggest that at the Gal gene promoters the histone acetyltransferase (HAT) complex can stabilize SWI/SNF binding to nucleosomes and that acetylated histones are preferentially displaced by the SWI/SNF remodeling complex (16–18). In addition, HAT complexes such as NuA4 and SAGA increase the ability of RSC to stimulate Pol II elongation and the stimulatory effect of SAGA on RSC is enhanced by the addition of acetyl-CoA in vitro (19). This observation suggests that SAGA- and NuA4-mediated histone modifications may be a direct target for binding of RSC via its bromodomains to the nucleosomes, although further in vivo evidence is still lacking. Similar interplay between SAGA and SWI/SNF is observed at the yeast *PHO5* gene upon phosphate starvation (20, 21), suggesting the prevalence of interdependency of PTMs of histones and the chromatin-remodeling complex.

To dissect such complex processes effectively, a combination of biophysical, biochemical and genetic methods are essential. We specifically focus here on genetic approaches aimed at discovering which processes individual histone residues participate in, and even more specifically on a recently described library of histone point mutants in H3 and H4.

We describe here a set of relatively generic protocols to systematically identify new histone mutations which are closely related to your favorite genes/phenotypes using this publicly available histone mutant library (22).

2. Materials

2.1. Construct the Reporter Strain

1. 2× yeast extract/peptone (YEP) medium: 20 g/L Bacto-yeast extract, 40 g/L Bacto-peptone in water and autoclaved.
2. 4% agar: Bacto-agar: 10 g of agar is added into 250 mL of water and autoclaved.
3. 20% (w/v) dextrose stock solution, sterile.
4. Petri plates: 90 × 15 mm.
5. Deionized (Mega-ohm resistance) water, to be used for all media preparations.
6. 1 M LiOAc stock solution: Add 51 g LiOAc·2H_2O into 400 mL ddH_2O to dissolve, and fill to 500 mL final volume. The pH should be between 8.4 and 8.9. Filter-sterilize or autoclave. This can also be made by dissolving LiOH to 1 M and adjusting the pH with acetic acid.
7. TE: 10 mM Tris acetate pH 8, 1 mM EDTA.
8. 0.1 M LiOAc in TE.
9. Salmon sperm DNA (sheared, 10 mg/mL): boiled for 5 min and kept on ice before use.
10. 50% (W/V) PEG 3350 stock solution: Add 125 g PEG 3350 to 120 mL ddH_2O, dissolve at low heat on a stirring plate. Fill to 250 mL final volume and filter-sterilize.
11. 40% PEG 3350 (W/V) in 0.1 M LiOAc/TE.

12. DMSO.
13. Nourseothricin (ClonNAT): make 100 mg/mL stock (1,000×) solution in sterile water and store at 4°C.
14. YPD plus clonNAT plates: Add 10 g agar to 250 mL 2× YEP medium, 50 mL 20% dextrose stock solution, and 200 mL sterile H_2O. Autoclave, let cool to about 55–60°C (you should be able to hold the flask without much discomfort), add nourseothricin (clonNAT) to 100 μG/mL. Pour into petri plates, uncovering plates only briefly to pour. If bubbles are present, carefully flame with a bunsen burner to break them. This will not work if the medium has cooled too much.
15. High-fidelity Taq polymerase such as ExTaq from Takara.
16. Manufacturer supplied 10× buffer for high-fidelity Taq polymerase.
17. dNTP mix (2.5 mM each dNTP).
18. Triton X-100, 20% solution.
19. Primers to knock out *HHT1-HHF1* locus and confirm correct knockout:

    JB2775: 5′-AATCCCAAATATTTGCTTGTTGTTACCGTTTTCTTAGAATTAGCCAGCTGAAGCTTCGTACGC-3′

    JB2776: 5′-GATTTATATTTTATTGTGTTTTTGTTCGTTTTTTACTAAAACTGATGACAATCAACAAAGCATAGGCCACTAGTGGATCTG-3′

    JB9608: 5′-CTCTTCCTGCTCTCAGATATTAATG-3′

    JB9610: 5′-CCAAAAGAAGGATATCAAGTTGGC-3′

    JB9611: 5′-GTAAGCCAGAAAACCAATTTGCCTTC-3′

    JB4677: 5′-AGGTGGTAAAGGTCTAGGTAA-3′

    KanB: 5′-CTGCAGCGAGGAGCCGTAAT-3′

    KanC3: 5′-CCTCGACATCATCTGCCCAGAT-3′

2.2. Prepare Histone Mutants for Integration

1. 2× LB media: 20 g/L Bacto-tryptone, 10 g/L Bacto-yeast extract, 20 g/L sodium chloride in water and autoclaved. 1× LB media can be made by diluting the 2× LB with sterile water.
2. Carbenicillin.
3. LB agar plates containing 50 μG/mL carbenicillin.
4. 96-pin replicator (e.g., BOEKEL, Model 140500).
5. Disposable sterile 96-well pin replicator (e.g., Genetix 5051).
6. 1.3 mL deep-well 96-well plate.
7. Aluminum foil lids for 96-well plates.
8. 96-well filter plate (1.3 mL).
9. Buffer P1: 50 mM Tris–HCl pH 8.0, 10 mM EDTA, and 100 μg/mL RNase A; store at 4°C.

10. Buffer P2: 200 mM NaOH and 1% (w/v) SDS; store at room temperature.
11. Buffer P3: 3.0 M potassium acetate, pH 5.5; store at room temperature.
12. 100% Isopropanol.
13. 70% ethanol.
14. 96-well flat bottom plate with lid (e.g., Costar 3370).
15. 10× NEBuffer 4: 500 mM potassium acetate, 200 mM Tris acetate, 100 mM magnesium acetate, 10 mM dithiothreitol pH 7.9.
16. *Bci*VI.
17. BSA, 10 mg/mL.

2.3. Integrate Mutants into Reporter Strain

1. 1× YPD media: Add 50 mL of 20% glucose and 200 mL of sterile water to 250 mL of 2× YEPD.
2. Glass tube with cap: 25 mL.
3. 0.1 M LiOAc in TE.
4. Transformation solution: 6.24 mL 50% PEG3350, 822 μL 1 M LiOAc, 958 μL DMSO, 500 μL 10 mg/mL salmon sperm DNA.
5. 5 mM $CaCl_2$ solution, filter-sterilized.
6. Omnitray: NUNC 242811 or equivalent.
7. 2× SC–Ura broth: 3.4 g/L yeast nitrogen base without amino acids and ammonium sulfate; 10 g/L ammonium sulfate; 4 g/L amino acid mixes without uracil.
8. SC–Ura plate: Mix 250 mL 4% agar (melted), 200 mL of 2× SC–Ura broth and 50 mL of 20% glucose. Aliquot 35 mL per Omnitray. Let the plate dry at room temperature for 3 days before use.

2.4. Confirm Correct Integration

1. High-fidelity Taq polymerase such as ExTaq from Takara.
2. Manufacturer supplied 10× buffer for high-fidelity Taq polymerase.
3. dNTP mix (2.5 mM each dNTP).
4. Triton X-100, 20% solution.
5. Primers to confirm correct integration by PCR.

   P1 (JB9612): 5′-GGAGCCATTTGTTAATATACCGTCTC ATC-3′

   N1 (JB7257): 5′-GAATCCGTCGAAGCATACTTA-3′

   P2 (JB12311, for H3 mutants): 5′-CTGTGCTCCTTCCT TCGTTC-3′

   P2′ (JB12312, for H4 mutants): 5′-CATCCGCTCTAACCGAA AAG-3′

3. Methods

This protocol aims to construct a systematic histone mutant library in a desired *S. cerevisiae* reporter strain to test the behavior of each histone mutant. As illustrated in Fig. 1, each histone mutant construct is prepared systematically from bacteria as a plasmid, digested using a restriction enzyme to release the synthetic fragment and integrated into yeast genome by homologous recombination after transformation. Integration at the target locus can be verified both by testing for the loss of a marker gene at the target locus and/or by yeast colony PCR. Each mutant links to the *URA3* selective marker that confers growth on yeast minimal medium lacking uracil

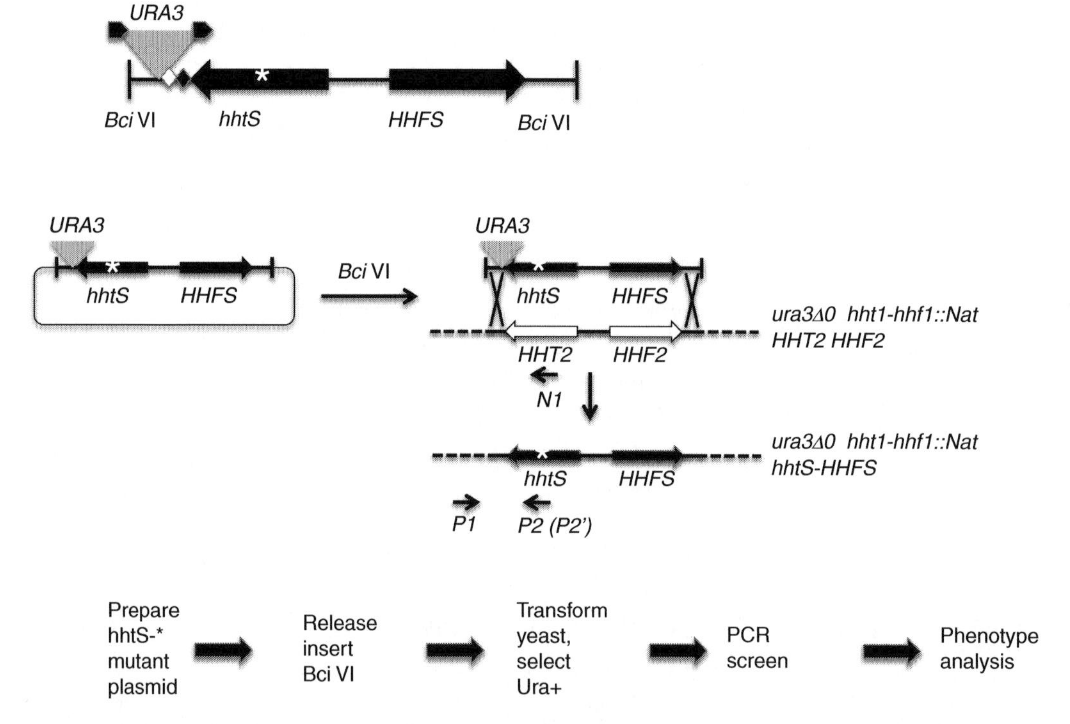

Fig. 1. Manipulation and integration of synthetic histone mutant libraries. The structure of the base construct for histone H3 mutants is shown in the *top panel*. The diagram indicates the mutant H3 gene (*hhtS*) and the wild-type H4 gene (*HHFS*) as divergent *black block arrows* cloned in the form of a synthetic DNA fragment consisting of HHT2-HHF2 flanking sequences (*solid black horizontal lines at left and right ends*) and terminal unique *BcN*I sites (*black vertical lines*). The mutation is marked by the * symbol. Other important features include pairs of molecular barcodes (*black and white diamonds*) unique to each mutant and loxP sites (*black arrowheads*) flanking the selectable marker, *URA3* (*gray triangle*), facilitating selectable marker removal or swapping. For H4 mutants, cloned in a distinct base construct, the *URA3* marker and tags are located adjacent to the 3′ end of *HHFS*. The *middle panel* shows how the liberated *BcN*I fragment can target and replace the endogenous *HHT2-HHF2* locus (*white block arrows* indicate resident native H3 and H4 genes) by homologous recombination (tall "X" symbols). A series of positive (P1/P2) and negative (P1/N1) primers (*small arrows*) can be used to confirm appropriate targeting. The genotypes at the *right end* indicate the relevant genotypes of the starting and swapped strains; the *ura3Δ0* genotype presumes the *URA3* marker used to select transformants has been removed by counterselection. The *lower panel* indicates the overall flow chart for the process of swapping mutants into the chromosome of a strain of interest.

(SC–Ura). To select for the mutant, the recipient yeast strain must carry mutations that prevent the synthesis of uracil and to further avoid off-target integration, the *URA3* gene should be completely deleted (including both 5′ and 3′ untranslated regions) to obtain the *ura3Δ0* strain. For details on constructing *ura3Δ0* strains, see ref. 23.

In order to use the histone mutant library as the sole source of histone H3 and H4, at least one copy of the endogenous histone loci should be knocked out before integrating the mutants. Since the histone mutants in the described collection target the *HHT2-HHF2* locus, we typically delete the *HHT1-HHF1* locus with the drug resistance marker *NatMX4* by one-step PCR replacement method (23) (see Note 1).

3.1. Construct the Reporter Strain

1. Inoculate your favorite strain in 5 mL of YPD media, grow at 30°C overnight in a roller drum.
2. Set up a PCR reaction to amplify the DNA fragment to knock out *HHT1-HHF1* using primers JB2775 and JB2776 using the plasmid pAG25 (24) or genomic DNA from a ClonNAT resistant strain as template (see Note 2).
3. Next morning, measure the cell density at A_{600} and inoculate the cells from overnight culture into 10 mL of YPD with a starting A_{600} of 0.1.
4. When the cells grow to an A_{600} of 0.6, spin them down, wash once with 10 mL sterile water and then with 10 mL 0.1 M LiOAc/TE. Resuspend the cells in 250 μL of 0.1 M LiOAc/TE and add 12.5 μL salmon sperm DNA to cells.
5. Split the cells into two 1.5 mL tubes with 100 μL cells each. To one tube, add 10 μL of sterile water as a "no DNA" control. In the other tube, add 10 μL of PCR mixture (~200 ng). Incubate both tubes at 30°C for 30 min.
6. Add 500 μL 40% PEG in 0.1 M LiOAc/TE. Mix well and incubate at 30°C for 45 min.
7. Add 70 μL 100% DMSO, mix immediately and heat-shock the cells at 42°C for 15 min.
8. Spin down the cells at 1600 × *g* for 1 min in a microfuge and wash with 1 mL of sterile water.
9. Resuspend the cells in 1 mL of YPD medium and incubate at 30°C for 5 h or 22°C overnight without shaking.
10. Spin down the cells and resuspend in 100 μL of water. Plate the entire tube of cells onto an YPD plate containing 200 μg/mL ClonNAT. Incubate the plate at 30°C until colonies appear (3–5 days).
11. Pick eight colonies from the plate, patch them onto a new YPD + ClonNAT plate and grow overnight at 30°C.
12. Set up four sets of colony PCR reactions to confirm the correct knockout of *HHT1-HHF1*: Primer JB9608 anneals to the

upstream of *HHT1-HHF1* locus and primer KanB is *NatMX4* specific. A PCR product of 824 base pair (bp) indicates the correctly integrated marker gene. On the contrary, JB4677 is specific to the native *HHT1-HHF1* locus and therefore JB9608 and JB4677 will not amplify anything in the correct knockout. The remaining two pairs (JB9611/KanC3 and JB9611/JB9610) are similar but specific to the downstream of *HHT1-HHF1* locus. Transfer about 2 μL of yeast cells into 30 μL of ddH_2O, mix well, and use 5 μL of for each PCR reaction.

Reagent	Volume (μL)
10× Taq buffer	1.5
dNTP mix	1.2
20% Triton X-100	0.75
primers (20 μM)	0.04 each
Taq polymerase	0.06
ddH_2O	6.41

3.2. Prepare Histone Mutants for Integration

1. The day before starting the experiment, the plasmid-containing *E. coli* strains should be grown on LB agar plates containing 50 μG/mL carbenicillin. We typically take the strains out of freezer, thaw the cells completely, replicate them onto agar plates using a 96-well pin replicator and grow the cells overnight at 37°C.
2. In 96-well deep well plates, aliquot 1 mL of 1× LB liquid media containing 50 μG/mL carbenicillin into each well. Inoculate the *E. coli* strains using a disposable sterile 96-pin replicator. Mix gently and avoid cross-well contamination. Seal the plate with Air-Pore membrane and allow the cells to grow for about 24 h at 37°C in an environmental shaker (220 rpm). Duplicate the culture to give 2 mL of cells per construct for plasmid preparation.
3. Spin the deep well plates at 3,000 × *g* for 5 min at room temperature to collect the cells. Discard the medium by pouring the liquid directly into sink. Drain the remaining medium by flipping the plates onto paper towel for a few seconds.
4. Resuspend the cells in 150 μL P1 buffer in one plate and transfer the cells into the other plate. Vortex briefly to completely resuspend the cells. Add 150 μL of P2 buffer into each well and mix the buffers by pipetting up and down several times. Add 150 μL of P3 buffer and pipette to mix. White precipitate will appear.

5. Centrifuge the samples at 4,000×*g* for 20 min at room temperature. Meanwhile, get a set of new 2 mL deep-well plates with a filter plate on top. Carefully transfer the supernatant onto the filter plate. Try to avoid taking the pellet but get as much liquid as possible. Centrifuge the deep well plates with filter plate on top at 2,500×*g* for 5 min at room temperature to collect the liquid. After spinning, discard the filter plate.
6. Add 500 μL of 100% isopropanol into each well. Mix by pipetting up and down several times, seal the plate with aluminum foil lids and put into −80°C freezer for 30 min to precipitate the plasmids. Centrifuge the plates at 4,000×*g* for 20 min at 4°C. Discard the liquid by pouring directly into sink. Drain remaining liquid by flipping the plates on paper towel for a few seconds.
7. Wash DNA pellet with 500 μL 70% ethanol. Shake the plate on plate shaker for 1 min and spin down at 4,000×*g* for 20 min at 4°C. Discard the liquid as above and dry the plasmids at room temperature. Resuspend DNA in 100 μL of ddH_2O and transfer plasmids into round bottom 96-well plates with lid. Seal with aluminum foil lid and store at −20°C (see Note 3).
8. Set up digestion in 96-well plates as follows. Seal the plates with aluminum foil lids and incubate the digestion mixture overnight at 37°C. The DNA will be ready to transform yeast. The Table indicates amounts per reaction, but it is best to make a master mix (110× volume) and aliquot 35 μL from this mix into each well (see Note 4).

Reagent	Volume (μL)	110× (μL)
Plasmid DNA (~100–200 ng/μL)	15	–
NEB buffer 4 (10×)	5	550
BSA (100×)	0.5	55
*Bci*VI	0.5	55
ddH_2O	29	3,190

3.3. Integrate the Mutants into the Reporter Strain

1. Two days before transforming the yeast strain, the yeast should be streaked onto agar plates containing selective medium (e.g., YPD+clonNAT). If taken from the freezer, two consecutive streaks are recommended.
2. Prepare the reporter strain for transformation. Inoculate 10 mL of liquid YPD medium in a 25 mL sterile glass tube with a single colony. Grow overnight at 30°C in either a roller drum or a shaking incubator.

3. Measure the cell density of the overnight culture (A_{600}). Subculture the cells in 100 mL of YPD medium with a starting A_{600} of 0.1. Grow to early log phase (A_{600} of ~0.6) (see Note 5).
4. Pellet cells 5 min at 1,600 × *g* at room temperature. Wash the cells with 20 mL ddH_2O first, and then with 20 mL of 0.1 M LiOAc/TE buffer. Pellet cells and resuspend in 2 mL of 0.1 M LioAc/TE. Let the cells sit at room temperature.
5. While preparing the cells, prepare the transformation plates by adding 10 μL of digested plasmid into each well (~300 ng). The remainder can be used to assess digestion by agarose gel electrophoresis.
6. Aliquot 20 μL cell suspension into each of the 96 wells in the microtiter plate.
7. Add 80 μL of transformation solution and mix with cells by pipetting up and down several times.
8. Incubate the plate at 30°C for 30 min (The incubation can be longer but not shorter than 30 min. We have incubated the plate up to 4 h without significantly affecting transformation efficiency).
9. Heat shock by placing in a 42°C water bath for 20 min (see Note 6).
10. Pellet the cells by centrifuging at 1600 × *g* for 6 min at room temperature.
11. Dump the supernatant into sink and blot off excess liquid by tapping on wad of paper towels for a few times. The pellet will be tightly on the bottom.
12. Wash the cells with 100 μL 5 mM $CaCl_2$. Shake the plates on a plate shaker for 5 min to resuspend. Pellet the cells (1600 × *g*, 6 min) and remove the supernatant.
13. Add 20 μL of 5 mM $CaCl_2$, mix well on the plate shaker to suspend the cells.
14. Plate cells by dripping row by row using 12-channel pipette onto the selective plates (Fig. 2) (see Note 7).
15. Incubate at 30°C for 2–3 days or until the colonies appear.

3.4. Confirm Correct Integration

1. Pick two independent colonies per construct and patch onto new selective plates. Arrange the strains in 96-well format. Incubate overnight at 30°C.
2. In a 96-well plate, aliquot 25 μL ddH_2O. Using 12-channel pipette to pick up yeast cells and put them into the water.
3. Make master mixture for yeast colony PCR, using primers indicated in Subheading 2.4. Use P1/P2 and P1/N1 for histone H3 mutants and P1/P2′ and P1/N1 for histone H4 mutants (see Note 8).

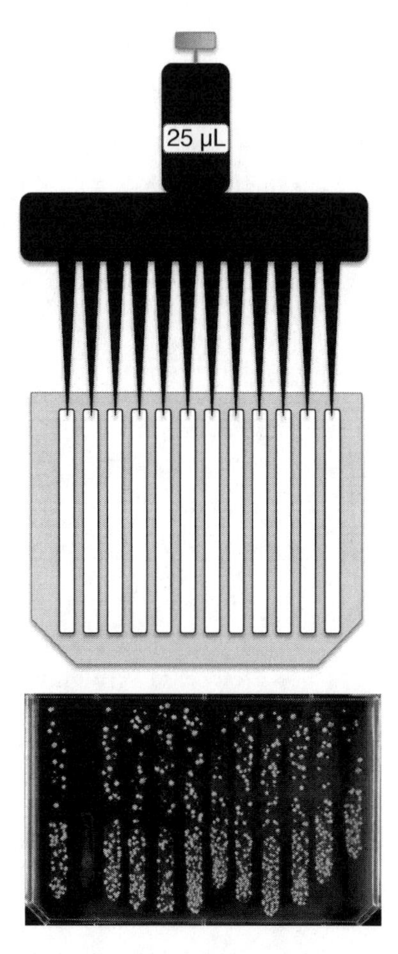

Fig. 2. Isolating single colonies from 96-well plate transformations. A 12-well pipettor is used to flow streaks of 25 μL of medium down the surface of a selective plate made in a Nunc Omnitray. After streaks run to the bottom of the plate the plate is adjusted to a horizontal position, the plates are incubated until single colonies form (*photo at bottom*). Using this method, 12 transformations can be consolidated per Omnitray, and position of the transformants is maintained within a row-of-12 format.

Reagent	Volume (μL)	One 96-plate (μL)	One 384-plate (μL)
10× Taq buffer	1.5	180	660
dNTP mix	1.2	144	528
20% Triton X100	0.75	90	330
Primers (20 μM)	0.04 each	4.8 each	17.6 each
Taq polymerase	0.06	7.2	26.4
ddH_2O	6.45	774	2,838

4. Aliquot 10 μL of PCR reaction mixture into each well in a 96- or 384-well PCR plate. Add 5 μL of yeast cells. Mix by pipetting

for several times. Seal the plate and run PCR reaction using the following program

94°C	5 min	
94°C	20 s	
55°C	20 s	
72°C	45 s	×35 cycles
72°C	7 min	
4°C	hold	

5. After PCR is done, load the entire PCR reaction mixture to check amplification on 1.5% agarose gel. Correctly integrated strains will have a 700 bp band in the positive reaction wells and blank in negative reaction wells. If the synthetic fragment is not integrated correctly, a PCR product of 660 bp will present in the negative reaction well.
6. Cherry-pick the correct strains onto a new selective plate. Grow the cells overnight at 30°C and store at 4°C until testing.

3.5. Carry Out the Assay

After constructing and verifying all of the mutants, array all of the strains in 96-well format and spot them on YPD plate, grow at 30°C for 1–2 days before picking the cells for serial dilution and testing the phenotypes you are interested in.

4. Notes

1. The method described here can only be used to identify nonessential mutations. This limitation can be overcome by the following strategy: During the process of integrating the histone mutants, lethal mutants can be identified because they cannot grow or can only grow poorly on the SC–Ura plate after transformation. Such mutants can be integrated directly into a strain containing both copies of the histone H3 and H4 genes. Then, follow the same assay to test whether those mutants may have a dominant phenotype.
2. Since both NatMX4 on pAG32 and HygMX4 on pAG29 are high GC content, DMSO (10% v/v) must be added to the PCR reaction.
3. The plate can be put on top of a 37°C heating block to speed the drying process. The plasmid can also be dissolved in 1× TE. The final DNA concentration will be around 100 ng/μL. A few rows of samples from each plate can be tested on an agarose gel to check quality and yield.

4. 50 μL of digestion is enough for five to six transformations. We typically use 8 μL of mixture for each transformation. There is no need for DNA purification after digestion. Scale up the volume based on the amount of plasmid you need to digest. To further eliminate recombination due to the *TRP1* fragment, *Xba*I can be added in the digestion, which cuts in the middle of *TRP1*. However, *Xba*I sites may not be unique in all of the synthetic constructs. If a synthetic construct contains an *Xba*I site due to codon change, it can be easily identified since no cells will grow after transformation. Overnight digestion is to reduce the consumption of restriction enzyme and there is no obvious DNA degradation or star reaction observed after overnight digestion. For quality control, we usually run a few rows of samples from each plate on an agarose gel to check completeness of digestion.
5. The cells can be grown to a higher density but normally A_{600} values should be below 1.0 for optimal transformation efficiency. We also don't recommend transforming the cells at A_{600} less than 0.5. Depending on strains, the culture time may vary from 6 to 8 h.
6. This can be done by sealing the plate well with parafilm and submerging under water. Put a weight on top of the plate to keep it from floating out of water. Do not let the incubation time exceed 20 min. Longer periods of heat shock will reduce the transformation efficiency dramatically.
7. The selective plates have to be dry enough to prevent merging of cells from adjacent columns. We normally make the plates 3 days before use and let them dry at room temperature. If plates are required more quickly the newly prepared plates can be dried at 37°C in an air incubator for 20 min with lids open after solidification. The plates should be tilted when spotting the cell suspensions, allowing them to form a straight streak, and then set flat right after. Alternatively, less volume can be used to suspend the cells for plates that are not dry enough.
8. We normally set up two PCR reactions for each mutant. One is the positive reaction, which uses one primer upstream of the *HHT2-HHF2* locus and the other primer specific to the synthetic construct. The other PCR reaction is the negative reaction, which is to check the loss of whatever DNA sequences (either the native *HHT2-HHF2* or a marker gene previously used to knock out the *HHT2-HHF2* locus, depending on parental strain genotype) at the *HHT2-HHF2* locus in the target strain.

References

1. Kouzarides T (2007) Chromatin modifications and their function. *Cell* 128(4):693–705.
2. Klose RJ & Zhang Y (2007) Regulation of histone methylation by demethylimination and demethylation. *Nat Rev Mol Cell Biol* 8(4): 307–318.
3. Altaf M, *et al.* (2007) Interplay of chromatin modifiers on a short basic patch of histone H4 tail defines the boundary of telomeric heterochromatin. *Mol Cell* 28(6):1002–1014.
4. Norris A & Boeke JD (2010) Silent information regulator 3: the Goldilocks of the silencing complex. *Genes Dev* 24(2):115–122.
5. Berger SL (2007) The complex language of chromatin regulation during transcription. *Nature* 447(7143):407–412.
6. Morillon A, Karabetsou N, Nair A, & Mellor J (2005) Dynamic lysine methylation on histone H3 defines the regulatory phase of gene transcription. *Mol Cell* 18(6):723–734.
7. Cheung WL, *et al.* (2005) Phosphorylation of histone H4 serine 1 during DNA damage requires casein kinase II in S. cerevisiae. *Curr Biol* 15(7):656–660.
8. Utley RT, Lacoste N, Jobin-Robitaille O, Allard S, & Cote J (2005) Regulation of NuA4 histone acetyltransferase activity in transcription and DNA repair by phosphorylation of histone H4. *Mol Cell Biol* 25(18):8179–8190.
9. Furuyama S & Biggins S (2007) Centromere identity is specified by a single centromeric nucleosome in budding yeast. *Proc Natl Acad Sci USA* 104(37):14706–14711.
10. Guillemette B, *et al.* (2005) Variant histone H2A.Z is globally localized to the promoters of inactive yeast genes and regulates nucleosome positioning. *PLoS Biol* 3(12):e384.
11. Weber CM, Henikoff JG, & Henikoff S (2010) H2A.Z nucleosomes enriched over active genes are homotypic. *Nat Struct Mol Biol* 17(12):1500–1507.
12. Whittle CM, *et al.* (2008) The genomic distribution and function of histone variant HTZ-1 during C. elegans embryogenesis. *PLoS Genet* 4(9):e1000187.
13. Barski A, *et al.* (2007) High-resolution profiling of histone methylations in the human genome. *Cell* 129(4):823–837.
14. Kumar SV & Wigge PA (2010) H2A.Z-containing nucleosomes mediate the thermosensory response in Arabidopsis. *Cell* 140(1):136–147.
15. Weake VM & Workman JL (2010) Inducible gene expression: diverse regulatory mechanisms. *Nat Rev Genet* 11(6):426–437.
16. Hassan AH, Neely KE, & Workman JL (2001) Histone acetyltransferase complexes stabilize swi/snf binding to promoter nucleosomes. *Cell* 104(6):817–827.
17. Hassan AH, *et al.* (2002) Function and selectivity of bromodomains in anchoring chromatin-modifying complexes to promoter nucleosomes. *Cell* 111(3):369–379.
18. Chandy M, Gutierrez JL, Prochasson P, & Workman JL (2006) SWI/SNF displaces SAGA-acetylated nucleosomes. *Eukaryot Cell* 5(10): 1738–1747.
19. Carey M, Li B, & Workman JL (2006) RSC exploits histone acetylation to abrogate the nucleosomal block to RNA polymerase II elongation. *Mol Cell* 24(3):481–487.
20. Reinke H & Horz W (2003) Histones are first hyperacetylated and then lose contact with the activated PHO5 promoter. *Mol Cell* 11(6):1599–1607.
21. Steger DJ, Haswell ES, Miller AL, Wente SR, & O'Shea EK (2003) Regulation of chromatin remodeling by inositol polyphosphates. *Science* 299(5603):114–116.
22. Dai J, *et al.* (2008) Probing nucleosome function: a highly versatile library of synthetic histone H3 and H4 mutants. *Cell* 134(6):1066–1078.
23. Brachmann CB, *et al.* (1998) Designer deletion strains derived from Saccharomyces cerevisiae S288C: a useful set of strains and plasmids for PCR-mediated gene disruption and other applications. *Yeast* 14(2):115–132.
24. Goldstein AL & McCusker JH (1999) Three new dominant drug resistance cassettes for gene disruption in Saccharomyces cerevisiae. *Yeast* 15(14):1541–1553.

Chapter 2

Measuring Dynamic Changes in Histone Modifications and Nucleosome Density during Activated Transcription in Budding Yeast

Chhabi K. Govind, Daniel Ginsburg, and Alan G. Hinnebusch

Abstract

Chromatin immunoprecipitation is widely utilized to determine the in vivo binding of factors that regulate transcription. This procedure entails formaldehyde-mediated cross-linking of proteins and isolation of soluble chromatin followed by shearing. The fragmented chromatin is subjected to immunoprecipitation using antibodies against the protein of interest and the associated DNA is identified using quantitative PCR. Since histones are posttranslationally modified during transcription, this technique can be effectively used to determine the changes in histone modifications that occur during transcription. In this paper, we describe a detailed methodology to determine changes in histone modifications in budding yeast that takes into account reductions in nucleosome.

Key words: Activated transcription, RNA polymerase II, Histone modifications, Acetylation, Gcn4, Gal4, *Saccharomyces cerevisiae*, Histone acetyltransferase, Histone deacetylase complexes

1. Introduction

The packaging of the eukaryotic genome into chromatin imposes a physical barrier for all DNA-dependent processes, including gene transcription. During transcription activation, the barrier imposed by nucleosomes, the fundamental unit of chromatin is relieved through sliding, repositioning, or evicting the nucleosomes that occlude regulatory DNA sequences (1). Similarly, histones are transiently evicted from the coding regions to facilitate transcription elongation by RNA Polymerase II (Pol II) (2–5). Acetylation of lysine residues in the amino-termini of histones is a major determinant of nucleosome occupancies in the promoter region and across the coding sequences of genes during transcription activation (5–7). Histone acetylation is

Randall H. Morse (ed.), *Chromatin Remodeling: Methods and Protocols*, Methods in Molecular Biology, vol. 833,
DOI 10.1007/978-1-61779-477-3_2, © Springer Science+Business Media, LLC 2012

dynamically regulated by histone acetyltransferase (HAT) and histone deacetylase complexes (HDACs) (5, 7–11). Additionally, the recruitment and function of these complexes, especially in coding regions is regulated by phosphorylation of the C-terminal domain (CTD) of Rpb1, the largest subunit of Pol II, and methylation of histone H3 (7, 12–15).

Genome-wide studies have revealed that histone occupancies at promoters are lower as compared to their corresponding coding regions (16, 17), where the occupancy of histones is inversely correlated with the rate of transcription (16). In the budding yeast *Saccharomyces cerevisiae*, activated transcription is associated with cotranscriptional loss of nucleosomes and this loss is more prominent at promoter regions. Transcription activation by Gcn4 or Gal4 is accompanied by a massive loss of nucleosomes from the promoters of their regulated genes, and the extent of histones lost from adjacent coding regions depends on both the rate of transcription and histone acetylation (5, 7, 9, 16). Our recent studies show that histone acetylation modulates histone occupancy in coding regions (5, 7). Deletion of *GCN5* or a mutation in *ESA1*, encoding the catalytic subunits of HAT complexes SAGA and NuA4, respectively, severely impairs histone acetylation and reduces histone eviction from the coding regions, and this reduced histone eviction is accompanied by lower transcription, which involves reductions in both the rate and processivity of Pol II during elongation (5, 9). Conversely, deletion of HDACs is accompanied by increased histone acetylation and greater histone eviction in the coding regions (7, 9). Thus, it appears that histone loss from the coding regions during transcription activation is regulated through modulation of histone acetylation by multiple HATs and HDACs and influences the elongation phase of transcription.

Chromatin immunoprecipitation (ChIP) has been extensively used to detect histone modifications that are associated with transcribed genes, as well as to detect the complexes that regulate histone modifications (8, 18). In this chapter, we provide a detailed experimental procedure for ChIP to detect both histone acetylation as well as recruitment of complexes that regulate modification and chromatin remodeling in *S. cerevisiae*. In this procedure, outlined in Fig. 1, the protein–protein and protein–DNA interactions are stabilized through use of formaldehyde (HCHO) as a cross-linking agent. The chromatin is then sheared into small pieces (~300–500 base pairs) and is subjected to immunoprecipitation using antibodies recognizing histones, a specific histone modification or a protein of interest that may be present as part of a larger complex. The detailed methodology provided is for the detection of histone modifications present at *ARG1*, a Gcn4-regulated gene, or at the Gal4-regulated gene *GAL1* under activating conditions.

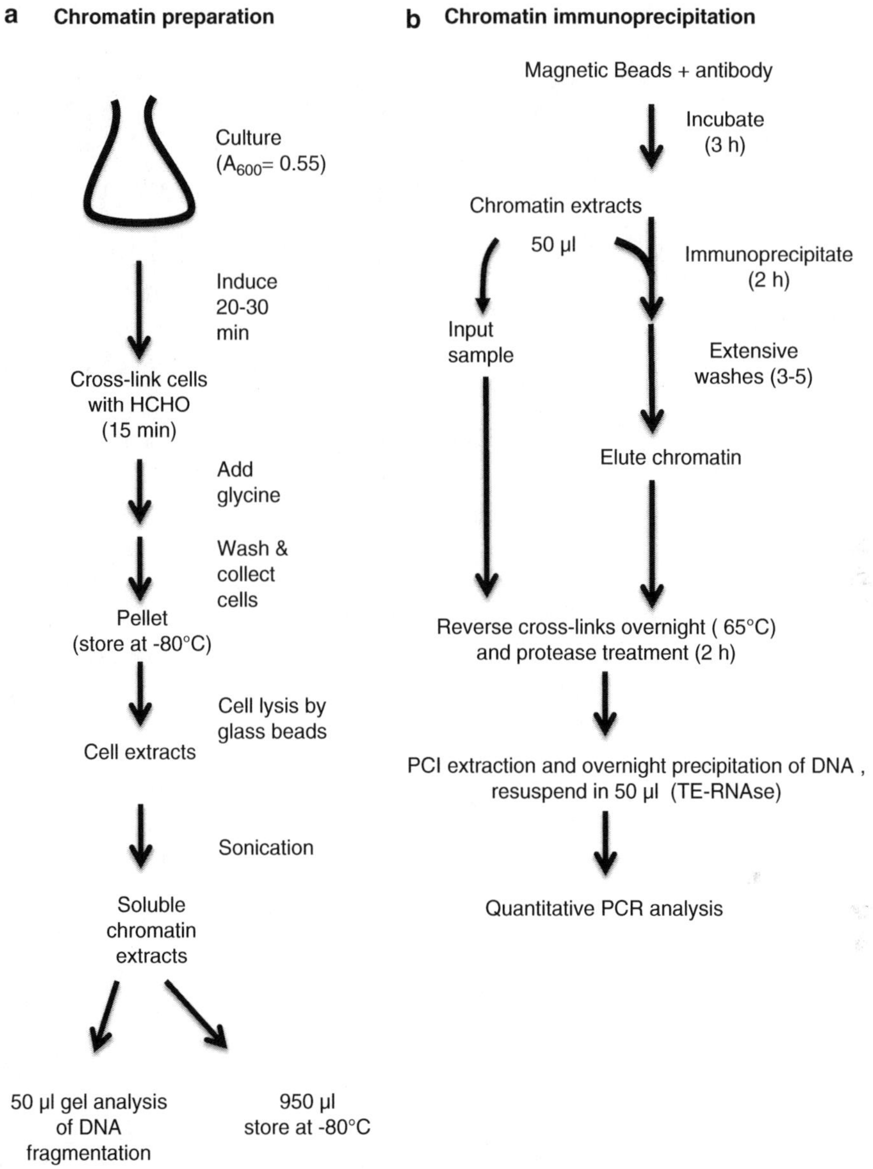

a **Chromatin preparation**

Culture
(A_{600}= 0.55)
Induce
20-30
min
Cross-link cells
with HCHO
(15 min)
Add
glycine
Wash &
collect
cells
Pellet
(store at -80°C)
Cell lysis by
glass beads
Cell extracts
Sonication
Soluble
chromatin
extracts
50 µl gel analysis
of DNA
fragmentation
950 µl
store at -80°C

b **Chromatin immunoprecipitation**

Magnetic Beads + antibody
Incubate
(3 h)
Chromatin extracts
50 µl
Immunoprecipitate
(2 h)
Input
sample
Extensive
washes (3-5)
Elute chromatin
Reverse cross-links overnight (65°C)
and protease treatment (2 h)
PCI extraction and overnight precipitation of DNA ,
resuspend in 50 µl (TE-RNAse)
Quantitative PCR analysis

Fig. 1. Schematic diagram of entire ChIP protocol: (**a**) chromatin preparation and (**b**) chromatin immunoprecipitation.

2. Materials

2.1. Media

1. YPD: Bacto-yeast extract (1%), Bacto-peptone (2%), dextrose (2%).
2. YPR: Bacto-yeast extract (1%), Bacto-peptone (2%), raffinose (2%).

3. Synthetic complete (SC): For 1 L of SC media, add 2 g of amino acid mix, 2.25 g of yeast nitrogen base without ammonium sulfate and amino acids and 5 g of ammonium sulfate in 900 mL of distilled water. Autoclave the media and add 50 mL of 40% dextrose solution (filter sterilized) for a final volume of 1 L.
4. 20% galactose solution, sterilized (for induction of Gal4 targets).

2.2. Reagents, Solutions, and Other Materials

1. Sulfometuron methyl (SM): 5 mg/mL in dimethyl sulfoxide (DMSO) (for induction of Gcn4 targets).
2. Formaldehyde (37% solution).
3. Cross-linking solution: 1 mM EDTA, 100 mM NaCl, 70 mM HEPES-KOH (pH 7.5). This can be stored at room temperature for up to 6 months. 37% formaldehyde is added to the required amount of the above solution to a final concentration of 11%.
4. Glycine stop solution: 2.5 M glycine (molecular biology grade) in distilled water heated to 70°C and stirred until completely dissolved. Filter-sterilize and store at room temperature.
5. FA lysis buffer*: 50 mM HEPES-KOH (pH 7.5), 150 mM NaCl, 1 mM EDTA, 0.1% sodium deoxycholate, 1% Triton X-100 in water. Filter-sterilize and store at room temperature, for up to 6 months.
   *Before use, take the required amount of FA-lysis buffer and add the following protease inhibitors to the indicated final concentrations: PMSF (1 mM; from a 100 mM stock solution in isopropanol), leupeptin (1 μg/mL; stock 10,000 μg/mL) pepstatin A (1 μg/mL; stock 1,000 μg/mL), aprotinin (10 μg/mL; stock 10,000 μg/mL).
6. Acid-washed glass beads (0.4–0.6 mm).
7. 26 gauge needle.
8. Wash buffer II: 50 mM HEPES-KOH (pH 7.5), 500 mM NaCl, 1 mM EDTA, 0.1% sodium deoxycholate, 1% Triton X-100.
9. Wash-buffer III: 10 mM Tris–Cl (pH 8.0), 250 mM lithium chloride, 1 mM EDTA, 0.5% sodium deoxycholate, 0.5% NP-40 substitute (Igepal CA-630; see Note 2 in Chapter 3).
10. Elution buffer: 50 mM Tris–HCl (pH 8.0), 10 mM EDTA, and 1% sodium dodecylsulfate (SDS).
11. Elution wash buffer: 10 mM Tris–HCl (pH 8.0), 1 mM EDTA, and 0.7% SDS.
12. Proteinase K, 10 mg/ml.
13. Phenol:chloroform:isoamyl alcohol (25:24:1, v/v).
14. Chloroform:isoamyl alcohol (24:1, v/v).
15. Glycogen, 20 mg/ml.
16. 4 M LiCl.

17. 100% ethanol.
18. Phosphate-buffered saline 1×: 137 mM NaCl, 2.7 mM KCl, 100 mM Na_2HPO_4 and 2 mM KH_2PO_4.
19. PBS/BSA: 0.5% bovine serum albumin (BSA) in 1× PBS.
20. RNase A, 10 mg/ml.
21. 6× ChIP dye solution: 4 mg bromophenol blue in 100 mL of 15% Ficoll solution prepared in 1× TBE (89 mM Tris–borate, 89 mM Boric acid, 2 mM EDTA).
22. 6% TBE-polyacrylamide gel: Use 1.0 mm 12 or 15 well combs.
23. Magnetic beads: Pan anti-mouse IgG and sheep anti-rabbit IgG Dynabeads or equivalent.
24. 1× TBS: 50 mM Tris–HCl pH 7.5, 150 mM NaCl.
25. ChIP antibodies: Rabbit polyclonal anti-H3 (0.7 μL; ab1791: Abcam), rabbit anti-acetyl histone H4 (0.5 μL; 06-866, Upstate Biotechnology), rabbit monoclonal anti-trimethyl (Lys4) histone H3 (1.0 μL; 05-745: Upstate Biotechnology), anti-acetyl histone H3 (0.7 μL; 06-599; Upstate Biotechnology), anti-Gal4p antibodies (1.0 μL SC577X; Santa Cruz Biotechnology), mouse monoclonal anti-Rpb3 (1.0 μL; Neoclone), and anti-phospho-Ser5 Rpb1 (1.0 μL; H14; Covance). If other sources are used for specific antibodies, it will be necessary to determine experimentally the optimal amount to use for ChIP.
26. PCR master-mix: (prepared on ice, 13 μL per reaction). Combine 6.32 μL water with 1.5 μL each of 10× PCR buffer (generally provided with hot-start Taq polymerase) and 15 mM $MgCl_2$, 1.5 μL of each of the specific and internal control primer sets (including forward and reverse primers; see Table 1 for a list of primers used for ChIP at the *ARG1* and *GAL1* promoters together with control primers), 0.3 μL of dNTPs (10 mM) and hot-start Taq polymerase (added last), and 0.08 μL of [α-^{33}P]-dATP.
27. PCR tubes (0.2 mL thin-walled strip tubes).
28. Thermal cycler.
29. Whatman 3 MM paper.
30. Ethidium bromide, 5 μg/mL.

3. Methods

3.1. Cell Culture and Transcription Induction

1. Cell culture: Inoculate a single colony of wild-type (WT) cells in 5 mL of SC media lacking isoleucine and valine (SC^{-ILV}) (for experiments involving induction of genes by Gcn4;

Table 1
Primers employed for ChIP analysis to study *ARG1* and *GAL1*

Gene	Location (relative to ATG)	Sequence
POL1	+2,477/+2,707 (ORF)	5′-GACAAAATGAAGAAAATGCTGATGCACC-3′ 5′-TAATAACCTTGGTAAAACACCCTG-3′
TEL VI-R	+51/+307	5′-GCTGAGTTTAACGGTGATTATT-3′ 5′-CCAGTCCTCATTTCCATCAAT-3′
ARG1	−376/−213 (UAS)	5′-ACGGCTCTCCAGTCATTTAT-3′ 5′-GCAGTCATCAATCTGATCCA-3′
	−197/−51 TATA	5′-TAATCTGAGCAGTTGCGAGA-3′ 5′-ATGTTCCTTATCGCTGCACA-3′
	+23/+186 (5′ ORF)	5′-TGGCTTATTCTGGTGGTTTAG-3′ 5′-ATCCACACAAACGAACTTGCA-3′
	+1,091/+1,258 (3′ ORF)	5′-TTCTGGGCAGATCTACAAAGA-3′ 5′-AAGTCAACTCTTCACCTTTGG-3′
GAL1	−408/−261 (UAS)	5′-TGTTCGGAGCAGTGCGGCGC-3′ 5′-ACGCTTAACTGCTCATTGCT-3′
	−185/−57 (TATA)	5′-GGTTATGCAGCTTTTCCATT-3′ 5′-CGAATCAAATTAACAACCATAGGA-3′
	+422/+567 (MID/5′ ORF)	5′-CCAGTTGGTACATCACCCTCA-3′ 5′-ATCCTTCTGTGTCGGACTGG-3′
	+1,233/1,355 (3′ ORF)	5′-CGTTCATCAAGGCACCAAAT-3′ 5′-TCAGAGGGCTAAGCATGTGT-3′

or appropriate media depending on experiment) and grow the culture overnight (O/N) with shaking at 30°C to late log phase or to saturation. Dilute the O/N culture in 100 mL of the same medium such that the starting absorbance at 600 nm (A_{600}) is ~0.1 and grow the cells at 30°C in an incubator shaker to A_{600} of 0.5–0.55 (see Note 1).

2. (1) Induction of Gcn4 target genes: Just before the culture reaches the desired A_{600}, thaw an aliquot of sulfometuron methyl (SM). When the culture reaches an A_{600} of 0.55, add 12 μL of SM solution to each flask and quickly place the flasks back into the incubator shaker for an additional 20–30 min (see Notes 2 and 3).

   (2) Induction of Gal4 regulated genes: Grow cells in YP-raffinose (YPR) to A_{600} of 0.55 and induce transcription for 20–30 min by adding galactose to a final concentration of 2% (10 mL of 20% galactose in 100 mL culture).

3. Add 11 mL of freshly prepared 11% formaldehyde solution by mixing 37% HCHO and cross-linking solution to 100 mL of induced culture (see Note 4). Mix gently by swirling the flask and incubate for 15 min at room temperature with intermittent mixing every 4 min.

4. Add 15 mL of 2.5 M glycine to quench the formaldehyde and mix gently (see Note 5).
5. Subsequent steps are carried out at 4°C or on ice with ice-cold solutions unless stated otherwise. Transfer the culture to 50 mL tubes (2 per 100 mL culture) and collect the cells by centrifugation at 1,500 × *g* for 5 min.
6. Decant the culture medium and wash two times with 1× TBS and collect the cells by centrifugation as in step 5. Decant the solution and add 1 mL of TBS, resuspend and transfer the cells from the 2 × 50 mL tubes into one 2 mL microcentrifuge (Eppendorf) tube.
7. Pulse-spin cells in a refrigerated microfuge to collect the pellet, which can be stored at –80°C until further use.

3.2. Preparation of Soluble Chromatin

1. Resuspend the cell pellet in 500 μL of FA lysis buffer with protease inhibitors and add ~500 μL of acid-washed glass beads (see Note 6). Place the tubes in a vortex mixer (e.g., Vortex Genie) and vortex for 40 min in a cold room. It is important to check the tubes for any leakage during this process. A large amount of frothing occurs at this step.
2. Place the tubes in an ice-bucket and carefully puncture both the cap and bottom of the tube using a 26 G red-hot needle. Place the tube in an un-capped 15 mL Falcon tube (save the cap) and spin at 300 × *g* for 1 min to collect cell lysate. Add an additional 500 μL of FA-lysis buffer to the tube and collect the lysate by centrifugation at 300 × *g* for 1 min. Discard the Eppendorf tube and place the 15 mL Falcon tube on ice and cap it.
3. Sonicate the cell lysate to shear the chromatin into ~300 bp fragments. The proper settings on a particular sonicator must be experimentally determined (see Note 7). For example, the Branson 450 sonifier fitted with the tapered tip (1/8 in., cat # 101 148 062) is set at 1.8 output with 60% duty cycle. Using these settings, sonicate the samples 10 times, for 30 s with 30 s intervals on ice. To prevent overheating of samples, the tube is placed in a glass beaker filled with ice during the entire sonication step.
4. After sonication, transfer the lysate to a 1.5 mL Eppendorf tube and spin at 16,000 × *g* for 30 min in a refrigerated microcentrifuge. Carefully remove the supernatant and transfer to a new 1.5 mL tube. Snap freeze by placing the tube in dry ice and store at –80°C if not required immediately.

3.3. ChIP and Reversal of Cross-Linking

1. Prepare antibody-bead conjugate. We use anti-mouse, anti-rabbit, and anti-IgM-conjugated magnetic beads (Dynabeads, Invitrogen). The particular type of magnetic beads employed depends on the specificity of the antibody (see Note 8).

For ChIP of acetylated H3, place 40 μL of bead suspension in each 1.5 mL tube and spin for ~8 s in a refrigerated microcentrifuge to pellet the beads. Place the tubes in a magnetic stand and remove the supernatant using suction. Wash beads twice with 1 mL of PBS containing BSA (5 mg/mL; PBS-BSA). Resuspend beads in 200 μL of PBS-BSA and add 0.7 μL of rabbit anti-acetyl histone H3 antibody and rotate the tubes in a cold room for 3 h. The antibody-coated beads can be prepared for any number of ChIPs.

2. Remove unbound antibody by washing beads twice with 1 mL of PBS-BSA. Resuspend beads in mixture of 30 μL of PBS-BSA, 20 μL of FA-lysis buffer and add 50 μL of thawed chromatin extract, reserving an identical chromatin aliquot as an "input sample (In)" on ice. It is important that the extracts are thawed on ice. Rotate the tubes (beads and extracts) for 2 h in a cold room. Overnight binding often increases the background and thus should be avoided; although for low affinity antibodies, overnight binding may provide more complete immunoprecipitations.
3. Collect the beads by "pulse-spin" (~8 s) centrifugation and resuspend the beads in 1 mL PBS-BSA. Agitate the tubes vigorously by flicking the bottom of the tube (4–5 times), collect the pellet by centrifugation and aspirate the buffer under vacuum. Repeat the washes with 1 mL each of FA-lysis buffer and wash buffer I and II (once when using anti-rabbit antibody-conjugated beads and twice for anti-mouse antibody conjugated beads). Wash gently with 1 mL 1× TE, discard the supernatant, and pulse-spin again to remove any remaining TE.
4. To elute DNA from the immune complexes, add 100 μL of elution buffer, vortex the tubes and incubate at 65°C in a water bath for 15 min. Collect the beads by centrifugation and transfer the supernatant to a new 1.5 mL tube. Add 150 μL of elution wash buffer to the beads, vortex, and incubate for 10 min at 65°C. Collect the supernatant and combine it with the elution buffer eluate. This is marked "IP" sample.
5. Add 200 μL of elution wash buffer to input tubes and leave both IP and In tubes at 65°C overnight to reverse the cross-linking.
6. On the next day, add 5 μL of proteinase K to each tube and incubate in a 37°C water bath for 2 h to degrade the chromatin proteins.
7. The samples (IP and In) are then processed for DNA isolation, as follows.
8. Extract DNA from the samples using 250 μL of phenol:chloroform:isoamyl alcohol (PCI) (25:24:1, v/v), vortexing for 15 s, centrifuging at 16,000×*g* for 10 min and

transferring the upper aqueous phase to a fresh 1.5 mL Eppendorf tube. Residual DNA is recovered from the chromatin samples by adding 130 μL of water to the PCI and repeating the extraction procedure just described. The aqueous phases from the two extractions are combined, and further extracted sequentially with 380 μL each of PCI and chloroform:isoamyl alcohol (24:1, v/v), exactly as above. To the final aqueous extract, add 1.0 μL of glycogen and 50 μL of 4 M LiCl. Fill the tube with 100% ethanol and incubate overnight at −80°C to precipitate the DNA.

9. Collect the precipitated DNA by centrifugation at 16,000 × *g* for 30 min, decant the supernatant by inverting the tube, add 70% ethanol and centrifuge at 16,000 × *g* for 5 min. Decant the supernatant by inverting the tube and gently blotting with paper towel to remove the residual liquid. Dry the DNA pellet in a Speed-vac (~ 5 min) and resuspend in 30–50 μL of TE containing RNAse. DNA can be stored at −20°C for PCR analysis at a later time.

3.4. PCR Analysis of Precipitated DNA

1. The concentration of the DNA sequence of interest in the precipitated DNA is determined by PCR analysis. The primer sets are designed to amplify ~150 bp segments. A primer set designed to amplify ~250 bp is included in the reaction as an internal control (see Notes 9 and 10).
2. Thaw all reaction components on ice and dilute the In sample (1:300) in distilled water. Mix all samples well using a vortex mixer. For each reaction, use 2 μL each of IP and diluted In samples.
3. Dispense 2 μL of IP samples in triplicate and In samples in duplicate into PCR tubes.
4. Add 13 μL PCR Master Mix to tubes containing IP and In samples, and pulse-spin in a microcentrifuge using adaptors for PCR-strips to collect the reaction mixture at the bottom of the tube.
5. The samples are amplified using the following settings: Initial denaturation at 94°C for 4 min, followed by 25 cycles of 94°C for 30 s, 52°C for 30 s, 65°C for 1 min, and a final extension for 5 min at 65°C.
6. After completion of PCR, add 3 μL of 6× ChIP loading dye and briefly spin the tubes to collect the reaction mixture at the bottom of the tube.
7. Separate PCR products on a 6% TBE-polyacrylamide gel by electrophoresis at 100 V until the dye reaches the bottom of the gel. This allows for sufficient separation of the ~250 bp control amplicon from the ~150 bp experimental amplicon. The gel is transferred to a small tray containing a solution of

ethidium bromide (5 µg/mL) in water and incubated for 2 min. The gel is then transferred on a plastic sheet and DNA fragments are visualized under UV illumination. After ensuring that the reaction worked, the gel is trimmed such that 1–2 cm on either side of the band are included. After trimming, place a Whatman sheet of 3 MM paper about the size of the trimmed gel over the gel. Take the "gel-Whatman-plastic sheet" sandwich and flip it over so that the plastic sheet is facing up and carefully remove the plastic sheet. The gel is dried on a gel dryer for 45 min. The dried gels are wrapped in a plastic sheet and exposed to a phosphorscreen overnight.

8. The next day, the radioactivity in each DNA band is quantified using a phosphorimager. First, the ratio of intensity of the experimental to the control band is calculated for both IP and input samples. The resulting IP_{exp}/IP_{con} ratio is divided by the corresponding In_{exp}/In_{con} ratio to obtain the "occupancy" of the experimental over the control sequences in the IP versus In DNA samples, which is equated with the occupancy of the experimental sequences by the protein immunoprecipitated from chromatin. The occupancy values are averaged and standard error mean is calculated from ChIP experiments conducted in duplicate using two replicate cultures (four independent ChIPs in total) and PCRs in triplicate. Mean occupancy values of unity indicate no sequence-specific association of the protein of interest with the experimental sequences in chromatin. Although dependent on the magnitude of the SEM relative to the mean, occupancy values of two or more are generally statistically significant. The fold occupancy differs over a wide range for different chromatin-binding proteins. An activator like Gcn4 or Pol II that directly contacts DNA can exhibit occupancies of 10 or more for highly expressed genes. Coactivators typically exhibit occupancies between 2 and 10 (see Note 11).

4. Notes

1. It is best to start early in the morning to allow sufficient time for cell growth and subsequent processing steps, which include induction (20–30 min), cross-linking (15 min), and washing of the cells (30 min). These steps usually take about 2 h after the cultures have reached the required A_{600} of 0.5–0.55. We have noticed that cultures grown beyond A_{600} of 0.6 frequently produce high nonspecific background signals (high IP_{con} values) and thus display decreased occupancies of the proteins of interest in chromatin.
2. We have noticed that optimal results are obtained for cells induced for 20–30 min.

3. After induction, all steps should be carried out as quickly as possible.
4. The 11% formaldehyde solution should be prepared fresh, just before the induction is complete.
5. We find that the addition of glycine does not completely inhibit HCHO action. It is therefore advisable to proceed to the next step immediately after the addition of glycine.
6. To prepare acid washed glass beads, take about 500 mL of glass beads in a 1,000 mL glass beaker. Carefully pour concentrated hydrochloric acid (in a fume hood) into the beaker to cover the glass beads. Cover the beaker with a glass plate (not with aluminum foil) and let it sit overnight. The following day, carefully decant the acid into a bottle and wash the beads with distilled water (500 mL each wash) at least five times. The beads must be washed with running distilled water until the pH is neutralized.
7. Efficient and complete fragmentation of the chromatin is crucial for obtaining a high degree of resolution and for preventing artifactual results. Optimal conditions for sonication must be empirically determined for each sonicator. We use a 450 Branson sonicator and a probe with a 1/8 in. tapered tip. The output knob is set to 1.8 with a 60% duty cycle. Care should be taken to prevent frothing of the sample during sonication, by keeping the tip of the probe submerged in the lysate and preventing any contact with the tube. It is important to check the extent of sonication by determining the length distribution of the sheared DNA fragments. One quick method is to reverse cross-linking ~50 μL of chromatin sample, extracting the DNA with phenol: chloroform: isoamyl alcohol (PCI), resolving the fragments by agarose gel electrophoresis, and staining the DNA with ethidium bromide. Additionally, we generally analyze Gcn4 occupancies in the 5′ end of the coding sequences in addition to the UAS of *ARG1*. Since Gcn4 is recruited exclusively to the UAS, significant Gcn4 occupancy of the 5′ ORF is indicative of inefficient sonication. Alternately, antibodies against TBP or other promoter-restricted general transcription factors can be used to determine the efficiency of fragmentation.
8. The antibody used for ChIP depends on the histone modification or the chromatin protein under investigation. The beads utilized in ChIP depend on the species in which the antibody was raised as well as the isotype of the antibody. For example, if the histone antibodies are generated in rabbit, choose beads coated with anti-rabbit IgG. If the antibody belongs to an isotype other than IgG, beads coated with the relevant isotype should be used. Thus, the mouse monoclonal antibody against serine 5 phosphorylated Rpb1 (ser5P), the largest subunit of Pol II, belongs to the IgM isotype, hence beads coated with anti-IgM should be used for ChIP analysis of ser5P-Pol II occupancy.

9. We find that Platinum Taq (Invitrogen) gives the best results and linearity. However, other hot start polymerases can be standardized empirically.
10. The primers, with a Tm of 55°C, are designed to amplify a ~150 bp region, and the primer sets are selected to anneal with sequences separated by at least 300–500 bp in the DNA sequence of interest. The primer concentration should be experimentally tested to determine the linear range of the PCR reaction. This is determined as follows. The DNA is purified from the chromatin extracts, diluted 1:300, and PCR reactions are set up with varying amounts of DNA (1, 2, 4, and 6 μL) and using 1–6 μM of primer. The primer concentration that reveals a linear increase in In_{exp} band intensity with constant In_{exp}/In_{cont} ratios is chosen to conduct ChIP PCRs. These conditions should be established for each primer set to insure that the yield of the amplicon obtained in the PCR reaction is proportional to the amount of IP or In sample over the complete range of IP and In DNA concentrations represented in the samples used in the experiments. This condition must be fulfilled to achieve quantitative measurements of the concentrations of experimental and control DNA sequences present in the In and IP samples. Real-time PCR can be employed as an alternative means of quantification of DNA concentrations in the In and IP samples. We generally use sequences from the coding sequences of the *POL1* gene, or an intergenic region of chromosome V as an internal control, to determine the occupancy of factors that are recruited to coding regions. Alternately, the right arm of telomere VI (TELVIR) is used as an internal control for measuring changes in histone acetylation. This region is hypoacetylated and hence small changes in histone acetylation can be detected using TELVIR as an internal control.
11. Transcription activation is accompanied by varying levels of histone eviction at the promoters and coding regions of a transcribed gene. The absolute histone acetylation levels do not measure accurate changes in histone acetylation that occur during activated transcription across the gene of interest. The occupancy ratios (IP_{exp}/IP_{TELVIR} to In_{exp}/In_{TELVIR}) for acetylated H3 and H4 should therefore be normalized to ratios obtained for total histone (H3 or H4) occupancy from the same chromatin sample. Thus, the changes in histone acetylation are presented as the fold change in acetylation relative to total H3 or H4. We have found that a "pan"-antibody against H3 provides the most reliable measurements of histone occupancy and, thus, we generally calculate the fold change in acetylation of H4 relative to total H3.

For analyzing the effects of HAT mutants (*gcn5Δ* or *esa1-ts*), the histone acetylation is measured by calculating the IP_{exp}/In_{exp}

ratios rather than occupancy values as defined above. This change in analysis is necessary because it is almost impossible to identify a control sequence for which IP_{cont}/In_{con} ratio is not substantially reduced by these HAT mutations. In such cases, it is necessary to carry out a large number of replicate immunoprecipitations, at least 4–6, to minimize variations in the amount of DNA that coimmunoprecipitates nonspecifically.

References

1. Workman JL, Kingston RE (1998) Alteration of nucleosome structure as a mechanism of transcriptional regulation. Annual Review of Biochemistry 67: 545–579.
2. Williams SK, Tyler JK (2007) Transcriptional regulation by chromatin disassembly and reassembly. Curr Opin Genet Dev 17: 88–93.
3. Li B, Carey M, Workman JL (2007) The role of chromatin during transcription. Cell 128: 707–719.
4. Schwabish MA, Struhl K (2004) Evidence for eviction and rapid deposition of histones upon transcriptional elongation by RNA polymerase II. Mol Cell Biol 24: 10111–10117.
5. Govind CK, Zhang F, Qiu H, Hofmeyer K, Hinnebusch AG (2007) Gcn5 promotes acetylation, eviction, and methylation of nucleosomes in transcribed coding regions. Mol Cell 25: 31–42.
6. Kouzarides T (2007) Chromatin Modifications and Their Function. Cell 128: 693–705.
7. Govind CK, Qiu H, Ginsburg DS, Ruan C, Hofmeyer K, et al. (2010) Phosphorylated Pol II CTD Recruits Multiple HDACs, Including Rpd3C(S), for Methylation-Dependent Deacetylation of ORF Nucleosomes. Molecular Cell 39: 234–246.
8. Govind CK, Yoon S, Qiu H, Govind S, Hinnebusch AG (2005) Simultaneous recruitment of coactivators by Gcn4p stimulates multiple steps of transcription in vivo. Mol Cell Biol 25: 5626–5638.
9. Ginsburg DS, Govind CK, Hinnebusch AG (2009) NuA4 Lysine Acetyltransferase Esa1 Is Targeted to Coding Regions and Stimulates Transcription Elongation with Gcn5. Mol Cell Biol 29: 6473–6487.
10. Selth LA, Sigurdsson S, Svejstrup JQ (2010) Transcript Elongation by RNA Polymerase II. Annual Review of Biochemistry 79: 271–293.
11. Shahbazian MD, Grunstein M (2007) Functions of Site-Specific Histone Acetylation and Deacetylation. Annual Review of Biochemistry 76: 75–100.
12. Carrozza MJ, Li B, Florens L, Suganuma T, Swanson SK, et al. (2005) Histone H3 methylation by Set2 directs deacetylation of coding regions by Rpd3S to suppress spurious intragenic transcription. Cell 123: 581–592.
13. Keogh MC, Kurdistani SK, Morris SA, Ahn SH, Podolny V, et al. (2005) Cotranscriptional set2 methylation of histone H3 lysine 36 recruits a repressive Rpd3 complex. Cell 123: 593–605.
14. Kim T, Buratowski S (2009) Dimethylation of H3K4 by Set1 Recruits the Set3 Histone Deacetylase Complex to 5′ Transcribed Regions. Cell 137: 259–272.
15. Smith E, Shilatifard A (2010) The Chromatin Signaling Pathway: Diverse Mechanisms of Recruitment of Histone-Modifying Enzymes and Varied Biological Outcomes. Molecular Cell 40: 689–701.
16. Lee C-K, Shibata Y, Rao B, Strahl BD, Lieb JD (2004) Evidence for nucleosome depletion at active regulatory regions genome-wide. Nat Genet 36: 900–905.
17. Sekinger EA, Moqtaderi Z, Struhl K (2005) Intrinsic Histone-DNA Interactions and Low Nucleosome Density Are Important for Preferential Accessibility of Promoter Regions in Yeast. Molecular Cell 18: 735–748.
18. Kuo M-H, Allis CD (1999) In Vivo Cross-Linking and Immunoprecipitation for Studying Dynamic Protein:DNA Associations in a Chromatin Environment. Methods 19: 425–433.

Chapter 3

Monitoring the Effects of Chromatin Remodelers on Long-Range Interactions In Vivo

Christine M. Kiefer and Ann Dean

Abstract

In metazoans transcriptional enhancers and their more complex relatives, locus control regions, are often located at great linear distances from their target genes. In addition, these elements frequently activate different members of gene families in temporal sequence or in different tissues. These issues have complicated understanding the mechanisms underlying long-range gene activation. Advances in primarily technical approaches, such as chromosome conformation capture (3C) and its derivatives have now solidified the idea that distant regulatory elements achieve proximity with their target genes when they are activating them. Furthermore, these approaches are now allowing genome-wide views of chromosome interactions that are likely to include regulatory, structural, and organization aspects from which we will be able to understand more about nuclear structure. At the base of these advances are experimental approaches to localize protein-binding sites in chromatin, to assess remodeling of chromatin and to measure interaction frequency between distant sites. Examples of these approaches comprise this review.

Key words: Long-range gene activation, Locus control regions, Chromatin immunoprecipitation, Chromatin remodeling, DNase I hypersensitivity, Chromosome conformation capture

1. Introduction

Long-range interactions between distal regulatory DNA elements and their associated genes have been implicated in the regulation of a wide variety of genetic loci (1). Perhaps the best characterized of these is the β-globin locus which contains five genes that are expressed during different stages of development (2). The developmental timing and high level of β-globin gene expression is coordinately regulated by the distal locus control region (LCR)

Randall H. Morse (ed.), *Chromatin Remodeling: Methods and Protocols*, Methods in Molecular Biology, vol. 833,
DOI 10.1007/978-1-61779-477-3_3, © Springer Science+Business Media, LLC 2012

located between 6 and 50 kb away from the globin genes. The LCR is defined by 4 erythroid-specific DNase I hypersensitive sites that are occupied by a number of erythroid-specific and ubiquitous transcription factors. Data obtained by using techniques such as chromatin conformation capture (3C) and RNA-trap to identify the stage-specific interactions between the LCR and the globin genes led to the proposal of the "chromatin hub" model, in which the LCR preferentially interacts with specific β-globin gene promoters at the appropriate stage of development to enhance transcriptional activity of those genes (3–5). Several factors, including erythroid-specific EKLF and GATA-1, and more widely expressed NLI/Ldb1, are required for long-range interactions between the LCR and the globin gene promoters (6–8). Although a number of published examples support this model, there is still much about the mechanism of LCR interactions and loop formation that is not well understood (9). How these interactions change during development and the factors involved remain an area of intense study.

Chromatin remodeling has been directly implicated in β-globin loop formation, and it is possible that this correlation extends to other loci involved in long-range transcriptional regulation. The chromatin remodeling protein BRG-1, an ATPase component of the SWI/SNF chromatin remodeling complex, was shown to be required for loop formation between the LCR and β-globin gene promoters (10). Brg-1 mutant cells do not form a loop between the LCR and the murine β-major gene, even though transcription factors required for looping and gene activation still occupy the LCR and globin gene promoter. This suggests that chromatin remodeling by Brg-1 is an early event in β-globin locus loop formation. In the related α-globin locus, Brg-1 was shown to mediate GATA-1-dependent chromatin looping by regulating DNase I sensitivity and histone modifications in the LCR and at globin gene promoters (11). The role of chromatin remodeling proteins on long-range interactions at other loci remains elusive, and methods to monitor the effects of chromatin remodeling on long-range interactions are important tools for understanding these relationships. The methods described herein can examine the in vivo occupancy of chromatin remodeling proteins and transcription factors and the changes in histone modification related to chromatin remodeling activity (ChIP), the effects of chromatin remodeling proteins on DNase I sensitivity, and the interaction frequency between widely separated DNA elements (3C). Although the specific example described below provides only a snapshot of one cell type, these techniques can be applied to experimental systems comparing two or more states of gene activation and can provide insight into the role of chromatin remodeling activities on long-range interactions.

2. Materials

2.1. Chromatin Immunoprecipitation

1. RPMI medium containing 10% fetal bovine serum (FBS).
2. 37% formaldehyde.
3. 1.25 M glycine (4.69 g glycine dissolved in 50 mL H_2O, filter sterilized) (see Note 1).
4. 1× phosphate-buffered saline (PBS): 137 mM NaCl, 2.7 mM KCl, 10 mM Na_2HPO_4 (dibasic, anhydrous), 2 mM KH_2PO_4 (monobasic, anhydrous) (pH 7.4).
5. Cell lysis buffer (50 mL): 10 mM Tris–HCl pH 8.0, 10 mM NaCl, 0.2% Igepal (see Note 2). Filter sterilize. Aliquot the amount needed per experiment. Protease inhibitors (see item 6) at a 1× concentration should be added just before use.
6. Protease inhibitors (see Note 3):
   (a) 50 mM 4-(2-aminoethyl) benzenesulfonyl fluoride hydro chloride(AEBSF) – Distribute ~1 mL aliquots as a 100× stock solution (final concentration is 500 μM).
   (b) 2 mM pepstatin – Distribute 200 μL aliquots as a 1,000× stock solution (final concentration is 2 μM).
   (c) 2 mg/mL aprotinin – Distribute 200 μL aliquots as a 1,000× stock solution (final concentration is 2 μg/μL).
   (d) 1 mM leupeptin – Distribute 200 μL aliquots as a 1,000× stock solution (final concentration is 1 μM).
7. 100 mM $CaCl_2$, filter sterilized.
8. MNase (20 U/μL).
9. 0.5 M EDTA pH 8.0.
10. Nuclei lysis buffer (50 mL): 50 mM Tris–HCl pH 8.0, 10 mM EDTA, 1% SDS. Filter sterilize. Aliquot the amount needed per experiment. Protease inhibitors (see item 6) should be added to 1× concentration just before use.
11. IP dilution buffer (100 mL): 20 mM Tris–HCl pH 8.0, 150 mM NaCl, 2 mM EDTA pH 8.0, 0.01% SDS, 1% Triton X-100. Filter sterilize. Aliquot the amount needed per experiment. Protease inhibitors (see item 6) should be added to 1× concentration just before use.
12. Sonicator – we have used the Misonex 3000 and the Diagenode Bioruptor.
13. RNase Cocktail (Invitrogen).
14. DynaBeads Protein G (Invitrogen) and magnetic rack. Beads should be washed 3 times and re-suspended in an equal volume of IP dilution buffer (see item 11) before use.
15. Bovine serum albumin (BSA) lyophilized powder.

16. ChIP-grade antibody (see Note 4).
17. Normal IgG: Used as an isotype control (anti-goat, anti-rabbit, or anti-mouse, as appropriate for the specific antibody being used).
18. 10% sodium deoxycholate, filter sterilized (see Note 5).
19. Modified RIPA buffer (500 mL): 10 mM Tris–HCl pH 7.5, 1 mM EDTA pH 8.0, 0.5 mM EGTA, 1% Triton X-100, 0.1% SDS, 0.1% sodium deoxycholate, and 14 mM NaCl. Filter sterilize.
20. Elution buffer (3 mL) (must be made fresh): 20 mM Tris–HCl pH 7.5, 5 mM EDTA pH 8.0, 0.5 M NaCl, 1% SDS, and 50 μg/mL proteinase K.
21. Thermomixer-R (Eppendorf).
22. Phenol:chloroform:isoamyl alcohol (25:24:1 v/v) pH 8.0–8.5.
23. 100% ethanol.
24. 1× TE buffer pH 8.0: 10 mM Tris–HCl pH 8.0, 1 mM EDTA pH 8.0.
25. TaqMan Universal PCR Master Mix (see Note 6).
26. ABI 7900HT or similar machine for real-time quantitative PCR (RT-qPCR).

2.2. DNase I Sensitivity by RT-qPCR

1. RPMI medium containing 10% FBS.
2. 1× phosphate-buffered saline (PBS): (137 mM NaCl, 2.7 mM KCl, 10 mM Na_2HPO_4 [dibasic, anhydrous], 2 mM KH_2PO_4 (monobasic, anhydrous) (pH 7.4).
3. Nuclei isolation buffers (100 mL) see Table 1 below. Final concentrations are shown in brackets. All solutions are filter sterilized prior to use and should be kept on ice at all times.
4. 0.05% trypan blue.
5. 100 mM $CaCl_2$, filter sterilized.
6. DNase I.
7. Phenol:chloroform:isoamyl alcohol (25:24:1 v/v) pH 8.0–8.5.
8. 3 M sodium acetate (NaOAc). Filter sterilize.
9. 100% ethanol.
10. 1× TE buffer pH 8.0: 10 mM Tris–HCl pH 8.0, 1 mM EDTA pH 8.0.
11. Stop solution (50 mL): 125 mM EDTA pH 8.0, 1 mg/mL proteinase K, 10% SDS. Filter sterilize.
12. TaqMan Universal PCR Master Mix (see Note 6).
13. ABI 7900HT or similar machine for RT-qPCR.

2.3. Chromosome Conformation Capture (3C)

1. RPMI medium containing 10% FBS.
2. 37% formaldehyde.
3. 1.25 M glycine, filter sterilized (see Note 1).

Table 1
Nuclear isolation buffers

Stock solution	Homogenization (5% sucrose)	Cushion (10% sucrose)	Wash (8.5% sucrose)
1 M Tris–HCl pH 7.4	1 mL (10 mM)	1 mL	1 mL
0.5 M EDTA	0.2 mL (1 mM)	0.2 mL	–
0.05 M EGTA	0.2 mL (0.1 mM)	0.2 mL	–
5 M NaCl	0.3 mL (15 mM)	0.3 mL	0.3 mL
1 M KCl	5 mL (50 mM)	5 mL	6 mL (60 mM)
0.1 M Spermine	0.15 mL (0.15 mM)	0.15 mL	0.15 mL
0.2 M Spermidine	0.25 mL (0.5 mM)	0.25 mL	0.25 mL
IGEPAL	0.2 mL (2%)	–	–
Sucrose	5 g (5%)	10 g (10%)	8.5 g (8.5%)
H_2O	~92.7 mL	~92.9 mL	~92.3 mL
Total	100 mL	100 mL	100 mL

4. 1× phosphate-buffered saline (PBS): (137 mM NaCl, 2.7 mM KCl, 10 mM Na_2HPO_4 (dibasic, anhydrous), 2 mM KH_2PO_4 (monobasic, anhydrous) (pH 7.4).
5. Cell lysis buffer (50 mL): 10 mM Tris–HCl pH 8.0, 10 mM NaCl, 0.2% Igepal (see Note 2). Filter sterilize. Aliquot the amount needed per experiment. Protease inhibitors (see item 6) at a 1× concentration should be added just before use.
6. Protease Inhibitors (see Note 3):
   (a) 50 mM 4-(2-aminoethyl) benzenesulfonyl fluoride hydro chloride(AEBSF) – Distribute ~1 mL aliquots as a 100×stock solution (final concentration is 500 μ M).
   (b) 2 mM pepstatin – Distribute 200 μ L aliquots as a 1,000×stock solution (final concentration is 2 μ M).
   (c) 2 mg/mL aprotinin – Distribute 200 μ L aliquots as a 1,000×stock solution (final concentration is 2 μ g/ μ L).
   (d) 1 mM leupeptin – Distribute 200 μ L aliquots.
7. EcoRI digestion buffer (5 mL): 1× EcoRI digestion buffer (New England Biolabs) (see Note 7) with 0.2% Igepal. Protease inhibitors (see item 6) at a 1× concentration should be added just before use.
8. 20% SDS.
9. 10% Igepal.

10. 20% Triton X-100.
11. EcoRI at 100,000 U/mL (high concentration).
12. Ligation buffer (5 mL): 1× T4 DNA ligase buffer (New England Biolabs) with 1% Triton X-100. Protease inhibitors (see item 6) at a 1× concentration should be added just before use.
13. T4 DNA ligase at 400,000 U/mL.
14. 5 M NaCl.
15. Thermomixer-R.
16. Proteinase K 10 mg/mL.
17. Phenol:chloroform:isoamyl alcohol (25:24:1 v/v) pH 8.0–8.5.
18. 100% ethanol.
19. 1× TE buffer pH 8.0: 10 mM Tris–HCl pH 8.0, 1 mM EDTA pH 8.0.
20. TaqMan Universal PCR Master Mix (see Note 6).
21. ABI 7900HT or similar machine for RT-qPCR.

3. Methods

For all methods described herein, the appropriate personal protective equipment should be worn at all times. Additionally, since all methods rely on RT-qPCR, the use of filter tips is highly recommended. Aliquot only the amount of solution needed from stocks to avoid contamination. The examples shown use K562 cells, a human erythroleukemia cell line that expresses predominantly the fetal γ-globin genes with lower levels of the embryonic ε-globin gene (Fig. 1).

3.1. ChIP (Results Are Shown in Fig. 1)

Day 1

1. *Cross linking*: Grow and harvest 2.5–5 × 10^7 K562 cells in 40 mL RPMI medium containing 10% FBS (see Note 8). For cross linking, add 1.1 mL of 37% formaldehyde (final concentration = 1%). Incubate for 10 min at room temperature with gentle agitation (see Note 9). To terminate the cross linking and quench the formaldehyde, add 4.6 mL of 1.25 M glycine (final concentration = 125 mM). Incubate for 5 min at room temperature with gentle agitation (see Note 9).
2. *Wash*: Centrifuge the cells at 4°C for 5 min at ~300 × *g*. Remove the supernatant and gently tap the bottom of the tube to loosen the cell pellet. Add 40 mL PBS and carefully rock by hand 3–4 times to resuspend and wash cells. Centrifuge the

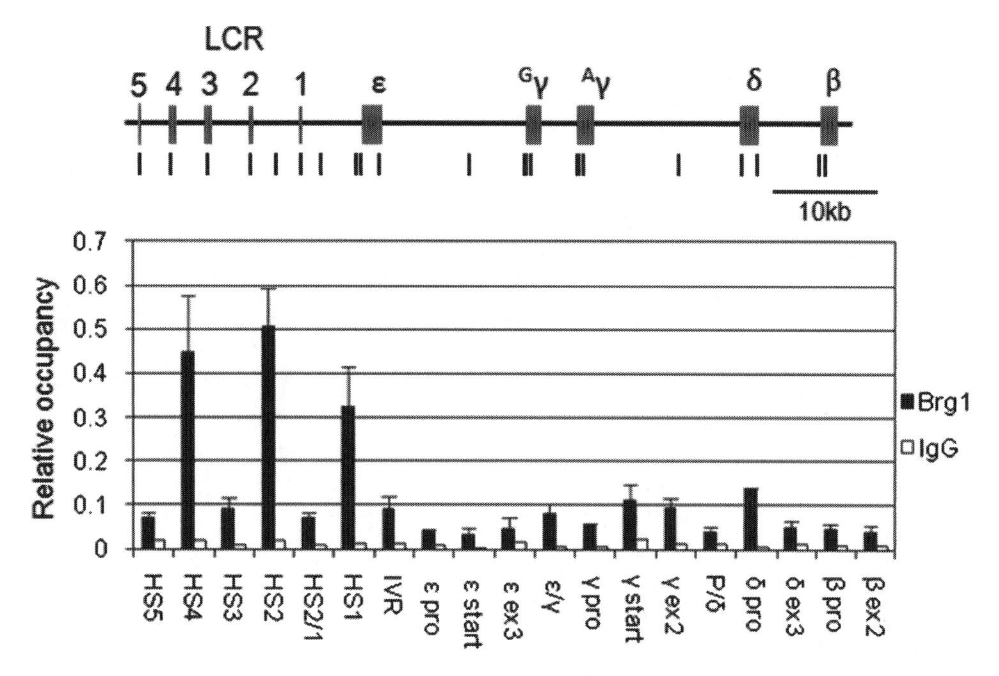

Fig. 1. BRG-1 occupancy in the human β-globin locus. A map of the β-globin locus appears at the top of the figure. Locations of PCR primers used for ChIP are indicated by *lines* below the map. ChIP was performed using a BRG-1 antibody (Santa Cruz) to determine occupancy in K562 erythroleukemia cells. qPCR primers were specific to the regions indicated on the *x*-axis. Data was calculated against the INPUT sample using the comparative Ct method and is shown as "Relative occupancy" (*y*-axis). Solid black bars indicate BRG-1 occupancy in K562 cells. White bars indicate values for an isotype control antibody (IgG).

cells at 4°C for 5 min at ~300 × *g*. Gently tap the bottom of the tube to loosen the cell pellet.

3. *Cell lysis*: Add protease inhibitors (AEBSF, leupeptin, aprotinin, pepstatin A) to 1 mL of cell lysis buffer at a 1× concentration. Use 400 μL to resuspend the cells and incubate for 10 min on ice. Centrifuge the nuclei at 4°C for 5 min at ~700 × *g*. Remove the supernatant and gently tap the bottom of the tube to loosen the pellet.
4. *MNase digestion*: Add 10 μL of 100 mM $CaCl_2$ to 1 mL of cell lysis buffer (final concentration of $CaCl_2$ = 1 mM). Use 500 μL to resuspend the nuclei. Incubate for 5 min at 37°C. Add 10 μL (200 U) of MNase and incubate at 37°C for an additional 10 min. MNase digestion reduces the overall fragment size to between 150 and 250 bp. This step helps to facilitate the analysis of histone modifications and transcription factor binding at a much higher resolution than may be achieved when using sonication alone.
5. *Nuclei lysis*: Add 10 μL of 0.5 M EDTA (final concentration = 10 mM) to stop the enzymatic reaction. Centrifuge the nuclei at 4°C for 5 min at ~700 × *g*. Remove the supernatant

and gently tap the bottom of the tube to loosen the pellet. Resuspend the nuclei in 1 mL of nuclei lysis buffer. Incubate on ice for 10 min.

6. *Sonication*: Increase the volume of the sample to 2.0 mL with IP dilution buffer. This step helps to dilute out the detergent in the sample before sonication where bubbles may form. Using the Misonex 3000, sonicate the lysate for a total process time of 5 min, cycling on for 30 s and off for 60 s, at 4.5 power (12 W). Alternatively, the Diagenode Bioruptor may be used for sonication. Settings for the Bioruptor are cycles of 30 s on and 30 s off on high power for a total of 15 min. For other sonicators, test by varying intensity and duration to obtain average DNA fragment size of 150–250 bp as outlined below.

   Note: After sonication, remove 100 μL and process separately to check fragment size:

   (a) Incubate at 65°C for 2 h to reverse cross-links. Perform an organic extraction using an equal volume of the phenol:chloroform:isoamyl alcohol mix (25:24:1). Use care to collect only the aqueous layer.

   (b) Add 5 μL RNase cocktail. Incubate for 30 min at 37°C. Microcentrifuge at ~ 13,000 × *g* for 1 min. No precipitation is necessary.

   (c) Run two volumes of the sample (generally 10 μL and 20 μL) on a 1.5% agarose gel. The MNase digestion combined with sonication should yield fragments that range between 150 and 250 bp.

7. *Preclearing*: Microcentrifuge the remaining sonicated chromatin at 4°C for 10 min at ~13,000 × *g* to pellet any cellular debris. Transfer supernatant to a clean 15 mL conical tube, and dilute the soluble chromatin with IP dilution buffer to 5 mL. Add 250 μL of protein G DynaBeads to preclear the chromatin of any proteins that may bind nonspecifically to the beads. For additional information on how to prepare and work with magnetic beads, see Subheading 2.1, item 14 and Note 10. Incubate on a rotating shaker at 4°C for 2 h.

8. *Capture bead preparation*: Add 30 mg BSA to 1 mL of IP dilution buffer to make a 3% blocking solution. Aliquot 60 μL of protein G Dynabeads to use for capturing chromatin complexes and remove IP dilution buffer. Dilute the beads in 300 μL blocking buffer. Incubate at 4°C for a *minimum* of 2 h on a rotating shaker.

9. *Immunoprecipitation*: Immobilize the beads in the precleared chromatin and remove 160 μL (20%) of precleared chromatin to serve as the INPUT sample. Store INPUT at 4°C until Day

2. Distribute remaining precleared chromatin into 800 μL aliquots. Add antibody (typically 2.5–20 μg depending on the quality of the antibody; we use 20 μg of anti-Brg1) to one aliquot and 6.25 μL of normal anti-goat IgG (2.5 μg) (or other IgG isotype, as appropriate) to another aliquot for an isotype control. Incubate at 4°C overnight on a rotating shaker. The remaining chromatin aliquots can be stored for future use at −80°C for up to 3 months. Beads that were used for preclearing may be discarded. Capture beads from step 8 should be stored at 4°C overnight (they may remain on the rotating shaker) for use on Day 2.

Day 2

10. *Capture immunoprecipitated chromatin*: Microcentrifuge the blocked Dynabeads from step 8, immobilize the beads, and remove 240 μL of blocking buffer to return the beads to the correct concentration. Mix gently by tapping the tube until the beads are evenly distributed and add 30 μL of beads to each sample. Incubate at 4°C for a minimum of 2 h on a rotating shaker.
11. *Wash and elute*: Wash beads three times with 500 μL of modified RIPA buffer followed by two times with 500 μL of 1× TE buffer (see Note 10). Remove the last wash. Add 150 μL of elution buffer to the beads. Add 140 μL of elution buffer to the INPUT sample (total volume of INPUT = 300 μL). Incubate IP, control IgG, and INPUT samples at 65°C for 2 h with shaking (1,300 rpm) using a Thermomixer (Eppendorf). After a 5–10 s spin in a microcentrifuge at <2,400 × *g* immobilize the magnetic beads and transfer the elution buffer from the IP and IgG samples to a clean tube. Add 150 μL of fresh elution buffer to the beads for a second round. Incubate at 65°C for an additional 30 min with shaking (1,300 rpm) using a Thermomixer. After a brief spin in a microcentrifuge at <2,400 × *g*, immobilize the magnetic beads and transfer the elution buffer from the IP and IgG samples to the tubes from the previous elution step (total elution volume = 300 μL). The INPUT sample should remain at 65°C throughout this step.
12. *DNA cleanup*: Perform an organic extraction on all samples using an equal volume (300 μL) of the phenol:chloroform:isoamyl alcohol mix (25:24:1). Use care to only remove the aqueous layer. Precipitate the DNA by adding 2.5 volumes (750 μL) of 100% ethanol. Salt (NaCl) is already present in sufficient concentrations for precipitation. If DNA concentrations are low, glycogen (1–2 μL of a 20 mg/mL stock) may be added as a carrier for precipitation and cleanup. Resuspend the pellet in

300 μL 1×TE and then dilute the INPUT sample 1:25 for qPCR analysis.

13. *qPCR setup*: Quantitate the amount of DNA in the diluted INPUT, IP, and IgG samples. Perform qPCR on equal amounts of DNA, generally 10–50 ng per 25 μL reaction. All samples should be done in duplicate or triplicate. Primers and TaqMan probes (if TaqMan chemistry is used) can be designed using Primer Express (ABI) or another equivalent primer design program. If SyBr Green chemistry is used, include a dissociation curve at the end of the qPCR cycle to ensure only one specific product was generated. PCR products may be additionally verified by running them on an agarose gel to look for the presence of a single band.

14. *Data analysis*: All data are calculated against the INPUT sample. Our lab routinely analyzes qPCR data using the comparative Ct method to express real-time PCR results for ChIP experiments as "relative occupancy" (12). Alternatively, if primer efficiencies are not equal then data analysis may be based on the Pfaffl method (13) or a standard curve. For an excellent review of real-time PCR analysis calculations and the advantages/disadvantages of choosing one method over another, please see The University of South Carolina School of Medicine's Real-Time PCR tutorial at http://pathmicro.med.sc.edu/pcr/realtime-home.htm.

3.2. DNase I Sensitivity by RT-qPCR (Results Are Shown in Fig. 2)

1. *Harvest cells*: Grow K562 cells in RPMI medium containing 10% FBS. Count cells and remove 5×10^6 cells (1×10^6 for each concentration of DNase I). Centrifuge the cells at 4°C for 5 min at ~300 × *g*. Gently tap the bottom of the tube to loosen the cell pellet. Add 10 mL ice-cold PBS to wash any remaining medium from the cells. Centrifuge the cells at 4°C for 5 min at ~300 × *g*. Gently tap the bottom of the tube to loosen the cell pellet.

2. *Cell lysis*: Add 0.6 mL of ice-cold 5% sucrose homogenization buffer. Incubate for 3 min on ice. Verify that cell lysis is complete by transferring 5 μL of the nuclei to a tube containing 50 μL of 0.05% trypan blue. Wait 1–2 min and then count the number of blue nuclei to confirm >90% of cells are lysed.

3. *Isolation of nuclei*: Layer 0.6 mL of nuclei from step 2 over 0.35 mL of 10% ice-cold sucrose cushion buffer. Centrifuge at 4°C for 20 min at 1,600 × *g*. Remove buffer and resuspend the nuclei pellet in 500 μL of ice-cold 8.5% sucrose wash buffer (1×10^6 cells/100 μL). Keep the nuclei on ice.

4. *DNase I digestion*: Aliquot nuclei into five 1.5 mL microfuge tubes (100 μL each). Add the following to

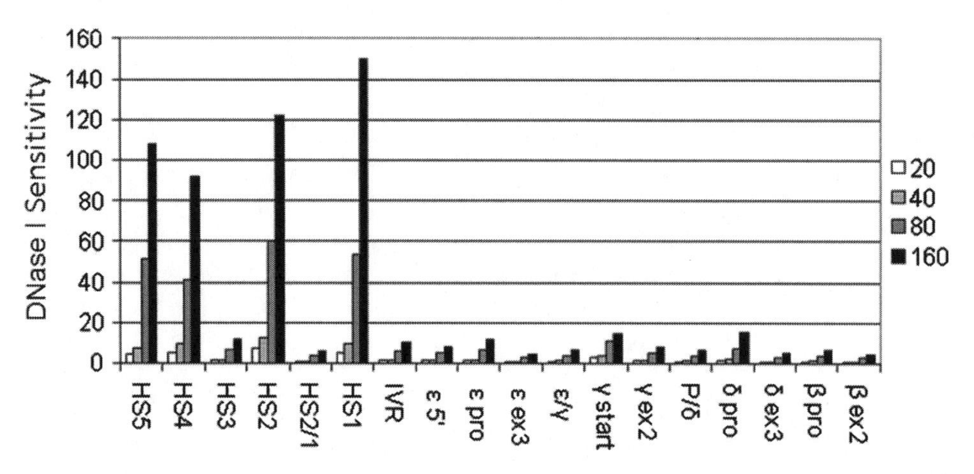

Fig. 2. DNase I sensitivity in the human β-globin locus. DNase I sensitivity analysis by RT-qPCR was performed to determine regions of general sensitivity and hypersensitivity to DNase I. qPCR primers were specific to the regions indicated on the *x*-axis. Data was calculated against an untreated control using the comparative Ct method and is plotted as "DNase I sensitivity" (*y*-axis). *White, light gray, dark gray,* and *black bars* represent DNase I sensitivity using 20 U, 40 U, 80 U, or 160 U of DNase I for digestion, respectively.

each tube (see Note 11) and incubate for 10 min at room temperature. Stop the digestion by adding 10 μL of stop solution and 10 μL of 5 M NaCl.

(a) *Tube 1*: 3 μL of 0.1 M $CaCl_2$ (serves as an untreated control).

(b) *Tube 2*: 3 μL of 0.1 M $CaCl_2$, 20 U DNase I.

(c) *Tube 3*: 3 μL of 0.1 M $CaCl_2$, 40 U DNase I.

(d) *Tube 4*: 3 μL of 0.1 M $CaCl_2$, 80 U DNase I.

(e) *Tube 5*: 3 μL of 0.1 M $CaCl_2$, 160 U DNase I.

5. *DNA cleanup*: Incubate DNase I treated samples at 55°C for 3 h OR 37°C overnight.

6. Perform an organic extraction on all samples using an equal volume of the phenol:chloroform:isoamyl alcohol mix (25:24:1). Use care to only remove the aqueous layer. Precipitate the DNA with 100% ethanol and 3 M sodium acetate. Resuspend the pellet in 200 μL 1× TE and measure concentration.

7. *qPCR setup*: Perform qPCR on an equal amount of DNA, generally 10–50 ng per 25 μL reaction. All samples should be done in duplicate or triplicate. Primers and TaqMan probes (if TaqMan chemistry is used) can be designed using Primer Express (ABI) or another equivalent primer design program. If SyBr Green chemistry is used, include a dissociation curve at the end of the qPCR cycle to ensure only one specific product was generated. PCR products may be additionally verified by running them on an agarose gel to look for the presence of a single band.

8. *Data analysis*: All data are calculated against the control untreated sample. Our lab routinely analyzes qPCR data using the comparative Ct method to determine DNase I sensitivity (12). Alternatively, if primer efficiencies are not equal then data analysis may be based on the Pfaffl method (13) or a standard curve. For an excellent review of real-time PCR analysis calculations and the advantages/disadvantages of choosing one method over another, please see The University of South Carolina School of Medicine's Real-Time PCR tutorial at http://pathmicro.med.sc.edu/pcr/realtime-home.htm.

3.3. Chromosome Conformation Capture (3C) (Results Are Shown in Fig. 3)

Day 1

1. *Cross linking and wash*: Grow K562 cells in RPMI medium containing 10% FBS. Complete steps 1 and 2 as described in the ChIP protocol (cross linking and wash). Resuspend cells in a small volume of PBS and split into aliquots of ~8×10^6 cells per 1.5 mL tube. Microcentrifuge cells at 800–1,000 rpm for 5 min and remove PBS. Gently tap the bottom of the tube to loosen the cell pellet.
2. *Cell lysis*: Incubate 8×10^6 cells in 1 mL of Cell Lysis buffer at 37°C for 20 min with gentle shaking (400 rpm) in a Thermomixer (see Note 12). Microcentrifuge nuclei at 800–1,000 rpm for 5 min. Remove buffer from pellet and add another 1 mL of cell

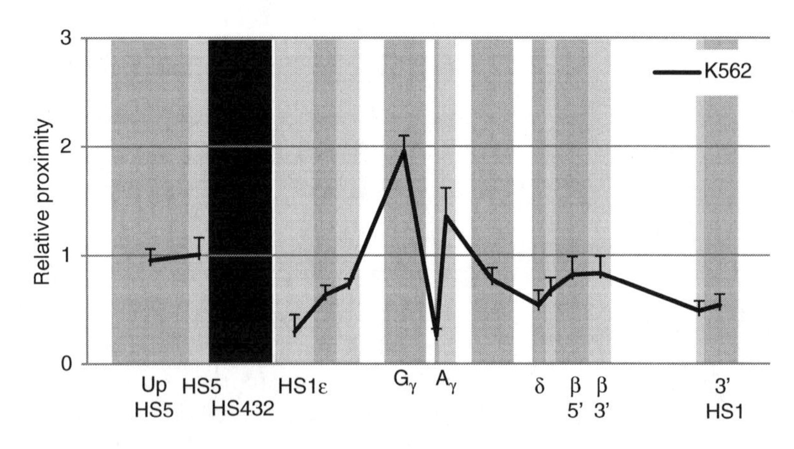

Fig. 3.3 Locus-wide interactions with the β-globin LCR. 3C analysis was performed on K562 cells. Cross-linked chromatin was digested with EcoRI and ligation products were quantitated by qPCR using a primer specific to an anchor fragment (containing LCR hypersensitive sites HS4, HS3, and HS2) with a second primer specific to 1 of the 15 other EcoRI fragments analyzed. Ligation frequencies were calculated using a conserved gene desert region on chromosome 16 (ENCODE region ENr313) as a control. EcoRI fragments that were analyzed by 3C are represented by alternating *dark* and *light gray bars* and the anchor fragment is shown as a *black bar*. The fragments containing hypersensitive sites and globin genes are labeled below the *x*-axis. The size of each fragment and the distance between fragments are shown to scale. Data are expressed as "Relative proximity" (*y*-axis). The *black line* represents data from K562 cells.

lysis buffer. Incubate at 37°C for 20 min with gentle shaking (400 rpm) in a Thermomixer. Microcentrifuge at 800–1,000 rpm for 5 min and remove buffer.

3. *Wash*: Resuspend the nuclei in 1 mL of EcoRI digestion buffer and gently rock the tube several times by hand (see Note 13). Microcentrifuge at 800–1,000 rpm for 5 min and remove buffer. Resuspend the nuclei in 0.8 mL of EcoRI digestion buffer. Remove 20 μL for use as an undigested control and store at −20°C for use in step 10. Aliquot the remaining nuclei into four 1.5 mL microcentrifuge tubes (~2×10^6 cells per tube) and increase the volume of each to 400 μL with EcoRI digestion buffer. Each of the four tubes serves as one technical replicate (see Note 14).
4. *Nuclei lysis*: Add 2 μL of 20% SDS to each tube (final concentration = 0.1%) and incubate at 65°C for 30 min to lyse the nuclei. Add 20 μL of 20% Triton X-100 (final concentration = 1%) to sequester the SDS.
5. *Restriction enzyme digestion*: Add 400 U of EcoRI (4 μL of 100 U/μL) and incubate at 37°C overnight with shaking (900 rpm) in a Thermomixer (see Note 15).

Day 2

6. Remove 40 μL (10 μL from each tube) for use as a digested control and store at −20°C for use in step 10. Incubate the remainder of each sample at 65°C for 20 min to inactivate EcoRI.
7. *Intramolecular ligation*: Dilute each sample with 600 μL of 1× ligation buffer (total volume of each tube = 1 mL). Incubate at 37°C for 30 min. Add 1,200 U of T4 DNA ligase and incubate at 16°C overnight with shaking (900 rpm) in a Thermomixer.

Day 3

8. *Intramolecular ligation (continued):* Stop the ligation reaction by adding 40 μL of 5 M NaCl (1/25 volume) and proteinase K and incubate at 65°C for 3–5 h with shaking (900 rpm) in a Thermomixer.
9. *DNA cleanup*: Perform several rounds of organic extraction (2× phenol, 2× phenol:chloroform:isoamyl alcohol, and 2× chloroform). Use care to only remove the aqueous layer. Precipitate the DNA by adding 2.5 volumes of 100% ethanol. Salt (NaCl) is already present in sufficient concentrations for precipitation. Resuspend in 200 μL 1×TE and measure concentration.
10. *Digestion efficiency test*: Samples removed as undigested and digested controls (steps 3 and 6) should be purified as described in step 9 and resuspended at a concentration of 10 ng/μL.

qPCR is performed using primers from either side of the EcoRI digestion sites that create each fragment to be analyzed by 3C along with a primer pair that does not span an EcoRI restriction site (control). Efficiency (the percentage of digestion) should be >85% and can be calculated using the following formula:

$$100-(2^{\Delta \text{Ct digested}}/2^{\Delta \text{Ct undigested}})\times 100$$

ΔCt is calculated by subtracting the Ct of the digested or undigested sample from the control Ct value.

11. *qPCR setup*: Dilute DNA to ~100–200 ng/μL and use 5 μL per qPCR reaction (see Note 16). All samples should be done in duplicate or triplicate. Proximity to the LCR was determined using a primer for the fragment containing LCR HS4, HS3, and HS2 as an anchor with a second primer for 1 of the 15 other beta globin locus EcoRI fragments analyzed. Because of the complexity and number of fragments analyzed, we have generally chosen to use SYBR Green chemistry, which is significantly less costly than TaqMan, for all 3C qPCR. Primers can be designed using Primer Express (ABI) or another equivalent primer design program. With SYBR Green chemistry, a dissociation curve should be included at the end of the qPCR cycle to ensure only one specific product was generated. qPCR samples should be additionally run on an agarose gel to look for the presence of a single band and the DNA extracted and sequenced to confirm the expected ligation products.
12. *Data analysis*: All data are calculated against the ligation frequency of two fragments in an unrelated locus that are separated by a third 1–2 kb fragment. For analysis in K562 cells, we have used α-tubulin and a conserved gene desert region on chromosome 16 (ENCODE region ENr313). Our lab routinely analyzes qPCR data using the comparative Ct method to determine relative proximity of two fragments (12). qPCR products are also sequenced to confirm expected ligation products. Primer efficiencies for each pair of primers (anchor + query fragment) should be determined using a control template (see Note 17). If primer efficiencies are not equal then data analysis may be based on the Pfaffl method (13) or a standard curve to correct for differences between primer pairs. For an excellent review of real-time PCR analysis calculations and the advantages/disadvantages of choosing one method over another, please see The University of South Carolina School of Medicine's Real-Time PCR tutorial at http://pathmicro.med.sc.edu/pcr/realtime-home.htm.

4. Notes

1. Unless otherwise indicated, H_2O is double distilled or molecular biology grade.
2. Igepal is the trade name for the detergent octylphenoxypolyethoxyethanol and is described as being "chemically indistinguishable" from Nonidet P-40. Either of these detergents should work equally well for this step.
3. Protease inhibitors may be made individually as described or purchased as a ready-made protease inhibitor cocktail.
4. Antibodies to any number of chromatin-remodeling proteins, transcription factors, or histone modifications may be used in this assay, but it is important that they have been validated for use in ChIP.
5. 10% stock of sodium deoxycholate should be protected from light.
6. Alternatively, we have used iQ SYBR Green Supermix (Bio-Rad).
7. Buffer 3 from New England Biolabs has been used as an equivalent substitution.
8. The protocol herein has been optimized for suspension tissue culture cells and utilizes ~4 million cells per antibody. We have been successful using this ChIP protocol for as few as 500,000 cells per antibody in situations where cell numbers were limited.
9. Generally, we use a nutator for gentle agitation, but a rotating shaker on the lowest speed is also appropriate.
10. Protein G DynaBeads are magnetic beads and require the use of a DynaMag-2 rack (or equivalent) to immobilize the beads. For each wash, beads should be immobilized and the buffer removed. After new wash buffer is added, the beads should be rocked gently back and forth by hand until completely resuspended. A brief spin in a microcentrifuge at <2,400 × *g* helps to remove all buffer and beads from the tube cap before immobilization. Beads that are bound to chromatin should always be mixed gently (without vortexing) and all microcentrifuge steps that include beads should be performed at <2,400 × *g* to avoid sample loss. For additional tips on the proper handling of DynaBeads, see the manufacturer's instructions.
11. The DNase I concentrations used here have been empirically determined for analysis of the human β-globin locus in K562 cells and could serve as a starting point for other applications. To determine the correct concentrations of DNase I to be used in a particular experimental system, aliquots of DNA treated with increasing concentrations of DNase I can be run on a 1.5% agarose gel. The size of DNA fragments should appear as

a smear that steadily decreases in size as DNase I concentration increases until the majority of high molecular weight DNA disappears.

12. Before spinning down the cells, a small aliquot can be removed and stained with trypan blue to determine efficiency of cell lysis (for instructions, see DNase I analysis step 2). Generally, K562 cells lyse well (>95%) in the first 20 min so the second incubation is unnecessary; however, we have observed other cell types that require both 20 min incubations to fully lyse.
13. EcoRI is used in this example of a 3C protocol; however, a number of other restriction enzymes have been shown by us and other groups to be appropriate for 3C, including HindIII, BglII, and ApoI. A restriction enzyme should be chosen that isolates the specific regions of interest (enhancers, promoters, etc.) on separate fragments. Any enzyme that has not been validated for 3C should be tested for digestion efficiency, which should be >85% after an overnight digestion and have no altered specificity (star activity) under these conditions. For additional details on 3C experimental design, refer to Dekker (14).
14. Although we routinely use ~2 million tissue culture suspension cells per 3C reaction, we have been successful using this 3C protocol for as few as one million cells per 3C reaction in situations where cell numbers were limited.
15. In certain circumstances, we have achieved higher digestion efficiencies by using a two-step digestion. For this approach, add 2 μL of EcoRI in the evening and then spike the reaction with another 2 μL of EcoRI first thing in the morning. After 3–4 h, proceed with step 6.
16. We have found the best results using this concentration, presumably because there is a high level of RNA contamination in the samples. Treatment with RNase and dilution of the sample to use 10–50 ng per 25 μL qPCR reaction has produced similar results; however, this extra step seems to be unnecessary as our results are highly reproducible without RNase treatment.
17. A control template may be generated using one of the following methods: *Method 1*: Bacterial artificial chromosomes (BACs) that include the entire region analyzed and the control region (e.g., α-tubulin or a conserved gene desert region) can be digested with EcoRI. Individual fragments are isolated, mixed in equimolar amounts, and ligated together to create a control template. *Method 2*: Genomic DNA is digested with EcoRI and ligated in a small volume to ensure all possible ligation products are formed. PCR across every ligation event to be analyzed is performed, and the PCR products are gel purified. The isolated PCR products are mixed in equimolar amounts to create a control template. In both cases, the control template

is used to test the efficiency of each primer pair using a standard curve. Any differences in efficiency are incorporated into calculations for relative proximity in the 3C samples.

Acknowledgments

We would like to thank members of our lab past and present for sharing their experimental experience and for their helpful comments on the manuscript. Work in the authors' laboratory is supported by the Intramural Program, NIDDK, NIH.

References

1. Palstra, R. J. (2010) Close encounters of the 3 C kind: long-range chromatin interactions and transcriptional regulation, 8 ed., pp 297–309.
2. Stamatoyannopoulos, G. (2005) Control of globin gene expression during development and erythroid differentiation, *Exp. Hematol. 33*, 259–271.
3. Tolhuis, B., Palstra, R. J., Splinter, E., Grosveld, F., and de Laat, W. (2002) Looping and interaction between hypersensitive sites in the active β-globin locus, *Mol. Cell 10*, 1453–1465.
4. Carter, D., Chakalova, L., Osborne, C. S., Dai, Y., and Fraser, P. (2002) Long-range chromatin regulatory interactions in vivo, *Nat. Genet. 32*, 623–626.
5. Palstra, R. J., Tolhuis, B., Splinter, E., Nijmeijer, R., Grosveld, F., and de Laat, W. (2003) The β-globin nuclear compartment in development and erythroid differentiation, *Nat. Genet. 35*, 190–194.
6. Vakoc, C. R., Letting, D. L., Gheldof, N., Sawado, T., Bender, M. A., Groudine, M., Weiss, M. J., Dekker, J., and Blobel, G. A. (2005) Proximity among distant regulatory elements at the β-globin locus requires GATA-1 and FOG-1, *Mol. Cell 17*, 453–462.
7. Drissen, R., Palstra, R. J., Gillemans, N., Splinter, E., Grosveld, F., Philipsen, S., and de Laat, W. (2004) The active spatial organization of the β-globin locus requires the transcription factor EKLF, *Genes Dev. 18*, 2485–2490.
8. Song, S.-H., Hou, C., and Dean, A. (2007) A positive role for NLI/Ldb1 in long-range β-globin locus control region function, *Mol. Cell 28*, 810–822.
9. Kadauke, S. and Blobel, G. A. (2009) Chromatin loops in gene regulation, *Biochim. Biophys Acta 1789*, 17–25.
10. Kim, S. I., Bultman, S. J., Kiefer, C. M., Dean, A., and Bresnick, E. H. (2009) BRG1 requirement for long-range interaction of a locus control region with a downstream promoter, *Proc Natl Acad Sci USA 106*, 2259–2264.
11. Kim, S. I., Bresnick, E. H., and Bultman, S. J. (2009) BRG1 directly regulates nucleosome structure and chromatin looping of the alpha globin locus to activate transcription, *Nucleic Acids Res 37*, 6019–6027.
12. Livak, K. J. and Schmittgen, T. D. (2001) Analysis of relative gene expression data using real-time quantitative PCR and the 2(–Delta Delta C(T)) Method, *Methods 25*, 402–408.
13. Pfaffl, M. W. (2001) A new mathematical model for relative quantification in real-time RT-PCR, *Nucleic Acids Res 29*, e45.
14. Dekker, J. (2006) The three 'C' s of chromosome conformation capture: controls, controls, controls, *Nat. Methods 3*, 17–21.

Chapter 4

Measuring Nucleosome Occupancy In Vivo by Micrococcal Nuclease

Gene O. Bryant

Abstract

Eukaryotic genomes are wrapped in nucleosomes. These nucleosomes could be a barrier or could help facilitate the binding of transcription or replication factors. To understand what biological role nucleosomes play, an accurate and reliable method for measuring not only the position of a nucleosome but the fraction of the population that is bound by a nucleosome is needed. Here is described a method for determining nucleosome occupancy that takes advantage of the difference in the rate of digestion of DNA by micrococcal nuclease when naked DNA is compared to the same DNA bound by a nucleosome. Curve fitting to a function that describes the amount of DNA remaining following a series of digestions over a broad range of micrococcal nuclease allows the calculation of nucleosome occupancy anywhere in the genome under many different conditions in vivo.

Key words: Nucleosome occupancy, Micrococcal nuclease, Yeast, In vivo assay

1. Introduction

In vivo eukaryotic DNA is wrapped around complexes of proteins containing eight histone subunits, which form nucleosomes over most of the genome (1). Nucleosomes, unlike the DNA binding domains (DBD) of transcription factors, do not bind exclusively to a specific DNA sequence but can bind to almost any underlying sequence (although some sequence preference is seen (2, 3)). Despite this fact it has been shown that in *Saccharomyces cerevisiae* nucleosomes are not randomly distributed but do have favored positions (4, 5). To understand how nucleosome position and occupancy (i.e., the percent of the population bound by a nucleosome at a given position) affect and are affected by transcription and/or replication, a quantitative assay to measure these properties is needed.

Randall H. Morse (ed.), *Chromatin Remodeling: Methods and Protocols*, Methods in Molecular Biology, vol. 833,
DOI 10.1007/978-1-61779-477-3_4, © Springer Science+Business Media, LLC 2012

A nucleosome substantially reduces the ability of micrococcal nuclease (MN) to digest the 147 bp of DNA that is wrapped around the histone octomer (6, 7). This reduction of digestion rate is utilized here to determine the percent of cells within a population that is occupied by a nucleosome. The amount of naked DNA that remains uncut following a micrococcal nuclease digestion is described by the one-state first-order decay function: $e^{-(km)}$ (see Fig. 1) where m is the amount of micrococcal nuclease and k is the rate of digestion (for the conditions of the reaction). So if several samples of naked DNA are digested with differing amounts of MN, the digestion rate is the value that minimizes the difference between the above function and the amount of DNA left uncut in the reactions. If occupied by a nucleosome, we found this same sequence of DNA would have a digestion rate some 50- to 300-fold slower. So in the case of a mixed population where some members are occupied and some are naked the two-state first-order decay function describes the amount of DNA remaining following a digest: $(1-\text{o})\text{e}^{-(\text{k}_1 m)}+\text{oe}^{-(\text{k}_2 m)}$ (see Fig. 2), where k_1 is the digestion rate of naked DNA, k_2 is the digestion rate of occupied DNA with $k_1 \gg k_2$ and 0 is the fractional occupancy with $0 \le o \le 1$. By curve fitting using this two-state function, the occupancy of a nucleosome can be calculated at any desired position within the genome even though the digestion rate of the underlying DNA sequence can differ from location to location (8, 9).

2. Materials

2.1. Cell Growth and Harvest

The cell growth and harvest conditions described below are for *S. cerevisiae* (see Note 1) grown in synthetic complete media using glucose as a carbon source. The growth media, carbon source,

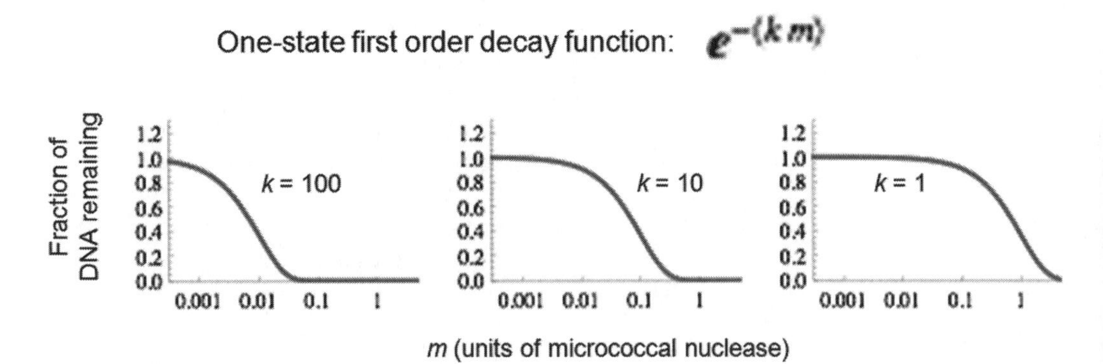

Fig. 1. Plots of the one-state first-order decay function where k is the digestion rate and m is the amount of micrococcal nuclease used in the reaction. Three separate plots of this decay function are shown with differing values for k.

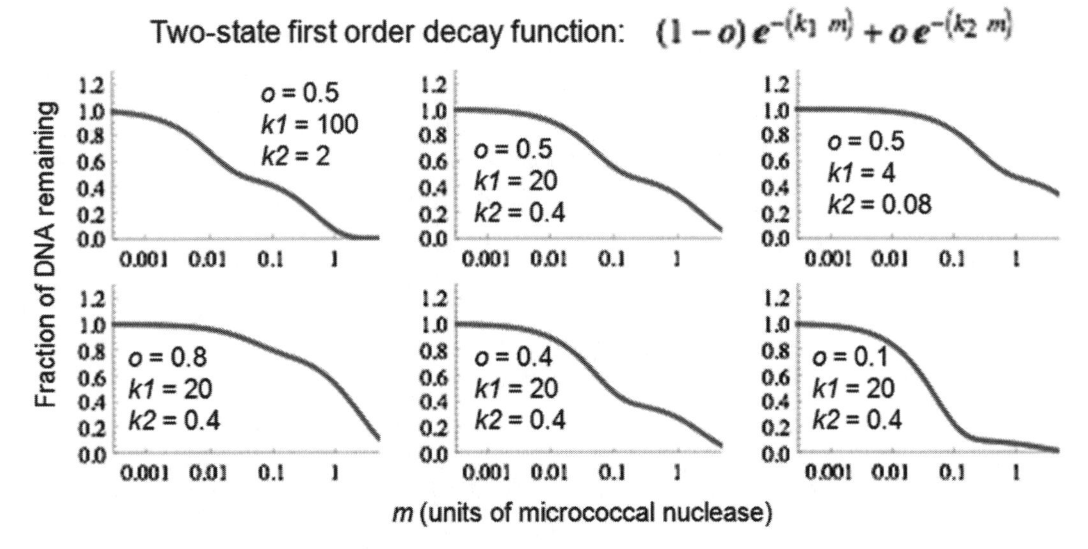

Fig. 2. Plots of the two-state first-order decay function where *o* is the fraction of the population bound by a nucleosome, *k1* is the rate of digestion of naked DNA and *k2* is the rate of digestion nucleosome bound DNA. Six separate plots of this decay function are shown with differing values for *o*, *k1*, *k2*.

growth temperature, etc., can all be adjusted for the particular experiment being performed.

1. SC media without sugar: 6.7 g/l Difco™ Yeast Nitrogen Base w/o Amino Acids, 1.8 g/l Complete Supplement Mixture (Bio 101 Systems). Mix with double distilled H_2O, autoclave and store under sterile conditions.
2. Glucose: 20% (w/v) D-(+)-Glucose. Mix with double distilled H_2O, autoclave, and store under sterile conditions.
3. Formaldehyde: Formaldehyde solution, 37%. Store undiluted at room temperature.
4. Glycine: 1 M Glycine. Mix with double distilled H_2O and use within a few days (this solution is easily contaminated so if long-term storage is desired it should be sterilized).
5. Centrifuge bottles.
6. Microcentrifuge tubes (1.7 ml).
7. 15-ml centrifuge tube.
8. Fixed angle centrifuge with a rotor compatible with centrifuge bottles.
9. Swing bucket centrifuge that can spin 15-ml centrifuge tubes.

2.2. Micrococcal Nuclease Digest

1. FA lysis buffer w/o EDTA: 50 mM Hepes–KOH (pH 6.5 see Note 2), 140 mM NaCl, 1% Triton X-100, 0.1% sodium deoxycholate. Add to double distilled H_2O and store at room temperature.

2. Micrococcal nuclease (MN): 1 U/μl Nuclease micrococcal from *Staphylococcus aureus*. Mix 500 U MN with 500 μl double distilled H_2O and store at –20°C.
3. $CaCl_2$ solution: 2 mM calcium chloride in double distilled H_2O; store at room temperature.
4. EDTA solution: 0.5 M EDTA, pH 7.0. Add to double distilled H_2O and store at room temperature.
5. Protease buffer: 200 mM Tris–HCl (pH 7.4), 4 M NaCl. Add to double distilled H_2O and store at room temperature.
6. Protease K: Protease K, recombinant PCR grade. Store at 4°C as it arrives from manufacturer.
7. Sonicator: Ultrasonic Cell Disruptor, Sonifier 250, (Branson Ultrasonics Corporation, CT) equipped with a Double Stepped Microtip (part number 101-063-212, Branson Ultrasonics Corporation, CT) (see Note 3).
8. 8 (or 12) channel pipettes: At least two separate pipettes are needed that will measure well in the 200, 15, and 5 μl volume ranges (see Note 4).
9. Collection microtubes: Collection microtube 96-well format (Qiagen Sciences, MD) (see Note 4).
10. DNA Purification Kit: QIAquick 96 PCR Purification Kit (Qiagen Sciences, MD) (see Note 5).

2.3. DNA Measurement

1. 2× QPCR buffer: 20 mM Tris–HCl (pH 8.3), 13 mM $MgCl_2$, 100 mM KCl, 400 μM dNTPs, 4% DMSO, 2× SYBR Green I (Invitrogen, Eugene, OR), 0.01% Tween 20, 0.01% NP40. Add to double distilled H_2O, store at 4°C.
2. Taq Polymerase at 5 U/μl. Store at –20°C.
3. Primer pairs: DNA oligomers for PCR should have a melting temperature of 59°C and create an amplicon of 40–60 bp in size (see Note 6). A different primer pair is needed for each location to be measured for occupancy.
4. Standard Curve samples: A dilution series of chromosomal DNA from the same strain that is being measured (see Note 7). The dilution series should contain 7 separate 3.33× serial dilutions plus an additional sample without any DNA (see Note 8).
5. Real time PCR machine such as the LightCycler 480/(384 or 96) (Roche Diagnostics Roche Applied Science, Mannheim, Germany) or 7900HT Fast Real-Time PCR System 384 or 96 wells (AB Applied Biosystems) (see Note 9).
6. QPCR plate: LightCycler Multi-well Plate 384, white with Sealing Foils (Roche, Mannheim, Germany).
7. 96-well mixing plate, such as BioExcell 96-well Ultra Rigid PCR plate Full Skirt (World Wide Medical Products, Inc., Hamilton NJ).

2.4. Data Analysis

1. A computer and software package that can calculate nonlinear least squares fit with constraints (e.g., Mathematica, Matlab, R, etc.).

3. Methods

The purpose of this assay is to measure, in vivo, under any growth condition, the fractional occupancy of a nucleosome, at any particular position within the genome. By cross-linking the cells with formaldehyde, the cells are effectively "frozen" in their state at the time of harvest. So by digesting the cross-linked chromatin we see the nucleosome state when the formaldehyde was added.

The fractional occupancy is determined by measuring the DNA remaining (at a particular location) after differing amounts of MN digestion. This means that the accuracy of the answer is directly dependent on the accuracy of the DNA measurement itself. QPCR can measure DNA to the needed accuracy (we typically get S.D. of <10% and a linear range >3,000×) but care must be taken to eliminate spurious measurements (one bad measurement can ruin 50 good measurements). Another issue that can affect the accuracy of the occupancy measurement is the slight differences in the amount of DNA lost during the purification process. If some DNA is lost from a particular sample there would be less DNA detected at all loci measured. This difference can be used to "normalize" the data thereby reducing this error.

3.1. Cell Culture and Harvest

1. Grow strain(s) of interest in media overnight. Use 10 ml of overnight media for every 300 ml of media that will be harvested (100 ml per harvest).
2. Add the appropriate amount of cells from the overnight to a pre-warmed and aerated harvest flask so the starting OD_{600} is 0.1–0.15. This flask should have 100 ml of media for each sample being harvested (e.g., if 10 samples will be harvested then 1,000 ml of media will be needed).
3. Grow cells under the desired conditions (~3–6 h depending on strain and conditions) and harvest cells when they reach an OD_{600} of 0.5–0.8. Harvest by transferring 100 ml of growing cells to the centrifuge tube containing 1.4 ml of 37% formaldehyde (final concentration of 0.5%) and mix to crosslink the histones to the DNA (see Note 10). Let this reaction proceed for 5–15 min at room temperature.
4. Add 14.5 ml of 1 M glycine to the cross-linked cells (resulting in 0.125 M glycine) to inactivate the formaldehyde and stop this reaction.
5. Spin down the cells in a centrifuge at 5,000 × g for 5 min, then decant off the supernatant.

6. Resuspend cells with 10 ml of H_2O, transfer to a 15-ml centrifuge tube, spin the cells down at 5,000 × *g* for 5 min, and again decant off supernatant.
7. Resuspend cells with 1 ml of H_2O, transfer to a 1.5-ml microcentrifuge tube, spin cells down for 1 min at full speed and again decant off supernatant.
8. Freeze down cell pellet at −70°C or proceed on to the micrococcal nuclease digest.

3.2. Micrococcal Nuclease Digest

1. Resuspend cells in 500 μl FA lysis buffer w/o EDTA in a microcentrifuge tube. Avoid bubbles and keep on ice.
2. Sonicate cells using a microtip at power 4 for 10 s and put back on ice (see Note 11).
3. Repeat step 2 after the temperature is back down to ~0°C.
4. Spin down sonicated cells for 5 min at 15,000 rpm (18,000 × *g*) and then transfer the supernatant to a new tube. The chromatin will be in the supernatant.
5. Measure the resulting volume and split the chromatin evenly between 16 separate collection microtubes. Then add FA lysis w/o EDTA to bring the volume up to 150 μl in each tube.
6. Make a twofold dilution series of micrococcal nuclease (MN) containing 14 separate concentrations with the highest concentration of 0.4 U/μl and extending to 0.0000488 U/μl plus two additional tubes containing no micrococcal nuclease (see Note 12).
7. Add 10 μl of each concentration of micrococccal nuclease to the 16 separate microtubes of chromatin. To start the reaction add 5 μl of 2 mM $CaCl_2$ to each tube and mix, then incubate at 37°C for 1.5 h.
8. Add 9 μl of 0.5 M EDTA to stop the reaction.
9. To each tube add 10 μl of Protease K mixed with protease buffer (1 μl Protease K per 50 μl protease buffer) and incubate at 42°C for 1 h.
10. Incubate cells at 65°C for 4 h or overnight to reverse the crosslink.
11. Cleanup the DNA using QIAquick 96-well plate as per manufacturer's recommendation. Elute with 200 μl H_2O or elution buffer into collection microtubes (see Note 5).

3.3. DNA Measurement

This protocol describes DNA measurement in a 384-well QPCR machine. Adjustments will need to be made for 96-well machines or other formats of QPCR machines.

1. Arrange the DNA to be measured in a 96-well mixing plate containing 12.5 μl of DNA for each location to be measured.

The first and last column of this 96-well DNA template should each contain standard curve samples. The remaining 80 wells should contain diluted DNA samples of the digested chromatin (see Notes 13 and 14).

2. Add Taq Polymerase to the 2× QPCR buffer (~ 5 μl Taq per ml buffer; see Note 15).
3. Add ~25 μl of each oligo (0.2 μg/μl) in the primer pair to 1.5 ml of 2× QPCR buffer w/Taq (see Notes 16 and 17). Distribute 12.5 μl of this solution to each well of a 96-well mixing plate (see Note 18).
4. Transfer 12.5 μl from each well of the DNA template (step 1) to the corresponding well in the 96-well mixing plate containing 2× QPCR buffer w/Taq and oligos (step 3). Mix these solutions by pipetting up and down without introducing bubbles.
5. Transfer from the 96-well plate (the DNA mixed with QPCR buffer) 5 μl four times to a 384-well QPCR plate where each individual well from the 96-well plate corresponds to 4 wells in the 384-well plate (see Note 19). Seal the 384-well plate with a QPCR compatible sealer (see Note 20).
6. Read this plate on a QPCR machine. Run the reaction for 40–50 cycles (depending on the primer pair) at 95°C for 4 s, 59°C for 26 s, and 72°C for 4 s with the fluorescence of the SYBR Green being read at the 72°C step.
7. After the QPCR machine has finished its run, analyze, clean up, and export the data as per manufacturer's instructions. Average together the four quadruplicate measurements for each of the up to 80 samples of DNA being measured (see Notes 21 and 22).

3.4. Data Analysis

1. Using the data from the standard curve samples, convert the measured QPCR values into relative concentrations ($a_{h,s,l}$ see Note 23). Then, using (1) and (2), convert these concentrations into the fractional amount of DNA remaining after digestion ($b_{h,s,l}$). First calculate the amount of DNA present before digestion for each harvest and locus measured using (1).

$$u_{hl} = \frac{1}{n}\sum_{s=1}^{n} a_{hsl}, \qquad (1)$$

where $a_{h,s,l}$ is the measured QPCR value, n is the number of samples that are (or appear to be; see Note 24) undigested and $u_{h,l}$ is the average undigested value.

Calculate the fractional DNA concentration ($b_{h,s,l}$) by dividing the measured QPCR values ($a_{h,s,l}$) by the undigested concentration ($u_{h,l}$) for all harvests, samples, and loci measured (2).

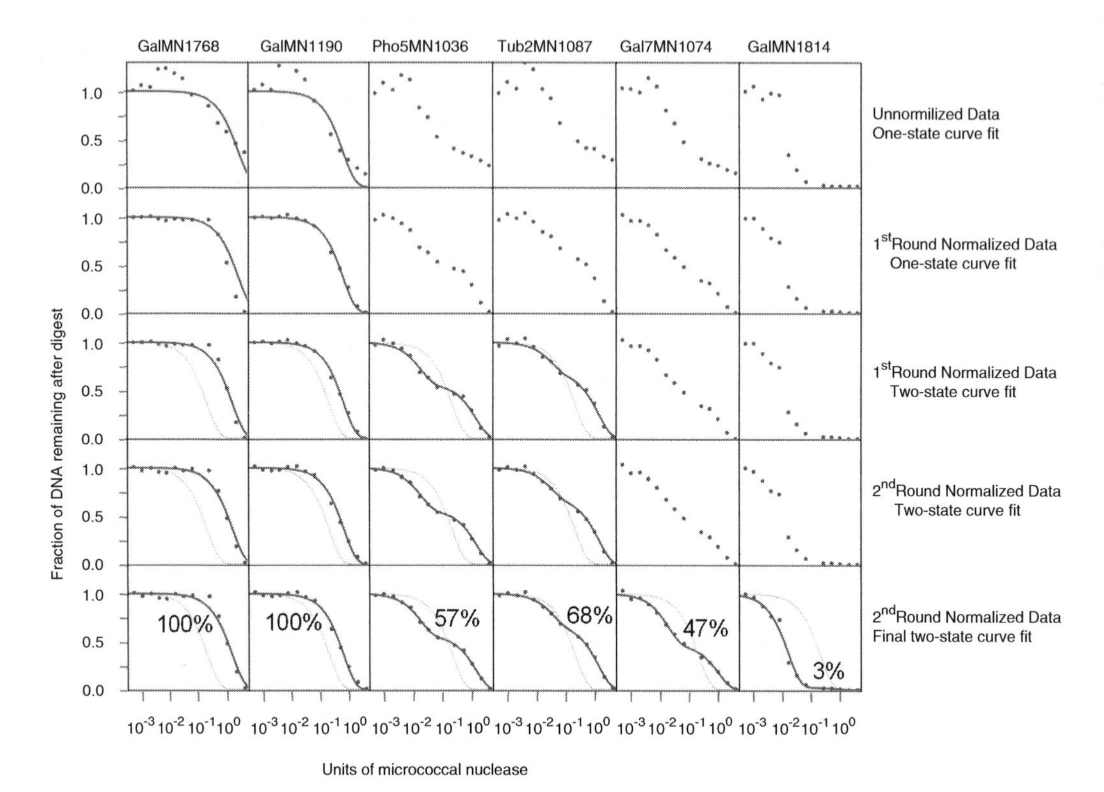

Fig. 3. The measured DNA concentrations and the optimal curve fits at each stage of the data analysis. Each column in this figure shows data and curve fits from six separate locations in the yeast genome in which the column title is the name of the primer pair used (see Table 1). Each row in this figure shows the data and curve fits at successive stages of the data analysis. The first row shows the fractional amount of DNA remaining after digestion ($b_{h,s,l}$) at each locus and the one-state fit on the first two loci, which are bound in 100% of the population. The second row shows the first round normalized values ($c_{h,s,l}$) and the same curves as in row one. The third row shows the first round normalized values ($c_{h,s,l}$) and the two-state fit on two loci bound 100% and two loci bound between 50 and 80% of the population. In all of the two-state curve fits, the faint curve indicates the threshold curve, which is a plot of one-state curve where k is set to *th*. The fourth row shows the second round normalized values ($d_{h,s,l}$) and the same curves as in row three. The last row shows the second round normalized values ($d_{h,s,l}$) and the final two-state fit. The percentages indicate the percent occupancy derived from the cure fit.

$$b_{hsl} = \frac{a_{hsl}}{u_{hl}}, \tag{2}$$

where $b_{h,s,l}$ fractional DNA and $a_{h,s,l}$ and $u_{h,l}$ are described in (1).

Examples of data and curve fits, at several stages in the data analysis, are shown in Fig. 3.

2. Perform the first round of normalizations for each harvest by first doing a curve fit using a one-state first-order decay function at loci that are known to be protected in 100% of the population (see Note 25). This is done by searching for the digestion rates ($k_{h,l}$) that minimize (3) (see Note 26).

$$\min \sum_{s=1}^{16} \left(b_{hsl} - e^{-(k_{hl} m_s)}\right)^2, \tag{3}$$

where $k_{h,l}$ is the adjustable parameter representing the one-state first-order digestion, and m_s is the amount of micrococcal nuclease for sample s (this should be the same for all harvests and loci).

For each sample, calculate the ratio of the sample's fractional DNA concentration over the value expected from the curve fit then take the average of this ratio for all loci.

$$q_{hs} = \frac{1}{n} \sum_{l=1}^{n} b_{hsl} e^{m_s k_{hl}}, \tag{4}$$

where $q_{h,s}$ is the average ratio of measured value divided by the expected value, n is the number of loci that were curve fit in this first round of normalization.

Finally, calculate the first round normalized values ($c_{h,s,l}$) by dividing the fractional DNA concentration ($b_{h,s,l}$) by the ratio just calculated ($q_{h,s}$) using (5).

$$c_{hsl} = \frac{b_{hsl}}{q_{hs}}, \tag{5}$$

where $c_{h,s,l}$ is the first round normalized values.

3. Do the second round of normalizations for each harvest by doing a curve fit using the two-state first-order decay function at the loci used in the first round plus a similar number of loci with an expected occupancy between 50 and 80% of the population (see Note 27). This is done by searching for the values of $k1_{h,l}$ $k2_{h,l}$ and $o_{h,l}$ that minimizes (6) (see Note 26).

$$\min \sum_{s=1}^{16} \left(c_{hsl} - (1 - o_{hl})e^{-(k1_{hl} m_s)} - o_{hl} e^{-(k2_{hl} m_s)}\right)^2, \tag{6}$$

where $0 \le o_{hl} \le 1$, $30t_h \ge k1_{hl} \ge t_h$, $k1_{hl} \div 30 \ge k2_{hl} \ge kl_{hl} \div 300$. There are three adjustable parameters $k1_{h,l}$ that represents the digestion rate of the naked, $k2_{h,l}$ that represents the digestion rate of the protected DNA, $o_{h,l}$ that represents the fractional occupancy of the. t_h is a threshold set between the $k1_{h,l}$ and $k2_{h,l}$ for all loci (see Note 28).

Now calculate (7) the amount that each sample's first round normalized value differs from the value expected from the curve fit for each locus used in this second round normalization and average them together ($q_{h,s}$).

$$r_{hs} = \frac{1}{n} \sum_{l=1}^{n} \frac{c_{hsl}}{(1 - o_{hl})e^{-m_s k1_{hl}} - o_{hl} e^{-m_s k2hl}}, \tag{7}$$

where $r_{h,s}$ is the average ratio of measured value divided by expected value, n is the number of loci that were curve fit in this second round of normalization.

Finally, calculate the second round normalized values ($d_{h,s,l}$) by dividing the first round normalized values ($c_{h,s,l}$) by the ratio just calculated ($r_{h,s}$) using (8).

$$d_{hsl} = \frac{c_{hsl}}{r_{hs}}, \tag{8}$$

where $d_{h,s,l}$ is the second round normalized values.

4. The data is now ready to calculate the occupancy and the two digestion rates for all loci. Search for the values of $k1_{h,l}$ $k2_{h,l}$ and $o_{h,l}$ that minimizes

$$\min \sum_{s=1}^{16} \left(d_{hsl} - (1 - o_{hl}) \mathrm{e}^{-(k1_{hl} m_s)} - o_{hl} \mathrm{e}^{-(k2_{hl} m_s)} \right)^2, \tag{9}$$

where $0 \le o_{hl} \le 1$, $30t_h \ge k1_{hl} \ge t_h$, $k1_{hl} \div 30 \le k2_{hl} \le k1_{hl} \div 300$.

5. Do a $\log(m_s)$ vs. DNA concentration plot of the data: $d_{h,s,l}$, along with the derived curve: $(1 - o_{hl})\mathrm{e}^{-(k1_{hl} m_s)} + o_{hl}\mathrm{e}^{-(k2_{hl} m_s)}$ and the threshold curve: $\mathrm{e}^{-(t_h m_s)}$, for each harvest and loci. Visually inspect the data for obviously spurious data points (see Note 29). Eliminate these points and recalculate from step 1.

6. Error is calculated by adjusting each parameter ($k1_{h,l}$ $k2_{h,l}$ and $o_{h,l}$) individually (while optimizing the other two parameters) above and below its optimal value until it increases (9) by 10% above its optimal value.

4. Notes

1. This assay has also been successfully applied to cultured mouse cells. The protocol after cell growth is essentially the same as described above except for a few modifications made to the digestion reaction. For mouse cells, 15-fold more micrococcal nuclease is used in the digestion series. 5 μl of 5 mM $CaCl_2$ is used to start the reaction and protease inhibitors (e.g., Protease Inhibitor Cocktail, Sigma-Aldrich Co.) are added to the reaction buffer.

2. The pH of the Hepes buffer has a significant effect on the micrococcal digestion reaction. The difference in the rate of digestion between naked and occupied DNA is lower at higher pH thereby reducing the accuracy of the occupancy measurement.

3. The sonication tip has a large effect on the amount of DNA sheared and therefore the amount of DNA found in the

supernatant after sonication. Large tips or damaged microtips will result in low DNA yields.

4. This protocol assumes the use of 8 (or 12)-channel pipettes and microtubes in a 96-well format. It is possible to do these steps with single-channel pipettes and microcentrifuge tubes but this will increase the workload and reduce the number of samples assayed.
5. The DNA purification kit can be replaced with a traditional EtOH precipitation step but again this will increase the workload and reduce the number of samples assayed.
6. Oligos can be placed almost anywhere in the genome, except at duplicated or highly repetitive sequences. Software that designs PCR oligos often limit the locations for oligo placement, and our experience suggests that this is not helpful. Oligo temperature can be calculated using:

$$T_{m} = 78.9 + \frac{41G - 820}{L},$$

where L is the oligo length in bp, G is the number of G and C residues, and T_{m} is the melting temperature.

7. The standard curve can be from a different strain then that being measured if the sequence is identical at all of the primer pair locations measured.
8. The standard curve is used in QPCR to calculate the relative concentrations of the unknown samples. It also shows what range of concentrations can be accurately measured. The concentration range used in the standard curve should be adjusted so that the highest concentration in the curve gives a Cp (or Ct) between 20 and 22.
9. In our hands the LightCycler 480 gave much better results than the 7900HT. The LightCycler had lower standard deviations of quadruplicate measurements but more significantly the same concentration of DNA gave the same value independent of its well location in the plate.
10. When doing timelines relative to an inducing event (e.g., heat shock or sugar addition) the time of harvest is the time at which the cells are mixed with the formaldehyde since this is a quick reaction and cross-linking stops further binding reactions.
11. The temperature of the solution following sonication should increase from the initial ~0°C to ~25–37°C if the sonication is done properly.
12. The MN dilution series should have 16 separate tubes with the concentrations of: 0.0, 0.0, 0.0000488, 0.0000977, 0.000195, 0.000391, 0.000781, 0.00156, 0.00312, 0.00625, 0.0125, 0.025, 0.05, 0.1, 0.2, and 0.4 U/μl.

13. All 16 samples from the same harvest must be diluted the same amount. Although 1:5 is a commonly used dilution this can vary from experiment to experiment. The optimal dilution can be determined empirically. An optimal dilution for a harvest will result in the undigested samples having a concentration close to the second highest concentrated DNA sample in the standard curve.
14. Since each harvest has 16 individually digested samples, 5 separate harvests can be fit into 80 wells of the 96-well plate. It is convenient to arrange each harvest in two neighboring columns ordered from the lowest level of digest to the highest level.
15. Unfortunately the activity of Taq Polymerase can change significantly from batch to batch so the amount added must be determined empirically. If the activity is too high then linearity is lost for the lower concentrated DNA samples (i.e., different concentrations of DNA give similar Ct (Cp) values). If the activity is too low then more cycles of PCR are needed to detect the same concentration of DNA. If a large number of measurements are to be performed it is convenient to combine several different batches of Taq and test the combined Taq once.
16. Similarly to the case of Taq (see Note 15), using the correct oligo concentration is critical to getting good DNA measurements. Each separate primer pair has its own optimal volume to be added (typically between 10 and 50 μl). Like the case for Taq, too much oligo results in a loss of linearity for lower concentrated samples and too little oligo increases the number of cycles needed to detect the DNA.
17. The time between the mixing of the oligos in step 3 to the addition of the DNA template in step 4 should be minimized (to less than 30 min) to prevent primer dimer formation. The effect of rapid primer dimer formation only occurs when the oligos are in the 2× QPCR buffer. When this solution is diluted twofold by addition of DNA, primer dimer formation is slowed considerably.
18. This step is much easier using 8-channel pipettes, by first distributing 185 μl into 8 separate wells in a column of a 96-well mixing plate and then using an 8-channel pipette to distribute the 12.5 μl to each well of a new 96-well mixing plate.
19. For example, the contents of well A1 from the 96-well plate corresponds to well A1, A2, B1 and B2 from the 384-well plate.
20. After the plate is sealed it can be read by the QPCR machine immediately or stored for up to a week at 4°C before reading. Depending on the primer pair, sealed plates can be left at room temperature for 10–24 h without adverse effects.

21. QPCR can be a very accurate method for measuring DNA but it does occasionally give spurious measurements. If a measurement is spurious then the standard deviation of the quadruplicate measurements will be high (e.g., >~25% S.D.) and eliminating this value will reduce the standard deviation of the remaining values significantly (e.g., > ~2× and < ~25% S.D.). If the standard deviation of a quadruplicate measurement is high but eliminating any individual value does not result in a low enough standard deviation then this quadruplicate measurement should just be eliminated from the data analysis.
22. Another way in which QPCR can give inaccurate results is when the DNA concentration being measured is too low. The lowest acceptable measurement can be determined by examining a Log(concentration) vs. Cp (Ct) plot of the standard curve (this should be a standard display in the QPCR machine's software). This plot should be linear over the entire concentration range of the standard curve. If the lower concentrated samples of the standard curve start to deviate from linearity (i.e., they give similar or the same Cp (Ct) values even though they have differing concentrations of DNA) then the concentration at which this happens is the lowest acceptable concentration and any values at or near this concentration should be eliminated from analysis.
23. For all of the equations shown in the data analysis section, subscripts are used to distinguish between each separate harvest, sample and locus, by using the subscript h, s and l respectively (e.g., the symbol $a_{h,s,l}$ represents the average QPCR measurement from a particular harvest, sample and locus, i.e., the primer pair used in the QPCR measurement). The samples are ordered from lowest level of digest to the highest so sample 1 was digested with 0 Units of MN and sample 16 was digested with 4 Units of MN.
24. Strictly speaking, only the first two samples are undigested but often the lowest levels of micrococcal nuclease show no detectable loss of DNA at any location measured. When this is the case these samples can be treated as undigested.
25. In *S. cerevisiae* we use two locations in the *GAL1 UASg*, which are 100% protected, for the first round of normalization (GalMN1768 and GalMN1190; see Table 1). If no locus is measured which is protected at 100% then do both the first and second round normalization as described in step 3 (although the normalization does work better if the first round is done with the one-step decay function on 100% occupied loci).
26. Curve fitting algorithms for nonlinear functions (that allow constraints on the parameter range) are available in many different software packages. In Mathematica the function is

Table 1
This table lists some example primer pairs showing their name, the sequence of each oligo and their expected occupancy

Primer pair	Oligo 1	Oligo 2	Occupancy (%)
GalMN1768	TCCTCCGTGCGTCCTCG	TTTCAGGAACGCGACCGGT	100
GalMN1190	GCGACAGCCCTCCGACGG	GGAACGCGACCGGTGAAG	100
Pho5MN1015	GTCTTAGCCAGACTGACAGTA	AGCAACTGCAAATGGTTGGTAG	57
Tub2MN1087	ACAATGTGCAAACTGCTGTGTG	AAAGTAGCAGCCATGTCCAAAC	64
Pho5MN1036	TCGGCTAGTTTGCCTAAGGG	ATTCAATTTTAGCCGCTTCTTTGG	56
GalMN2049	AAGGAATTACCAAGACCATTGGC	TAATTATGCTCGGGCACTTTTCG	62

ArgMin, in MatLab the function is lsqnonlin, and in R it is nlminb. Although when using (3) no parameter constraint is needed, it is very important that when using (6) and (9), the stated constraints are used. Without these constraints nonsensical values can be derived; for example, occupancy values of >100% or <0% or digestion rates of naked DNA lower than that for protected DNA.

27. In *S. cerevisiae* we have used several loci for the second round normalizations: Pho5MN1015, Tub2MN1087, Pho5MN1036, and GalMN2049; see Table 1.

28. The threshold (*th*) is used to distinguish the rate of digestion of the naked DNA from that of protected DNA. Without this threshold a locus that is near 100% occupancy is indistinguishable from a locus that is near 0% occupancy. A good initial guess for the threshold would be a value some 50-fold higher than the digestion rate found for primer pair GalMN1768 (see Table 1). Visually inspect the plots described in step 5 to verify that the threshold is set appropriately. The quickly digesting part of the curve should be below (higher rate) the threshold and the slowly digesting region should be above (lower rate) at all loci. If the threshold does need to be changed, redo all calculations from (6) onward.

29. A more objective method for eliminating bad data points would be to calculate how much the sum of squares of the differences between observed and expected is reduced when each sample is eliminated from the curve. If dropping an individual sample reduces the sum of squares by a large amount it should be eliminated.

References

1. Kornberg RD and Klug A. (1981) The nucleosome. *Sci. Am.* **244**, 52–64.
2. Satchwell SC, Drew HR and Travers AA. (1986) Sequence periodicities in chicken nucleosome core DNA. *J. Mol. Biol.* **191**, 659–675.
3. Segal, E, Fondufe-Mittendorf Y, Chen L, Thåström A, Field Y, Moore IK, Wang JZ, and Widom J. (2006) A genomic code for nucleosome positioning. *Nature* **442**, 772–778.
4. Yuan GC, Liu YJ, Dion MF, Slack MD, Wu LF, Altschuler SJ, Rando OJ. (2005) Genome-scale identification of nucleosome positions in *S. cerevisiae*. *Science* **309**:626–30.
5. Lee W, Tillo D, Bray N, Morse RH, Davis RW, Hughes TR, Nislow C. (2007) A high-resolution atlas of nucleosome occupancy in yeast. *Nat Genet.* **39**:1235–44
6. Hewiah, D.R., and Burgoyne, L.A. Chromatin sub-structure. (1973) The digestion of chromatin DNA at regularly spaced sites by a nuclear deoxyribonuclease. *Biochem Biophys Res Commun.* **52**:504–10.
7. Luger K, Mader AW, Richmond RK, Sargent DF and Richmond TJ. (1997) Crystal structure of the nucleosome core particle at 2.8 Å resolution. *Nature* **389**, 251–260.
8. Bryant GO, Prabhu V, Floer M, Wang X, Spagna D, Schreiber D, Ptashne M. (2008) Activator control of nucleosome occupancy in activation and repression of transcription. *PLoS Biol.* **6**:2928–39.
9. Floer M, Wang X, Prabhu V, Berrozpe G, Narayan S, Spagna D, Alvarez D, Kendall J, Krasnitz A, Stepansky A, Hicks J, Bryant GO, Ptashne M. (2010) A RSC/nucleosome complex determines chromatin architecture and facilitates activator binding. *Cell* **141**: 407–18.

Chapter 5

Analysis of Nucleosome Positioning Using a Nucleosome-Scanning Assay

Juan Jose Infante, G. Lynn Law, and Elton T. Young

Abstract

The nucleosome-scanning assay (NuSA) couples isolation of mononucleosomal DNA after micrococcal nuclease (MNase) digestion with quantitative real-time PCR (qPCR) to map nucleosome positions in chromatin. It is a relatively simple, rapid procedure that can produce a high-resolution map of nucleosome location and occupancy and thus is suitable for analyzing individual promoters in great detail. The analysis can also quantify the protection of DNA sequences due to interaction with proteins other than nucleosomes and show how this protection varies when conditions change. When coupled with chromatin immunoprecipitation (ChIP), NuSA can identify histone variants and modifications associated with specific nucleosomes.

Key words: Nucleosomes, Nucleosome scanning, Micrococcal nuclease, Quantitative real-time PCR, Transcription

1. Introduction

The importance of nucleosome positioning has been debated since the discovery of the histone octamer as the basic unit of chromosome architecture (reviewed in ref. 1). The key questions have been "Do nucleosomes occupy specific locations on the underlying DNA, and if so, how is their position established and maintained?" Genome-wide analysis of nucleosome location has confirmed on a global scale what was initially observed at a local level: nucleosomes are present at defined locations throughout the genome (2–5). The global analysis also confirmed that most yeast promoter regions were deficient or even bereft of nucleosomes. The second question – how are nucleosome positions established and maintained? – is the subject of considerable debate (4, 6–11).

Randall H. Morse (ed.), *Chromatin Remodeling: Methods and Protocols*, Methods in Molecular Biology, vol. 833
DOI 10.1007/978-1-61779-477-3_5, © Springer Science+Business Media, LLC 2012

Chromatin is viewed now as a very dynamic structure, which shapes all cellular processes interacting with DNA (12). This picture raises new questions. How does the basic unit of chromatin, the nucleosome core particle (NCP), regulate the accessibility of DNA to the cellular machinery? Different dynamic properties of NCPs have been associated with promoting changes in DNA accessibility relevant to cellular processes. For instance, changes in histone charge or histone assembly dynamics lead to the transient exposure of regulatory regions (13–15). These regions can also be exposed by changes in the translational position of nucleosomes or by changes in chromatin higher-order structure, which are still poorly understood (16, 17). Regardless of the biophysics underlying these changes, the nucleosome-scanning assay (NuSA) protocol that will be discussed in this chapter is useful to locate nucleosomes and quantify the accessibility of chromatin-associated DNA sequences in different conditions.

The approximate location of nucleosomes on DNA in chromatin can be determined at low resolution using an indirect end-labeling protocol (18). However, the position of individual nucleosomes is poorly resolved. More precise localization can be achieved by isolating mononucleosome-sized fragments of DNA after extensive Micrococcal nuclease (MNase) treatment and determining the DNA sequence that was protected from digestion. Initially primer extension was used to determine at high resolution the borders of single nucleosomes (19, 20). Subsequent genomic approaches utilized cloning and sequencing (9), tiled oligonucleotide arrays (4, 5, 21), quantitative real-time PCR (qPCR) (3), and high-throughput DNA sequencing (22) to map nucleosome positions (reviewed in refs. 8, 23, 24).

In this section, we describe the procedure that we have used to map the position of nucleosomes at several yeast promoters (15, 25) using the "nucleosome-scanning" assay (NuSA) first described by Sekinger and Struhl (3). This procedure couples MNase digestion, isolation of mononucleosome-size DNA fragments, and qPCR to provide a quantitative map of nucleosome positions across a chromosomal region. The basic question that might be solved by performing the NuSA is whether a specific portion of a chromatin region promotes more or less accessibility to the DNA by proteins including histones. Also, it might reveal whether a change in accessibility affects a single nucleosome or a subset of them over a chromatin region. The advantage over other chromatin analysis techniques is that the answer to that question is quantitative, fitting better with the dynamic equilibrium of histone–DNA interaction.

As examples we show the results of NuSA of the *ADH2* promoter (*ADH2prm*) for several reasons. First, it was used to evaluate various experimental parameters, such as extent of MNase digestion, use of unfractionated MNase-digested DNA, and different normalization procedures. Second, its chromosomal architecture and the changes that accompany gene activation have been characterized by indirect end-labeling (26–28), high-resolution primer extension (19), NuSA (15, 25), and minichromosome supercoiling (15, 25). Thus, we can compare various aspects of nucleosome organization as determined by several assays. Third, the *ADH2prm* nucleosomes appear to represent two types that are frequently found throughout the yeast genome: one that is strongly positioned and two others covering the TATA box and RNA initiation site that are less well positioned.

Figure 1 illustrates the major steps of the NuSA procedure. Figure 2 represents hypothetical NuSA data from a promoter with two nucleosomes (ovals) and a nucleosome-free region (nfr). The left-hand side of the figure assumes that both nucleosomes are present at the indicated locations in all of the cells. The right-hand side of the figure illustrates the same chromatin but in this case it is imagined that 50% of the time nucleosome N(−1) is either missing

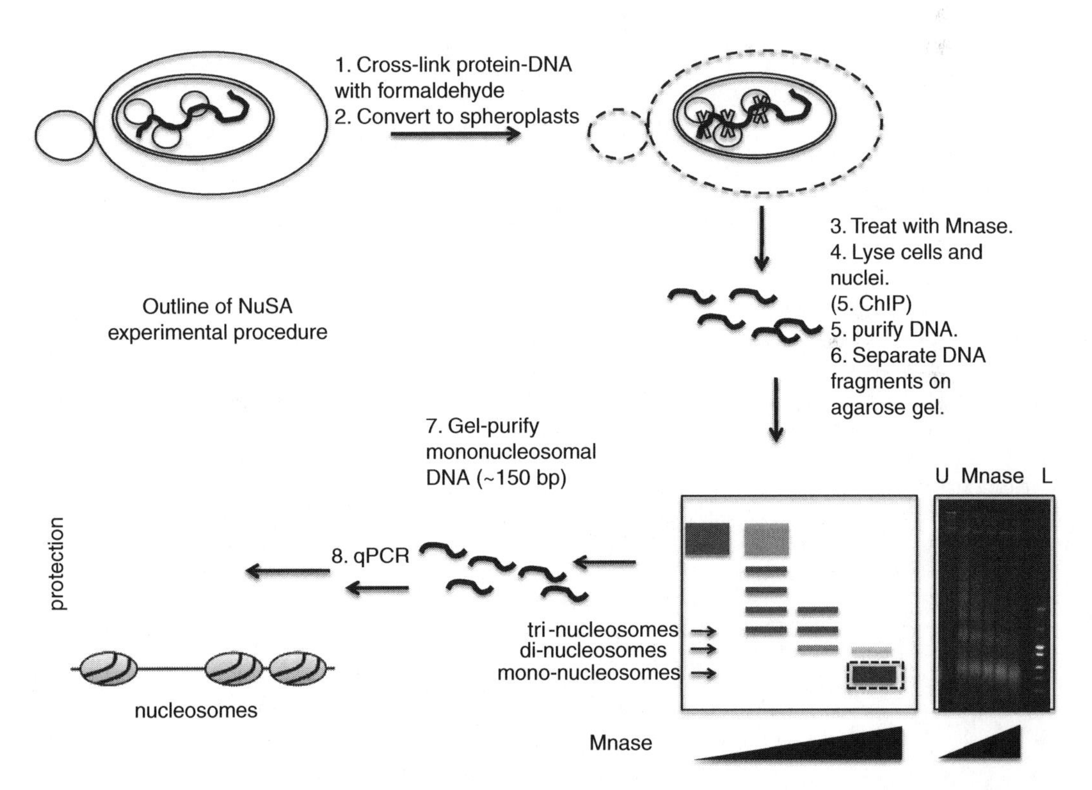

Fig. 1. Outline of steps used in the procedure to perform a nucleosome-scanning assay (NuSA). The picture shows a hypothetical 2% analytical agarose gel used to analyze the DNA products of MNase digestion of chromatin. The gel shows the ladder of nucleosomes produced at low levels of digestion, and production of mostly mononucleosomal DNA (*highlighted band*) at high MNase concentrations. A reproduction of a MNase titration of chromatin with increasing amounts of MNase (5, 10, 15, 20, and 40 units) is shown in the photograph at the *bottom*. The undigested DNA sample is labeled as U. L designates a molecular weight ladder. The fold-enrichment of protected sequences in the mononucleosomal DNA over undigested DNA is quantified by quantitative real-time PCR (qPCR) for a subset of overlapping amplicons covering the sequence of interest. An example of a typical NuSA result is illustrated in the *lower left*. This diagram shows the location of protected sequences and the relative degree of protection from a typical assay of *ADH2* promoter nucleosomes.

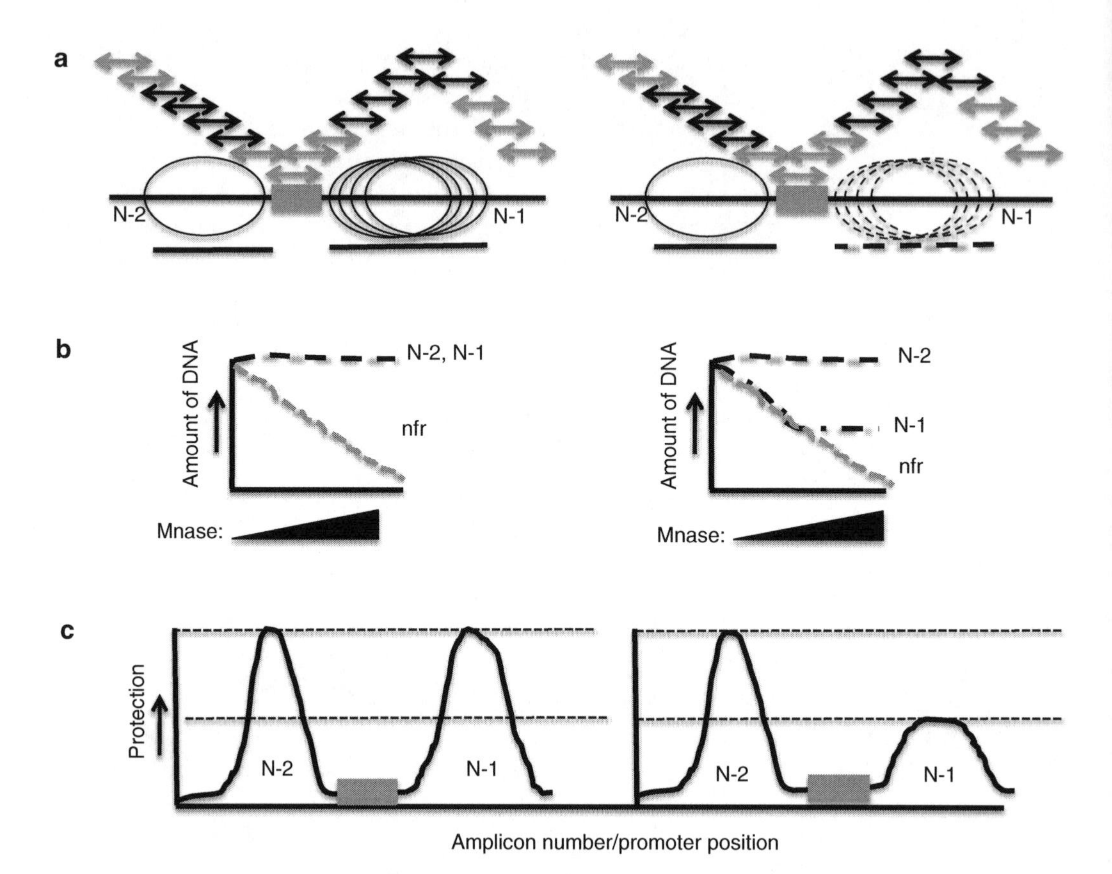

Fig. 2. Diagram of a hypothetical promoter with two nucleosomes showing position of primer pairs. (**a**) Two hypothetical scenarios are shown. On the *left*, all of the genomic DNA is protected by two nucleosomes in the region show. The N − 1 nucleosome comprises a family that occupies different translational positions on the DNA. On the *right half* is diagrammed a situation in which half of the population is lacking the N − 1 nucleosome. (**b**) The graphs represent the relative amount of DNA protected by N–1 and N–2 nucleosomes in the two situations illustrated in (**a**). The relative amount of an amplicon that lies entirely within the nucleosome (*black double-headed arrows in* (**a**)) is shown by *black dashed lines*; the relative amount of an amplicon that lies within the nucleosome free region (nfr, *gray double-headed arrows* in (**a**)) is shown by *gray dashed lines*. The relative amount of MNase used for digestion is shown by the *gray triangle* below the graphs. (**c**) NuSA patterns expected in the two scenarios shown in (**a**). As in panel (**a**) the *solid gray bar* represents the nucleosome-free (or nucleosome-depleted) region.

or bound in a manner that allows access to MNase. In Fig. 2a, the ~150 bp DNA fragments isolated after MNase digestion (see Fig. 1) are indicated by lines below the ovals. Primer-pairs are indicated by arrowheads at the end of the amplicons that would result from PCR of genomic DNA. N(−2) represents a well-positioned nucleosome and N(−1) represents a nucleosome that can occupy any one of four positions. The black arrows represent regions of the DNA that would be fully protected because both primers lie completely within the nucleosome. Gray arrows represent regions of DNA that would not be protected because at least one of the primers lies outside of the nucleosome-protected region. The relative protection from increasing amounts of MNase digestions using primer-pairs within or outside of the nucleosome borders are illustrated in Fig. 2b and a NuSA assay is shown in Fig. 2c.

When NuSA is preceded by ChIP to enrich for DNA sequences associated with specific histone variants, modifications, or other DNA-bound proteins, a particularly powerful analysis of chromatin-associated proteins can be achieved (3, 29). A protocol for combining ChIP with NuSA is described and an example of the promoter of the *S. cerevisiae FBP1* gene (*FBP1prm*) is shown that illustrates its utility for mapping nucleosome alterations accompanying gene activation.

2. Materials

2.1. Harvesting Cells for NuSA

1. Media for growth of cells (30).
2. Formaldehyde, 1% solution (Stock is 37%. Formaldehyde is toxic; take appropriate precautions).
3. 2.5 M glycine.

2.2. Spheroplasting

1. Buffer Z: 1 M sorbitol, 50 mM Tris–HCl, pH 7.4, 10 mM ß-mercaptoethanol (added just before use).
2. Zymolyase, 10 mg/mL (Seikagaku America, Falmouth MA, USA). Dissolve in 40% glycerol, 60% Buffer Z. The solution should be stored at –20°C in small aliquots.

2.3. Micrococcal Nuclease Digestion

1. NPS buffer: 0.5 mM Spermidine, 0.075% NP-40, 50 mM NaCl, 10 mM Tris–Cl, pH 7.5, 5 mM $MgCl_2$, 1 mM $CaCl_2$, 1 mM ß-mercaptoethanol (always add fresh from stock 14.3 M).

   To prepare 300 mL of NPS buffer: 22 mg of solid Spermidine, 2.25 mL of 10% NP-40, 3 mL of 5 M NaCl, 3 mL of 1 M Tris–Cl pH 7.5, 1.5 mL of 1 M $MgCl_2$, 0.3 mL of 1 M $CaCl_2$, 20 μl of 14.3 M ß-mercaptoethanol (14.3 M is the usual concentration of this chemical in supplied stock solutions, which should be stored at 4°C) and 290 mL of water.

1. MNase. Dissolve in MNase buffer: 10 mM Tris–HCl, pH 7.5, 10 mM NaCl, 100 μg/mL BSA. Store in small aliquots of 15 units MNase/μl at –80°C.
2. 500 mM EDTA solution. Small volumes per sample would be used, therefore adjust the volume of the stock solution to the predicted working schedule to minimize problems derived from working with old stock solutions.
3. 200 mM EGTA. To prepare the stock solution, follow the recommendation made for EDTA above.

2.4. Protein Degradation and DNA Purification

1. Proteinase K solution: 10 mM Tris–HCl pH 7.5, 1 mM EDTA, 0.4 mg/mL glycogen, 1 mg/mL proteinase K.
2. RNase A solution: 10 mg/mL, DNase-free in TE.
3. 0.1 M Tris–HCl, pH 7.5-saturated phenol. Extreme precaution is required when working with phenol. Always work under the hood, wearing gloves and glasses.
4. Chloroform.
5. 3 M Sodium Acetate (NaAc) pH 5.3.
6. 100% Ethanol.
7. TE buffer: 10 mM Tris–HCl pH 8.0, 0.5 mM EDTA.
8. Agarose (high-quality molecular biology grade).
9. 10× TBE buffer: 890 mM Tris base, 890 mM Boric Acid, 20 mM EDTA pH 8.0.
10. Loading dye: 0.25% Bromophenol Blue, 18% Ficoll 400, plus 0.3% xylene cyanol.
11. DNA molecular weight markers. Since it is important to check the digested DNA for mono-, di-, tri-nucleosomal DNA, use appropriate DNA size markers for interpretation of DNA fractionation. Markers with 100 bp-increments between 100 bp and 1 kb are recommended.
12. 10 mg/mL stock Ethidium Bromide solution. Final concentration for gel staining is 0.5 μg/mL. Always wear gloves when handling Ethidium Bromide solutions and avoid contamination of the room with these solutions. Nitrile gloves are recommended because latex gloves do not protect efficiently against Ethidium Bromide. Dispose of Ethidium Bromide contaminated-solutions in an environmentally correct manner.
13. Sterile razor blade or scalpel.
14. UV-Transilluminator.
15. Gel-purification kit for DNA such as Qiagen QIAquick Gel Purification Kit.
16. Picofluor™ Fluorometer with blue optical configuration.
17. 10 mm × 10 mm methacrylate fluorescence cuvettes.
18. PicoGreen dsDNA Quantitation Reagent, (Molecular Probes, Inc., Eugene OR, USA).

2.5. Quantitative Real-Time PCR Analysis of Isolated DNA

1. 0.2-mL siliconized tubes.
2. 96-well PCR-plates and sealers.
3. Primers for quantitative real-time PCR (qPCR). Primer pairs for the region of interest are designed to produce overlapping PCR products (amplicons) of 60–80 bp with a 10–30 bp overlap (see Note 1).

4. Syber Green (SYBRG) mix.
5. PCR Thermal Cycler for Real-time quantitative PCR and computer equipped with analysis software.

2.6. Chromatin Immunoprecipitation-NuSA Protocol

1. Protease Inhibitors Cocktail (Sigma).
2. 100 mM PMSF in isopropanol.
3. Buffer L. Final working concentrations are 50 mM HEPES–KOH, pH 8.0, 140 mM NaCl, 1 mM EDTA, 1% Triton X-100, 0.1% Sodium Deoxycholate. For working samples of 800 μL, make a 4× solution of buffer L by adding these volumes from the following stock solutions: 56 μl water, 40 μl 1 M HEPES–KOH pH 8.0, 22.4 μl 5 M NaCl, 1.6 μl 0.5 M EDTA, 80 μl 10% Triton X-100, and 8 μl 10% Sodium deoxycholate.
4. Braun (or equivalent) sonicator.
5. TBS buffer: 140 mM NaCl, 2.5 mM KCl, 50 mM Tris–HCl pH 7.4.
6. Chromatin immunoprecipitation (ChIP) lysis buffer: 50 mM HEPES–HCl pH 7.5, 140 mM NaCl, 1% Triton X-100, 0.1% Sodium deoxycholate.
7. ChIP lysis buffer high salt: 50 mM HEPES–HCl pH 7.5, 500 mM NaCl, 1% Triton X-100, 0.1% Sodium deoxycholate.
8. ChIP wash buffer: 50 mM Tris–HCl pH 7.5, 250 mM LiCl, 0.5% NP-40, 0.5% Sodium deoxycholate, 1 mM EDTA.
9. ChIP elution buffer: 50 mM Tris–HCl, pH 8.0, 0.1% SDS, 1 mM EDTA.
10. PCR product purification Kit (such as Qiagen QIAquick PCR purification kit).

3. Methods

3.1. Harvesting Cells for NuSA

1. Grow 220 mL of *Saccharomyces cerevisiae* to an OD_{600} ~ 0.6.
2. (Harvest 20 mL for RNA preparation if required).
3. Add 5.4 mL of 37% formaldehyde (final concentration 1%). Incubate for 15 min at room temperature, swirling occasionally.
4. Add 1/19 volume of 2.5 M glycine (final concentration 125 mM) to quench the formaldehyde. Incubate cells for 5 min at room temperature.
5. Pellet cells (3,000 × *g* in Sorvall SS34 rotor or equivalent) for 5 min at 4°C and wash twice with cold sterile water.

3.2. Spheroplasting (31)

1. Resuspend cell pellets in 19.5 mL Buffer Z and transfer to 125-mL flasks.
2. Add fresh 14 μl of 14.3 M β-mercaptoethanol (final concentration 10 mM) to each flask.
3. Add 0.5 mL of zymolyase solution. Incubate cells in 125-mL flasks at 30°C with shaking for approx. 15 min (see Note 2).
4. Pellet spheroplasts at 3,000 × *g*, 10 min, 4°C.

3.3. Micrococcal Nuclease Digestion

1. Resuspend spheroplasts gently in 1.5 mL NPS buffer. Make three aliquots of 600 μl each for two MNase digestions and one undigested control (see Note 3).
2. There will be two digestions with MNase and one mock digestion (to make genomic DNA) for each strain or condition (see Note 4).
   (a) Digest with 15 U MNase for 30 min at 37°C.
   (b) Digest with 15 U MNase for 40 min at 37°C.
   (c) No digestion-incubate for 30 min at 37°C.
3. Stop digestions by shifting the tubes to 4°C and adding 18 μl 500 mM EDTA and 7 μl 200 mM EGTA. At this point, samples can be frozen on dry ice and stored at –20°C until further use. (If chromatin immunoprecipitation (ChIP) is to be performed go to Subheading 3.7 below. Be sure not to add SDS, as this will inhibit the antibody in the ChIP procedure).

3.4. Protein Degradation and DNA Purification

1. Add 60 μl 10% SDS, 10 μl of 10 mg/mL Proteinase K and 10 μl of 10 mg/mL DNase-free RNase. Incubate at 37°C for ~2 h.
2. Extract twice with an equal volume phenol saturated with 0.1 M Tris–HCl, pH 7.5 and once with an equal volume of chloroform.
3. Precipitate with 0.1 vol. of 3 M NaAc, pH 5.3 and 2.5 vol. 100% ice-cold ethanol. The samples can be left indefinitely at –20°C at this point.
4. Pellet DNA by centrifugation at 1,500 × *g* in a tabletop microfuge for 15 min at 4°C.
5. Dry pellet and resuspend pellet in 40 μl TE. If using a speed vac, do not overdry.
6. Measure the total DNA concentration using Quant-iT PicoGreen dsDNA Assay Kit.
7. Analyze 2 μl of each digested sample on a 1% agarose gel in 1×TBE buffer to check whether the MNase treatment yielded mononucleosomal DNA (see Note 4). Run the undigested genomic DNA as a control.

8. Electrophorese 25–50 μl on a 2% preparative agarose gel in 1×TBE buffer to isolate mononucleosome DNA. Leave an empty lane between samples. If necessary, reduce the volume using the speed vac before electrophoresis.
9. For the MNase-digested samples cut out the band between the 100 bp and 200 bp DNA size markers using a sterile razor blade or scalpel. Purify the DNA using a Qiagen Qiaquick Gel Extractions kit following the manufacturer's recommendations except elute the DNA using 30–50 μl of 50°C Elution Buffer for 10 min at 50°C with shaking at 700 rpm. Measure the DNA concentration using PicoGreen (see Note 5).

3.5. Quantitative Real-Time PCR Analysis of Isolated DNA

1. Using siliconized tubes for qPCR dilutions:
   (a) Dilute mononucleosome DNA samples 1:100 (6 μl into 600 μl of 1 mM Tris–HCl, pH 8.0) (see Note 6).
   (b) Undigested genomic DNA: dilute 1:500 (2 μl into 1 mL of 1 mM Tris–HCl, pH 8.0) (see Note 6).
2. qPCR assays. For each primer set, run MNase-treated mononucleosome and undigested DNA samples in triplicate. Include a no-DNA control for each primer set. Run a standard curve for each primer set using a series of tenfold dilutions of undigested DNA (1:50 to 1:500,000 dilution series). If the analysis method is based on a standard curve, this must be determined in each run for primers with identical efficiencies. If a relative analysis method, such as the Pfaffl method (32), is used, the standard curve might be determined only once to calculate the primer efficiency. Determine the relative amount (RA) of each amplicon in the undigested genomic DNA and the gel-purified mononucleosome DNA. Standards for performing, analyzing, and interpreting qPCR are discussed in detail in ref. 33.

3.6. Data Analysis and Interpretation (see Note 7)

1. A relative protection value is calculated for each amplicon. The protection value set for a given amplicon corresponds to the fold-enrichment of the target sequence in the mononucleosomal DNA sample over the undigested DNA ("genomic DNA") sample. If the quantification method chosen for the qPCR data is based on a standard curve, the relative protection value would result from dividing the absolute value obtained for mononucleosomal DNA by that obtained for the undigested sample. However, since the NuSA analysis is relative, the Pfaffl method of qPCR quantification could be used for simplicity. With this option, primer efficiency for each amplicon must be known. Primer efficiency has to be determined at least once with a standard curve or an equivalent method.
2. Normalize the relative protection values for each amplicon to differences in DNA concentration among different samples.

The simplest normalization method is to divide the relative amount of each amplicon, determined from the qPCR of the mononucleosomal DNA by the amount of total mononucleosomal DNA measured by PicoGreen.

3. An alternative normalization method uses a reference amplicon instead of the amount of total DNA measured by PicoGreen. The reference amplicon should correspond to the location of a nucleosome that shows 100% occupancy under all of the conditions tested. In this case the relative amount determined by qPCR for each experimental amplicon is divided by the relative amount of the reference amplicon. The normalization is not strictly necessary and its use depends on the object of the NuSA assay. However comparison to a reference amplicon such as a centromeric nucleosome (e.g., *CEN3*) is recommended, since it helps to determine the best experimental conditions, most importantly the optimal parameters of MNase digestions. In addition, it can be used to check for the quality of the DNA samples. And finally, it is relevant to interpret variations in relative protection values caused by a change in the cellular conditions (see Note 7). If relative protection values in two experimental conditions would be compared, they would be normalized to the relative protection of a sequence protected by a fixed nucleosome in both experimental conditions (see Note 8).

3.7. Chromatin Immunoprecipitation-NuSA Protocol (see Note 9)

From Subheading 3.3 in the preceding protocol: Before proceeding with ChIP, verify that the MNase digestion gave mostly mononucleosomes because the DNA will not be size-fractionated.

1. Thaw MNase-digested samples from Subheading 3.3 step 3 on ice. Add 1 μl protease inhibitor cocktail (Sigma) and 10 μl 100 mM PMSF to each 600 μl sample. Add 200 μl of 4× Buffer L components to bring samples to 800 μl.
2. Sonicate (4 × 20 s pulses with low, constant power). Spin 10 min at 13,000 rpm in a tabletop microfuge at 4°C. Transfer supernatant to a new tube and add another 10 μl of PMSF.
3. Remove a 20 μl sample and store at 4°C for a no-antibody input DNA control.
4. To the remainder add antibody and agitate overnight or for an appropriate amount of time at 4°C.
5. Spin 10 min at 13,000 rpm in a microfuge at 4°C.
6. Transfer supernatant to a new tube with protein A sepharose (60 μl for 2 μg of primary antibody; 90 μl for 4 μg antibody) and incubate 1 h at 4°C with rocking.
7. Wash with quick vortexing and 30 s centrifugation in a tabletop microfuge at 2,000 rpm, 4°C, using spin columns:

2× 1 mL ChIP lysis buffer.

2× 1 mL ChIP lysis buffer high salt.

2× 1 mL ChIP wash buffer.

2× 1 mL TE.

8. Spin 10 min at 2,000 rpm in a microfuge at 4°C.
9. Carefully aspirate and discard supernatant.
10. Add 100 μl ChIP elution buffer and incubate overnight 10 min at 65°C with shaking at 700 rpm. An Eppendorf Thermomixer is convenient for this step.
11. Spin at 2,000 rpm in a microfuge for 30 s, collect the supernatant and elute the pellet a second time with 100 μl ChIP elution buffer (10 min at 65°C with shaking at 700 rpm). Combine the two supernatant fractions (~200 μl).
12. Incubate the combined supernatant fractions overnight at 65°C with shaking at 700 rpm to reverse the cross-links.
13. Add 180 μl ChIP elution buffer to the input control samples (step 14) and incubate overnight at 65°C with shaking at 700 rpm.
14. Spin tubes briefly to collect all of the liquid. Add 1 mL Qiagen PCR purification kit PB buffer and mix briefly.
15. Add 500 μl to a spin column over a collection tube, spin 1 min at 10,000 rpm; discard flow-through and add remainder of sample to the spin column, spin 1 min at 10,000 rpm.
16. Discard flow-through , spin again to remove all liquid, add 750 μl ethanol wash buffer PE; spin 1 min at 10,000 rpm. Discard the flow-through and spin again 1 min at 10,000 rpm to dry.
17. Transfer the spin column to a new collection tube, elute with 50 μl of 50°C elution buffer EB added to the center of the filter, and incubate 10 min at 50°C.
18. Spin 1 min at 10,000 rpm. Store DNA at –20°C. Note that in the ChIP-NuSA protocol, the mononucleosomal DNA is not gel-purified.
19. qPCR assay (see Subheading 3.5 above)
    (a) For each primer set run Input, ChIP and genomic DNA samples in duplicate. Include a no-DNA control for each primer set. In principle, it might be necessary to run a standard curve for each primer set using a tenfold dilution series of genomic DNA. In practice, if primer sets have similar efficiencies, one standard curve can be used for all of them (see Note 6).
    (b) The recovery of DNA will generally be low after ChIP so it might not be reliably quantified by PicoGreen analysis. The dilutions needed for the ChIP samples should be

determined empirically. The dilution for the ChIP DNA is usually 1:10, 1:40 for the Input DNA, and 1:500 for the genomic DNA. Perform data analysis as described in Subheading 3.6.

4. Notes

1. The position of primer-pairs is chosen based on the efficiency of the resulting PCR reaction as well as their relative location on the promoter of interest. The program used for primer design was Primer 3 (http://frodo.wi.mit.edu/primer3/). The primers must be specific for the region (e.g., promoter) of interest. Amplicon length should be within the 60–200 bp range. The primers should have 20–80% GC content. Avoid repeats of G's and C's longer than 3 bp. There should be no predicted stable interaction between forward and reverse primers (i.e., avoid primer-dimer production). Place C's and G's on ends of primers, but no more than two in the last five bases on 3′ end. The T_m should be 60–62°C. By keeping all of the primers in this range of T_m one can use a standard PCR protocol with one annealing/elongating temperature of 60°C.

   Figure 3 shows an example of the amplicons used to analyze nucleosome-DNA interactions at the yeast *ADH2prm*. The expected positions of the three nucleosomes at the promoter based on previous studies and the location of relevant sequences for both nucleosomes and transcription initiation are indicated. The bar diagram shows the relative protection of the sequences of each amplicon calculated by applying the NuSA protocol to yeast cells growing on 5% glucose, therefore with the *ADH2prm* being repressed. The relative protection values represented by each bar were normalized to the value obtained for amplicon 4 of *CEN3* (Table 1) which was set to 100. The sequences and other properties of the primers are listed on Table 1. The sequence of the *ADH2prm* of the reference *S. cerevisiae* strain S288c can be obtained at the *Saccharomyces* Genome Database Web site (http://www.yeastgenome.org). As is evident from the figure, the position and overlap of the primers is critical to resolving the location of the nucleosomes. After preliminary NuSA analysis, it might become evident that additional primer-pairs are needed to achieve the desired resolution. In the example shown, amplicon 12.5 was added after initial analysis indicated that the 5′ border of the N(−1) nucleosome would be better resolved with an additional pair of primers between amplicon 12 and 13.

2. Zymolyase digestion can be followed in various ways to monitor the conversion of intact cells to osmotically fragile spheroplasts.

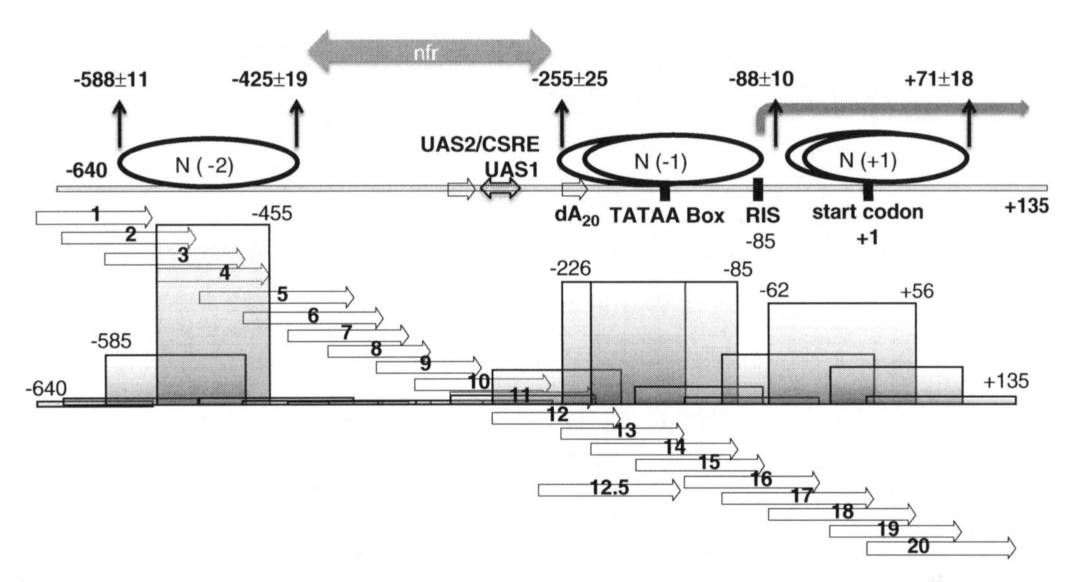

Fig. 3. Diagram of the *ADH2* promoter showing positions of nucleosomes, transcription factor binding sites, and primer pairs used for nucleosome scanning assays. Amplicons 1–20 used for NuSA of the *ADH2prm* are shown by *open arrows*. In the upper part of the figure the location of the three previously described nucleosomes are represented by *ovals* and the nucleosome-free region (nfr) is shown as a *gray double-headed arrow* above the figure. The location of the binding site for the transcription factors Cat8 (UAS2/CSRE) and Adr1 (UAS1), a poly dA string (dA$_{20}$), the TATA box, RNA initiation site (RIS) and the start codon are also represented. The *shaded vertical bars* show the relative protection values obtained for each amplicon by NuSA. The protection values were normalized to the value obtained for amplicon 4 (AMP4) of *CEN3* (Fig. 7 and Table 1). Above *some bars*, the relative positions of the amplicon 5′- or 3′-end with respect to the start codon are indicated (see Table 1 for location of *ADH2* primer pairs).

Mix one volume of spheroplast suspension (one drop) with one volume of water (one drop). Observe under the microscope at ~1,000× magnification. Swollen "ghosts" which have lost most of their refractive cell contents are an indication of effective spheroplasting. Add another volume of water and the spheroplasts should lyse, releasing cell wall material. If little cell wall material is visible, spheroplasting is complete. The ease with which the cell walls are digested is highly dependent on the growth stage of the cells. Cells in log phase are more easily converted to spheroplasts than cells in late log or stationary phase. The extent to which the cells have been converted to spheroplasts is a major determinant of the sensitivity to MNase, and thus the eventual recovery of mononucleosomal DNA.

3. To resuspend the spheroplasts use a rounded glass rod or a sterile rubber policeman and gently mix the pelleted spheroplasts before adding buffer. Then, add 1/10 vol. buffer and continue to resuspend until the mixture is homogeneous; then, add the remainder of the buffer. It is best to make a large number of small aliquots of MNase in the buffer described and store them at −80°C. They seem to be stable for several years stored in this manner.

Table 1
Primers used for nucleosome scanning assays (NuSA)

Gene	Amplicon	Primer	Sequence	Direction	Tm	Middle of amplicon (Promoter coordinate)	5′/3′ end	Size (bp)
ADH2	A1	LL31	GAAGAGACTAATCAAAGAATCGTTTTC	Forward	59	−594	−640	92
ADH2	A1	LL32	CGTTTGTTTGCCCCTACG	Reverse	60		−548	
ADH2	A2	LL33	CGTTTTCTCAAAAAAATTAATATCTTAAC	Forward	58	−567	−620	106
ADH2	A2	LL34	TCTTGGCATCAGAAAATTTGAG	Reverse	59		−514	
ADH2	A3	LL35	GTTTGATCAAAGGGGCAAAACG	Forward	65	−530	−585	111
ADH2	A3	LL36	CGGATCATAAGGCAATTTTTAGATAAG	Reverse	61		−475	
ADH2	A4	LL37	AAATCGTTTCTCAAATTTTCTGATG	Forward	60	−500	−545	89
ADH2	A4	LL38	CAGGCTGTAACCGGAGAGAC	Reverse	60		−455	
ADH2	A5	LL39	TCTAACCAGTCTTATCTAAAATTGCC	Forward	59	−450	−511	122
ADH2	A5	LL40	TGAAGACAAAATCCCTTAATTAAAAC	Reverse	58		−389	
ADH2	A6	LL41	CCGTCTCTCCGGTTACAGC	Forward	61	−424	−477	106
ADH2	A6	LL42	ATGAGCGAAAGCCGTTAATG	Reverse	60		−371	
ADH2	A7	LL43	CCTGCCTTTCTAATCACCATTC	Forward	60	−393	−440	95
ADH2	A7	LL44	GCGGGCAAAACGTCATAAC	Reverse	61		−345	
ADH2	A8	LL45	AATTAAGGGATTTTGTCTTCATTAACG	Forward	61	−370	−410	80

ADH2	A8	LL46	GGATGGTTTCCCGCCTG	Reverse	62		−330	
ADH2	A9	LL47	AAAATGTTATGACGTTTTGCCCG	Forward	64	−330	−371	82
ADH2	A9	LL48	AGATGCCCGGTGTTCCG	Reverse	63		−289	
ADH2	A10	LL49	GAAACCATCCACTTCACG-AGACTG	Forward	64	−287	−350	107
ADH2	A10	LL50	TTTTTTTTCATTCTCTCAAT-CTGAAAT	Reverse	60		−244	
ADH2	A11	LL51	CCTCTGCCGGAACACCG	Forward	64	−256	−313	114
ADH2	A11	LL52	CCATTTCTATGCTCTCCTCTGC	Reverse	60		−199	
ADH2	A12	LL53	AAGTTGGAGAAATAAGAGAATT-TCAGATTG	Forward	62	−229	−280	101
ADH2	A12	LL54	GCTTTACCAAAAAGTGAACCCC	Reverse	61		−179	
ADH2	A12.25	LL104	GAGAATTTCAGATTGAGAGAATGAA	Forward	58	−222		86
ADH2	A12.25	LL54	GCTTTACCAAAAAGTGAACCCC	Reverse	61			
ADH2	A12.5	LL106	AAAAAAAAAAAAAAAAAAAAG-GCAGAGG	Forward	64	−192		126
ADH2	A12.5	LL56	TGAAAAAAGTCGCTACTGGCAC	Reverse	62			
ADH2	A13	LL55	AAAAGGCAGAGGAGAGCATAGAAATG	Forward	64	−177	−225	97
ADH2	A13	LL56	TGAAAAAAGTCGCTACTGGCAC	Reverse	62		−128	
ADH2	A14	LL57	TGGGGTTCACTTTTTGGTAAAGC	Forward	63	−144	−202	116
ADH2	A14	LL58	TGATAAAAACAACAAGAGAG-CAGTAGTAA	Reverse	60		−86	
ADH2	A15	LL59	ATCACATATAAATAGAGTGC-CAGTAGCGAC	Forward	64	−116	−167	102
ADH2	A15	LL60	TTACCAAGAAGAAACAAGAAGT-GATAAA	Reverse	60		−65	

(continued)

Table 1 (continued)

Gene	Amplicon	Primer	Sequence	Direction	Tm	Middle of amplicon (Promoter coordinate)	5′/3′ end	Size (bp)
ADH2	A16	LL61	CACTCGAAATACTCTTACTACT GCTCTC	Forward	60	−76	−129	106
ADH2	A16	LL62	GTTGATAGTTGATTGTATGCT TTTTG	Reverse	58		−23	
ADH2	A17	LL63	GTTGTTTTTATCACTTCTTG TTTCTTC	Forward	58	−39	−99	120
ADH2	A17	LL64	TTGAGTTTCTGGAATAGAC ATTGTG	Reverse	60		+21	
ADH2	A18	LL65	GAATATCAAGCTACAAAAAGC ATACAATC	Forward	61	−3	−62	117
ADH2	A18	LL66	ACTTGCCGTTGGATTCGTAG	Reverse	60		+56	
ADH2	A19	LL67	TATCGTAATACACAATGTCT ATTCCAGAAA	Forward	61	39	−14	106
ADH2	A19	LL68	GGCTTTGGCTTTGGAACTG	Reverse	61		+92	
ADH2	A20	LL69	CTCAAAAAGCCATTATCTTCTACG	Forward	59	76	+16	119
	A20	LL70	GTGGCAGACACCAGAGTAC	Reverse	58		+136	
CEN3	AMP1	LL111	TATACATTTCATAAACAT GGCATGGC					
CEN3	AMP1	LL112	TGTGACTTATTTGTACTATTTTT TCAATGAAT					
CEN3	AMP2	LL113	CGCCAAACAATATGGAAAATCC					

CEN3	AMP2	LL114	AAAAATATAATAAAATCAAAT ATCATCATGTGAC
CEN3	AMP3	LL115	TTGAAAAAATAGTACAAAT AAGTCACATGA
CEN3	AMP3	LL116	CGGAAATCAAATACACTAAT ATTTTAAA
CEN3	AMP4	LL117	CATGATGATATTTGATTTTAT TATATTTTTAA
CEN3	AMP4	LL118	TTTCTTTTTAACTTTCGG AAATCA
CEN3	AMP5	LL119	TTAGTGTATTTGATTTCC GAAAGTT
CEN3	AMP5	LL120	CGTTTCATATATCCATTCA ATGAAA
CEN3	AMP6	LL121	TATTTCATTGAATGGATA TATGAAACG
CEN3	AMP6	LL122	ATTACTTCTATTGAATAATAA TATATGAGCAAA

4. Owing to variations in spheroplasting from experiment-to-experiment it is best to do at least two and sometimes more digestions using different amounts of MNase so that at least one sample yields the desired level of mononucleosomal DNA. Titrate the amount of MNase for a particular strain and set of conditions you will be using. The appropriate amount of MNase digestion will give about 80–90% mononucleosomal DNA and 10–20% dinucleosomal DNA. An example of a typical MNase titration is shown in Fig. 1, *bottom right.* More extensive digestion gives lower yields of nucleosome peaks in the NuSA assay, perhaps because the DNA has been nicked extensively within a nucleosome. Any mononucleosomal DNA that has been nicked on both strands will not be PCR-ampified.
5. PicoGreen is a very sensitive, specific and accurate way to measure the amount of double-stranded DNA. Because the DNA is likely to have a significant amount of contaminating small RNA, a method based on absorbance is not likely to be accurate.
6. If 1 mM Tris–HCl, pH 8.0 and siliconized tubes are used when diluting DNA samples for qPCR assays, the dilutions are stable for a few weeks. However, it is not recommended to using stocks that are older than 1 month. Adding glycogen or *E. coli* RNA seems to reduce the loss of DNA.
7. Different mononucleosomes could have different sensitivities to MNase. This might be due to different affinities of the histone octamer for a particular DNA sequence, different nucleosome composition (of, for example variant histones or histone modifications), different occupancies at any given time, or some combination of these and other reasons. Nucleosome-free regions (nfr's) will be much more MNase sensitive. This is illustrated in Fig. 4 showing the relative amount of different amplicons produced from the *ADH2prm* as a function of MNase concentration. At high levels of MNase the relative amount of protection in the promoter region varies significantly. As determined by the final level of protection, the N(−2) nucleosome is the most resistant (amplicon 4), followed by the N(−1) and N(+1) nucleosomes (amplicons 13, 14 and 18, respectively) with the region previously characterized as being nucleosome-free being the most sensitive (amplicons 6–11). However, within the nfr there is a region (amplicon 9) that shows greater resistance to MNase than its flanking regions, but less resistance than the region protected by nucleosomes N(−2), N(−1) and N(+1). A region of analogous partial resistance was recently reported in the *GAL1,10* promoter (29). It appears to be a partially unwound nucleosome bound to the Rsc complex. The greater MNase resistance of amplicon 9 in the *ADH2prm* could be due to an inherent feature of this DNA sequence and/or to partial protection by an unknown protein.

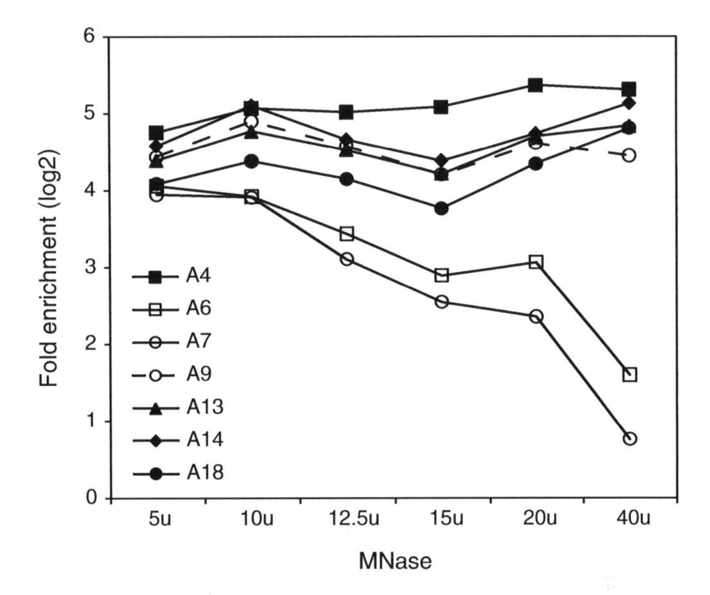

Fig. 4. Quantification of resistance to MNase digestion for different amplicons used in the NuSA assay of the *ADH2* promoter. Cross-linked samples were serially digested with different amounts of MNase for 30 min. The graph shows enrichment of each amplicon in a digested sample over the same amplicon in the undigested sample calculated as the qPCR ratios of digested/undigested DNA normalized to differences in DNA concentration. The data are expressed as the $\log_2$ ratio. The graph lines of DNA regions predicted to be protected by a nucleosome have solid labels (A4, A13, A14, A18), while the lines of DNA regions located at the nucleosome free region of the *ADH2* promoter have open labels (A6, A7). The *dashed line* shows the pattern of amplicon 9, which surprisingly was similar to patterns of protected sites although being located in the nucleosome free region (see Note 7).

The MNase titration curves might also reveal whether the nucleosomal DNA is uniformly protected. If some promoters were completely free of N(−1), for example, a biphasic curve of MNase sensitivity would be expected, as shown in the schematic diagram in Fig. 2, right. A variant of NuSA has been developed with the purpose of measuring values for the proportion of yeast cells in which a given sequence is protected by a nucleosome in any given moment at any given condition (see Chapter 4) (34). The novel aspect of this assay is that cross-linked chromatin is digested with MNase over a wide range of nuclease concentrations. Then a series of overlapping primer pairs, which generate amplicons of about 60 bp each, are used to quantify the reaction products by qPCR without any prior fractionation for the size of protected fragments. Figure 5 is an illustration of the effect of different extents of MNase digestion on the final NuSA pattern of the *ADH2prm* when cells are grown in repressing conditions (*ADH2prm* inactive). This pattern is consistent with the MNase titration curves in Fig. 4, showing the differential sensitivity of the nucleosomal DNAs, with N(−2) being the most resistant, followed by N(−1), N(+1), and the nfr.

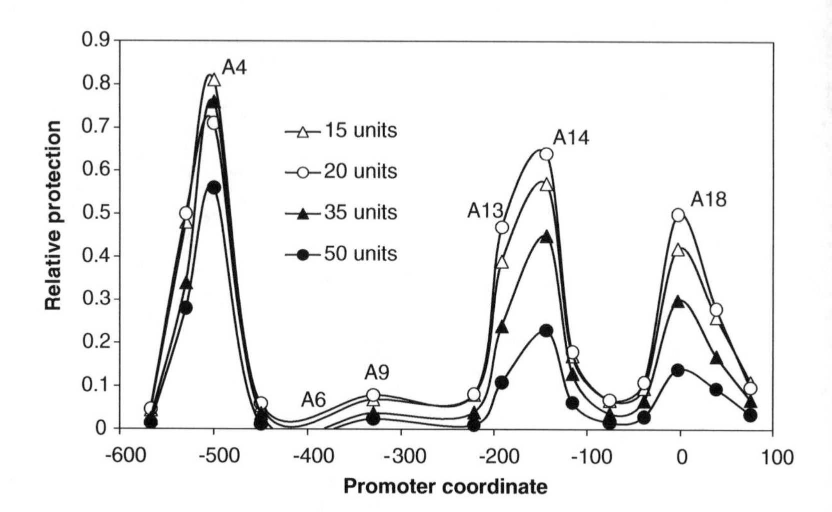

Fig. 5. NuSAs of the repressed *ADH2* promoter with samples treated with different amounts of MNase. The assays were performed as described in the Methods section but different amounts of MNase were used to digest the DNA for 30 min. A NuSA assay was performed for each digested sample. Relative protection was normalized to a centromeric nucleosome-associated amplicon (AMP4 of *CEN3*). The points corresponding to particular amplicons of Fig. 4 are indicated.

The isolation of size-fractionated DNA may not be necessary for some types of analysis. Figure 6 shows a NuSA of the *ADH2prm* using both size-fractionated, gel-purified DNA (mononucleosomal DNA), and an equivalent DNA sample that had not been size-fractionated (total DNA). The DNA was isolated from cultures of cells in which the *ADH2prm* was inactive (repressed growth conditions, high glucose) and active (derepressed growth conditions, low glucose). In conditions in which the promoter is inactive, the NuSA patterns using total and size-fractionated DNA are very similar. Using DNA from the active-promoter conditions, when all of the nucleosomes appear to be more sensitive to MNase, the patterns are dissimilar, with the non-size-fractionated DNA showing relatively poor signal-to-noise ratios for all nucleosome positions. This is likely due to a higher level of DNA representing di-, tri-, etc. nucleosomes in those preparations. This would lead to a higher relative level of amplification with primers located between nucleosomes.

8. If determining the position of nucleosomes is the only purpose of the procedure, the question of normalization is relatively unimportant. However, if the relative level of nucleosomal protection in different experimental conditions is being addressed, it is important to normalize the protection values. The simplest normalization method is to divide the relative amount of each amplicon, determined from the qPCR of the mononucleosomal DNA by the amount of total mononucleosomal DNA measured

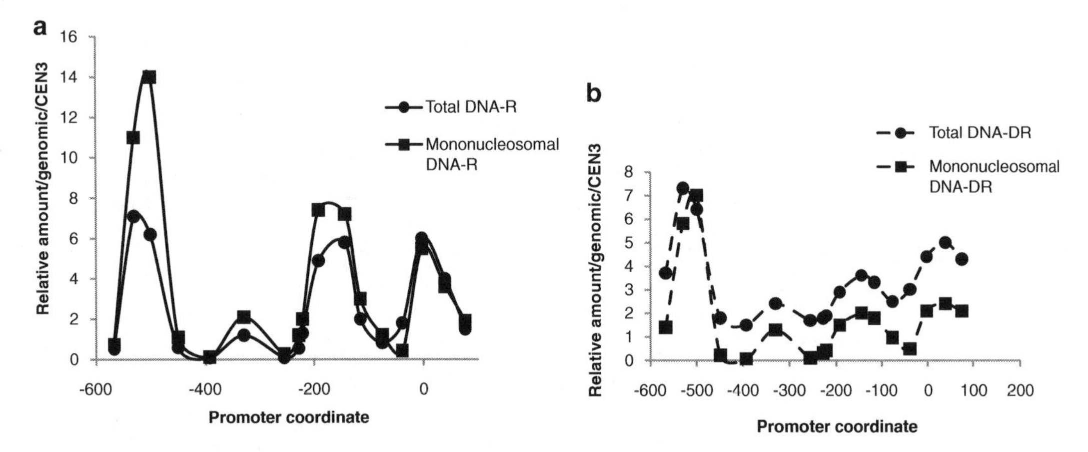

Fig. 6. *ADH2* promoter NuSA performed with size-fractionated mononucleosomal DNA and non-size-fractionated DNA. Spheroplasts from cultures grown in high (**a**) and low (**b**) glucose were prepared and digested with MNase as described in the protocol. After MNase digestion the samples were divided and one portion was size-fractionated to isolate *mononucleosomal DNA*. DNA was isolated from the other portion without size fractionation (*total DNA*). The relative amount of each amplicon quantified by qPCR was normalized to the amount of amplicon 4 from the *CEN3* locus measured by qPCR in the same sample (Fig. 7 and Table 1). The high background in the NuSA representing chromatin from derepressed cultures is most likely due to under-digestion with MNase, leading to some di- and trinucleosomal DNA that protects amplicons from the nucleosome-free region and the regions between nucleosomes.

by PicoGreen. However, this does not take into consideration any confounding effects due to different levels of spheroplasting, different degrees of MNase digestion that might not be discernable on the gels, or other complications. Therefore, we often use an internal reference that attempts to correct for these potential differences. Internal references we have used are derived from the *PHO5* promoter (TATA nucleosome) (3), *CEN3*, and *ACT1*. *CEN3* contains a stably positioned nucleosome with H3 replaced by the histone CEN-PA. As shown in Fig. 7, the relative protection of the *CEN-PA*-containing nucleosome (amplicon 4) is similar in repressed and derepressed cells. In most NuSAs we have performed, the interpretation of the results is unaffected by the method of normalization, suggesting that in these experiments the reference (e.g., CEN-PA-containing) and experimental mononucleosomes experienced the same exposure to MNase. However, for different strains or growth media it may be necessary to find other nucleosomes whose sensitivity to MNase digestion are invariant in the specific conditions or strains being analyzed.

9. Figure 8 illustrates the use of NuSA ChIP to demonstrate the presence of histones in the protected regions of the promoter. It also demonstrates a change in nucleosome occupancy and position that accompanies activation of the glucose-repressed

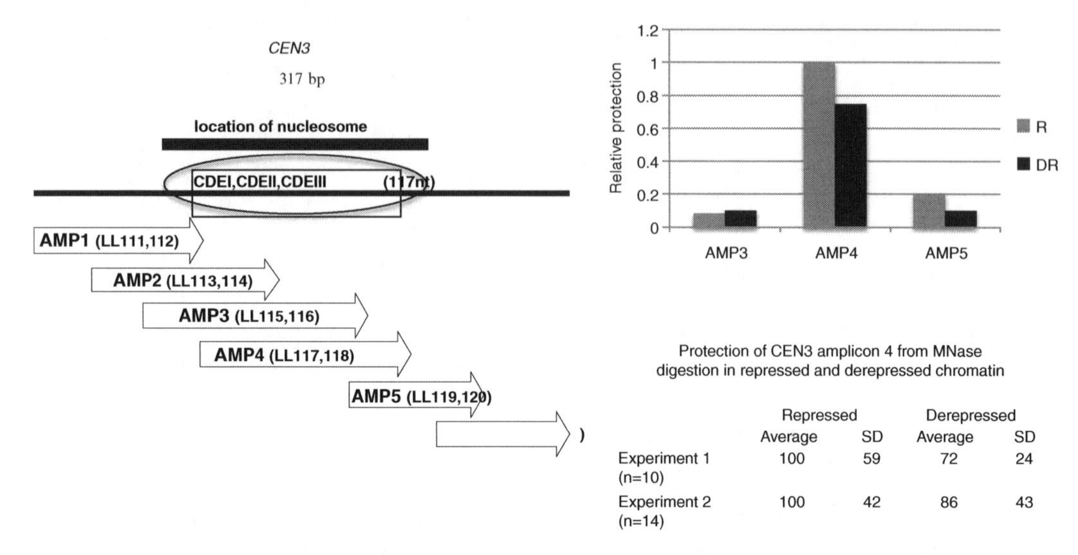

	Repressed		Derepressed	
	Average	SD	Average	SD
Experiment 1 (n=10)	100	59	72	24
Experiment 2 (n=14)	100	42	86	43

Fig. 7. Cartoon of the *CEN3* reference region and primer pairs used for normalizing NuSA data. Cells were grown in either high (repressing) or low (derepressing) conditions. After MNase digestion the samples were size-fractionated to isolate mononucleosomal DNA and the relative protection of the *CEN3* amplicons was measured by qPCR. The relative qPCR values for each amplicon obtained using mononucleosomal DNA isolated from repressed and derepressed cultures were normalized to the amount of mononucleosomal DNA measured by PicoGreen in each sample. Only the AMP4 amplicon was protected by a nucleosome, as predicted.

FBP1 gene. The *FBP1prm* is activated by glucose starvation by Cat8-dependent recruitment of chromatin modifying factors and coactivators (25). Activation is accompanied by Cat8-dependent nucleosome loss and/or translational alteration of nucleosomes N(−1) and N(−2) as shown in Fig. 8 and in ref. 25. The ChIP analysis, followed by NuSA, demonstrates that all of the regions of protection in the *FBP1prm*, in both the inactive and active state, are due to nucleosomes. The evidence for this assertion is the congruence of the curves for total mononucleosomal DNA (Input) and DNA isolated after ChIP for histone H3.

Because samples isolated from cells grown in different conditions often show dramatic differences in sensitivity to MNase digestion, it is important to have a reference gene that is unaffected by these changes. In this case the relative protection values for the *FBP1prm* amplicons were normalized to the amount of amplicon 4 of *CEN3* in the top panel to take the changes of MNase sensitivity into account between repressed and derepressed cell cultures. In the middle and bottom panels, the relative amounts were normalized to the total amount of mononucleosomal DNA and to a specific amplicon within that panel because it is only the comparison between ChIP and total input DNA that is important.

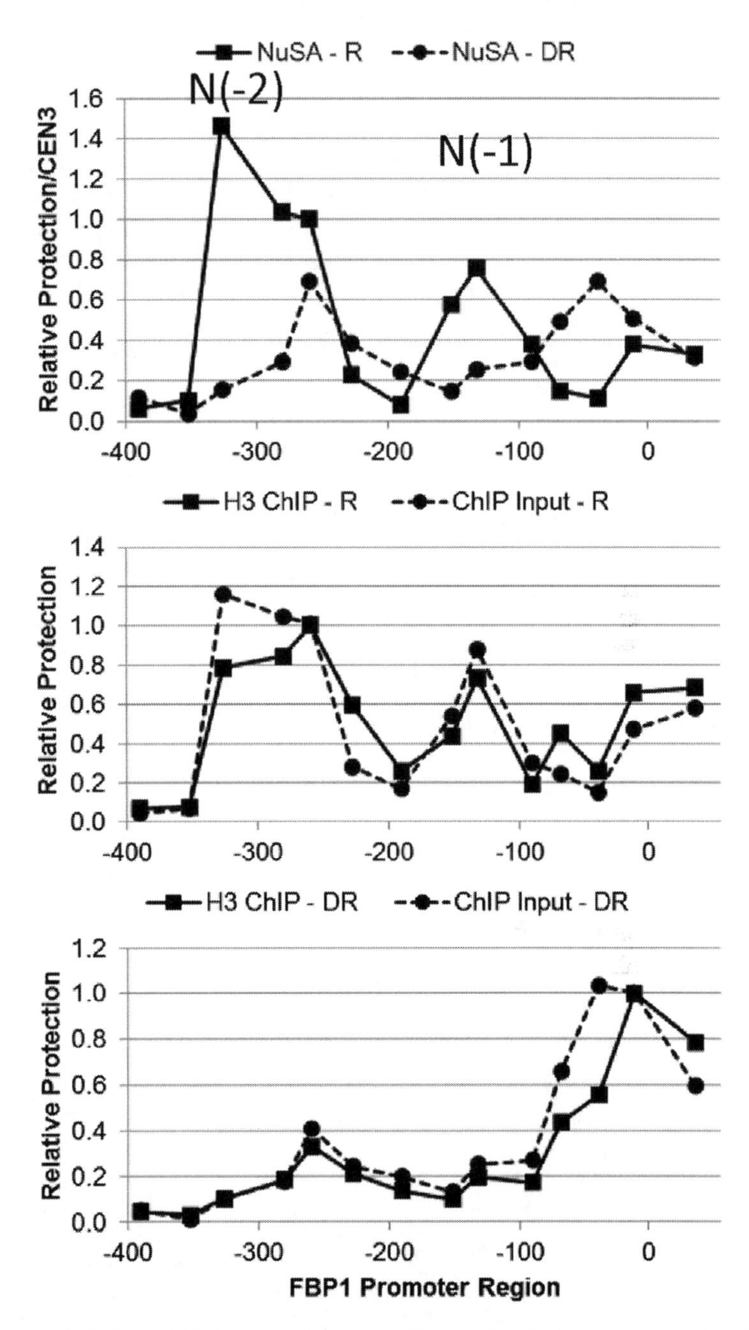

Fig. 8. ChIP-NuSA of *FBP1prm*. NuSAs of the *FBP1prm* (25) illustrating the use of ChIP-NuSA to demonstrate that peaks of presumed nucleosomal protection from MNase digestion contain histones. The *upper panel* represents a standard NuSA using size-fractionated mononucleosomal DNA. The relative protection was normalized to the protection of AMP4 from *CEN3*. The *bottom two panels* show both a standard NuSA and a NuSA that was performed with DNA isolated after ChIP for histone H3. The samples were from cells growth either in high glucose (*middle panel*) or low glucose (*lower panel*). In these two panels the relative protection was normalized to a specific amplicon within each curve.

Acknowledgements

Research was supported by a grant from the General Medical Sciences Institute of the National Institutes of Health, GM26079 to E.T.Y. J.J. Infante was the recipient of a postdoctoral fellowship from the Ministry of Science and Innovation of Spain.

References

1. Kornberg, R. D., and Lorch, Y. (1999) Twenty-five years of the nucleosome, fundamental particle of the eukaryote chromosome, *Cell 98*, 285–294.
2. Lee, C. K., Shibata, Y., Rao, B., Strahl, B. D., and Lieb, J. D. (2004) Evidence for nucleosome depletion at active regulatory regions genome-wide, *Nat Genet 36*, 900–905.
3. Sekinger, E. A., Moqtaderi, Z., and Struhl, K. (2005) Intrinsic histone-DNA interactions and low nucleosome density are important for preferential accessibility of promoter regions in yeast, *Mol Cell 18*, 735–748.
4. Peckham, H. E., Thurman, R. E., Fu, Y., Stamatoyannopoulos, J. A., Noble, W. S., Struhl, K., and Weng, Z. (2007) Nucleosome positioning signals in genomic DNA, *Genome Res 17*, 1170–1177.
5. Lee, W., Tillo, D., Bray, N., Morse, R. H., Davis, R. W., Hughes, T. R., and Nislow, C. (2007) A high-resolution atlas of nucleosome occupancy in yeast, *Nat Genet 39*, 1235–1244.
6. Ioshikhes, I. P., Albert, I., Zanton, S. J., and Pugh, B. F. (2006) Nucleosome positions predicted through comparative genomics, *Nat Genet 38*, 1210–1215.
7. Kaplan, N., Moore, I. K., Fondufe-Mittendorf, Y., Gossett, A. J., Tillo, D., Field, Y., LeProust, E. M., Hughes, T. R., Lieb, J. D., Widom, J., and Segal, E. (2009) The DNA-encoded nucleosome organization of a eukaryotic genome, *Nature 458*, 362–366.
8. Radman-Livaja, M., and Rando, O. J. (2010) Nucleosome positioning: how is it established, and why does it matter? *Dev Biol 339*, 258–266.
9. Segal, E., Fondufe-Mittendorf, Y., Chen, L., Thastrom, A., Field, Y., Moore, I. K., Wang, J. P., and Widom, J. (2006) A genomic code for nucleosome positioning, *Nature 442*, 772–778.
10. Segal, E., and Widom, J. (2009) What controls nucleosome positions? *Trends Genet 25*, 335–343.
11. Caserta, M., Agricola, E., Churcher, M., Hiriart, E., Verdone, L., Di Mauro, E., and Travers, A. (2009) A translational signature for nucleosome positioning in vivo, *Nucleic Acids Res 37*, 5309–5321.
12. Chakravarthy, S., Park, Y. J., Chodaparambil, J., Edayathumangalam, R. S., and Luger, K. (2005) Structure and dynamic properties of nucleosome core particles, *FEBS Lett 579*, 895–898.
13. Adkins, M. W., Williams, S. K., Linger, J., and Tyler, J. K. (2007) Chromatin disassembly from the PHO5 promoter is essential for the recruitment of the general transcription machinery and coactivators, *Mol Cell Biol 27*, 6372–6382.
14. Schermer, U. J., Korber, P., and Horz, W. (2005) Histones are incorporated in trans during reassembly of the yeast PHO5 promoter, *Mol Cell 19*, 279–285.
15. Tachibana, C., Biddick, R., Law, G. L., and Young, E. T. (2007) A poised initiation complex is activated by SNF1, *J Biol Chem 282*, 37308–37315.
16. Kim, Y., McLaughlin, N., Lindstrom, K., Tsukiyama, T., and Clark, D. J. (2006) Activation of Saccharomyces cerevisiae HIS3 results in Gcn4p-dependent, SWI/SNF-dependent mobilization of nucleosomes over the entire gene, *Mol Cell Biol 26*, 8607–8622.
17. Luger, K. (2006) Dynamic nucleosomes, *Chromosome Res 14*, 5–16.
18. Wu, C. (1980) The 5′ ends of Drosophila heat shock genes in chromatin are hypersensitive to DNase I, *Nature 286*, 854–860.
19. Buttinelli, M., Di Mauro, E., and Negri, R. (1993) Multiple nucleosome positioning with unique rotational setting for the Saccharomyces cerevisiae 5 S rRNA gene in vitro and in vivo, *Proc Natl Acad Sci USA 90*, 9315–9319.
20. Fragoso, G., John, S., Roberts, M. S., and Hager, G. L. (1995) Nucleosome positioning on the MMTV LTR results from the frequency-biased occupancy of multiple frames, *Genes Dev 9*, 1933–1947.
21. Bernstein, B. E., Liu, C. L., Humphrey, E. L., Perlstein, E. O., and Schreiber, S. L. (2004) Global nucleosome occupancy in yeast, *Genome Biol 5*, R62.

22. Shivaswamy, S., Bhinge, A., Zhao, Y., Jones, S., Hirst, M., and Iyer, V. R. (2008) Dynamic remodeling of individual nucleosomes across a eukaryotic genome in response to transcriptional perturbation, *PLoS Biol* *6*, e65.
23. Jiang, C., and Pugh, B. F. (2009) Nucleosome positioning and gene regulation: advances through genomics, *Nat Rev Genet* ***10***, 161–172.
24. Rando, O. J., and Chang, H. Y. (2009) Genome-wide views of chromatin structure, *Annu Rev Biochem* *78*, 245–271.
25. Biddick, R. K., Law, G. L., and Young, E. T. (2008) Adr1 and Cat8 mediate coactivator recruitment and chromatin remodeling at glucose-regulated genes, *PLoS One* *3*, e1436.
26. Di Mauro, E., Verdone, L., Chiappini, B., and Caserta, M. (2002) In vivo changes of nucleosome positioning in the pretranscription state, *J Biol Chem* *277*, 7002–7009.
27. Verdone, L., Camilloni, G., Di Mauro, E., and Caserta, M. (1996) Chromatin remodeling during Saccharomyces cerevisiae ADH2 gene activation, *Mol Cell Biol* ***16***, 1978–1988.
28. Verdone, L., Cesari, F., Denis, C. L., Di Mauro, E., and Caserta, M. (1997) Factors affecting Saccharomyces cerevisiae ADH2 chromatin remodeling and transcription, *J Biol Chem* *272*, 30828–30834.
29. Floer, M., Wang, X., Prabhu, V., Berrozpe, G., Narayan, S., Spagna, D., Alvarez, D., Kendall, J., Krasnitz, A., Stepansky, A., Hicks, J., Bryant, G. O., and Ptashne, M. (2010) A RSC/nucleosome complex determines chromatin architecture and facilitates activator binding, *Cell* ***141***, 407–418.
30. Sherman, F. (1991) Getting started with yeast, *Methods Enzymol* ***194***, 3–21.
31. Liu, C. L., Kaplan, T., Kim, M., Buratowski, S., Schreiber, S. L., Friedman, N., and Rando, O. J. (2005) Single-nucleosome mapping of histone modifications in S. cerevisiae, *PLoS Biol* *3*, e328.
32. Pfaffl, M. W. (2001) A new mathematical model for relative quantification in real-time RT-PCR, *Nucleic Acids Res* *29*, e45.
33. Bustin, S. A., Benes, V., Garson, J. A., Hellemans, J., Huggett, J., Kubista, M., Mueller, R., Nolan, T., Pfaffl, M. W., Shipley, G. L., Vandesompele, J., and Wittwer, C. T. (2009) The MIQE guidelines: minimum information for publication of quantitative real-time PCR experiments, *Clin Chem* ***55***, 611–622.
34. Bryant, G. O., Prabhu, V., Floer, M., Wang, X., Spagna, D., Schreiber, D., and Ptashne, M. (2008) Activator control of nucleosome occupancy in activation and repression of transcription, *PLoS Biol* *6*, 2928–2939.

Chapter 6

Assaying Chromatin Structure and Remodeling by Restriction Enzyme Accessibility

Kevin W. Trotter and Trevor K. Archer

Abstract

The packaging of eukaryotic DNA into nucleosomes, the fundamental unit of chromatin, creates a barrier to nuclear processes, such as transcription, DNA replication, recombination, and repair. This obstructive nature of chromatin can be overcome by the enzymatic activity of chromatin remodeling complexes, which create a more favorable environment for the association of essential factors and regulators to sequences within target genes. Here, we describe a detailed approach for analyzing chromatin architecture and remodeling by restriction endonuclease hypersensitivity assay. This procedure uses restriction endonucleases to characterize changes in chromatin that accompany nucleosome remodeling. The specific experimental example described in this article is the BRG1 complex-dependent chromatin remodeling of the steroid hormone-responsive mouse mammary tumor virus promoter. Through the use of these methodologies one is able to quantify changes at specific nucleosomes in response to regulatory signals.

Key words: Chromatin, Restriction enzyme, Hypersensitivity, BRG1, SWI/SNF, Transcription

1. Introduction

In the eukaryotic nucleus, DNA is packaged with histone and non-histone proteins to form a highly organized chromatin structure (1–2). This DNA–protein assembly represents a significant barrier to a number of nuclear processes, such as DNA repair, recombination, replication, and transcription (3). Often the structural changes in chromatin that accompany transcriptional activation require multiprotein enzymatic complexes that manipulate nucleosomal architecture (4). Alteration in chromatin structure by ATP-dependent remodeling complexes is considered a significant initial step in transcriptional regulation (5). Chromatin remodeling complexes use the energy of ATP-hydrolysis to alter local chromatin

Randall H. Morse (ed.), *Chromatin Remodeling: Methods and Protocols*, Methods in Molecular Biology, vol. 833,
DOI 10.1007/978-1-61779-477-3_6, © Springer Science+Business Media, LLC 2012

structure and facilitate a more "open/active" conformation conducive for the recruitment and binding of essential nuclear factors. A number of chromatin remodeling complexes have been identified that modulate the arrangement and stability of nucleosomes in a non-covalent manner (5). Generally, these ATP-dependent remodeling machines are divided into four major subfamilies, characterized by the identity of their central catalytic subunit, which include BRG1 (or hBrm), ISWI, Mi-2, and Ino80 of their respective complexes SWI/SNF, ISWI, NuRD, and INO80 (6, 7).

The SWI/SNF remodeling complex was originally identified in yeast, by mutations in mating type switching (SWI) and sucrose fermentation (SNF), and its function is highly conserved in eukaryotes with homologs found in *Drosophila* and mammals. Human SWI/SNF is a large multiprotein complex that possesses either BRG1 or hBrm, two ATPases related to yeast SWI2/SNF2 and STH1, and is composed of about 10 BRG1-associated factors (BAFs) subunits, most of which are orthologous to those found in yeast SWI/SNF and RSC (5).

This chapter provides a detailed approach for analyzing chromatin architecture and remodeling by the BRG1-SWI/SNF complex upon nuclear receptor signaling (Fig. 1). Using the human SW-13 cell line, containing stably integrated mouse mammary tumor virus (MMTV) promoter, as a model system we are able to evaluate chromatin changes upon hormone signaling (8). This procedure allows the use of restriction endonucleases to characterize alterations in the chromatin structure that accompany nucleosome remodeling and transcriptional activation (9). The methodology presented focuses on analyzing nucleosomal organization using isolated nuclei, restriction endonucleases, and reiterative primer extension. All methods described here have been employed extensively to study the activity of BRG1 on chromatin structure and transcriptional regulation from the MMTV promoter (10). Although the focus of this article is on the MMTV promoter, these techniques have been applied to a variety of inducible and developmentally responsive gene promoters in higher eukaryotes to analyze changes within chromatin (11–15).

2. Materials (see Note 1)

2.1. Cell Culture

1. Dulbecco's modified Eagle's medium (DMEM) high glucose, supplemented with 10% normal fetal bovine serum, 2 mM L-glutamine, 10 mM HEPS, 100 μg/ml penicillin, and 100 μg/ml streptomycin.
2. Humidified incubator for growth of cultured cells at 37°C under 5% CO_2.
3. 150 mm tissue culture plates.

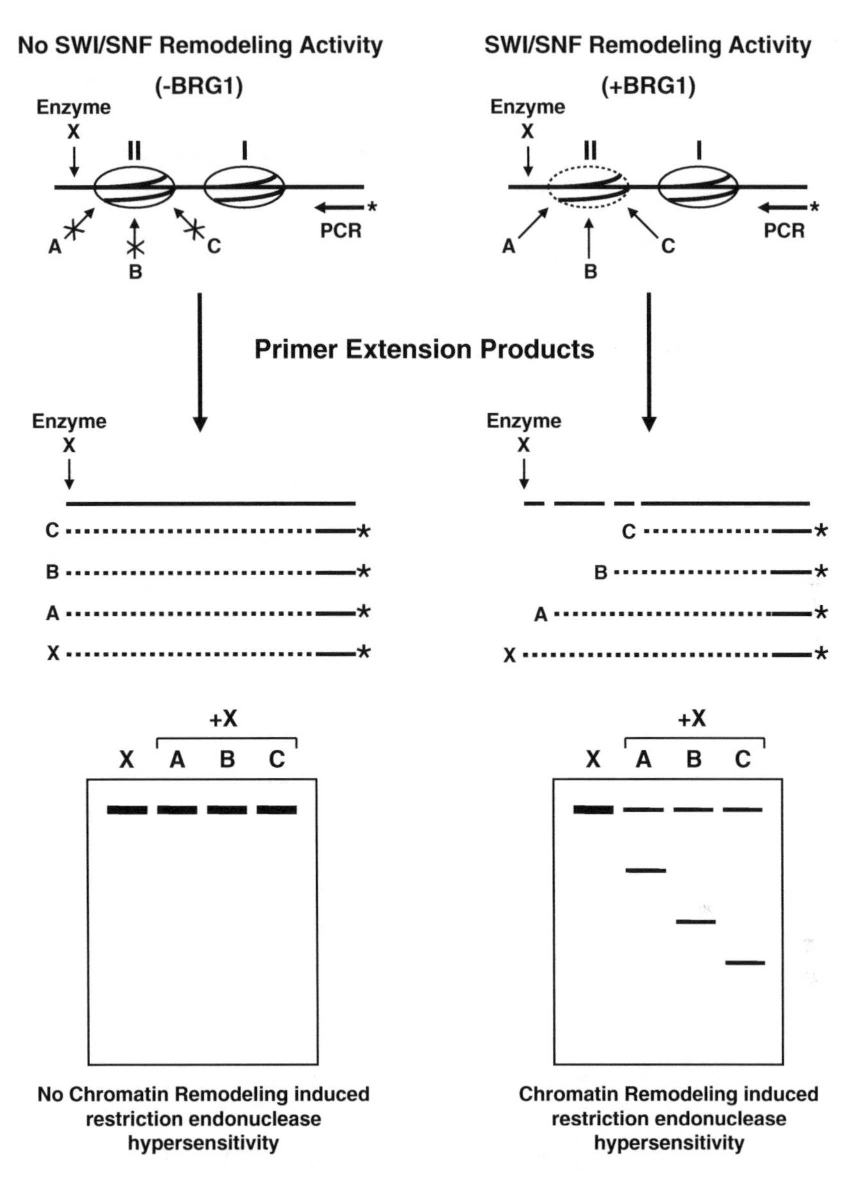

Fig. 1. Schematic of the restriction enzyme hypersensitivity assay used to evaluate chromatin structure and detect SWI/SNF-mediated chromatin remodeling activity. Nuclei are isolated from naïve or hormone-treated cells with (+BRG1) or without (–BRG1) expression of a functional SWI/SNF chromatin remodeling complex. The chromatin is partially digested with various restriction endonucleases (A, B, or C) that target the predicted hypersensitive region. After purification, the genomic DNA is digested to completion with a second restriction enzyme which cuts outside the hypersensitive region (enzyme X). The resulting DNA is purified and analyzed by reiterative primer extension using *Taq* polymerase and a ^{32}P-labeled oligonucleotide specific for the target of interest. The amplified products are resolved on a denaturing polyacrylamide gel and exposed to a PhosphorImager screen to detect the presence of bands that correspond, in length, to the predicted in vivo restriction endonuclease cleavage pattern. Redigestion with the second restriction enzyme (in vitro digest) serves as an internal standard to ensure equal amounts of template DNA are loaded in each reaction. The prediction in this model would suggest that the SWI/SNF complex targets regions within nucleosome II where it remodels the chromatin structure to allow access to transcription factors or, in this case, restriction endonucleases to the genomic DNA. In the absence of BRG1, SWI/SNF is unable to remodel nucleosome II where it acts to block restriction enzyme access to the target DNA.

4. LipofectAMINE 2000 (Invitrogen, Carlsbad, CA) or equivalent transfection system, to be used according to manufacturer's recommendations (see Note 2).
5. 100 nM dexamethasone (Dex) (for glucocorticoid induction) (see Note 2).
6. Vehicle (100% ethanol) for control induction 1 h prior to nuclei isolation.

2.2. Nuclei Isolation (see Note 3)

1. Phosphate-buffered saline (PBS): 137 mM NaCl, 2.7 mM KCl, 10 mM Na_2HPO_4 (dibasic, anhydrous), 2 mM KH_2PO_4 (monobasic, anhydrous) (pH 7.4).
2. Homogenization buffer: 10 mM Tris–HCl, pH 7.4, 15 mM NaCl, 60 mM KCl, 1 mM EDTA, 0.1 mM EGTA, 0.1% Nonidet P-40 (NP-40), 5% sucrose, 0.15 mM spermine, and 0.5 mM spermidine.
3. Sucrose pad: 10 mM Tris–HCl, pH 7.4, 15 mM NaCl, 60 mM KCl, 1 mM EDTA, 0.1 mM EGTA, 10% sucrose, 0.15 mM spermine, and 0.5 mM spermidine.
4. Wash buffer: 10 mM Tris–HCl, pH 7.4, 15 mM NaCl, 60 mM KCl, 1 mM EDTA, 0.1 mM EGTA, 0.15 mM spermine, and 0.5 mM spermidine.
5. 7.5-ml Dounce tissue grinder/homogenizer designed for fine particle size reduction without damage to cell nuclei.

2.3. In Vivo Digestion by Restriction Endonuclease

1. Restriction endonuclease(s) which cleavages the target genomic DNA sequence within close proximity of a transcription factor binding site or region of chromatin to be assayed for remodeling.
2. 1,000× DTT stock: 1 M DTT.
3. 100× proteinase K stock: 20 mg/ml proteinase K in deionized water.
4. Restriction enzyme digestion buffer: 10 mM Tris–HCl, pH 7.4, 15 mM NaCl, 60 mM KCl, 5 mM $MgCl_2$, 0.1 mM EDTA, 5% glycerol, and 1 mM dithiothreitol (DTT) (see Note 4).
5. Proteinase K buffer: 10 mM Tris–HCl, pH 7.6, 10 mM EDTA, 0.5% sodium dodecyl sulfate (SDS), and 0.2 mg/ml proteinase K (see Note 4).
6. Phenol:Chloroform:Isoamyl alcohol (25:24:1) saturated with 10 mM Tris–HCl, pH 8.0, 1 mM EDTA.
7. Chloroform.

2.4. In Vitro Redigestion and Purification of DNA

1. Restriction endonuclease which recognizes a cleavage site 200–300 bp upstream of the in vivo restriction enzyme site.
2. Reaction buffer supplied with enzyme.

3. Phenol:Chloroform:Isoamyl alcohol (25:24:1) saturated with 10 mM Tris, pH 8.0, 1 mM EDTA.
4. Chloroform.

2.5. End Labeling of Oligonucleotide (see Note 5)

1. 5 pmol gene-specific oligonucleotide.
2. ATP, (γ-32P)- 6,000 Ci/mmol 10 mCi/ml.
3. T4 polynucleotide kinase (PNK).
4. T4 PNK buffer: 70 mM Tris–HCl, 10 mM $MgCl_2$, 5 mM DTT, pH 7.6 (supplied with enzyme).
5. MicroSpin Sephadex G-25 columns.
6. Liquid scintillation counter.

2.6. Reiterative Primer Extension

1. PCR reaction mix: 1× PCR buffer (10 mM Tris–HCl, pH 8.3, 50 mM KCl, 2–4 mM $MgCl_2$, 0.5% Tween 20), 200 μM deoxynucleotides, 2.5 U Taq DNA polymerase.
2. PCR stop buffer: 10 mM Tris–HCl, pH 7.4, 200 mM sodium acetate, pH 7.0, 5 mM EDTA, 0.1 μg/μl yeast tRNA.

2.7. Analysis of Primer Extension Products (see Note 6)

1. Acrylamide stock solution: 38% acrylamide and 2% bis-acrylamide (w/v).
2. For 8% denaturing polyacrylamide gel: 25.2 g urea, 6 ml 10× TBE, 112 ml acrylamide stock solution, water to 60 ml, 10 μl TEMED and 200 μl ammonium persulfate (10%). Pour gel immediately.
3. Sample loading buffer: 80% formamide, 0.01 M NaOH, 1 mM EDTA, 0.04% bromophenol blue, 0.04% xylene cyanol.
4. Sequi-Gen GT nucleic acid sequencing system; 38 × 30 cm gel (Bio-Rad, Hercules, CA).
5. PhosphorImager screen, scanner, and image analysis software (e.g., Imagequant).

3. Methods

The methodology presented is suitable for the analysis of chromatin structure in vivo and can assist in the mapping of DNA–protein interactions and to elucidate their mechanistic implications regarding various nuclear processes, including transcriptional regulation, DNA repair, and replication. The requirements for this procedure are that the genomic DNA of interest (1) contains sequence-specific binding sites for and responds to a specific transcription factor or factors and (2) has a restriction enzyme cleavage site located proximal to the binding region and is contained within the confines of a nucleosome or is organized as chromatin.

3.1. Cell Culture

The experimental procedure outlined below is based on studies in human carcinoma cells that were stably transformed to contain multiple copies of the nuclear receptor-dependent MMTV promoter. For the purpose of this protocol, we have selected the human adrenal carcinoma cell line, SW-13, which does not express BRG1 protein (10). The use of this multicopy promoter cell line provides a strong signal-to-noise ratio for the MMTV sequence, which enhances the ability to define the chromatin structure of the promoter. The use of this cell line has greatly enhanced our ability to study nuclear receptor-initiated cascades that lead to structural changes within chromatin upon receptor binding to specific hormone response elements within target promoters (16, 17). Therefore, these general growth conditions have been optimized for our established SW-13/MMTV cell line. Specific growth requirements may vary from this description depending on cell type used.

3.2. Isolation of Nuclei

This protocol outlines the standard procedure for nuclei isolation from tissue cultured cells. All steps are performed on ice with prechilled equipment and solutions at 4°C. Cells in our illustrative example were treated with hormone or vehicle prior to nuclei isolation to stimulate glucocorticoid receptor signaling.

1. Treat cells with 100 nM Dex or vehicle for 1 h prior to nuclei isolation.
2. Rinse cells with cold PBS and detach from 150 mm plates by scraping cells into 10 ml cold PBS. Transfer cells to a prechilled 15-ml conical centrifuge tube.
3. Pellet cells by centrifugation at 500 × *g* for 5 min at 4°C. Remove PBS from cell pellet.
4. Add cold homogenization buffer (5 ml) and gently to dislodge cell pellet by pipetting. Transfer cells and buffer to a prechilled 7-ml Dounce tissue grinder/homogenizer and incubate on ice for 2 min.
5. Lyse cells by gently using four complete strokes of the Dounce pestle (tight pestle). Transfer lysate to a prechilled 15-ml conical tube (see Note 7).
6. Gently add 1 ml sucrose pad directly to the bottom of the tube using a micropipette.
7. Sediment nuclei through sucrose pad by centrifugation at 1,400 × *g* for 20 min at 4°C. Delicately remove supernatant from nuclei pellet.
8. Add 1 ml wash buffer to nuclei and gently resuspend pellet using a micropipette. Once nuclei are fully in solution, add an additional 4 ml wash buffer and centrifuge at 750 × *g* for 5 min at 4°C.
9. Carefully remove all traces of wash buffer and store nuclei on ice.

3.3. In Vivo Digestion by Restriction Endonuclease

The choice of restriction endonuclease to use for this accessibility assay is dependent upon the availability of cleavage sites within the genomic region of interest. Most analysis requires the testing of numerous restriction enzymes found to cleave within the target sequence until an optimal enzyme(s) is identified. The cleavage buffer selected for this assay was chosen because it maintains the structural integrity of the nuclei and is compatible with a broad range of restriction enzymes (see Note 8). The quantity of enzyme to use should be derived empirically and will depend on the efficiency at which the enzyme cleaves DNA when using the buffer recommended for hypersensitivity assays.

1. Gently resuspend nuclei in cold restriction enzyme digestion buffer (use a 3:1 ration of buffer to nuclei pellet, i.e., resuspend a 50 μl compact nuclei pellet in 150 μl buffer). Transfer aliquots of 100 μl nuclei to prechilled 5-ml polypropylene tubes.
2. Digest nuclei with appropriate restriction endonuclease (100–1,000 U/ml) at 30°C for 15 min (in vivo digest). Use 100 μl resuspended nuclei for each digest.
3. Stop reactions by adding 1 ml proteinase K buffer. Mix each sample by inverting five times and incubate overnight at 37°C.
4. Purify total DNA by four extractions with phenol/chloroform/isoamyl alcohol (PCI- 25:24:1, v/v) and two extractions with chloroform. The first two extractions should be carried out with twice the volume PCI (2.0 ml) and each subsequent extraction with one volume (1.0 ml). Mix samples by vigorous shaking and centrifuge at 10,000×*g* for 5 min at room temperature.
5. Precipitate the DNA by addition of 1/10th volume of 1 m NaCl and 3 volumes ice-cold 95% ethanol. Incubate samples at –20°C for 1 h then pellet the DNA by centrifugation at 12,000×*g* for 30 min at 4°C.
6. Wash DNA pellet with 1 ml cold 70% (v/v) ethanol and centrifuge at 12,000×*g* for 10 min at 4°C.
7. Allow DNA pellet to air-dry for 1 h at room temperature, resuspend in 100 μl sterile water, and transfer to 1.5 ml microfuge tube (see Note 9).

3.4. In Vitro Redigestion and Purification of DNA

The purified in vivo digested genomic DNA is cut to completion with a second restriction endonuclease which recognizes a cleavage site upstream or downstream of the in vivo restriction enzyme site depending on which template strand is to be extended. This in vitro digestion is performed overnight usually with 100 U of endonuclease to ensure complete digestion of the target sequence. The in vitro digest serves as an internal control to ensure equal amounts of DNA template was used in the reiterative primer extension reaction.

1. Digest DNA to completion with a second restriction enzyme recognizing a site upstream of the in vivo restriction enzyme site. Use 100 U of enzyme with manufacture's provided buffer according to manufacturer's instructions (in vitro digest). Incubate reaction overnight at 37°C (see Note 10).
2. Purify digested DNA by two extractions with phenol/chloroform/isoamyl alcohol (PCI- 25:24:1, v/v) and one extraction with chloroform. Mix samples by vigorous shaking and centrifuge at high speed (>18,000 × g) for 5 min at room temperature.
3. Precipitate the DNA by addition of 625 μl ice-cold 95% ethanol and 5 μl 5 M NaCl. Incubate samples at –20°C for 30 min and pellet the DNA by centrifuge at high speed (>18,000 × g) for 30 min at 4°C.
4. Allow DNA pellet to air-dry for 1 h at room temperature, resuspend in 100 μl of sterile water, and determine DNA concentration by $A_{260/280}$ absorbance reading.

3.5. End-Labeling of Sequence-Specific Oligonucleotides

End-labeling is a rapid and sensitive method for radioactively labeling DNA fragments, such as oligonucleotide, and is useful for visualizing small amounts of DNA. All of the enzymes employed are specific to either the 3′ or 5′ termini of DNA and will, consequently, only incorporate one radiophosphate per oligonucleotide. The most common method of radio-labeling oligonucleotides uses PNK to transfer a single radioactive phosphate group from γ-32P-ATP to the 5′-end of the oligonucleotide.

1. In a 1.5-ml microfuge tube add the following in the order indicated; combine PCR-grade water for a total volume of 20 μl, 2 μl 10× T4 PNK buffer (supplied with enzyme), 2 μl target-specific oligonucleotide (5 μM), 50 μCi ATP, (γ-32P)-6,000 Ci/mmol, and 20U T4 PNK.
2. Mix the reaction by vortex and briefly centrifuge.
3. Incubate the reaction mixture for 10 min at 37°C.
4. Prepare MicroSpin G-25 columns according to manufacturer's guidelines.
5. Pass labeling reaction over readied G-25 column as directed by manufacturer's protocol.
6. Determine incorporation of radiolabel (cpm/μl) by liquid scintillation counter measurement.

3.6. Reiterative Primer Extension

The extent of restriction endonuclease hypersensitivity for a given target sequence is determined using reiterative primer extension with *Taq* polymerase and a template-specific ^{32}P-labeled oligonucleotide. The primer selected to detect restriction endonuclease hypersensitivity regions should be located either upstream or downstream of the transcription factor binding site, ideally 200–300 bp from the in vivo cleavage site.

The concentration of Mg and oligonucleotide selected, for primer extension, greatly influences the specificity and yield of the reaction; therefore, the amounts of both should be titrated to achieve optimal results. The primer selected should anneal downstream of the in vivo restriction endonuclease cleavage site and be greater than 18 bases in length with melting temperatures ranging between 45 and 70°C. The reiterative primer extension/PCR reaction, described above, has been optimized for studies involving the MMTV promoter (10). These conditions may be employed for other steroid-responsive promoters although optimization should be performed (see Note 11).

1. Amplify 10–20 μg purified in vivo/in vitro digested genomic DNA in 30 μl PCR reaction mix using 1–5 × 10^6 cpm ^{32}P-labled sequence-specific oligonucleotide (see Note 12).
2. Thermocycler program should include an initial cycle of denaturation at 94°C for 4 min, annealing for 60°C for 2 min, and primer extension at 72°C for 2 min. Follow initial cycle with 29 cycles of 2 min denaturation at 94°C, 2 min annealing at 60°C, and 2 min primer extension at 72°C with final extension for 10 min.
3. Stop each PCR reaction with addition of 150 μl PCR stop buffer.
4. Purify extended products by extraction with two rounds of phenol/chloroform/isoamyl alcohol (500 μl) and one round chloroform (500 μl). Mix samples by vortexing for 5 s and centrifuge at high speed (>18,000 × g) for 5 min at room temperature.
5. Precipitate extension products by addition of 625 μl cold 95% ethanol and 5 μl 5 M NaCl followed by centrifugation at high speed (>18,000 × g) for 30 min at 4°C. Precipitated products were washed with 70% cold ethanol, recovered by centrifugation (>18,000 × g) for 10 min at 4°C and allow pellet to air-dry.

3.7. Analysis of Polymerase Extension Products

Primer extension products are resolved on denaturing polyacrylamide gels. Samples should be allowed to electrophorese to yield maximal separation between bands corresponding to the in vivo restriction endonuclease cleavage site and the in vitro extension terminal cleavage site. For the purposes of this procedure, samples were resolved on 38 × 30 cm gels using Sequi-Gen GT nucleic acid sequencing system (see Note 13).

1. Resuspend DNA pellets in 7 μl sample loading buffer, vortex at high speed for 10 s, briefly centrifuge, heat for 5 min at 95°C, vortex again, and recentrifuge to collect sample.

2. Pour denaturing polyacrylamide gel and prerun according to manufacturer's specifications.
3. Load primer extension products and separate on denaturing polyacrylamide gel.
4. After electrophoresis, allow gel to cool then transfer to filter paper.
5. Dry gel under vacuum for 1 h at 80°C.
6. Expose to PhosphorImager Screen, scan and analyze suing ImageQuant or other image analysis software (see Note 14).

3.8. Concluding Remarks

The authors have described the use of restriction enzyme hypersensitivity/ reiterative primer extension for assaying nucleosome remodeling which is often a prerequisite for transcriptional activation (15). This method allows for a high-resolution analysis of changes in the chromatin architecture upon promoter stimulation. The individual example presented here demonstrates the use of restriction enzymes to determine steroid-responsive SWI/SNF-mediated promoter hypersensitivity (Fig. 2). This method has also been employed to characterize chromatin remodeling and induction of hypersensitivity regions within endogenous nuclear receptor-dependent promoters. Restriction endonuclease hypersensitivity assays were used to identify differences in the chromatin structure of the IκBα and MMTV promoters upon induction by the glucocorticoid and progesterone receptors (18). Nuclease hypersensitivity has been used to demonstrate that estrogen induction of ERRα involves chromatin remodeling around the multiple hormone-response element (19).

Genome-wide strategies can be used in concert with restriction endonuclease hypersensitivity to identify regulatory domains and elucidate specificity and function. DNase I hypersensitivity assays can be used to map discrete sites in the genome where changes in chromatin conformation yields DNA hypersensitivity to digest by the endonuclease DNase I. These DNase I hypersensitivity

Fig. 2. BRG1-dependent restriction endonuclease chromatin accessibility within the MMTV promoter. (**a**) Nucleosomal organization of the MMTV promoter. When stably integrated into the host genome the MMTV promoter is organized into a phased array of six positioned nucleosomes (**a**–**f**). The hormone inducible hypersensitivity region is encompassed by nucleosome B (Nuc-B). The expanded schematic of the proximal portion of the MMTV promoter shows the location of hormone-response elements (HREs), nuclear factor 1 (NF1), octamer transcription factors (OTFs), and the TATA-binding protein (TBP) and identifies cleavage sites for restriction enzymes to test for induced hypersensitivity and chromatin remodeling (21). (**b**) SW-13/MMTV cells, expressing GR, were transiently transfected with empty vector (–BRG1) or BRG1 expression plasmid (+BRG1). Nuclei, from untreated (–) or dexamethasone treated (10^{-7} M) cells, were isolated and digested with restriction endonucleases targeting sequences within nucleosome-B (in vivo: Sst I, Mbo I, or Fok I). Purified genomic DNA was digested to completion with Hae III (In vitro). Fifteen micrograms of each sample was analyzed using linear *Taq* polymerase reiterative primer extension with a ^{32}P-labeled single-stranded specific for MMTV. Extension products were analyzed on a 6% polyacrylamide denaturing gel and exposed to PhosphorImage screen.

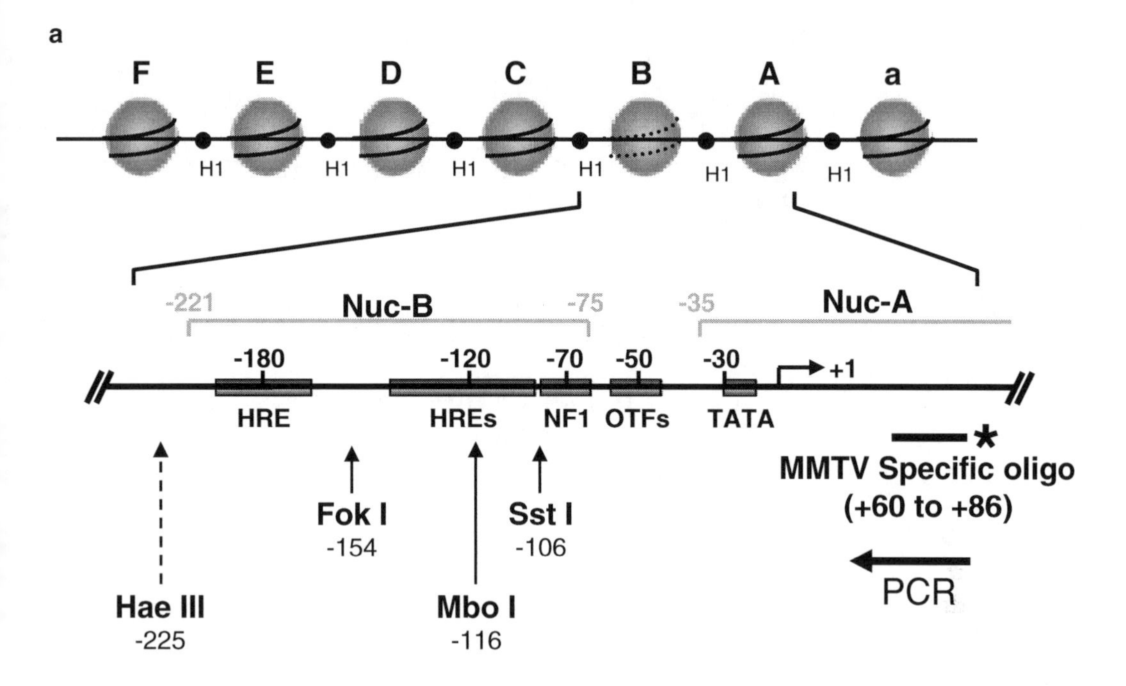
a
F
E
D
C
B
A
a
H1
H1
H1
H1
H1
H1
-221
Nuc-B
-75
-35
Nuc-A
-180
-120
-70
-50
-30
+1
HRE
HREs
NF1
OTFs
TATA
MMTV Specific oligo
(+60 to +86)
Fok I
-154
Sst I
-106
PCR
Hae III
-225
Mbo I
-116

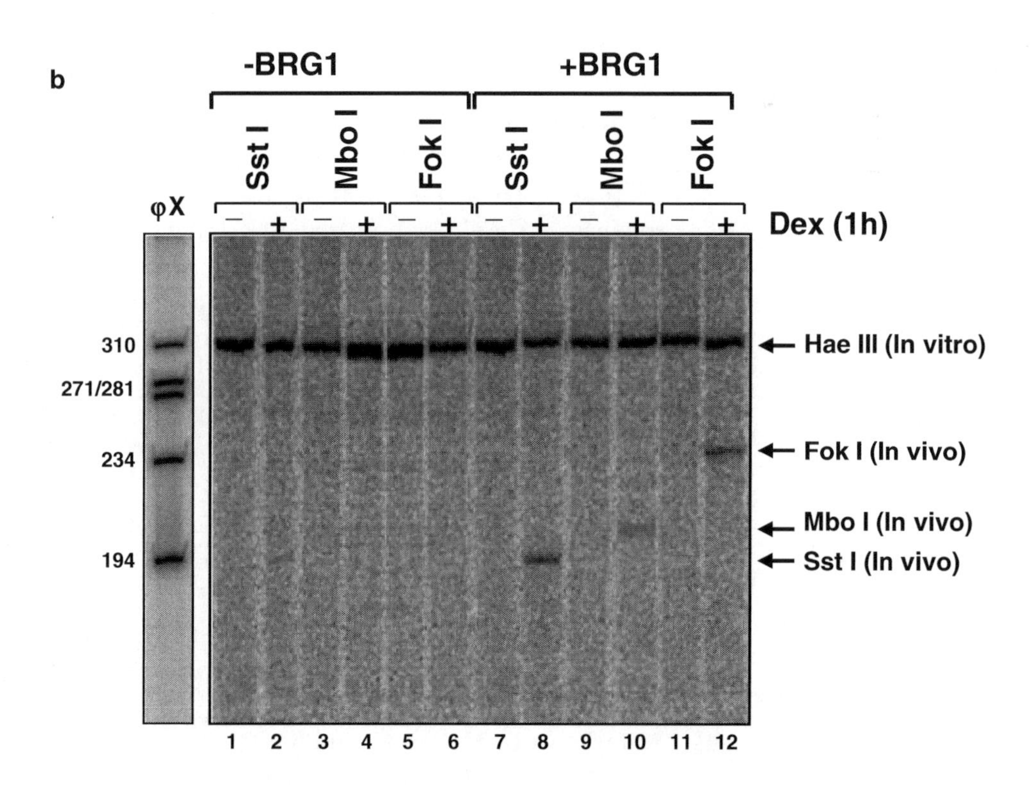
b
-BRG1
+BRG1
Sst I
Mbo I
Fok I
Sst I
Mbo I
Fok I
φX
Dex (1h)
310
271/281
234
194
Hae III (In vitro)
Fok I (In vivo)
Mbo I (In vivo)
Sst I (In vivo)
1 2 3 4 5 6 7 8 9 10 11 12

sites have been found in the regulatory domains of promoters, enhancers, silencers, insulators, and locus control regions (20). Taken together, the use of these in vivo chromatin analysis techniques can provide insight into the roles structured chromatin and remodeling complexes play in the regulation of nuclear processes, such as transcription.

4. Notes

1. The authors do not endorse any products or manufacturers listed in this protocol.
2. Transfection of BRG1 expression plasmids was carried out in our example in order to determine the dependence of SWI/SNF chromatin remodeling on restriction enzyme accessibility. Depending on cell line being used transfection may not be necessary. Similarly, induction of glucocorticoid receptor by treatment with Dex is indicated here for illustrative purposes and may be eliminated if not needed for the user's application.
3. Generally, buffers used for nuclei isolation are not maintained at 4°C for more than 2 months. The following buffer stock solution may be used in preparation of these solutions. 10× salt buffer: 100 mM Tris–HCl, pH 7.4, 150 mM NaCl, and 600 mM KCl (store at room temperature). Spermine (0.15 M) and spermidine (0.5 M) stock solutions (1,000×) should be prepared and added to each buffer just prior to isolation. Store 1,000× stock solutions at −20°C.
4. Prepare DTT as a 1 M (1,000×) stock and add prior to use. Proteinase K should be prepared as a 20 mg/ml (100×) stock and added prior to use.
5. Sequence specific real-time primer pairs may substitute for 32P-labeled oligonucleotide for real-time PCR analysis of cleavage products.
6. Our laboratory uses the following premixed sequencing solutions: SequaGel Sequencing System (National Diagnostics, Atlanta, GA).
7. The number of Dounce homogenizer pestle strokes needed for efficient nuclei isolation is cell type specific. It is recommended, after using four pestle strokes, to evaluate the extent of nuclei isolation under a light microscope.
8. The restriction enzyme buffer supplied by the manufacture may be used for in vivo digest.
9. If using a Speed-Vac, be careful not to over dry DNA pellet.
10. The recognition sequence for this second restriction enzyme should located upstream of the in vivo restriction enzyme site.

This digest serves as an internal standard to ensure equal loading of DNA into the *Taq* polymerase reiterative primer extension assay. This second digest also makes the sample less viscous and the genomic DNA easier to pipette.

11. The PCR conditions described here have been optimized for the analysis of stably integrated MMTV promoter. However, these conditions have been used to evaluate in vivo restriction enzyme hypersensitivity from other nuclear receptor-dependent promoters and should prove to be good starting point for analyzing other target sequences.
12. Sequencing reactions can be performed to map predicted restriction enzyme cleavage sites using the PCR reactions mix and conditions described above with three deoxynucleotides at 200 μM and the fourth, including dideoxynucleotide as follows: 100 μM ddGTP/20 μM dGTP, 200 μM ddATP/5 μM dATP, 200 μM ddCTP/ 20 μM dCTP, or 200 μM ddTTP/20 μM dTTP.
13. Detection of multiple transcription factor-binding sites can be visualized more effectively suing longer denaturing gels, such as 40–50 cm.
14. Kodak X-OMAT Blue film can be used for gel exposure if PhosphorImager screen is unavailable.

References

1. Wolffe, A. P. (1994) Transcription: in tune with the histones, *Cell 77*, 13–16.
2. Kornberg, R. D., and Lorch, Y. (1999) Twenty-Five Years of the Nucleosome, Fundamental Particle of the Eukaryote Chromosome, *Cell 98*, 285–294.
3. Felsenfeld, G., and Groudine, M. (2003) Controlling the double helix, *Nature 421*, 448–453.
4. Weake, V. M., and Workman, J. L. Inducible gene expression: diverse regulatory mechanisms, *Nat Rev Genet 11*, 426–437.
5. Trotter, K. W., and Archer, T. K. (2008) The BRG1 transcriptional coregulator, *Nucl Recept Signal 6*, e004.
6. Sif, S. (2004) ATP-dependent nucleosome remodeling complexes: enzymes tailored to deal with chromatin, *J Cell Biochem 91*, 1087–1098.
7. Eberharter, A., and Becker, P. B. (2004) ATP-dependent nucleosome remodelling: factors and functions, *J Cell Sci 117*, 3707–3711.
8. Trotter, K. W., and Archer, T. K. (2007) Nuclear receptors and chromatin remodeling machinery, *Molecular and cellular endocrinology 265–266*, 162–167.
9. Archer, T. K., Lefebvre, P., Wolford, R. G., and Hager, G. L. (1992) Transcription factor loading on the MMTV promoter: a bimodal mechanism for promoter activation, *Science 255*, 1573–1576.
10. Trotter, K. W., and Archer, T. K. (2004) Reconstitution of glucocorticoid receptor-dependent transcription in vivo, *Molecular and cellular biology 24*, 3347–3358.
11. Almer, A., and Horz, W. (1986) Nuclease hypersensitive regions with adjacent positioned nucleosomes mark the gene boundaries of the PHO5/PHO3 locus in yeast, *The EMBO journal 5*, 2681–2687.
12. Boyes, J., and Felsenfeld, G. (1996) Tissue-specific factors additively increase the probability of the all-or-none formation of a hypersensitive site, *The EMBO journal 15*, 2496–2507.
13. Okino, S. T., and Whitlock, J. P., Jr. (1995) Dioxin induces localized, graded changes in chromatin structure: implications for Cyp1A1 gene transcription, *Molecular and cellular biology 15*, 3714–3721.
14. Verdin, E., Paras, P., Jr., and Van Lint, C. (1993) Chromatin disruption in the promoter of human immunodeficiency virus type 1 during

transcriptional activation, *The EMBO journal 12*, 3249–3259.

15. Archer, T. K., Cordingley, M. G., Wolford, R. G., and Hager, G. L. (1991) Transcription factor access is mediated by accurately positioned nucleosomes on the mouse mammary tumor virus promoter, *Molecular and cellular biology 11*, 688–698.
16. Archer, T. K., Deroo, B. J., and Fryer, C. J. (1997) Chromatin modulation of glucocorticoid and progesterone receptor activity, *Trends Endocrinol Metab 8*, 384–390.
17. Deroo, B. J., and Archer, T. K. (2001) Glucocorticoid receptor-mediated chromatin remodeling in vivo, *Oncogene 20*, 3039–3046.
18. Deroo, B. J., and Archer, T. K. (2001) Glucocorticoid receptor activation of the I kappa B alpha promoter within chromatin, *Mol Biol Cell 12*, 3365–3374.
19. Hu, P., Kinyamu, H. K., Wang, L., Martin, J., Archer, T. K., and Teng, C. (2008) Estrogen induces estrogen-related receptor alpha gene expression and chromatin structural changes in estrogen receptor (ER)-positive and ER-negative breast cancer cells, *J Biol Chem 283*, 6752–6763.
20. Pipkin, M. E., and Lichtenheld, M. G. (2006) A reliable method to display authentic DNase I hypersensitive sites at long-ranges in single-copy genes from large genomes, *Nucleic Acids Res 34*, e34.
21. Hebbar, P. B., and Archer, T. K. (2003) Chromatin remodeling by nuclear receptors, *Chromosoma 111*, 495–504.

Chapter 7

Generation of DNA Circles in Yeast by Inducible Site-Specific Recombination

Marc R. Gartenberg

Abstract

Site-specific recombinases have been harnessed for a variety of genetic manipulations involving the gain, loss, or rearrangement of genomic DNA in a variety of organisms. The enzymes have been further exploited in the model eukaryote *Saccharomyces cerevisiae* for mechanistic studies involving chromosomal context. In these cases, a chromosomal element of interest is converted into a DNA circle within living cells, thereby uncoupling the element from neighboring regulatory sequences, obligatory chromosomal events, and other context-dependent effects that could alter or mask intrinsic functions of the element. In this chapter, I discuss general considerations in using site-specific recombination to create DNA circles in yeast and the specific application of the R recombinase.

Key words: Site-specific recombination, R Recombinase, *RS* site, DNA circle, *Saccharomyces cerevisiae*

1. Introduction

The properties of any given chromosomal element are subject to context effects. Neighboring sequences can directly impact a property of interest or simply mask that property. One option is to relocate the element to a new chromosomal position. This, however, only exchanges one set of context effects for another. Furthermore, the properties of the element might be influenced by obligatory chromosomal events (e.g., DNA replication) that cannot be avoided in any context. A powerful approach to bypass context effects is to convert chromosomal elements into extrachromosomal circles by site-specific recombination. Often referred to as "pop-outs" or "excisions," these reactions can be remarkably efficient and rapid. The requirements are minimal: an inducible expression system for the recombinase and modification of the chromosomal element to include a pair of small recombinase target

Randall H. Morse (ed.), *Chromatin Remodeling: Methods and Protocols*, Methods in Molecular Biology, vol. 833,
DOI 10.1007/978-1-61779-477-3_7,

sites. This chapter discusses general considerations and practical aspects in using the R site-specific recombinase to create DNA circles in the yeast *Saccharomyces cerevisiae* (hereafter referred to as budding yeast).

A few examples from the budding literature illustrate the utility of the site-specific recombination approach in addressing biological consequences of chromosomal context. First, Raghuraman et al. excised a late-firing replication origin to study the impact of chromosomal context on replication origin firing time. They found that uncoupling erased the context effect, but only if excision occurred within a certain cell cycle window (1). This showed that the replication-timing program is determined at a specific cell cycle stage (early G1). Second, Megee and Koshland combined excision reactions with fluorescence microscopy in a visual hunt for DNA elements that mediate sister chromatid cohesion of minichromosomes (2). This study showed that sequences necessary for kinetochore assembly were critical. Third, the Broach laboratory and my own used excision reactions to study the role of *cis*-acting regulatory elements (termed silencers) in maintenance of yeast heterochromatin domains. When the silencers were uncoupled from heterochromatin, transcriptional silencing was lost during cell cycle progression in one instance, and even without cell cycle progression in another (3–6). These findings indicated that silencers act continuously to maintain the silent state. Fourth, the Rine laboratory and my own generated nonreplicating DNA circles to study the role of DNA replication in establishment of transcriptional silencing (7, 8). In contrast to the prevailing view at that time, we discovered that passage of a replication fork was not necessary to acquire a repressive chromatin state. More recently, the heterochromatic domain known as *HMR* was excised to study the determinants of heterochromatic cohesion and nuclear localization, both features that would have been masked at the native chromosomal location (9, 10).

To the biochemist, the site-specific recombination approach offers additional advantages. Converting sequence elements in linear chromosomal templates into closed-circular forms permits analysis of DNA supercoiling, a sensitive metric of chromatin structure and nucleosome number (5). Moreover, when coupled with affinity purification the technique can be used to isolate select chromatin fragments (11) (see also chapter by Unnikrishnan et al., this volume). The Kornberg laboratory refined this approach to determine how transcriptional activation in vivo influenced the nucleosome content and position of a single promoter (12, 13). Conceivably, the material isolated by such purification schemes could be used as native templates for biochemical analyses of transcription and other processes. With ever improving affinity purification methods, it should be possible to couple the site-specific recombination approach with mass spectrometry to identify all of the constituents of any given chromosomal landmark on a well-defined segment of DNA (see chapter by Unnikrishnan et al. for progress toward this goal).

2. Materials

2.1. General Considerations

2.1.1. The Recombinases

Three members of the tyrosine family of site-specific recombinases (Cre, Flp, and R) have been used for various programmed genome rearrangements in budding yeast. Cre and Flp have been used more broadly to revolutionize genetic manipulation in mice and flies (14, 15). The enzymes cut and rejoin DNA via a covalent-protein DNA intermediate. DNA ends are not exposed during the concerted reaction, and therefore they are not recognized as DNA damage. No cofactors or accessory proteins are required, simplifying the matter of using these enzymes in heterologous organisms. Cre is encoded by the bacteriophage P1, whereas Flp and R are encoded by native plasmids of *S. cerevisiae* and *Zygosaccharomyces rouxii*, respectively.

In budding yeast, the recombinase genes are typically linked to a strong and inducible promoter like *GAL1p* that provides a greater than 1,000-fold increase in expression over very low basal levels. Added control has been achieved by fusing the recombinases to the human estrogen-binding domain ((EBD), (16, 17)). EBD chimeras are sequestered in an inactive form until exposed to β-estradiol. The most recent generation of Cre-EBD chimeras are so tightly controlled that they can be expressed from strong constitutive promoters without concern for hormone-independent recombination (18). Application of Flp in budding yeast requires that the native plasmid encoding the recombinase (the 2 μm plasmid) first be evicted. This is now a simple task (19). One disadvantage, however, is that the 2 μm plasmid also encodes trans-acting factors that stabilize retention of common 2-μm based, high-copy vectors (20). The R recombinase can be used in strains bearing the 2 μm plasmid because the enzyme does not cross-react with Flp (21). Based on these practical considerations and the availability of *GAL1p-R* constructs that regulate recombination tightly, we selected the R recombinase for our earliest experiments with DNA circles (22).

2.1.2. Recombinase Target Sites and Excision Reactions

The target sites for Cre, Flp, and R range from 31 to 34 bp. One target site is placed on either side of the locus to be excised, and thus only minimally invasive changes must be made to generate an excision cassette. Each target site is composed of a pair of inverted repeats that flank an asymmetric 7–8 bp core (Fig. 1a). The core sequences between pairs of targets must be identical for recombination to occur. Multiple independent recombination events can be driven simultaneously by the same recombinase if each pair of

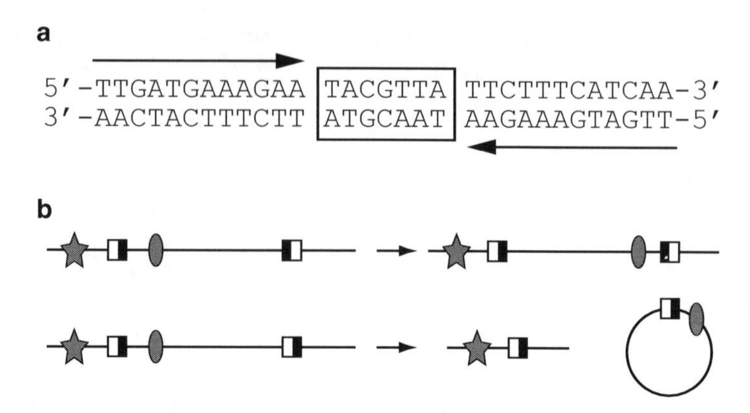

Fig. 1. (**a**) The *RS* target site of the R recombinase. The asymmetric core is boxed. The inverted repeats are marked with *arrows*. This minimal sequence is embedded within a larger 58 bp fragment that is used commonly in construction of recombination cassettes for budding yeast (see Note 4). (**b**) The relative orientation of *RS* sites (*half-filled boxes*) determines the outcome of the recombination reaction. An inverted orientation yields inversion whereas a tandem orientation yields excision. The star and oval serve are landmark features.

target sites contains a unique core sequence. The relative orientation of the core sequences determines the outcome of the recombination event (Fig. 1b). When two target sites (shown as half-filled boxes) are inverted relative to one another, the intervening DNA will be inverted by the recombinase. When the target sites are oriented in a tandem fashion, then the intervening DNA will be excised into a DNA circle. Recombination reactions by Cre, Flp, and R are reversible events. Therefore, excision within a population of cells may approach but never reach 100% completion until the excised circles are lost by out-growth of the culture (see Note 1).

Facile excision reactions of DNA segments spanning from several hundred base pairs to over 30 kb have been observed (1, 2). For example, using a multicopy R recombinase expression vector and 2.5 kb excision cassette, the reaction proceeded to over 90% completion in 60 min, well within the budding yeast cell cycle (11). In each scenario, the rate of excision depends on the copy number of the recombinase expression vector, the distance between the recombinase target sites, local chromatin effects, and environmental factors (see Note 2).

2.2. Generating Chromosomal Recombination Cassettes

A two-step procedure can be used to convert almost any nonessential chromosomal locus into a cassette for excision (Fig. 2; see Note 3). In the first step, PCR-mediated gene replacement is used to replace the locus of interest with a counter-selectable marker (23, 24). All sequences that will ultimately be flanked by recombinase sites should be eliminated at this time. The *URA3* genes of *C. albicans* or *K. lactis* are ideal replacements because they lack sufficient homology to recombine efficiently with the endogenous *ura3* gene. Accurate integration of the replacement gene and loss of the original locus should be confirmed by PCR. In the second

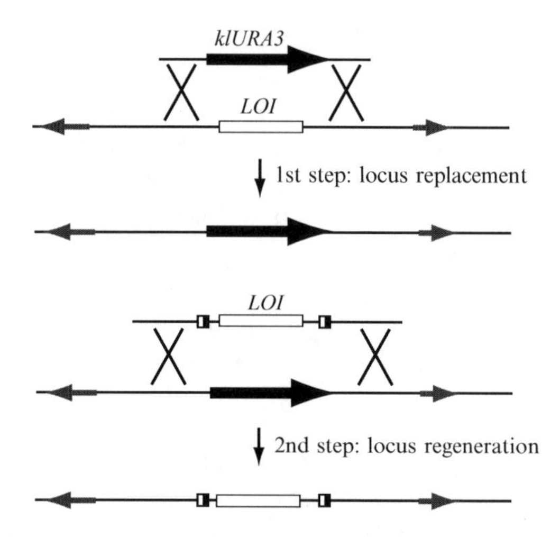

Fig. 2. A two-step procedure to generate chromosomal excision cassettes. In the first step, all of the sequences of the designated chromosomal landmark (*white* bar labeled *LOI*) are removed by gene replacement using a PCR fragment bearing the *K. lactis URA3* gene. In the second step, a synthetic construct bearing the chromosomal landmark flanked by *RS* sites is used to replace the *K. lactis URA3* gene.

step, the *URA3* replacement gene is substituted with an engineered locus that bears flanking *RS* recombinase sites. Counter-selection for loss of *URA3* can be achieved with 5-fluoro-orotic acid (5-FOA, (25)). The design and construction of each excision cassette depends on details specific to the locus of interest and the materials available. In the simplest case, *RS* sites could be inserted at convenient restriction enzyme sites of a cloned fragment by standard procedures (see Note 4). However, it should also be possible to introduce recombinase sites into synthetic constructs with cloning-free PCR methods (26). The regenerated locus should be confirmed with PCR.

2.3. Generating a Recombination-Proficient Strain

GAL1p-R recombinase gene fusions are available as integrative vectors, as well as single and multicopy episomes. Each of these three options has distinct merits that must be weighed given the demands of the experiment. Multicopy episomes produce faster excision rates when compared to their single-copy counterparts. However, the cumulative effect of leaky expression from multicopy constructs occasionally results in excision without induction. While these "pre-excision" reactions may seem insignificant by Southern hybridization, they are likely to be present within a cell population and measurable by PCR. Single-copy expression vectors produce less frequent pre-excision. Given this small but inherent instability, we recommend that strains bearing excision cassettes be stored as frozen glycerol stocks both with and without recombinase vectors whenever possible.

Integrative vectors assure that every cell in a population bears an identical complement of recombinase genes. Episomal vectors, by contrast, are absent from a fraction of cells (5–10%) even when cultured under selective pressure. Integrative vectors would thus be preferable in analyses involving single cells. These strains can also be grown in rich media, which is advantageous for experiments requiring microtubule inhibitors, like nocodazole, that function poorly in minimal media. Collectively, the benefits gained by integrating the recombinase may balance the additional effort necessary to confirm proper integration (see Note 5). For the sake of simplicity, the induction protocol described below uses the single-copy vector pRS415-RecR.

2.4. Itemized Reagents

2.4.1. Reagents for Production of DNA Circles in Budding Yeast

1. A duplex oligonucleotide *RS* site or the pHM401 plasmid that bears a single *RS* site within a 58 bp fragment flanked by *Sal*I sites (21).
2. A vector bearing the *GAL1p-R* recombinase gene fusion: Multicopy, 2 μm vector pHM153 (21); single-copy, CEN vector pRS415-RecR (27); integrating vector pRINT (1). All bear the *LEU2* marker.
3. A budding yeast strain with a competent galactose induction pathway and mutations in the *URA3* and *LEU2* nutritional markers.
4. Synthetic complete media lacking leucine (SC-leu media): 0.67% (w/v) yeast nitrogen base with ammonium sulfate but without amino acids (Difco); 0.13% (w/v) leu- drop-out powder (28). After sterilization, supplement with dextrose or raffinose to a final concentration of 2%.
5. A 20% (w/v) stock solution of dextrose sterilized by autoclave and 20% (w/v) stock solutions of galactose and raffinose sterilized by filtration.
6. Toluene/EtOH/Tris cocktail: 95% (v/v) EtOH, 3% (v/v) toluene (v/v), 2% (v/v) 1 M Tris–HCl pH 8.0.

2.4.2. Reagents to Isolate DNA by Glass Beads Lysis

1. Lysis buffer: 100 mM NaCl, 10 mM Tris–HCl pH 8.0, 10 mM EDTA, 2% Triton X-100, 1% SDS (add Triton X-100 and SDS last).
2. 0.45 mm acid-washed glass beads.
3. Phenol/Chloroform/Isoamyl Alchohol (49.5/49.5/1) pH 8.0.
4. 100% EtOH, 70% EtOH.
5. TE: 10 mM Tris–HCl pH 8.0, 1 mM EDTA.
6. 10 mg/ml DNAse-free RNAse.
7. 7.5 M NH_4OAc.

3. Methods

3.1. Time Course of DNA Circle Production

The goal of this procedure is to measure the rate of recombination of an untested excision cassette. The culture volumes were chosen to obtain enough DNA for Southern blot hybridization. Smaller volumes could be used if recombination were to be measured by a quantitative PCR procedure.

1. Obtain fresh transformants of pRS415-RecR in a strain that carries an excision cassette or revive prior transformants from frozen glycerol stocks.
2. Inoculate the transformed strain into 3 ml of SC-leu media supplemented with dextrose. Grow at 30°C from morning to evening (at least 6 h) with good aeration.
3. Adjust the cell density of the initial culture to 0.3–0.7 (OD_{600}) and inoculate 100 ml of SC-leu media supplemented with raffinose. Use a 1/200 dilution. Culture overnight with good aeration (see Note 6).
4. When the culture reaches mid-log growth phase the next morning (OD_{600} = 0.3–0.7) add galactose to a final concentration of 2%. Remove a 20 ml aliquot of culture prior to galactose addition and save as the 0 h time point. Remove additional 20 ml aliquots of culture at 1, 2, and 4 h (see Note 7).
5. Terminate cell growth in each aliquot by adding EDTA to a final concentration of 20 mM followed by adding an equal cell culture volume of ice-cold toluene/EtOH/Tris cocktail. Invert to mix. Store on ice (see Note 8).
6. Collect cell pellet by tabletop centrifugation (1,000 × *g*) for 5 min at room temperature. Discard supernatant and wash pellet once with water in a 1.5 ml microcentrifuge tube. Pellet can be stored at −20°C until DNA isolation.

3.2. Isolation of DNA for Analysis by Restriction Digestion

This is a glass-beads protocol to isolate DNA for subsequent enzymatic digestion and analysis by Southern blotting. Traditional spheroplasting protocols have been used to isolate high-quality DNA samples for the analysis of DNA supercoiling. Commercially available kits might now substitute (e.g., Gentra Puregene from Qiagen).

1. Resuspend the cell pellet in 500 μl of lysis buffer and then add 500 μl of phenol/chloroform. Add approximately 200 μl of acid-washed 0.45 mm glass beads. Be certain that no beads interfere with the seal of the tube or the sample will spill. Vortex vigorously for 3 min.
2. Separate phases by microcentrifugation (16,000 × *g*) for 5 min. Collect the upper aqueous phase, carefully avoiding precipitate

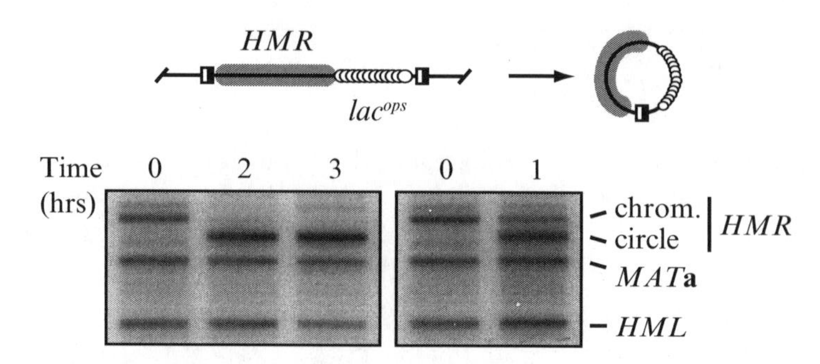

Fig. 3. DNA circle production as a function of time. Two induction protocols were performed with strain MRG2201 (9): one spanned 1 h and the other spanned 3 h. The silenced *HMR* mating-type locus in this strain (*gray* bar labeled *HMR*) contains an array of lac operators (*open circles* labeled *lac*^{ops}) and a pair of *RS* recombinase target sites that are separated by 16.8 kb. The strain also contains two tandem copies of the R site-specific recombinase integrated at the *leu2-3,112* locus. Purified DNA was digested with *Eco*RI. A radiolabeled probe to *HMR* also hybridized to the chromosomal *MAT***a** and *HML* loci.

at the interface. Repeat phenol/chloroform extraction at least once to increase purity.

3. Transfer aqueous phase to a new tube containing 1 ml of 100% EtOH. Vortex. Collect the precipitate by microcentrifugation (16,000 × *g*) for 5 min. Pour off supernatant and remove recalcitrant drop by pipet.
4. Resuspend pellet in 200 μl of TE. Add 1 μl of 10 mg/ml DNAse-free RNAse and incubate for 5 min at 37°C.
5. Add 200 μl of phenol/chloroform, vortex, and separate the phases by microcentrifugation (16,000 × *g*) for 1 min.
6. Collect the aqueous phase and add 100 μl of 7.5 M NH_4OAc, vortex, and then add 750 μl of 100% EtOH. Vortex. Collect the precipitate by microcentrifugation (16,000 × *g*) for 5 min. Rinse pellet with ice-cold 70% EtOH.
7. Resuspend pellet in 20 μl of TE and store at −20°C.
8. Digest the isolated DNA with enzymes that produce diagnostic fragment lengths for parental and recombined DNA, which can be detected with a single probe. Perform electrophoresis and Southern blotting according to standard protocols. Figure 3 shows the rate of an excision reaction that produces a 16.8 kb circle bearing the *HMR* locus.

4. Notes

1. The serine integrase *ϕ*C31 has also been exploited for a wide variety of genetic manipulations. Unlike the tyrosine recombinases, members of the serine integrase family require a specific protein cofactor to reverse recombination of small *attB* and

attP sites. Therefore, ϕC31 can be expressed in heterologous systems to catalyze irreversible recombination events (29).

2. Closely spaced *RS* sites recombine more rapidly than distantly spaced sites. Reactions driven by *GAL1p-R* proceed more rapidly at 30°C than at 37°C (also true for HO-mediated double strand breaks when the HO endonuclease is expressed from *GAL1p*). *RS* sites within a yeast heterochromatin domain recombine more slowly than sites located beyond the domain.
3. If the locus of interest is essential, like a centromere, a modification of this two-step protocol must be used (30).
4. In principle, the simple 31 bp *RS* site depicted in Fig. 1 should be sufficient in constructions. To my knowledge, however, all applications have used a larger 58 bp fragment listed here:

   5′-cgagatcatatcactgtggacgttgatgaaagaatacgttattctttcatcaaatcgt-3′
5. The integrative vector pRINT contains *GAL1p-R*, an *LEU2* marker and sequences for propagation in bacteria. The vector can be integrated at the *leu2* allele of many laboratory strains following linearization by *Bst*EII. In strains bearing the *leu2*Δ*0* deletion allele (31), integration might still be possible with a linearization by *Bsa*XI. Most transformants contain a single integrated copy of the vector. However, increasing the amount of transformed DNA favors multiple tandem integrations, perhaps due to iterative rounds of ends-in recombination (32). Diagnostic enzyme digests and Southern blotting can be used to distinguish these advantageous increases in *GAL1p-R* copy number. All of the strains for our cytology work bear two tandemly integrated copies of pRINT. When possible, we move the tandem pair from one strain to the next by genetic crosses.
6. Growth in raffinose removes glucose repression so that subsequent addition of galactose results in an immediate burst of R recombinase expression. The purpose of preliminary growth in liquid dextrose is to obtain a healthy cell population in mid-log phase for transfer to raffinose. Direct inoculation into raffinose from agar plates or from cultures grown to saturation in dextrose results in a prohibitively long lag phase. There is flexibility in the final density of the dextrose culture and the dilution into raffinose (avoid extreme dilutions) but the goal is to reach mid-log phase growth at the time of galactose induction the following morning. Some trial and error may be necessary to optimize this step.
7. Owing to the addition of galactose and growth during the procedure, the density of cells at each time point will differ. Compensate for this by adjusting the volume of each aliquot according to cell density at the time of removal.

8. This is a traditional cold-stop procedure that was used to rapidly terminate cell growth and inactivate nucleases that might interfere with isolation of supercoiled plasmids. We use the procedure here to ensure that the time points accurately reflect the extent of the recombinase reaction. Rapid termination of cell growth may not be necessary for most applications.

References

1. Raghuraman MK, Brewer BJ, Fangman WL (1997) Cell cycle-dependent establishment of a late replication program. Science 276: 806–809.
2. Megee PC, Koshland D (1999) A functional assay for centromere-associated sister chromatid cohesion. Science 285: 254–257.
3. Holmes SG, Broach JR (1996) Silencers are required for inheritence of the repressed state in yeast. Genes Dev 10: 1021–1032.
4. Bi X, Broach JR (1997) DNA in transcriptionally silent chromatin assumes a distinct topology that is sensitive to cell cycle progression. Mol Cell Biol 17: 7077–7087.
5. Cheng T-H, Li Y-C, Gartenberg MR (1998) Persistence of an alternate chromatin structure at silenced loci in the absence of silencers. Proc Natl Acad Sci USA 95: 5521–5526.
6. Cheng T-H, Gartenberg MR (2000) Yeast heterochromatin is a dynamic structure that requires silencers continuously. Genes Dev 14: 452–463.
7. Kirchmaier AL, Rine J (2001) DNA replication-independent silencing in *S. cerevisiae*. Science 291: 646–650.
8. Li Y-C, Cheng T-H, Gartenberg MR (2001) Establishment of transcriptional silencing in the absence of DNA replication. Science 291: 650–653.
9. Gartenberg MR, Neumann FN, Laroche T, Blaszczyk M, Gasser SM (2004) Sir-mediated repression can occur independently of chromosomal and subnuclear contexts. Cell 119: 955–967.
10. Wu CS, Chen YF, Gartenberg MR (2011) Targeted sister chromatid cohesion by Sir2. PLoS Genet 7: e1002000.
11. Ansari A, Cheng T-H, Gartenberg MR (1999) Isolation of selected chromatin fragments from yeast by site-specific recombination in vivo. Methods 17: 104–111.
12. Boeger H, Griesenbeck J, Strattan JS, Kornberg RD (2003) Nucleosomes unfold completely at a transcriptionally active promoter. Mol Cell 11: 1587–1598.
13. Griesenbeck J, Boeger H, Strattan JS, Kornberg RD (2004) Purification of defined chromosomal domains. Methods Enzymol 375: 170–178.
14. Branda CS, Dymecki SM (2004) Talking about a revolution: The impact of site-specific recombinases on genetic analyses in mice. Dev Cell 6: 7–28.
15. Bischof J, Basler K (2008) Recombinases and their use in gene activation, gene inactivation, and transgenesis. Methods Mol Biol 420: 175–195.
16. Logie C, Stewart AF (1995) Ligand-regulated site-specific recombination. Proc Natl Acad Sci USA 92: 5940–5944.
17. Cheng T-H, Chang C-R, Joy P, Yablok S, Gartenberg MR (2000) Controlling gene expression in yeast by inducible site-specific recombination. Nucleic Acids Res 28: E108.
18. Verzijlbergen KF, Menendez-Benito V, van Welsem T, van Deventer SJ, Lindstrom DL, et al. (2010) Recombination-induced tag exchange to track old and new proteins. Proc Natl Acad Sci USA 107: 64–68.
19. Tsalik EL, Gartenberg MR (1998) Curing *Saccharomyces cerevisiae* of the 2 Micron plasmid by targeted DNA damage. Yeast 14: 847–852.
20. Ghosh SK, Hajra S, Paek A, Jayaram M (2006) Mechanisms for chromosome and plasmid segregation. Annu Rev Biochem 75: 211–241.
21. Matsuzaki H, Nakajima R, Nishiyama J, Araki H, Oshima Y (1990) Chromosome engineering in *Saccharomyces cerevisiae* by using a site-specific recombination system of a yeast plasmid. J Bact 172: 610–618.
22. Gartenberg MR, Wang JC (1993) Identification of barriers to rotation of DNA segments in yeast from the topology of DNA rings excised by an inducible site-specific recombinase. Proc Natl Acad Sci USA 90: 10514–10518.
23. Goldstein AL, McCusker JH (1999) Three new dominant drug resistance cassettes for gene disruption in *Saccharomyces cerevisiae*. Yeast 15: 1541–1553.

24. Güldener U, Heck S, Fielder T, Beinhauer J, Hegemann JH (1996) New efficient gene disruption cassette for repeated use in budding yeast. Nucl Acids Res 24: 2519–2524.
25. Boeke JD, LaCroute F, Fink GR (1984) A positive selection for mutants lacking orotidine-5'-phosphate decarboxylase activity in yeast: 5-fluoro-orotic acid resistance. Mol Gen Genet 197: 345–346.
26. Reid RJ, Lisby M, Rothstein R (2002) Cloning-free genome alterations in *Saccharomyces cerevisiae* using adaptamer-mediated PCR. Methods Enzymol 350: 258–277.
27. Ansari A, Gartenberg MR (1999) Persistence of an alternate chromatin structure at silenced loci in vitro. Proc Natl Acad Sci USA 96: 343–348.
28. Ausubel FM, Brent R, Kingston RE, Moore DD, Seidman JG, et al., editors (2010) Current Protocols in Molecular Biology. New York: John Wiley & Sons.
29. Smith MC, Brown WR, McEwan AR, Rowley PA (2010) Site-specific recombination by phiC31 integrase and other large serine recombinases. Biochem Soc Trans 38: 388–394.
30. Weber SA, Gerton JL, Polancic JE, DeRisi JL, Koshland D, et al. (2004) The kinetochore is an enhancer of pericentric cohesin binding. PLoS Biol 2: E260.
31. Brachmann CB, Davies A, Cost GJ, Caputo E, Li J, et al. (1998) Designer deletion strains derived from *Saccharomyces cerevisiae* S288C: a useful set of strains and plasmids for PCR-mediated gene disruption and other applications. Yeast 14: 115–132.
32. Rothstein R (1991) Targeting, disruption, replacement, and allele rescue: integrative DNA transformation in yeast. Methods Enzymol 194: 281–301.

Chapter 8

An Efficient Purification System for Native Minichromosome from *Saccharomyces cerevisiae*

Ashwin Unnikrishnan, Bungo Akiyoshi, Sue Biggins, and Toshio Tsukiyama

Abstract

We have recently established a system for purifying minichromosomes in a native state from *Saccharomyces cerevisiae*. This system is extremely efficient, and a single-step purification yields samples with sufficient purity and quantity for mass spectrometry (MS) analysis of histones and non-histone proteins tightly associated with the minichromosome. The templates can also be used in various biochemical assays in vitro, such as transcription and recombination, and could be suitable for EM or other biophysical studies.

Key words: Chromatin, Minichromosome, Purification, Histone modifications, Episome, Mass spectrometry, DNA replication, Replication origin, TRP1-ARS1, FLAG tag

1. Introduction

The identification of chromatin-associated proteins and their modification status greatly facilitates the elucidation of the molecular mechanisms underlying DNA-dependent processes such as transcription, DNA replication, and chromosome segregation. However, it has been difficult to obtain chromatin templates of specific genomic regions in sufficient purity and quantity for these analyses. We have recently established an improved TRP1-ARS1 system (1) using multimerized *lac* operators and an epitope-tagged LacI that allows purification of the mini-circles from *Saccharomyces cerevisiae* in a native state (2) (see Note 1). The purified samples are suitable for direct mass spectrometry (MS) analysis for the identification of chromatin-associated proteins and the detection of post-translational modifications on these proteins, and also as chromatin templates in biochemical assays in vitro.

Randall H. Morse (ed.), *Chromatin Remodeling: Methods and Protocols*, Methods in Molecular Biology, vol. 833,
DOI 10.1007/978-1-61779-477-3_8, © Springer Science+Business Media, LLC 2012

2. Materials

2.1. Growing and Harvesting Cells

1. Yeast strain harboring pRS406-CMV-LacI-3FLAG and TALO8 (*TRP1-ARS1-Lac Operator 8*) (Fig. 1).
2. TALO8 is propagated in *Escherichia coli* as pUC-TALO8 as shown in Fig. 1a. In order to preserve eight copies of *lac* operator (8×*lacO*) sequences, Stbl2 (Invitrogen) is used as a bacterial host for plasmid preparation and 1 mM Isopropyl β-D-1-thiogalactopyranoside (IPTG) is added to all bacterial culture media at 30°C. After each preparation of pUC-TALO8, PCR is performed using primers surrounding 8×*lacO* (5′-CAGCT ATGAC-CATGATTACG and 5′-AATGCGAGATCCGT TTAAC) to ensure that the 8×*lacO* are not lost by recombination (with intact *lacO*, the PCR product should be about 300 bp). The plasmid backbone is digested with EcoRI enzyme, ~1.7-kb fragment corresponding to TALO8 is gel purified, ligated in vitro under conditions that promote intramolecular ligation, and then transformed into a yeast strain expressing epitope-tagged LacI. The sequence of the plasmid is available at http://www.labs.fhcrc.org/tsukiyama/protocols/TALO8_Protocol.pdf (a pdf file).
3. pRS406-CMV-LacI-3FLAG (LacI expression vector). As shown in Fig. 1b, LacI protein with three copies of FLAG epitope at the C-terminal end is expressed from the plasmid pRS406-CMV-LacI-3FLAG. This plasmid is linearized within the *URA3* gene by BstBI digestion and transformed into yeast. Integration of the plasmid is confirmed by detection of 3× FLAG-LacI by western blotting using FLAG M2 antibody, which should recognize a band about 45 kDa. The sequence of the plasmid is available at http://www.labs.fhcrc.org/tsukiyama/protocols/TALO8_Protocol.pdf (a pdf file).
4. Yeast media: synthetic media without tryptophan (3).
5. 50-ml Falcon tubes.

2.2. Preparation of Whole Cell Extract

1. 200 mM phenylmethanesulfonyl fluoride (PMSF) in 100% methanol.
2. Buffer H 150: 25 mM HEPES KOH pH 7.6, 2 mM $MgCl_2$, 0.5 mM EGTA, 0.1 mM EDTA, 10% glycerol, 150 mM KCl, 0.02% NP40, freshly supplemented with 2 mM DTT (see Note 2).
3. 100× Protease inhibitors: 100 mM PMSF, 200 μM pepstatin, 60 μM leupeptin, 200 mM benzamidine, 200 μg/ml chymostatin A in 100% methanol. Store at –20°C.
4. 100× Phosphatase inhibitors: 200 mM imidazole, 100 mM sodium fluoride, 115 mM sodium molybdate, 100 mM sodium

orthovanadate, 400 mM sodium tartrate dihydrate in H_2O. Store at –20°C.

5. 1,000× Phosphatase inhibitors: 2.5 mM (–)-*p*-bromotetramisole oxalate, 0.5 mM cantharidin, 500 nM microcystin in DMSO. Store at –20°C.
6. 1,000× Histone deacetylase inhibitors: 500 μM Trichostatin A (Sigma), 25 mM Sirtinol (Calbiochem) in DMSO. Store at –20°C.
7. Zirconia/silica beads (500 μm).
8. 2-ml screw cap tube.
9. Bead beater for cell breakage.
10. Syringe.

2.3. Coupling Anti-FLAG M2 Antibody with Magnetic Beads

1. Dynabeads Protein G (Invitrogen).
2. Anti-FLAG M2 antibodies (Sigma).
3. 0.1 M sodium phosphate pH 7.0.
4. 0.1 M sodium phosphate pH 7.0, 0.01% Tween-20.
5. 0.2 M triethanolamine pH 8.2.
6. 20 mM dimethyl pimelimidate, 0.2 M triethanolamine, pH 8.2. Freshly prepared.
7. 50 mM Tris–HCl pH 7.5.
8. PBST (Phosphate buffered saline with 0.01% Tween-20): (137 mM NaCl, 2.7 mM KCl, 10 mM Na_2HPO_4 (dibasic,

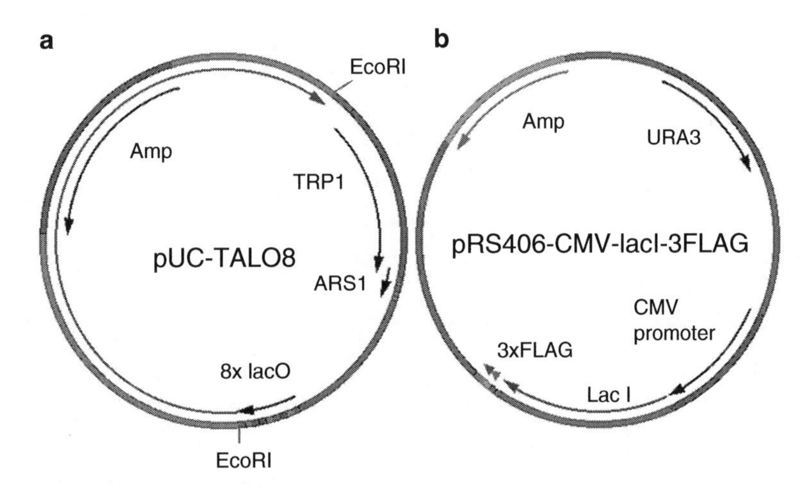

Fig. 1. Map of pUC-TALO8 and pRS406-CMV-LacI-3FLAG. (**a**) TALO8 is inserted into the pUC18 vector to form pUC-TALO8 plasmid. To preserve eight copies of Lac operators, special care is taken during propagation of pUC-TALO8. TALO8 is released from pUC18 vector by EcoRI digestion. (**b**) LacI protein followed by three copies of FLAG epitope is expressed from the CMV promoter.

anhydrous), 2 mM KH_2PO_4 (monobasic, anhydrous), 0.01% Tween-20).

9. Magnetic particle concentrator (MPC, Invitrogen).

2.4. Purification of TALO8 from Cell Extract

1. Buffer H 150: 25 mM HEPES KOH pH 7.6, 2 mM $MgCl_2$, 0.5 mM EGTA, 0.1 mM EDTA, 10% glycerol, 150 mM KCl, 0.02% NP40.
2. Buffer H 300: 25 mM HEPES KOH pH 7.6, 2 mM $MgCl_2$, 0.5 mM EGTA, 0.1 mM EDTA, 10% glycerol, 300 mM KCl, 0.02% NP40.
3. Rinse Buffer: 25 mM HEPES KOH pH 7.6, 2 mM $MgCl_2$, 10% glycerol, 150 mM KCl.
4. Elution Buffer: 50 mM Ammonium bicarbonate, 0.1% Rapigest (Waters Corporation).
5. 3× FLAG peptide (optional, for purification for biochemical assays, such as in vitro transcription).

3. Methods

3.1. Growing and Harvesting Cells

1. Grow yeast cells harboring TALO8 and pRS406-CMV-LacI-3FLAG to an appropriate cell density (OD_{660} = 0.7~1.2) in media lacking tryptophan.
2. Spin cells down at ~6,000 × g for 5 min at 4°C.
3. Suspend cells in ~20× packed cell volume of ice-cold water supplemented with 2 mM PMSF and pellet them as above.
4. Suspend cells in ~10× packed cell volume of Buffer H 150 freshly supplemented with 1× protease inhibitors, phosphatase inhibitors, and histone deacetylase inhibitors, and pellet them in 50-ml Falcon tubes at ~2,500 × g for 5 min at 4°C.
5. Whole cell extracts can be prepared immediately or the cell pellet can be frozen in liquid nitrogen and stored at −80°C.

3.2. Preparation of Whole Cell Extract

All the steps are done on ice or at 4°C.

1. Thaw cells in room temperature water, then add an equal volume of Buffer H 150 freshly supplemented with 1× protease inhibitors, phosphatase inhibitors, and histone deacetylase inhibitors.
2. Aliquot equal volumes of cell suspension and zirconia/silica beads to fill up screw capped 2-ml tubes. Beat cells for 3–5 min using Mini-Beadbeater-96 (BioSpec Products) or equivalent until majority of the cells are broken as assessed under a light microscope.

3. Puncture holes at the bottom and top of the tubes, and place them on 12×75-mm tubes using microfuge tube locks. Recover the cell extract by spinning the tubes at ~285×*g* for 3 min.
4. Alternatively, frozen cell pellet in Subheading 3.1, step 5 can be ground in a blender or coffee grinder in the presence of dry ice for 20 min. Frozen ground cells are then thawed in Buffer H 150 freshly supplemented with 1× protease inhibitors, phosphatase inhibitors, and histone deacetylase inhibitors.
5. Clarify the cell extract by centrifugation at ~125,000×*g* for 90 min in Beckman SW41 or equivalent at 4°C.
6. Soluble cell extract is drawn out through a syringe. Insert needle just above the top of precipitates and slowly draw whole cell extract. Avoid taking up soft, fluffy precipitates on the top of firmly packed precipitates.
7. The cell extract can be used immediately in purification or be frozen in liquid nitrogen and stored at −80°C.

3.3. Coupling Anti-FLAG M2 Antibody with Magnetic Beads

1. Typically, the antibody-conjugated beads are prepared immediately before use. Cross-linking of FLAG M2 antibody to beads is not essential for purification, but significantly reduces the amount of contaminating proteins in eluates. For each liter of cells from which extract was prepared, 25 μl of Dynabeads Protein G beads slurry and 11.5 μg of anti-FLAG M2 antibodies are used. Concentrate magnetic beads on an MPC, then suspend and concentrate beads twice in 0.5 ml of 0.1 M sodium phosphate pH 7.0.
2. Mix antibody and magnetic beads in 0.1 M sodium phosphate pH 7.0 and gently shake them at room temperature for 30 min.
3. Suspend and concentrate beads twice in 0.5 ml of 0.1 M Sodium Phosphate pH 7.0, 0.01% Tween-20.
4. Suspend and concentrate the beads twice in 1 ml 0.2 M triethanolamine pH 8.2.
5. Suspend the beads in 1 ml of 20 mM dimethyl pimelimidate in 0.2 M triethanolamine pH 8.2 (prepared fresh), and incubate them for 30 min at room temperature with constant rotational mixing.
6. Concentrate and suspend the beads in 1 ml 50 mM Tris–HCl pH 7.5 and incubate for 15 min at room temperature with constant rotational mixing.
7. Wash the beads three times in 1 ml PBST. The beads are ready for use in purification.

3.4. Purification of TALO8 from Cell Extract

1. Take small aliquots of the whole cell extract prior to mixing with beads for western blots and DNA analyses to monitor

purification efficiency. Incubate antibody-conjugated beads and cell extract at 4°C for 3 h with constant rotational mixing.

2. Concentrate beads on an MPC and transfer the supernatant (unbound material) into a fresh tube. Ensure that the vast majority of beads are concentrated to the wall to minimize losses. At the same time, do not leave the beads on MPC for more than a few minutes as the beads will clump up, leading to increased background. Save small aliquots of unbound material for western blots and DNA analyses to monitor purification efficiency, freezing the rest in liquid nitrogen and saving them for troubleshooting purposes if required. Suspend the beads in 1 ml Buffer H 150 and transfer them into a siliconized 1.7-ml microfuge tube.
3. Suspend and concentrate the beads three times in 1 ml Buffer H 150, freshly supplemented with protease inhibitors, phosphatase inhibitors, histone deacetylase inhibitors, and 2 mM dithiothreitol (DTT).
4. Suspend the beads in 1 m Buffer H 300 freshly supplemented with protease inhibitors, phosphatase inhibitors, histone deacetylase inhibitors, and 2 mM DTT and rotate them at 4°C for 5 min. Concentrate the beads on MPC, and repeat this wash step three more times, for a total of four times.
5. Suspend and concentrate the beads three times in 1 ml of Rinse Buffer.
6. Mix beads with 50 μl Elution Buffer and agitate them vigorously for 30 min at room temperature. Concentrate the beads on an MPC, transfer the supernatant into a fresh microfuge tube. Take small aliquots from the eluted samples for western blots and DNA analyses to monitor purification efficiency, then immediately freeze the rest in liquid nitrogen and store them at −80°C. Elution is performed a total of four times, with the first two done for 30 min each and the subsequent two for 15 min each (see Note 3). Freeze the beads in liquid nitrogen and store them at −80°C for troubleshooting purposes if required. The samples are ready for MS analyses.
7. Determine the yield and purity of the sample by DNA preparation and SDS-PAGE gel electrophoresis followed by silver staining. Typical yield of TALO8 from 10 l culture at $OD_{660} = 0.7$ is about 2–4 μg of chromatin (1–2 μg of core histones). see Notes 3 and 4 for troubleshooting.
8. If TALO8 is purified as a template for biochemical assays, such as in vitro transcription, it can be eluted by 3× FLAG peptide [(Met-Asp-Tyr-Lys-Asp-His-Asp-Gly-Asp-Tyr-Lys-Asp-His-Asp-Ile-Asp-Tyr-Lys-Asp-Asp-Asp-Asp-Lys), suspended at 5 mg/ml in Buffer H 0.1], which allows elution of the template in a native state. After the step 4 above, rinse beads

three times with Buffer H with a desired salt concentration (usually 100–150 mM KCl), then incubate the beads with 0.5 mg/ml 3× FLAG peptide in the same buffer for 30 min with constant and gentle shaking at 4°C. Repeat elution for 3–4 times totally. For analyses of non-histone proteins associated with the templates, see Note 5.

4. Notes

1. Affinity purification of TRP1-ARS1 minichromosome from *S. cerevisiae* using lacI was first reported by R. Simpson's lab to determine the stoichiometry of the Tup1 general transcriptional repressor to nucleosomes (1). In this system, point mutations were introduced within the ARS1 to create a single *lac* operator, and lacI was expressed and purified from *E. coli* as a recombinant protein to create an affinity resin in vitro. In our system, we have inserted eight copies of *lac* operators within a nucleosome-free region of the template (4), and expressed FLAG-epitope tagged lacI in vivo. These changes significantly increased the yield of the minichromosome. We have also tested Tet repressor and LexA as affinity modules to purify the TRP1-ARS1 mini-circle, but the lacI system was the most efficient in purification (data not shown).

   Another strength of the TALO8 system is the purity of the samples. Southern blotting (data not shown) and SDS-PAGE gel followed by silver staining (Fig. 2) showed that the level of contamination of non-specific DNA and proteins in eluted fractions is quite low. Using Rapigest in the elution buffer enables the direct use of eluted samples in the MS analyses.
2. All solutions are prepared in water that is highly purified (resistivity: ~18.2 ΩM-cm).
3. If the yield of TALO8 is too low, consider the following: low yield of TALO8 can be caused by inefficiency in either cell breakage, binding of TALO8 to antibody-coated magnetic beads, or elution of TALO8 from beads. To monitor cell breakage, examine cells under microscope before clarifying the cell extract by high-speed spin. The binding and elution efficiencies of TALO8 should be monitored in two ways: western blotting and DNA analysis using the starting materials, unbound extract, eluates and the beads after the final elution to determine how much lacI protein and TALO8, respectively, were bound and eluted during the process. If the efficiency of elution is low, increase the volume of Elution Buffer to 100 μl.

4. If the purity of the sample is too low (too much contaminant), consider the following. High levels of contamination can be caused by inefficient washing of beads before elution. Try washing beads more extensively or at a higher salt concentration. We have also found that excess levels of lacI lead to a large amount of proteins non-specifically co-purifying. If the level of free lacI in eluates is too high or the copy number of the template is too low, the expression construct for lacI (pRS406-CMV-LacI-3FLAG) needs to be modified to achieve optimal levels of expression. If the purification is successful, histone proteins should be the major bands when analyzed by silver stained gel (Fig. 2).
5. Insertion of model genes or specific *cis*-elements into TALO8 will enable identification of proteins and their modification status that are specifically associated with these modules. For example, a modified TALO8 with a centromere sequence has been used to identify a previously unknown kinetochore protein (5). This purification required decreased salt concentrations throughout the process. To identify non-histone proteins associated with TALO8 or its derivative, optimal salt

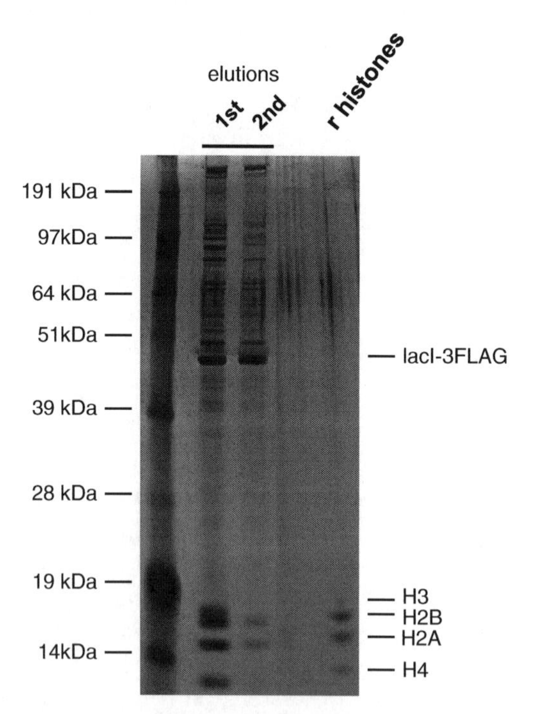

Fig. 2. Typical SDS-PAGE gel of proteins eluted from purified TALO8. Proteins eluted off TALO8 from ~700 ml culture were loaded onto NuPAGE 12% Bis–Tris gel and silver stained. “r histones” denotes 25 ng recombinant yeast histones.

concentrations for binding and washing should be carefully determined. If the protein of interest survives 250 mM KCl wash, highly clean template can be obtained. If a lower salt concentration is required to retain the protein of interest, quantitative MS analysis, with mock purification as a negative control, may be required for identification of proteins specifically associated with the template (5).

Acknowledgments

This work was supported in part by grants from NIGMS and Leukemia & Lymphoma Society to S.B. and T.T and a Beckman Young Investigator Award to S.B.

References

1. Ducker, C. E., and Simpson, R. T. (2000) The organized chromatin domain of the repressed yeast a cell-specific gene STE6 contains two molecules of the corepressor Tup1p per nucleosome, *Embo J 19*, 400–409.
2. Unnikrishnan, A., Gafken, P. R., and Tsukiyama, T. (2010) Dynamic changes in histone acetylation regulate origins of DNA replication, *Nat Struct Mol Biol 17*, 430–437.
3. Adams, A., Gottschling, D., and Stearns, T. (1997) *Methods in yeast genetics*, Cold Spring Harbor Laboratory Press, New York.
4. Thoma, F., Bergman, L. W., and Simpson, R. T. (1984) Nuclease digestion of circular TRP1ARS1 chromatin reveals positioned nucleosomes separated by nuclease-sensitive regions, *J Mol Biol 177*, 715–733.
5. Akiyoshi, B., Nelson, C. R., Ranish, J. A., and Biggins, S. (2009) Quantitative proteomic analysis of purified yeast kinetochores identifies a PP1 regulatory subunit, *Genes Dev 23*, 2887–2899.

Chapter 9

Simultaneous Single-Molecule Detection of Endogenous C-5 DNA Methylation and Chromatin Accessibility Using MAPit

Russell P. Darst, Carolina E. Pardo, Santhi Pondugula, Vamsi K. Gangaraju, Nancy H. Nabilsi, Blaine Bartholomew, and Michael P. Kladde

Abstract

Bisulfite genomic sequencing provides a single-molecule view of cytosine methylation states. After deamination, each cloned molecule contains a record of methylation within its sequence. The full power of this technique is harnessed by treating nuclei with an exogenous DNMT prior to DNA extraction. This exogenous methylation marks regions of accessibility and footprints nucleosomes, as well as other DNA-binding proteins. Thus, each cloned molecule records not only the endogenous methylation present (at CG sites, in mammals), but also the exogenous (GC, when using the *Chlorella* virus protein M.CviPI). We term this technique MAPit, methylation accessibility protocol for individual templates.

Key words: Chromatin, Chromatin remodeling, DNA methylation, DNA methyltransferases, Footprinting, *MLH1*, Nucleosomes, Single-molecule analysis, Transcription

1. Introduction

Heterogeneity of gene expression within clonal populations has been observed at every scale of life, from bacteria to human. Among single-celled microbes, transcriptional heterogeneity facilitates adaptation to changing environmental conditions (reviewed in ref. 1). Heritable transcriptional heterogeneity, which is termed epigenetic, underpins the development of multicellular organisms, and also plays a role in disease, especially cancer (reviewed in ref. 2). However, it remains difficult to study transcriptional heterogeneity at the molecular level, since many techniques for probing molecular interactions, such as chromatin immunoprecipitation,

Randall H. Morse (ed.), *Chromatin Remodeling: Methods and Protocols*, Methods in Molecular Biology, vol. 833, DOI 10.1007/978-1-61779-477-3_9, © Springer Science+Business Media, LLC 2012

measure population averages. Thus, minority molecular states that could be of biological significance go undetected.

One commonly used technique that does produce single-molecule information is bisulfite genomic sequencing (BGS) (reviewed in ref. 3). Treatment with bisulfite converts cytosine to uracil, but 5-methylcytosine is not affected (Fig. 1). Loci of interest can then be amplified with bisulfite and strand-specific primers

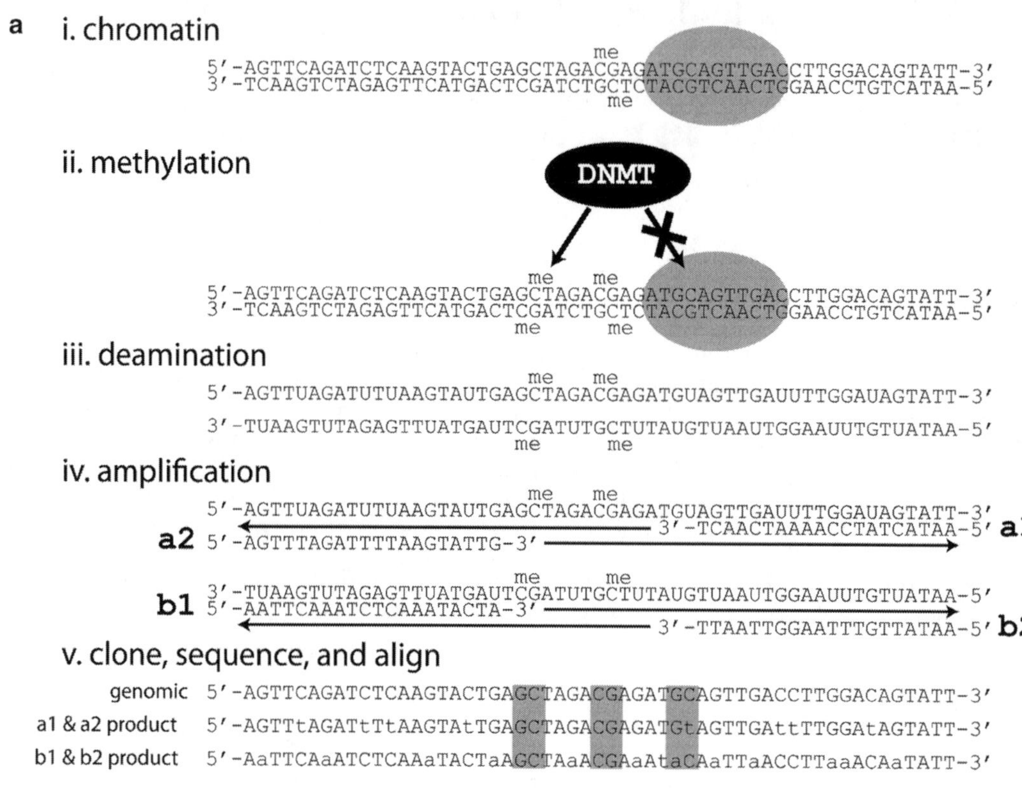

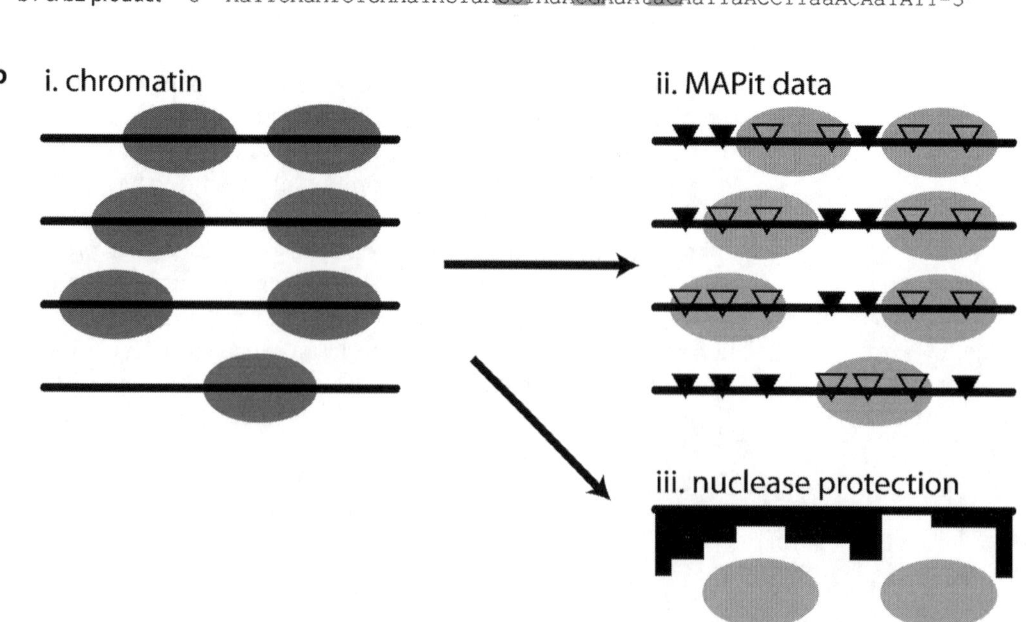

(strands cease to be complementary upon cytosine deamination), which replaces uracil with thymine. Of the several ways to analyze bisulfite-treated DNA, the "gold standard" is to clone and sequence a number of molecules. This approach is widely used to study DNA methylation, which is often associated with transcriptional silencing. Because each cloned sequence represents one molecule from the original population, subpopulations with different DNA methylation states are readily found, for instance at imprinted loci.

Recently, BGS has been married to DNA methyltransferase (DNMT) accessibility probing (reviewed in ref. (4). Of commercially available DNMTs, two are especially suited for this purpose, M.SssI (New England Biolabs) and M.CviPI (New England Biolabs and Zymo Research). Whereas M.SssI methylates cytosine in CG dinucleotides (5), the *Chlorella*-virus-derived M.CviPI methylates cytosine in GC dinucleotides (6). This gives M.CviPI two key advantages: it allows for discrimination between accessibility and endogenous CG methylation, and it increases footprinting resolution in parts of the genome depleted for CG dinucleotides. Just as for nucleases commonly used to probe chromatin structure, M.SssI and M.CviPI interact most readily with open DNA, and can be used to footprint both nucleosomes and DNA-bound proteins (7–9). However, unlike nucleases, DNMTs do not break the phosphodiester backbone. Therefore, multiple "hits" (methylations) occur along continuous molecules. After bisulfite treatment, cloned sequences record multiple hits on single molecules. This allows footprinting of "unpositioned" nucleosomes, whereas nuclease footprinting sees only nucleosomes that occur at fixed positions (Fig. 1). Thus, it is possible to map multiple chromatin conformations occurring at a single locus (10, 11). A second advantage is that, when a non-CG DNMT such as M.CviPI is used, both endogenous and probe methylation can be mapped (9, 12). Most commonly this is used to show that endogenous and probe methylation do not occur on the same molecules (e.g., see ref. 13).

Fig. 1. MAPit overview. (**a**) BGS readout of methylation. In a simplified chromatin structure (i), a protein binds a DNA site 3′ of a dimethylated CG dinucleotide. The presence of the protein protects a GC dinucleotide from attack by an exogenous DNMT, whereas a GC 5′ of the site becomes methylated (ii). Once deaminated and desulfonated (iii), unmethylated cytosines become uracils, and the two strands are no longer complementary. Either strand may then be amplified with bisulfite- and locus-specific primers (iv). The a1 and a2 primers amplify the top strand, whereas the b1 and b2 primers amplify the bottom. Next, DNA is cloned and sequenced. Comparison of sequence with known genomic sequence (v) indicates position of methylation (potential methylation sites are shaded, and base sequence altered by bisulfite treatment is indicated by lower-case letters). (**b**) Advantage of MAPit over nuclease footprinting. Suppose that, in a population of cells, there is a locus that has one fixed and one variably positioned nucleosome. A minority of cells lack either nucleosome, having instead an intermediately positioned nucleosome (i). By MAPit (ii), both chromatin states are detected, and it is seen that the variably positioned nucleosome and the fixed nucleosome occur on the same molecules. *Black triangles* represent methylated sites, and *gray ovals* are inferred nucleosome positions. By contrast, in nuclease footprinting (iii, nuclease accessibility is plotted under the line, which represents 100% protection) provides less information. In particular, nuclease footprinting does not see the minority population.

Although BGS has traditionally been used to study silencing, DNMT probing brings actively transcribed loci into relief. Genome-wide profiling in *Drosophila* indicates that transcription start sites are most accessible to M.SssI, whereas almost no accessibility occurs in heterochromatin (14). The technique has even been termed "M-SPA," for methylase-based single-promoter analysis assay (11). As nucleosomes can be footprinted outside CG islands (15), we see no need to limit application of this technology to transcription start sites. To point, we have instead set precedence with the acronym "MAPit," for methylation accessibility protocol for individual templates (9, 10). By whatever name, MAPit adds a new dimension to chromatin studies: characterization of heterogeneity within populations.

In this section, we describe two MAPit protocols. The first protocol leverages the GC methyltransferase M.CviPI to probe nuclei of cultured mammalian cells, mapping both endogenous DNA methylation and chromatin accessibility along the same molecules (9, 12). Although the method given works with mammalian tissue culture cells, it can be easily adapted to any source of intact nuclei, such as *Drosophila* embryos (14). It is also possible to express DNMTs in live yeast, which has been described elsewhere (16–20). The second protocol extends MAPit to chromatin reconstituted in vitro (21).

Analysis of chromatin structure at a transcription start site of the tumor suppressor gene *MLH1* is provided as an example of MAPit in cultured human cells (Fig. 2). The sequences reveal what appears to be a variably positioned nucleosome flanked by large (~150 bp) histone-depleted regions. Protected footprints of approximately 150 bp size that do not occupy rigidly fixed positions are usually to be interpreted as nucleosomes. MAPit footprinting of a nucleosome reconstituted in vitro is shown in Fig. 3. Smaller footprints may be interpreted as transcription factors (7, 8, 18, 19), particularly if they occupy invariant positions or known transcription factor-binding sites (11). Larger footprints may occur where nucleosomes are closely packed, where linkers are protected by linker histones, or perhaps in higher-order chromatin structures. Hypotheses derived from MAPit data can be tested, if necessary, by use of other techniques, such as chromatin immunoprecipitation or knock-down.

2. Materials

Equipment

1. Water baths set to 37, 42, and 50°C.
2. UV spectrophotometer, e.g., Nanodrop (Thermo Scientific).
3. Thermocycler.

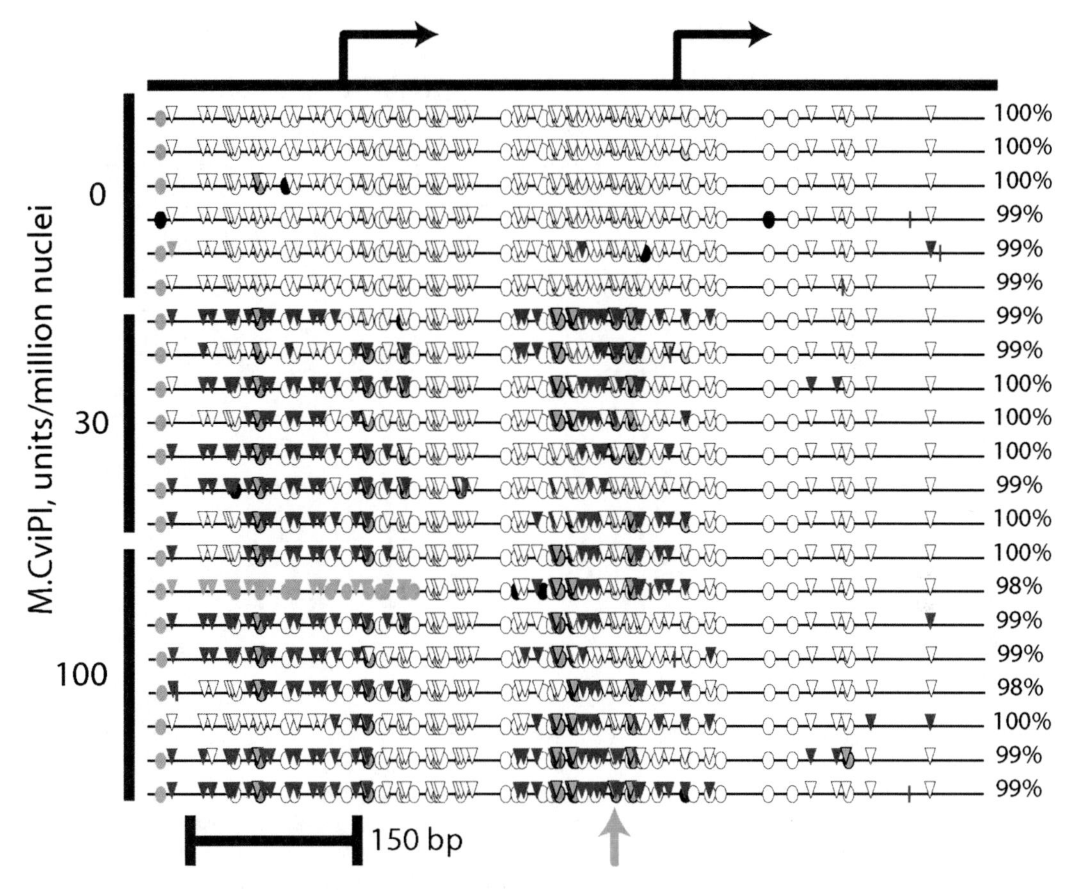

Fig. 2. Single-molecule analysis of *MLH1* transcription start sites (TSSs) in K562 cells. The *MLH1* gene, which encodes a key mismatch repair factor, is frequently silenced by DNA hypermethylation in cancer. Active *MLH1* promoters are characterized by open chromatin (13). Nuclei prepared from K562 cells, a CML-derived lineage, were treated with the indicated concentrations of M.CviPI. Following deamination, a 755 bp amplicon encompassing the *MLH1* TSS was cloned and sequenced. Sequences were processed in MethylViewer to image GC and CG methylation (12). Each row represents one cloned sequence. *Black circles mark* sites of HCG methylation and *red triangles* are methylated GCH sites, where H is A, C, or T. *Gray symbols mark* sites of GCG methylation. *Open symbols* indicate unmethylated sites. *Orange* and *yellow* symbols indicate missing and unaligned sequence, respectively. *Blue vertical lines* are HCH sites, indicative of incomplete conversion (conversion efficiency for each sequence is noted on *right*). Location of each *MLH1* TSS is indicated at *top*. Note that housed between the two TSSs is a loosely positioned footprint of nucleosomal size flanked by open DNA. Also, the cyan arrow labels a footprint of smaller size present in most M.CviPI-probed molecules.

4. Incubator set to 37°C.

Sterile Consumables

1. 1.7 ml microcentrifuge tubes.
2. Micropipet tips for volumes 1–250 μl.
3. Glass scintillation vials.
4. Bisulfite clean-up kit, such as EZ Bisulfite Cleanup Kit (Zymo Research).
5. DNA concentration and desalting kit.

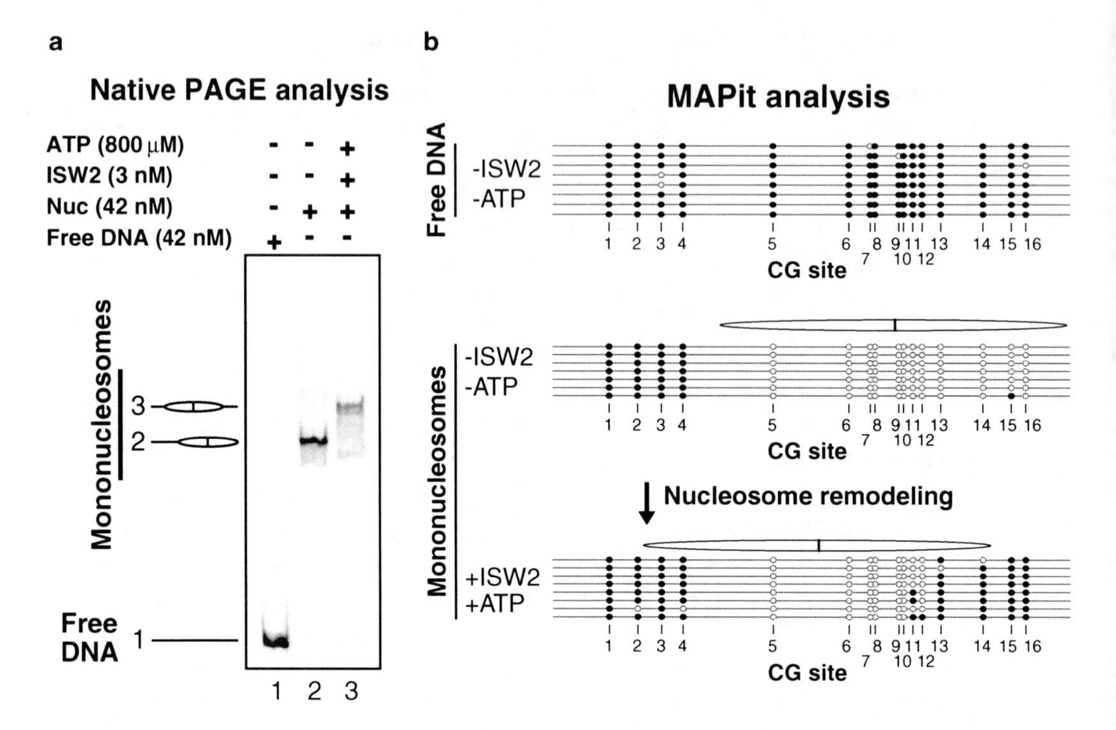

Fig. 3. MAPit view of chromatin remodeling in vitro. (**a**) Gel electrophoresis assay of nucleosome reconstitution and remodeling. In the presence of ATP, purified ISW2 complex moves the histone octamer to the center of the fragment (30–33), reducing mobility in the gel. (**b**) After terminating the remodeling reactions, the samples were probed with M.SssI for MAPit analysis. Each line graphically represents a sequence of a cloned, deaminated molecule. *Circles* indicate CG dinucleotides; *filled circles* mark sites of methylation. *Ovals* are nucleosome positions, inferred from footprints. Variability in apparent extent of remodeling may be due to spontaneous site exposure, a transient unraveling of nucleosome ends (28, 29).

6. Plastic petri dishes.
7. 96-well plates.
8. AirPore™ tape sheets (Qiagen).
9. Aluminum foil tape.

2.1. Nuclei Isolation

1. Phosphate-buffered saline (PBS): 137 mM NaCl, 5.37 mM Na_2HPO_4, 2.68 mM KCl, 1.76 mM KH_2PO_4, pH to 7.4.
2. Resuspension buffer: 20 mM HEPES (4-(2-hydroxyethyl)-1-piperazineethanesulfonic acid) pH 7.5, 70 mM NaCl, 0.25 mM EDTA (ethylenediaminetetraacetic acid) pH 8.0, 0.5 mM EGTA (ethylene glycol tetraacetic acid) pH 8.0, 0.5% glycerol. Store at 4°C for up to a month. Add DTT to 10 mM and PMSF to 0.25 mM immediately before use and place on ice.
3. Lysis buffer: As resuspension buffer, plus 0.19% Nonidet P-40 equivalent (octylphenoxypolyethoxyethanol), made immediately before use. Add DTT to 10 mM and PMSF to 0.25 mM immediately before use and place on ice.
4. Dithiothreitol (DTT), 1 M stock. Make 100 μl aliquots and store at –20°C. Do not reuse thawed aliquots.

5. Phenylmethanesulfonylfluoride (PMSF), 25 mM stock in ethanol. Store at –20°C.
6. Trypan blue, 0.4% (w/v).

2.2. Chromatin Reconstitution and Remodeling

1. Primers to amplify 601 or other nucleosome-positioning sequence, one Cy5-labeled.
2. Plasmid containing 601 or other nucleosome-positioning sequence, e.g., pGEM-3Z/601 (22).
3. Taq DNA polymerase (such as Sigma Jump Start Taq) and buffer, including $MgCl_2$: 10 mM Tris–HCl pH 8.3, 50 mM KCl, 1.5 mM $MgCl_2$.
4. dNTP mix, 2.5 mM each dATP, dCTP, dGTP, dTTP.
5. Refolded histone octamer (23).
6. Reconstitution buffer: 2 M NaCl, 10 mM Tris–HCl pH 7.5, 1 mM EDTA, 5 mM β-mercaptoethanol.
7. Salt dilution buffer: 10 mM Tris–HCl pH 7.5, 1 mM EDTA, 5 mM β-mercaptoethanol.
8. Phenol:chloroform:isoamyl alcohol (25:24:1). Phenol should be brought to pH 7 with Tris solution. Store at 4°C for up to a year.
9. Ethanol, both 100 and 70% solution with 0.1× TE.
10. Sodium acetate, 3 M pH 5.2.
11. TE, 0.1×: 1 mM Tris–HCl pH 8.0, 100 μM EDTA pH 8.0.
12. TBE: 10 mM Tris, 10 mM boric acid, 2 mM EDTA pH 8.0.
13. Native gel: 4% polyacrylamide in 1× TBE.
14. Gel loading buffer, 6×: 15% w/v Ficoll 400, 10 mg/ml Orange G, 10 mM Tris–HCl pH 7.5.
15. Purified ISW2 or other chromatin remodeler (24).
16. Reaction buffer: 25 mM HEPES-KOH pH 7.6, 5 mM $MgCl_2$, 40–50 mM KCl, 0.1 mg of bovine serum albumin/ml, 6 mM Tris–HCl (pH 8.0), 5% glycerol, 30 mM NaCl, 0–300 μM ATP (see Note 1).
17. Nonhydrolyzable ATP analog, e.g., ATP-γ-S, 100 mM.

2.3. Probing Chromatin

1. *S*-adenosylmethionine (SAM), 32 mM. Store aliquots at –80°C. Do not reuse thawed aliquots.
2. DNMT storage buffer: 15 mM Tris–HCl pH 7.4, 200 mM NaCl, 100 μM EDTA pH 8.0, 1 mM DTT, 200 μg/ml nuclease-free bovine serum albumin, 50% glycerol. Store at –20°C.
3. Stop buffer, 2×: 100 mM NaCl, 10 mM EDTA pH 8.0, 1% SDS (sodium dodecyl sulfate). Store at room temperature (20–25°C) no more than 1 day.

4. GC methyltransferase (M.CviPI) fused to either glutathione *S*-transferase (Zymo Research Corp.) or maltose-binding protein (New England Biolabs), obtained from commercial supplier at more than 20 U/μl, or other DNMT (see Note 2). DNMT should be aliquoted and stored at −20°C.
5. Proteinase K, 20 mg/ml. Do not vortex when dissolving in water. Store aliquots at −20°C. Aliquots can be reused up to five times.
6. Phenol:chloroform:isoamyl alcohol (25:24:1). Phenol should be brought to pH 7 with Tris solution. Store at 4°C for up to a year.
7. Ethanol, both 100 and 70% solution with 0.1× TE.
8. Ammonium acetate, 10 M.
9. TE, 0.1×: 1 mM Tris HCl pH 8.0, 100 μM EDTA pH 8.0.

2.4. Bisulfite Treatment

1. Degassed water: Degas 200 ml by stirring 20 min either while boiling or under vacuum. Fill 125 ml glass bottle to brim and cap. If boiled, let cool overnight before using.
2. Sodium hydroxide (NaOH), 3 N. Make fresh with degassed water and use within 6 h.
3. Hydroquinone, 100 mM. Make fresh with degassed water and use within 6 h.
4. Sample denaturation buffer: 3×: per sample, 6.5 μl degassed water, 3.0 μl 3 N NaOH, 0.5 μl 500 mM EDTA pH 8.0. Make fresh and use within 6 h.
5. Sodium metabisulfite. Prepare 5 g aliquots in an oxygen-free environment and store dessicated and in the dark.

2.5. Amplification and Cloning

1. Hot-start Taq DNA polymerase, such as HotStar Taq (Qiagen; see Note 3) and buffer, to amplify deaminated DNA.
2. Generic Taq DNA polymerase and buffer, to amplify cloned DNA.
3. $MgCl_2$, 25 mM.
4. dNTP mix, 2.5 mM each dATP, dCTP, dGTP, dTTP.
5. Locus- and bisulfite-specific primers (see Note 4).
6. TAE buffer, 50×: 2 M Tris, 5.71% v/v glacial acetic acid, 40 mM EDTA.
7. Agarose, molecular biology grade.
8. Ethidium bromide or other DNA stain.
9. Gel loading buffer, 6×: 15% w/v Ficoll 400, 10 mg/ml Orange G, 10 mM Tris–HCl pH 7.5.
10. DNA purification kit, such as Qiagen QIAEX II Gel Extraction Kit.
11. TA cloning kit, such as pGEM (Promega).

12. Primers that flank cloning site, such as M13 forward and reverse.
13. Calcium-competent bacteria, such as DH10B (see Note 5). Freeze in 200 μl aliquots and store at –80°C.
14. SOC medium: 20 g/l tryptone, 5 g/l yeast extract, 3.6 g/l dextrose, 500 mg/l NaCl, 186 mg/l KCl. Adjust pH to 7.0 and autoclave. Store at room temperature and add sterile $MgCl_2$ to 10 mM immediately before use.
15. Solid Luria Bertani (LB) medium: 2% w/v agar, 1% tryptone, 1% NaCl, 0.5% yeast extract, plus biologic agents appropriate to cloning vector used. For pGEM, add 100 mg/l ampicillin and 40 μg/ml X-gal (5-bromo-4-chloro-3-indolyl-β-D-galactopyranoside). X-gal can be top-spread on plates as needed, since it is not stable at 4°C.
16. Liquid LB medium: 1% w/v tryptone, 1% NaCl, 0.5% yeast extract, plus antibiotic appropriate to cloning vector used, e.g., 100 mg/l ampicillin.

3. Methods

3.1. Nuclei Isolation

This protocol describes M.CviPI fusion protein MAPit of nuclei from K562 cells, which grow in suspension. Adherent cells must first be detached by trypsin or scraping.

1. Pellet cells in clinical centrifuge for 5 min at 1,000 × *g* at 4°C. Aspirate medium and resuspend cells in 5 ml ice-cold PBS. Pellet cells and repeat PBS wash.
2. Mix 20 μl cells in PBS with an equal volume trypan blue solution. Count the cells either on a hemocytometer by light microscopy, or with an automated counter.
3. Aliquot 1.2×10^6 cells to an ice-cold 1.7 ml microcentrifuge tube per DNMT concentration to be tested (see Note 6).
4. Centrifuge 1,000 × *g* for 5 min at 4°C and resuspend cell pellet in 200 μl ice-cold resuspension buffer with freshly added DTT and PMSF. Centrifuge 1,000 × *g* for 5 min at 4°C.
5. To cell pellet, per 1.2×10^6 cells, add 42 μl ice-cold lysis buffer with freshly added DTT and PMSF. Resuspend the cells by *gently tapping* the tube. Keep tubes at 4°C as much as possible.
6. Incubate 10 min on ice to lyse cells (lysis time may need optimization) and then add 62.4 μl resuspension buffer per 1.2×10^6 nuclei. Leave nuclei on ice for now.
7. Stain 2 μl nuclei suspension by addition of 2 μl 0.4% trypan blue and examine by light microscopy. At this point, the nuclei should be blue, round, and granular, with no attached cytoplasmic debris.

3.2. Chromatin Reconstitution and Remodeling

1. To produce DNA template for chromatin reconstitution, amplify chosen sequence from plasmid using Taq polymerase and 1× manufacturer's supplied buffer, 1.5–3.5 mM $MgCl_2$ total, 200 nM dNTP mix and 500 nM each primer (one Cy5-labeled and one unlabeled). Melt DNA at 95°C for 4 min in thermocycler, then amplify with 40 cycles of: 95°C melt for 40 s, anneal at primer calculated $T_m \pm 3$°C for 40 s, and extend for 2 min, with an extra 10-min extension step at end. 20 or more microgram DNA will be needed, so adjust number and volume of reactions accordingly.
2. Bring amplified DNA to 200 μl total volume with 0.1× TE. Add an equal volume phenol:chloroform:isoamyl alcohol (25:24:1) and vortex 1 min. Centrifuge 5 min at 14,000 × *g* in microcentrifuge to separate organic and aqueous phases.
3. Extract aqueous layer to a new tube and add 0.1 volume 3 M sodium acetate pH 5.2, vortex, then add 3 volumes ethanol. Incubate 2 h or more at −20°C.
4. Pellet DNA 15 min at 14,000 × *g* in microcentrifuge at room temperature. Carefully withdraw supernatant by aspiration, keeping the aspirator tip well away from the presumed position of the pellet. Wash with 0.4 ml 70% ethanol and pellet again. Air-dry DNA 10 min.
5. Resuspend DNA in 20 μl 0.1× TE. Measure nucleic acid concentration by 260-nm absorbance and dilute to 1 mg/ml with 0.1× TE.
6. To reconstitute mononucleosomes, mix 10 μg histone octamer with purified DNA in molar ratios of 1:4, 1:2, 3:4, and 1:1 in reconstitution buffer, in the minimum volume necessary.
7. Incubate 25 min at 37°C, then dilute salt concentration serially to 1.5, 1, 0.7, and 0.3 M by addition of appropriate volumes of salt dilution buffer at 10 min intervals.
8. To a volume containing 1 μg DNA, add 0.2 volumes 6× gel loading dye. Analyze DNA by native 4% w/v polyacrylamide gel electrophoresis (PAGE) at 4°C buffered with 0.2× TBE. Image the gel for Cy5 fluorescence to assay DNA incorporation into nucleosomes. Use the sample with the least free DNA for further analysis.
9. To remodel chromatin with ISW2 complex, mix nucleosomes and purified complex in ISW2 reaction buffer. Addition of ISW2 complex should be staggered by 30 s to ensure precise reaction times. Incubate 30 min in 30°C water bath, then stop reaction by adding a nonhydrolyzable ATP analog, such as ATP-γ-S, to 5 mM final concentration (see Note 7).
10. Follow extent of nucleosome remodeling by electrophoresis of 1/2 each reaction by native PAGE on a 4% w/v gel at 4°C buffered with 0.2× TBE. Image the gel for Cy5 fluorescence.

3.3. Probing Chromatin

1. On ice, make desired dilutions of the DNMT (see Note 6). For example, to treat nuclei with 0, 30, and 100 U M.CviPI fusion protein per 10^6 cells:
   (a) 100 U/10 μl: 1 part 80 U/μl M.CviPI fusion protein plus seven parts resuspension buffer with freshly added DTT and PMSF.
   (b) 30 U/10 μl: 24 parts 100 U/10 μl M.CviPI fusion protein from step a, plus seven parts DNMT storage buffer and 49 parts resuspension buffer.
   (c) 0 U/10 μl: 1 part DNMT storage buffer plus seven parts resuspension buffer.

   The dilution scheme ensures that DNMT concentration is the only variable. For MAPit of reconstituted chromatin, replace resuspension buffer with reaction buffer (0 mM ATP), and use DNMT at 0–4 U/pmol DNA (including unlabeled and competitor DNA, if any).
2. Prewarm samples at 37°C for 5 min. Prewarm stop buffer at 50°C. Bring ice bucket with DNMT dilutions, timer, micropipet, and tips to 37°C water bath.
3. Add SAM to each sample at 160 μM final concentration as follows. If working with nuclei, add 0.5 μl 32 mM SAM per 10^6 nuclei and then withdraw a 90-μl aliquot for each methylation reaction. For in vitro reconstituted chromatin, bring sample to 90 μl volume in reaction buffer supplemented with 160 μM SAM.
4. To ensure that all samples receive identical DNMT treatment, addition of DNMT to each sample should be staggered. Therefore, set timer to 15 min. After 30 s, add 10 μl DNMT dilution to first sample. Mix by flicking the tube briefly and return to 37°C. Repeat every 30 s until all samples have been treated.
5. 15 min after addition of DNMT (30 s after alarm sounds), add 100 μl 2× stop buffer to first sample treated. Vortex 5 s and place tube at 50°C. Treat samples in the same order as before, one every 30 s.
6. Add proteinase K to 100 μg/ml to each sample (1 μl 20 mg/ml per 200 μl sample). Incubate 16–20 h at 50°C.
7. After proteinase K digestion, add an equal volume phenol:chloroform:isoamyl alcohol (25:24:1) and vortex 30 s. Centrifuge 5 min at 14,000 × *g* in microcentrifuge to separate organic and aqueous phases.
8. Extract aqueous layer to a new tube and add 0.25 volumes 10 M ammonium acetate (i.e., 40 μl ammonium acetate to 200 μl DNA solution). Vortex and add 2.5 volumes ethanol. Incubate 10 h or more at –20°C.

9. Pellet DNA 15 min at 14,000 × *g* in microcentrifuge. Carefully withdraw supernatant by aspiration, keeping the aspirator tip well away from the presumed position of the pellet. Wash with 0.4 ml 70% ethanol and pellet again. Air-dry DNA 10 min.
10. Resuspend DNA in 50 μl 0.1× TE. Measure nucleic acid concentration by 260-nm absorbance. This value will include both DNA and RNA.

3.4. Bisulfite Treatment (see Note 8)

1. Prepare degassed water, 3 N NaOH, 100 mM hydroquinone, and 3× sample denaturation buffer.
2. Suspend 0.2–4.0 μg total nucleic acid (DNA and RNA) in 20 μl 0.1× TE. Add 10 μl 3× sample denaturation buffer. Leave at room temperature.
3. Pipet 100 μl 100 mM hydroquinone into a glass scintillation vial with a small stir bar. Add quickly in order: 5 g sodium metabisulfite, 7 ml degassed water, and 1 ml 3 N NaOH. Stir to dissolve and then adjust pH to 4.95–5.05 with 3 N NaOH (may take 100–200 μl). Place at 50°C.
4. Denature DNA samples at 98°C in thermocycler. After 5 min, quickly add 200 μl metabisulfite solution to each sample and vortex. Keep samples at 50°C or above to prevent renaturation.
5. Incubate samples at 50°C in the dark for 6 h to deaminate DNA (see Note 9).
6. For desulfonation and DNA concentration, it is best to use a commercially available kit, such as EZ Bisulfite Cleanup Kit (Zymo Research). Elute DNA in 20 μl 0.1× TE and store at −20°C.

3.5. Amplification and Cloning (see Note 10)

1. Put 0.2–3 μl deaminated DNA (see Note 11) in a 20 μl reaction with 0.2 μl HotStar Taq (see Note 3) and 1× manufacturer's supplied buffer, 1.5–3.5 mM $MgCl_2$ total (see Note 12), 200 nM dNTP mix, and 500 nM each primer. Melt DNA at 95°C for 4 min in thermocycler, then amplify with 40 cycles of: 95°C melt for 40 s, anneal at primer calculated $T_M \pm 3°C$ for 40 s, and extend for 2 min, with an extra 10-min extension step at end. Keep at 16°C or below when done.
2. Microwave TAE plus 1.5% agarose mixture. Ethidium bromide may be added before pouring gel at 1 part per 15,000. Alternately, DNA can be stained after gel has been electrophoresed.
3. Add 0.2 volumes 6× loading dye to each 1 volume PCR reaction and load on gel. Electrophorese until product clearly separates from primers, as viewed by ethidium stain under ultraviolet light. Photograph if desired, and excise band.

4. A commercially available kit, such as Qiagen QIAEX II Gel Extraction Kit, may be used to desalt and concentrate DNA. Elute in 20 μl 0.1× TE. Afterward, assay DNA concentration by ultraviolet absorbance and electrophorese 2 μl on a 1.5% agarose gel to check quality of preparation.
5. DNA may be cloned using a TA cloning kit, such as pGEM (Promega) or TOPO-TA (Invitrogen). If using a T4 DNA ligase-based approach, mix 1–3 μl DNA with vector, ligase, and buffer per manufacturer's recommendation, then incubate overnight at 4°C.
6. To transform ligation into calcium-competent cells, such as DH10B (see Note 5), thaw 50 μl cells on ice 5–10 min, add 20–50% of a ligation, incubate 20 min on ice, and then 45 s at 42°C. Add 250 μl SOC and incubate 1–1.5 h at 37°C before plating on selective medium appropriate to the cloning vector used. Grow 12–16 h at 37°C in incubator.
7. To check cloning efficiency by colony PCR, suspend 5–10 colonies per plate in separate 100 μl LB aliquots. Amplify 1 μl colony suspension with generic Taq DNA polymerase and primers flanking the cloning site (e.g., M13 forward and reverse for pGEM or TOPO), using standard cycling conditions with at least a 5 min initial denaturation at 95°C to lyse cells.
8. Check cloning efficiency by running PCR product on a 1.5% agarose TAE gel as before. We send DNA from positive colonies and colonies from plates that test more than two-thirds positive off for Sanger sequencing.
9. Prepare colonies for sequencing. We currently use rolling circle amplification and automated BigDye sequencing (Applied Biosystems) performed at the Interdisciplinary Center for Biotechnology Research (ICBR) at the University of Florida. To prepare samples, we inoculate 1 colony (or 3 μl of colony diluted in LB used for colony PCR in step 7) in 100 μl LB + appropriate antibiotic in each well of a 96-well plate, sealing the plate with an AirPore™ (Qiagen) tape sheet. After growing bacteria 12–16 h at 37°C, we add 16 μl 50% v/v glycerol to each well, seal with aluminum tape, and incubate at –80°C for at least 30 min before delivering the plate for sequencing.

3.6. Mapping Methylation in Sequences

There are several programs that automate scoring of methylation from sequencing data. We recommend MethylViewer, as it is able to map and distinguish multiple methylation codes (i.e., CG versus GC) (12). We usually discard sequences with <97% conversion of cytosines outside potential methylation sites. Unaltered MethylViewer output is reproduced in Fig. 2.

4. Notes

1. Reaction buffer can vary, depending on the chromatin remodeler requirements, but should not destabilize or inhibit the DNMT (i.e., pH should be neutral; detergents should be nonionic, etc.). To verify that the DNMT is active in the buffer used, one can methylate plasmid DNA and test for protection from an appropriate restriction enzyme.
2. Different sources of nuclei or enzyme may affect methyltransferase activity. To control for such effects, we use either quantitative methylation-sensitive restriction enzyme digestion qMSRE (25) or methylation-specific PCR (MSP) (26). For the former assay, digest 20 ng unmethylated and methylated genomic DNA with restriction endonuclease R.HaeIII. Amplify digests, as well as mock-treated sample, by real-time PCR with primers to a known open region containing a HaeIII site, such as human *GAPDH* promoter (primers TACTAGCGGTTTT-ACGGGCG and TCGAACAGGAGGAGCAGAGAGCGA). R.HaeIII digestion is blocked by GC methylation, so the restriction enzyme should cut unmethylated but not methylated DNA. For MSP at human LINE, amplify 20 ng DNA with primer sequences AGGTATTGTTTTATTTGGGAAG-TGT and CCTTACAATTTAATCTCAAACTACTATA, which amplify non-GC-methylated or "U" LINE, and CATTGCT-TTATTTGGGAAGCGC and CTTGCAATTTAATCTCAAA-CTGCTATG, which amplify GC-methylated or "M" LINE. Compare abundance of the two products by agarose gel: the "M" product will be more abundant if the DNMT was active.
3. Because deaminated DNA contains uracil, it stalls archaeal family B DNA polymerases, such as Pfu. Taq DNA polymerase, however, works. A hot-start formulation is preferred as the deaminated DNA is single stranded; since deaminated DNA is also low complexity, extensive denaturation is necessary. Pfu Turbo Cx (Stratagene), engineered to lack uracil recognition, is another option, and is preferred for library construction (27). If using Pfu Turbo Cx, one must subsequently treat product with Taq or other polymerase for 10 min at 72°C in the presence of dNTPS, to A-tail the product for TA cloning. We also advise the use of phosphorothioate linkages in primers so that they cannot themselves be edited.
4. To protect against amplicon bias, we recommend that several primer pairs be used for each locus studied, preferably amplifying opposite strands. Each chromatin structure presents a unique sequence in MAPit, which could be toxic in the bacterium used, or could form secondary structure blocking DNA polymerase passage. By these means, some amplicons may be

under-represented in the data. Since opposite strands have radically different sequences following deamination (Fig. 1), one would not expect to see biases against the same chromatin structures. Therefore, if the arrays of chromatin structures seen on the two strands match, amplicon bias is likely not a problem.

5. In some cases, cloning efficiency may be improved by switching bacterial strains. We have found that even closely related strains can differ in their ability to clone particular deaminated loci.
6. We recommend a pilot experiment to determine which DNMT concentrations to use. Comparison of treated with untreated DNA is often useful for discriminating between endogenously and exogenously methylated GCG sites. A two-point titration of M.CviPI activity can show whether enzyme is limiting for methylation at specific probed sites or regions, or conversely, invading protein–DNA contacts as might occur at the edges of nucleosomes (see Fig. 3b) (28, 29). For these reasons, we generally set up three reactions per cell culture, to treat samples at 0, 30, and 100 U/10^6 cells.
7. Chromatin remodelers can be removed by addition of competitor DNA, to prevent their footprinting. Competitor DNA must not contain the (undeaminated) BGS amplicon sequence.
8. There are several commercially available kits for bisulfite treatment, desulfonation, and cleanup. We find that we can get better percent conversion of cytosines, using the deamination conditions given, than with any commercially available kit. However, we use EZ Bisulfite Cleanup Kit (Zymo Research) for desulfonation and cleanup.
9. To ensure that the metabisulfite solution is the correct pH, as some additional metabisulfite may become soluble over the course of the treatment, the scintillation vial containing the remaining metabisulfite can be left at 50°C. Ten minutes before the end of the 6 h incubation, remove the metabisulfite vial and measure pH, which should be in the range 4.95–5.05.
10. DNA can be stored at –20°C between any steps in this procedure. However, for best results, all steps from amplification to transformation of bacteria should be performed in succession without freezing DNA.
11. The amount of DNA can be critical. Because of the reduced complexity of deaminated DNA, primers have reduced specificity. The bulk of the genome, therefore, competes with the target locus for primer recruitment; and once primers have extended at another locus, they no longer work. If too much deaminated DNA is added to the reaction, neither product nor unincorporated primers will be visible on the gel (step 3).

12. Annealing temperature and magnesium concentration must be optimized for each amplicon, and sometimes, for each experiment, since different chromatin structures result in different sequences that could form different secondary structures.

Acknowledgments

We are grateful to Amber Delmas for advice on troubleshooting BGS. This work was supported by the National Institutes of Health (CA95525 to MPK) as well as the Department of Defense, Breast Cancer Research Program (BC062914, BC087311, and BC097648 to MPK).

References

1. Raj A, van Oudenaarden A. (2008) Nature, nurture, or chance: stochastic gene expression and its consequences. Cell 135:216–226
2. Jones PA, Baylin SB. (2007) The epigenomics of cancer. Cell 128:683–692
3. Darst RP, Pardo CE, Ai L, Brown KD, Kladde MP. (2010) Bisulfite sequencing of DNA. Curr Protoc Mol Biol Chapter 7:Unit 7.9.1–16
4. Pondugula S, Kladde MP. (2008) Single-molecule analysis of chromatin: changing the view of genomes one molecule at a time. J Cell Biochem 105:330–337
5. Renbaum P, Abrahamove D, Fainsod A, Wilson G, Rottem S, Razin A. (1990) Cloning, characterization, and expression in *Escherichia coli* of the gene coding for the CpG DNA from *Spiroplasma* sp strain MQ-1 (M.Sss I). Nucleic Acids Res 18:1145–1152
6. Xu M, Kladde MP, Van Etten JL, Simpson RT. (1998) Cloning, characterization and expression of the gene coding for cytosine-5-DNA methyltransferase recognizing GpC sites. Nucleic Acids Res 26:3961–3966
7. Kladde MP, Xu M, Simpson RT. (1996) Direct study of DNA-protein interactions in repressed and active chromatin in living cells. EMBO J. 15:6290–6300
8. Xu M, Simpson RT, Kladde MP. (1998) Gal4p-mediated chromatin remodeling depends on binding site position in nucleosomes but does not require DNA replication. Mol Cell Biol 18:1201–1212
9. Kilgore JA, Hoose SA, Gustafson TL, Porter W, Kladde MP. (2007) Single-molecule and population probing of chromatin structure using DNA methyltransferases. Methods 41: 320–332
10. Jessen WJ, Hoose SA, Kilgore JA, Kladde MP. (2006) Active *PHO5* chromatin encompasses variable numbers of nucleosomes at individual promoters. Nat Struct Mol Biol 13:256–263
11. Delmas AL, Riggs BM, Pardo CE, Dyer LM, Darst RP, Izumchenko E, Monroe M, Hakam A, Kladde MP, Siegel EM, Brown KD. (2011) *WIF1* is a frequent target for epigenetic silencing in squamous cell carcinoma of the cervix. *Carcinogenesis*, doi:10.1093/carcin/bgr193
12. Pardo CE, Carr IM, Hoffman CJ, Darst RP, Markham AF, Bonthron DT, Kladde MP. (2010) MethylViewer: computational analysis and editing for bisulfite sequencing and methyltransferase accessibility protocol for individual templates (MAPit) projects. Nucleic Acids Res 39:e5
13. Lin JC, Jeong S, Liang G, Takai D, Fatemi M, Tsai YC, Egger G, Gal-Yam EN, Jones PA. (2007) Role of nucleosomal occupancy in the epigenetic silencing of the *MLH1* CpG island. Cancer Cell 12:432–444
14. Bell O, Schwaiger M, Oakeley EJ, Lienert F, Beisel C, Stadler MB, Schubeler D. (2010) Accessibility of the *Drosophila* genome discriminates PcG repression, H4K16 acetylation and replication timing. Nat Struct Mol Biol 17:894–900
15. Wolff EM, Byun HM, Han HF, Sharma S, Nichols PW, Siegmund KD, Yang AS, Jones PA, Liang G. (2010) Hypomethylation of a LINE-1 promoter activates an alternate transcript of the *MET* oncogene in bladders with cancer. PLoS Genet 6:e1000917

16. Singh J, Klar AJS. (1992) Active genes in yeast display enhanced *in vivo* accessibility to foreign DNA methylases: a novel *in vivo* probe for chromatin structure of yeast. Genes Dev. 6:186–196
17. Kladde MP, Simpson RT. (1996) Chromatin structure mapping *in vivo* using methyltransferases. Methods Enzymol 274:214–233
18. Jessen WJ, Dhasarathy A, Hoose SA, Carvin CD, Risinger AL, Kladde MP. (2004) Mapping chromatin structure *in vivo* using DNA methyltransferases. Methods 33:68–80
19. Hoose SA, Kladde MP. (2006) DNA methyltransferase probing of DNA-protein interactions. Methods Mol Biol 338:225–244
20. Pardo C, Hoose SA, Pondugula S, Kladde MP. (2009) DNA methyltransferase probing of chromatin structure within populations and on single molecules. Methods Mol Biol 523:41–65
21. Dechassa ML, Sabri A, Pondugula S, Kassabov SR, Chatterjee N, Kladde MP, Bartholomew B. (2010) SWI/SNF has intrinsic nucleosome disassembly activity that is dependent on adjacent nucleosomes. Mol Cell 38:590–602
22. Lowary PT, Widom J. (1998) New DNA sequence rules for high affinity binding to histone octamer and sequence-directed nucleosome positioning. J Mol Biol 276:19–42
23. Dyer PN, Edayathumangalam RS, White CL, Bao Y, Chakravarthy S, Muthurajan UM, Luger K. (2004) Reconstitution of nucleosome core particles from recombinant histones and DNA. Methods Enzymol 375:23–44
24. Tsukiyama T, Palmer J, Landel CC, Shiloach J, Wu C. (1999) Characterization of the imitation switch subfamily of ATP-dependent chromatin-remodeling factors in *Saccharomyces cerevisiae*. Genes Dev 13:686–697
25. Hashimoto K, Kokubun S, Itoi E, Roach HI. (2007) Improved quantification of DNA methylation using methylation-sensitive restriction enzymes and real-time PCR. Epigenetics 2:86–91
26. Herman JG, Graff JR, Myohanen S, Nelkin BD, Baylin SB. (1996) Methylation-specific PCR: a novel PCR assay for methylation status of CpG islands. Proc. Natl. Acad. Sci. USA 93:9821–9826
27. Smith ZD, Gu H, Bock C, Gnirke A, Meissner A. (2009) High-throughput bisulfite sequencing in mammalian genomes. Methods 48: 226–232
28. Polach KJ, Widom J. (1995) Mechanism of protein access to specific DNA sequences in chromatin: a dynamic equilibrium model for gene regulation. J Mol Biol 254:130–149
29. Polach KJ, Widom J. (1996) A model for the cooperative binding of eukaryotic regulatory proteins to nucleosomal target sites. J Mol Biol 258:800–812
30. Fitzgerald DJ, DeLuca C, Berger I, Gaillard H, Sigrist R, Schimmele K, Richmond TJ. (2004) Reaction cycle of the yeast Isw2 chromatin remodeling complex. EMBO J 23:3836–3843
31. Kagalwala MN, Glaus BJ, Dang W, Zofall M, Bartholomew B. (2004) Topography of the ISW2-nucleosome complex: insights into nucleosome spacing and chromatin remodeling. EMBO J 23:2092–2104
32. Zofall M, Persinger J, Bartholomew B. (2004) Functional role of extranucleosomal DNA and the entry site of the nucleosome in chromatin remodeling by ISW2. Mol Cell Biol 24: 10047–10057
33. Zofall M, Persinger J, Kassabov SR, Bartholomew B. (2006) Chromatin remodeling by ISW2 and SWI/SNF requires DNA translocation inside the nucleosome. Nat Struct Mol Biol 13:339–346

Chapter 10

Analysis of Stable and Transient Protein–Protein Interactions

Stephanie Byrum, Sherri K. Smart, Signe Larson, and Alan J. Tackett

Abstract

The assembly of proteins into defined complexes drives a plethora of cellular activities. These protein complexes often have a set of more stably interacting proteins as well as more unstable or transient interactions. Studying the in vivo components of these protein complexes is challenging as many of the techniques used for isolation result in the purification of only the most stable components and the transient interactions are lost. A technology called transient isotopic differentiation of interactions as random or targeted (transient I-DIRT) has been developed to identify these transiently interacting proteins as well as the stable interactions. Described here are the detailed methodological approaches used for a transient I-DIRT analysis of a multi-subunit complex, NuA3, that acetylates histone H3 and functions to activate gene transcription. Transcription is known to involve a concert of protein assemblies performing different activities on the chromatin/gene template, thus understanding the less stable or transient protein interactions with NuA3 will shed light onto the protein complexes that function synergistically, or antagonistically, to regulate gene transcription and chromatin remodeling.

Key words: Chromatin, Protein–protein interactions, Mass spectrometry, Affinity purification, Isotope-labeling, I-DIRT

1. Introduction

To identify networks of stable and transient protein–protein interactions, technology termed transient isotopic differentiation of interactions as random or targeted (transient I-DIRT) was developed (Fig. 1) (1). This technique combines mild in vivo chemical cross-linking and isotopic-labeling with mass spectrometric readout to classify proteins co-purifying with an affinity-tagged "bait" protein as stable, transient, or nonspecific. There has been a multitude of protein–protein interaction studies using in vitro chemical cross-linking; however, there have only been a few using in vivo cross-linking with affinity purification followed by mass spectrometric analysis (1–3).

Randall H. Morse (ed.), *Chromatin Remodeling: Methods and Protocols*, Methods in Molecular Biology, vol. 833,
DOI 10.1007/978-1-61779-477-3_10,

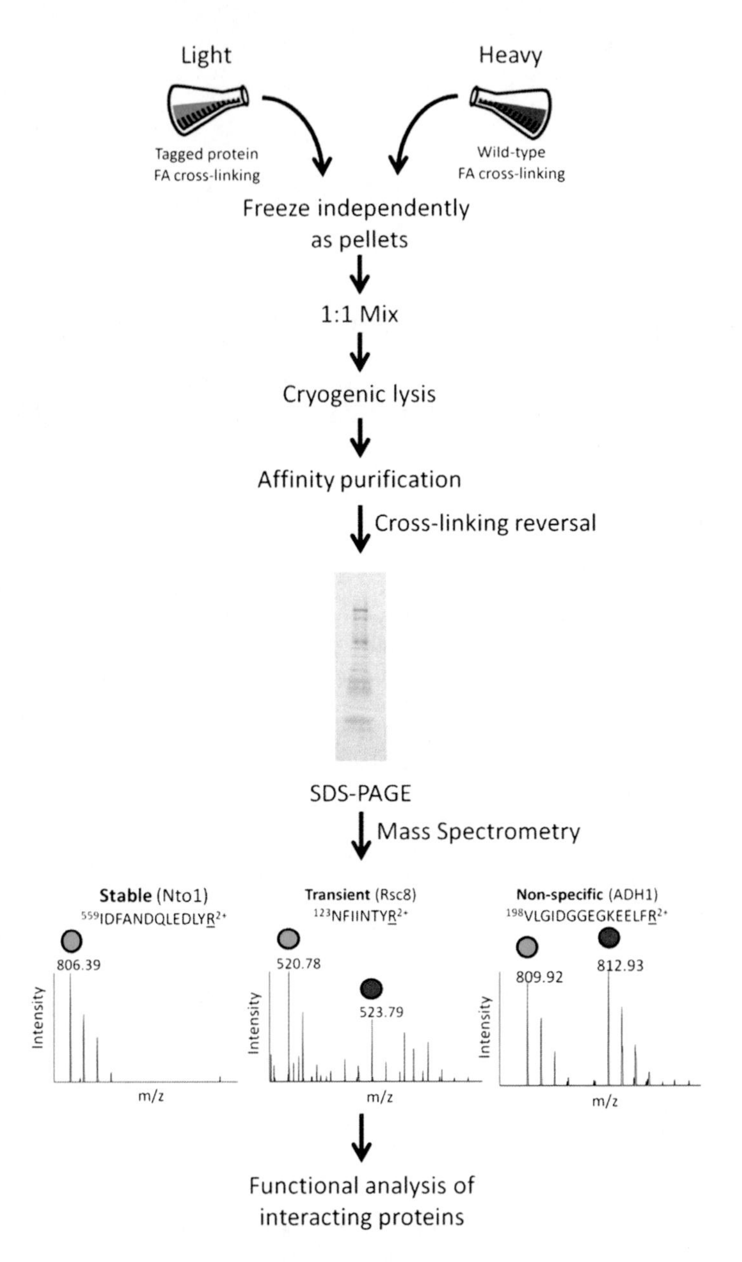

Fig. 1. Transient I-DIRT technology for identifying stable and less stable protein–protein interactions. One culture containing an affinity-tagged protein is grown in isotopically light media and cross-linked with formaldehyde (FA). A second culture of wild-type cells is grown in isotopically heavy media (e.g., $^{13}C_6$-arginine) and also cross-linked with formaldehyde. Following independent harvesting and freezing, the cultures are mixed 1:1 and cryogenically lysed under liquid nitrogen temperature. The affinity-tagged protein is purified with associating proteins and resolved by SDS-PAGE. The gel lane is sliced into 2 mm sections and proteins are identified with mass spectrometry. Proteins interacting stably with the affinity-tagged protein complex have ~100% isotopically light tryptic peptides (i.e., these protein interactions have been preserved from the original isotopically light culture), while nonspecific proteins co-purifying with the affinity-tagged protein complex will have ~50% isotopically light tryptic peptides (i.e., contamination occurs during the purification procedure and these nonspecific protein associations have an equal probability to be isotopically light or heavy). Transiently interacting proteins will have an intermediate level of isotopically light tryptic peptides. Examples are shown of each interaction from a purification of the NuA3 histone acetyltransferase. Following mass spectrometric analysis, the role of the specific protein interactions are explored with functional analyses.

Here, we describe the methodology for transient I-DIRT as it provides for a straightforward and quantifiable approach for defining the in vivo stable and less stable protein interactions. As an example of transient I-DIRT, we describe an analysis of the multi-subunit complex NuA3. NuA3 is a histone acetyltransferase in *Saccharomyces cerevisiae* that acetylates lysine 14 of histone H3 and activates gene transcription (4). NuA3 is a five member protein complex composed of Sas3, Nto1, Yng1, Eaf6, and Taf30 (4). Only Yng1 and Nto1 are found solely in the NuA3 complex. To identify the stable and transient protein–protein interactions with NuA3 specifically, the Yng1 protein was utilized as the purification "bait." Specifically a TAP-tagged version of Yng1 was used in the methods described; however, other affinity tags can be utilized with an appropriate antibody for enriching.

2. Materials

2.1. Cell Culture and Cryogenic Lysis

1. Synthetic complete media per liter: 6.7 g yeast nitrogen base without amino acids, 2 g synthetic drop-out media minus arginine, 900 mL dH_2O. Media is sterilized by autoclaving. After cooling to room temperature, add 1 mL arginine (80 mg/mL in dH_2O, sterile filtered) for isotopically light media, or add 1 mL of $^{13}C_6$-arginine (80 mg/mL in dH_2O, sterile filtered, Cambridge Isotope Laboratories CLM-2265) for isotopically heavy media. After adding light or heavy arginine, add 1 mL of ampicillin (100 mg/mL in 50% ethanol) and 100 mL of autoclaved 20% glucose (w/v).
2. Glycine is a 2.5 M stock, autoclaved.
3. 37% formaldehyde w/v.
4. 20 mM Hepes (pH 7.5)/1.2% polyvinylpyrrolidone.
5. Liquid nitrogen.
6. Retsch MM301 mixer mill with stainless steel cylinders (Retsch 25 mL screw top grinding jars) and ball bearings (Retsch 20 mm stainless steel).

2.2. Affinity Purification

1. Dynabeads M270-epoxy (Invitrogen 143-02D) are suspended in N,N-dimethylformamide at 30 mg/mL and stored at 4°C.
2. Purified rabbit IgG is resuspended at 17 mg/mL in dH_2O and stored at –80°C in 100 μL aliquots (only thaw an aliquot once).
3. 0.1 M sodium phosphate (pH 7.4) stock.
4. 3 M ammonium sulfate stock.
5. Solutions for washing IgG-coated Dynabeads: 100 mM glycine pH 2.5, 10 mM Tris pH 8.8, PBS, PBS + 0.5% tritonX-100.

6. Affinity purification buffer: 20 mM Hepes pH 7.5, 300 mM NaCl, 2 mM $MgCl_2$, 0.1% Tween-20, 1/100 protease inhibitor cocktail (stored at –20°C). Make buffer immediately prior to use and keep at 4°C.
7. Fisher Tissuemiser (or any available probe tissue homogenizer that can operate at 30,000 RPMs).
8. Branson Sonifier 450.
9. Rocking platform.
10. 0.5 N ammonium hydroxide/0.5 mM EDTA. Make immediately prior to use.
11. 2× SDS-PAGE loading buffer: 125 mM Tris–HCl pH 6.8, 20% glycerol, 4.1% SDS, 0.05% bromophenol blue, 4% 2-mercaptoethanol. Store at –20°C in 0.5 mL aliquots.

3. Methods

The methodological workflow for a transient I-DIRT analysis is shown in Fig. 1. The key components of this methodology are (1) a strain with an affinity-tagged protein and (2) a strain without the tag on the protein that is grown isotopically heavy. The tagged and nontagged strains are grown to equivalent densities, subjected to a *mild* in vivo chemical cross-linking with formaldehyde, and frozen independently. The mild cross-linking helps to partially stabilize the protein interactions. Particularly for chromatin-associated protein complexes, the cross-linking has to be mild as extensive cross-linking precludes the efficient purification of protein complexes (1, 5). Frozen strains are mixed 1:1 prior to cryolysis with a ball mill maintained under liquid nitrogen temperature, which provides a method for generating cell lysate without thawing. The ability to differentiate stable, transient, and nonspecific protein interactions with the affinity-tagged complex occurs at the point of thawing the cell lysate and extends for the duration of the purification procedure. During the course of the purification, the stable protein interactions (which are exclusively isotopically light) with the affinity-tagged complex are maintained, while the transient protein interactions (which are also isotopically light) are only somewhat maintained relative to their affinity for the tagged protein complex. Thus, for a transient protein interaction, some exchange of the isotopically light protein will occur with the isotopically heavy counterpart. For nonspecific protein associations with the affinity-tagged protein complex that occur during the purification procedure, there is an equal probability that the nonspecific protein will be isotopically light or isotopically heavy.

The ultimate readout for these stable, transient, and nonspecific protein interactions with the affinity-tagged protein complex is mass spectrometry. When peptides are assigned to a given protein co-purifying with the affinity-tagged complex, the type of protein interaction can be classified as one of the following: stable if the peptides are ~100% isotopically light, nonspecific if the peptides are ~50% isotopically light and ~50% isotopically heavy or transient if the peptides are in between 50 and 100% isotopically light. The following transient I-DIRT methodology is presented for an analysis of the NuA3 histone acetyltransferase, but the methodology can be extended to any affinity-tagged protein. Using the transient I-DIRT technique, we previously identified five stable (Sas3, Nto1, Yng1, Taf30, Eaf6), five transient (Pob3, Nap1, Spt16, Rsc8, Rsc7), and 278 nonspecific protein interactions with an NuA3 purification (1). The analysis of NuA3 demonstrates the high level of nonspecific associations with affinity purifications, and the need for a technique like transient I-DIRT to identify the subpopulation of specific and transient interactions for subsequent functional studies.

3.1. Cell Culture and Cryogenic Lysis

1. *S. cerevisiae BY4741 YNG1::TAP-HIS* is grown to midlog phase (~2.5×10^7 cells/mL) at 30°C with shaking at 200 RPM in isotopically light synthetic complete media. The arginine auxotroph strain *BY4741 arg4::KAN* is grown to ~2.5×10^7 cells/mL in isotopically heavy synthetic complete media with $^{13}C_6$-arginine (see Note 1). Four liters of each strain are typically grown to give ~5 g of wet cell pellet. Large-scale growths are inoculated from 3 mL stationary phase cultures (in the respective synthetic media) with an estimated doubling time of 2.5 h.
2. Protein–protein interactions are partially trapped with mild in vivo chemical cross-linking. When the cultures reach ~2.5×10^7 cells/mL, remove the flasks from the incubator and add formaldehyde (37% stock concentration) to a final amount of 0.05%. Swirl the flask to mix and leave at room temperature for 5 min. Quench the cross-linking by adding 2.5 M glycine to a final concentration of 125 mM. Swirl the flask to mix and leave at room temperature for 5 min.
3. Cells are collected by centrifugation (2,500 × *g*) at 4°C for 30 min in 1 L bottles, washed with 100 mL of ice cold dH_2O, and re-collected by centrifugation. Add 20 mM Hepes (pH 7.5)/1.2% polyvinylpyrrolidone to the wet cell pellet (1 mL buffer/10 g of wet cells) and mix by pipetting.
4. Cells are frozen as a suspension in liquid nitrogen. Use scissors to cut ~1 cm off the end of a 1 mL pipette tip – increasing the size of the opening will provide for easier pipetting. Next, fill a 50 mL polypropylene conical tube with liquid nitrogen. Slowly pipette the cell suspension drop-wise into the liquid nitrogen.

Liquid nitrogen should be added at intervals during the freezing procedure to keep the tube full. Once the cells are frozen as pellets, pour off the excess liquid nitrogen. Poke three holes in the cap of the conical tube with a needle and place it on the tube. The frozen cells are now stored at –80°C.

5. For cryolysis, the isotopically light *YNG1-TAP* and isotopically heavy *arg4::KAN* cells are mixed 1:1 by cell weight. Cells are kept at liquid nitrogen temperature as much as possible during this process to avoid thawing. Weigh each set of pellets, mix 1:1 by weight, and add to a 50 mL polypropylene conical tube. Shake this tube thoroughly to ensure mixing of the pellets.
6. Cryogenic lysis is performed with a Retsch MM301 mixer mill using stainless steel cylinders (Retsch 25 mL screw top grinding jars) and ball bearings (Retsch 20 mm stainless steel). Cell pellets and cell powder are kept at liquid nitrogen temperature as much as possible during this process to avoid thawing. Stainless steel cylinders and ball bearings are precooled in liquid nitrogen (the cylinders and ball bearings are cooled once the nitrogen stops "boiling"). Tongs should be used to retrieve the cylinders from the liquid nitrogen. Approximately 3 g of mixed cell pellets are added to a cylinder with a ball bearing and then placed into liquid nitrogen. Once the cylinder with ball bearing and yeast pellets is cooled, it is attached to the mixer mill, processed for 3 min at 30 Hz, and then returned to the liquid nitrogen. The cycle of cryolysing is repeated five times. Following the final cycle, the cylinder is opened and the cell powder is scooped out into a 50 mL polypropylene conical tube that is in a bath of liquid nitrogen (no liquid nitrogen in the tube). The tube is sealed with a cap containing three holes made with a needle and placed at –80°C for storage.

   Cryogenic lysis with a mixer mill is the preferred method for lysing and blending the cells. One should avoid methods, such as lysis, with glass beads as the samples will thaw during the procedure, which precludes uniform blending of the samples prior to thawing. If a mixer mill is not readily available, a reasonable alternative for cryogenic lysis is manual grinding of the cells in the presence of liquid nitrogen with a mortar and pestle. When manually grinding, the cells should be covered with liquid nitrogen during the lysis process. The cells should be ground into a fine powder. Grinding should continue until >75% lysis is visually observed with a light microscope. After the cells are ground, the cells are stored at –80°C as described above immediately after allowing the liquid nitrogen to evaporate.

3.2. Affinity Purification

1. For affinity purification via the TAP-tag on Yng1, 40 mg of M270-epoxy Dynabeads are coated with 3 mg of rabbit IgG (see Note 2). The following mixture is incubated overnight at 30°C with rocking: 40 mg of M270-epoxy Dynabeads, 3 mg

of rabbit IgG, 1 M ammonium sulfate, 60 mM sodium phosphate (pH 7.4). Following the overnight coupling, the beads are collected with a magnet and washed with: 1 mL of 100 mM glycine pH 2.5, 1 mL of 10 mM Tris pH 8.8, 4 times with 1 mL of PBS, 1 mL of PBS/0.5% tritonX-100, 1 mL of PBS. All washes are done quickly except for the PBS/tritionX-100 wash that should be done for 15 min. For each purification, the beads should be coupled fresh.

2. For a typical purification of a chromatin-associated protein complex like NuA3, we use 10 g of cryogenically lysed cell powder. All steps are performed at 4°C. To 10 g of cell powder containing isotopically light YNG1-TAP and isotopically heavy *arg4::KAN* cell lysates, add 50 mL of affinity purification buffer (see Note 3). The cell lysate is suspended by gentle inversion. Cell lysate is thoroughly blended with a handheld Fisher Tissuemiser (or any available probe tissue homogenizer) for 20 s at 30,000 RPMs. Chromatin is sheared to ~500 bp with a Branson Sonifier 450 with five 10-second cycles (optimal shearing occurs with a maximum of 3 mL of lysate per tube and at least 1 min of incubation on ice following each cycle). The number of cycles and duration of sonication will vary between sonicators, thus parameters for sonication of chromatin to ~500 bp should be empirically determined. The supernatant is collected by centrifugation at 2,500 × *g* for 10 min. The 40 mg of IgG-coated Dynabeads are added for ~16 h with gentle inversion on a rocking platform (see Note 4). Beads are collected with a magnet and washed five times with 1 mL of affinity purification buffer. Proteins are eluted from the beads with 0.5 mL of 0.5 N ammonium hydroxide/0.5 mM EDTA for 5 min at room temperature and the eluant is lyophilized. This elution procedure minimizes the amount of heavy and light chain antibody that is released from the resin. The lyophilized proteins are resuspended in 20 μL SDS-PAGE loading buffer and heated to 95°C for 20 min.

3. Proteins collected from the affinity purification are resolved by 4–20% SDS-PAGE. In our laboratory, SDS-PAGE is performed with the pre-cast Invitrogen Tris–glycine gel system, and the gel is stained with the Thermo GelCode Blue Stain (a colloidal Coomassie stain) prior to imaging for documentation. A colloidal Coomassie stain is recommended, as this stain is sensitive for imagine purposes and easily removed for mass spectrometry. All handling of the gel should be done with powder-free gloves and clean labware. Cleaning items with a commercial glass cleaner such as Windex is good for minimizing keratin contamination, which is the major source of contamination during processing. The gel lane is sliced into 2 mm bands in preparation for mass spectrometry. The easiest way to slice

the gel is to place it on a clean glass plate with a ruler underneath and use a clean razor to excise 2 mm sections. The excised gel bands are placed into 1.7 mL microcentrifuge tubes and stored at −20°C.

4. Gel bands are submitted to a proteomics facility for protein identification and high resolution analysis of tryptic peptides (see Note 5). The proteomics facility performs the following: destaining of the gel bands, in-gel trypsin digestion and tandem mass spectrometric identification of the tryptic peptides. A high resolution mass analyzer should be used to collect the mass spectra. Tandem mass spectra do not necessarily need to be collected with a high resolution mass analyzer. Proteins will be identified with database searching of the tandem mass spectrometric data. Peak areas of peptides corresponding to the assigned proteins will be extracted from the mass spectra. Note that peak areas only need to be extracted for peptides that will contain the heavy amino acid(s). The percent isotopically light peptide is calculated: (light area/(heavy area + light area)) × 100. Percent light peptide values are averaged together for a given protein and the standard deviation is calculated. A typical representation of this data is a bar graph with percent light on the *y*-axis and the given protein on the *x*-axis (1). Stable protein interactions are seen as ~100% isotopically light (e.g., Fig. 1 shows Nto1 as a stable component of NuA3). Nonspecific protein associations are seen as ~50% isotopically light (e.g., Fig. 1 shows Adh1 as a contamination during the purification of NuA3). Transient protein interactions are observed as an intermediate between the stable and nonspecific levels (e.g., Fig. 1 shows Rsc8 as a transiently interacting protein with NuA3). Proteins qualify as transient interactions if the percent isotopically light is >2 standard deviations above the contaminant threshold and less than ~100% isotopically light (1).
5. After identification of stable, transient, and nonspecific interactions, the investigator must select an appropriate system to explore the functional significance of the identified associating proteins. Systems for knocking out or knocking down a particular protein provide a good method for studying the significance of the protein interaction.

4. Notes

1. The methodology presented is for incorporation of isotopically heavy arginine using the *S. cerevisiae BY4741 arg4::KAN* strain. If one is using a different auxotroph, heavy amino acid(s) or organism, then the incorporation efficiency of the heavy amino

acid must be measured. For the transient I-DIRT procedure, one needs to approach complete incorporation of the heavy amino acid. To measure incorporation of heavy amino acid, the strain under study should be grown to mid-log phase and the proteins isolated. Proteins should be subjected to high resolution mass spectrometric analysis of tryptic peptides using a proteomics facility. Tryptic peptides containing the heavy amino acid should be completely labeled. It is always a good idea to label arginine or lysine as these are represented in tryptic peptides (since trypsin cuts after arginines and lysines). Additionally, it is a good idea to use isotopically heavy versions of carbon that are at least six Daltons heavy because (1) carbon isotopes (relative to deuterated) of amino acids do not affect peptide elution from reverse phase columns that are a key component of most mass spectrometric setups and (2) peptides with an amino acid that is less than six Daltons heavy result in isotopic overlap when compared to their isotopically light counterparts.

2. The procedure outlined for affinity purification was optimized for NuA3. NuA3 is a relatively low abundance chromatin-associated protein complex. The amount of cell lysate necessary for other protein complexes must be determined empirically. The amount of cell lysate used in this section is a good starting point for most protein complexes that have been studied with the transient I-DIRT technique. Note that all steps in this section are scalable to the amount of starting lysate (e.g., 4 mg of coupled Dynabeads are used for each gram of cell powder).
3. The affinity purification buffer described in this section is an empirically determined buffer that provides good yields for chromatin-associated protein complexes. The components of the buffer can be varied in accordance to the protein complex under study. Some of the components typically varied include: NaCl (200–400 mM) and TritonX-100 (0–1% v/v).
4. The time of incubation with coated Dynabeads is empirically determined. Shorter affinity isolation times (e.g., 1–4 h) can be explored and may result in fewer nonspecific interactions (6).
5. The mass spectrometric processing and data analysis are standard procedures for a proteomics facility. Proteomics facilities are available at many universities and are also available commercially.

Acknowledgments

Funding was provided by NIH grants P20RR015569, P20RR016460, and R01DA025755.

References

1. Smart, S. K., Mackintosh, S. G., Edmondson, R. D., Taverna, S. D., and Tackett, A. J. (2009) Mapping the local protein interactome of the NuA3 histone acetyltransferase. *Protein Sci.* **18**, 1987–1997.
2. Gingras, A. C., Gstaiger, M., Raught, B., and Aebersold, R. (2007) Analysis of protein complexes using mass spectrometry. *Nat Rev Mol Cell Biol.* **8,** 645–654.
3. Guerrero, C., Tagwerker, C., Kaiser, P., and Huang, L. (2006). An integrated mass spectrometry-based proteomic approach: quantitative analysis of tandem affinity-purified in vivo cross-linked protein complexes (QTAX) to decipher the 26 S proteasome-interacting network. *Mol Cell Proteomics* **5**, 366–378.
4. Taverna, S. D., Ilin, S., Rogers, R. S., Tanny, J. C., Lavender, H., Li, H., Baker, L., Boyle, J., Blair, L. P., Chait, B. T., Patel, D. J., Aitchison, J. D., Tackett, A. J., and Allis, C. D. (2006) Yng1 PHD finger binding to histone H3 trimethylated at lysine 4 promotes NuA3 HAT activity at lysine 14 of H3 and transcription at a subset of targeted ORFs. *Mol. Cell* **24**, 785–796.
5. Byrum, S., Mackintosh, S. G., Edmondson, R. D., Cheung, W. L., Taverna, S. D., and Tackett, A. J. (2010) Analysis of histone exchange during chromatin purification. *JIOMICS*, **1**, 61–65.
6. Cristea, I. M., Williams, R., Chait, B. T., and Rout, M. P. (2005) Fluorescent proteins as proteomic probes. *Mol Cell Proteomics* **4**, 1933–1941.

Chapter 11

Monitoring Dynamic Binding of Chromatin Proteins In Vivo by Fluorescence Recovery After Photobleaching

Florian Mueller, Tatiana S. Karpova, Davide Mazza, and James G. McNally

Abstract

Fluorescence recovery after photobleaching (FRAP) has now become widely used to investigate nuclear protein binding to chromatin in live cells. FRAP can be applied qualitatively to assess if chromatin binding interactions are altered by various biological perturbations. It can also be applied semi-quantitatively to allow numerical comparisons between FRAP curves, and even fully quantitatively to yield estimates of in vivo diffusion constants and nuclear protein binding rates to chromatin. Here we describe how FRAP data should be collected and processed for these qualitative, semi-quantitative, and quantitative analyses.

Key words: FRAP, Binding, Modeling, Chromatin, Microscopy

1. Introduction

In fluorescence recovery after photobleaching (FRAP), a region of a cell is selectively photobleached and then the rate at which fluorescence enters the bleached region is measured (Fig. 1a). The fluorescence recovery contains information about both the protein's diffusion rate plus any binding interactions with large, relatively immobile substrates that may be present in the bleached region (1, 2). Chromatin is a good example of such a substrate, as nuclear proteins that bind to it will be retarded as they enter a bleached region of the nucleus. The amount of retardation in the FRAP recovery rate depends on the residence time of the protein on chromatin. In this way, FRAP has been increasingly used as a tool to detect and quantify nuclear protein interactions with chromatin (3).

Randall H. Morse (ed.), *Chromatin Remodeling: Methods and Protocols*, Methods in Molecular Biology, vol. 833,
DOI 10.1007/978-1-61779-477-3_11, © Springer Science+Business Media, LLC 2012

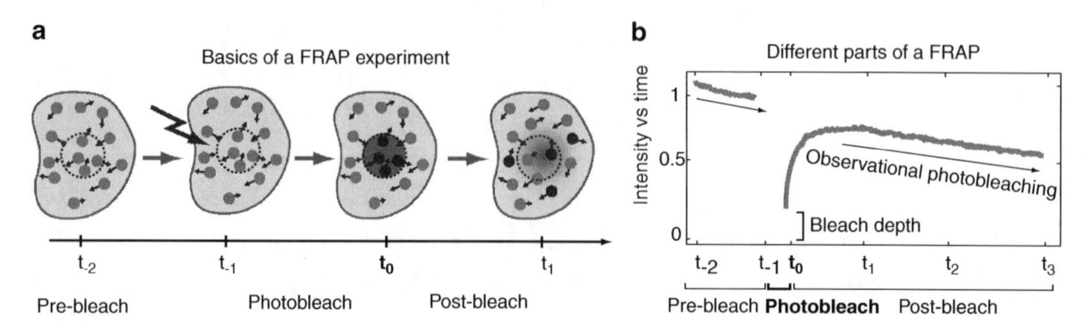

Fig. 1. Basics of FRAP. (**a**) Shown is a nucleus (*solid black outline*) containing fluorescently tagged proteins (*solid gray circles*). *Small arrows* indicate protein motion. FRAP is performed by photobleaching a region (*dashed circle*) of the nucleus. During the recovery phase, bleached molecules (*black circles*) move out of the bleach zone and unbleached molecules move in. The rate of fluorescence recovery can be measured by estimating the average fluorescent intensity inside the measurement area (*dashed circle*) as a function of time, generating a FRAP curve. (**b**) The FRAP curve shows observational photobleaching, i.e., the fluorescence intensity decreases over time due to unintentional photobleaching during all periods of observation. The bleach depth is defined here as the initial measured value of the FRAP curve immediately after the photobleach.

A typical FRAP experiment is composed of three parts (Fig. 1a, b): (1) the pre-bleach, which is the first part of the FRAP measurement to establish that the fluorescence concentration is at equilibrium; (2) the photobleach, which is the next part of the FRAP measurement where a specific region in the cell is intentionally photobleached; (3) the post-bleach, which is the last part of the FRAP measurement where the recovery of fluorescence in the bleached region is recorded. During all phases of image acquisition (pre-bleach and post-bleach), unintentional, low-level bleaching occurs. This is called observational photobleaching. It must be minimized when designing the FRAP protocol, and corrected for when computing the FRAP curve.

Here we describe procedures for three different types of FRAP experiment: qualitative FRAP, semi-quantitative FRAP, and quantitative FRAP, where the latter two procedures include additional processing steps to extract quantitative information.

In qualitative FRAP, the FRAP curve is generated and then conclusions are drawn about whether a protein is very stably bound (recoveries on a time scale of hours) or very transiently bound (recoveries on a time scale of seconds). Often, qualitative FRAP is used to compare two different biological conditions. For example, FRAP might be performed on the wild-type and on a mutant version of a nuclear protein to see if the FRAP curve changes (4). A different FRAP curve often indicates that chromatin binding interactions have been disrupted in the mutant.

Semi-quantitative FRAP extends qualitative FRAP by enabling a quantitative comparison of two FRAP curves (4, 5). Different metrics can be employed for this comparison, including the time it takes the FRAP recovery to reach 50% of its final value (the half time) or the fraction of fluorescent molecules that are permanently bound (the immobile fraction).

Quantitative FRAP goes beyond a semi-quantitative analysis by enabling estimates of the protein's diffusion constant and its association and dissociation rates with chromatin (1, 2, 6). These numbers can in turn be used to calculate the fraction of free and bound molecules in the nucleus, plus the time to diffuse from one binding site to the next and the dwell time on a chromatin target site. The quantitative estimates are obtained by fitting mathematical models for diffusion and binding to the FRAP data.

2. Materials

2.1. GFP-Tagged Proteins

To perform FRAP the molecule of interest must be fluorescently tagged. Fluorescent proteins are typically used for this purpose, with eGFP being the most common.

The fusion protein gene can be transiently transfected or stably incorporated into the cell of interest. Low expression levels are preferable to avoid over-expression artifacts, but expression must be high enough to provide a sufficient signal above background.

2.2. Microscope

The protocols below are for a specific microscope, the Zeiss LSM 5 Live confocal microscope. FRAP can be performed on most confocal microscopes, and also on two-photon microscopes or wide-field microscopes equipped with a photobleaching laser. The principles (although not all the details) described in the protocols below are applicable to these other systems. For more detailed information on the operation of a confocal microscope, consult the excellent Web site http://www.microscopyu.com/articles/confocal/.

2.3. Imaging Chamber

Temperature and CO_2 content during imaging should be regulated by an environmental chamber, such as the one from Precision Plastics Inc. (Beltsville, MD, USA).

2.4. Software

The microscope is typically equipped with software to collect the data and to do some of the early processing steps. If necessary, further image analysis can be done with other image processing software, such as ImageJ. A spreadsheet program such as Excel is required to tabulate and manipulate the initial measurements of the FRAP. For quantitative FRAP, we can provide Matlab software that performs the analysis. Alternatively, the equations described in the Notes can be implemented in any software program that permits data fitting.

3. Methods

Three FRAP protocols are described: qualitative, semi-quantitative, and quantitative FRAP. General guidelines are provided. Specific examples are also included at many points in the protocol and accompanying figures. These are all derived from FRAP experiments using an eGFP-tagged glucocorticoid receptor (GR) in the nuclei of mouse fibroblast cells.

3.1. Protocol for Qualitative FRAP

Use qualitative FRAP to compare FRAP curves. For example, compare wild-type and mutant versions of the same protein (7), or compare the same protein in the presence and absence of a specific cofactor (removed for example by siRNA knockdown) (4), or compare wild-type and drug-treated cells (8), or compare FRAPs of different proteins belonging to the same regulatory complex (4).

3.1.1. Set Up Time-Lapse Imaging

These conditions should be optimized in preliminary tests and then used without modification such that different FRAP curves can be compared (see Note 1).

1. Objective lens. For imaging of most nuclei, choose a 63× or 100× objective lens with high numerical aperture (NA = 1.2–1.4) (see Note 2).
2. Pixel size and zoom. For imaging of mammalian nuclei set the zoom such that the pixel-size is 0.1–0.2 μm (see Note 3).
3. Size of the imaged area. Typically the whole nucleus or a subregion of the nucleus is sufficient (see Note 4).
   Example: The entire nucleus is imaged using a region of 160×160 pix^2 = 17×17 μm^2 (Fig. 2a).
4. Confocal pinhole. For bleaching an arbitrary region of the nucleus in which the protein is homogenously distributed, set the pinhole to a value that gives an optical slice, which is somewhat less than the thickness of the nucleus in that region (see Note 5). In this way, the maximum amount of nuclear fluorescence is collected without any contribution of cytoplasmic fluorescence. For bleaching a specific nuclear structure, such as a nuclear organelle, adjust the pinhole such that the optical slice corresponds to the dimensions of the structure.

   Example: GR distribution is relatively homogeneous throughout the nucleus (with the exception of nucleoli) (Fig. 2a), so the photobleach is not designed for a specific structure. The pinhole is set to 2.5 Airy units, which yields an optical slice thickness of 2.0 μm. Note that the nucleus in these cells is an ellipsoid with a maximum height of about 4 μm.
5. Gain and offset. Determine first the range of cell intensities that are of interest (e.g., an intermediate intensity range might

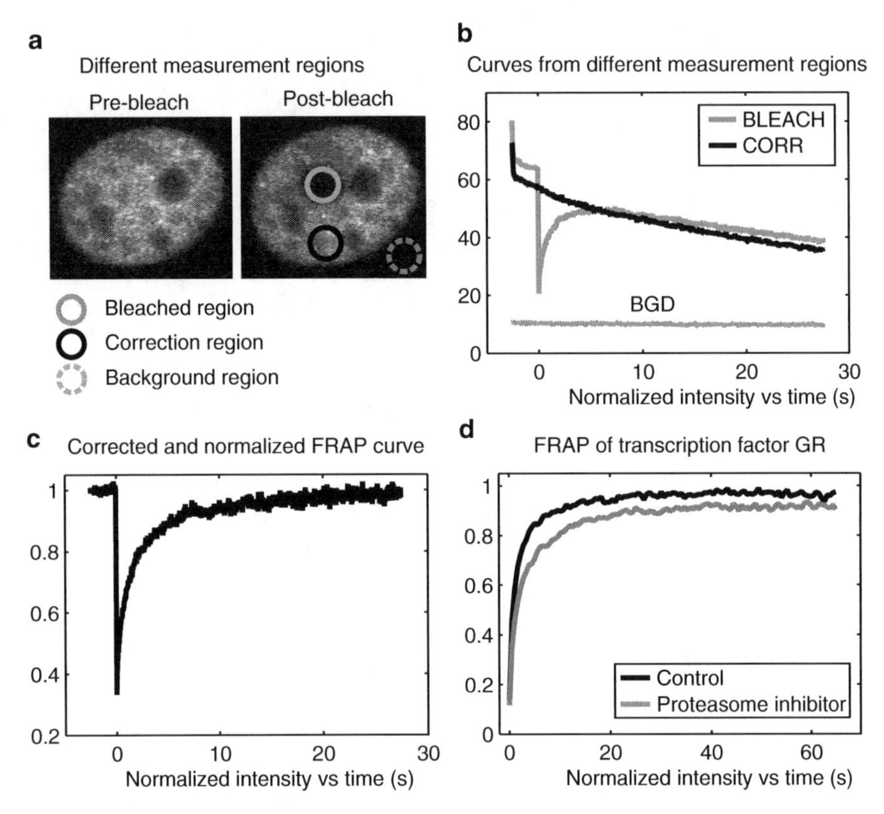

Fig. 2. Processing steps in qualitative FRAP. (**a**) Image of cells before and after the photobleach. *Circles* indicate the different measurement regions: *solid gray* – bleached region, *solid black* – correction region, *dashed gray* – background region. (**b**) Averaged fluorescence intensity from each of the regions in (**a**). Note that the bleached and correction regions have comparable brightness and that the background region does not show observational photobleaching. (**c**) The corrected FRAP curve is obtained by first correcting for the loss of fluorescence due to both intentional photobleaching and observational photobleaching, and then renormalizing the resulting curve by its pre-bleach intensity. (**d**) Application of qualitative FRAP to the glucocorticoid receptor (GR). Treatment of the cells with a proteasome inhibitor leads to a slowdown of the FRAP (8).

be of interest, thereby avoiding very bright cells, which might yield over-expression artifacts and very dim cells which might yield insufficient signal). Then set the gain (see Note 6) such that all cells within the desired intensity range can be recorded without further gain adjustments. Specifically, the brightest cells within the selected intensity range should not cause saturation of the detector. The offset should be set such that the dimmest cells within the selected intensity range should be above zero. Practically speaking, this means that a background measurement outside of the cell should yield intensities that are 5–10% of the total dynamic range of the detector.

Example: For an 8-bit image the dynamic range of the detector is from 0 to 255. To avoid saturation, the gain is then set such that cell intensities do not exceed 255. The offset is set such that background levels range from 12 to 25.

6. Acousto-optical tunable filter (AOTF) setting for imaging. The laser power used for imaging is modulated by the AOTF. It should be set as low as possible to minimize observational photobleaching, but high enough such that cell signals are well above background.

   Example: FRAP on the Zeiss LSM 5 Live is performed with a 488-nm/100-mW diode laser in the DuoScan mode. In this mode 80% of the laser power is used for the photobleach and 20% is used for imaging. The AOTF is set to use 6% of the imaging laser. Thus the final fraction of the 100 mW laser that is used for imaging is 6% of 20% or 1.2%.

7. Scan speed for imaging. Set the scan speed as fast as possible in order to record the first post-bleach frame as quickly as possible. We recommend the fastest possible setting regardless of how fast or slow the FRAP recovery is.

   Example: Imaging scan speed of 40 fps (frames per second).

8. Number of frames to collect before the photobleach. Optimally, collect pre-bleach images for at least 2–5 s before the photobleach. For fast scanning this can be up to 100 frames. Reduce this number to as low as five frames if fluorescent signal is rapidly lost due to observational photobleaching. This is often necessary for small nuclei, which have limited reservoirs of fluorescence and so lose signal rapidly during the imaging process.

   Example: 100 pre-bleach frames for the large 10 μm diameter mouse fibroblast nucleus (Fig. 2c).

9. Number of frames to collect after the photobleach. Optimally, collect around 1,000 frames after the photobleach. Reduce this number if the fluorescent signal has decayed close to background levels by the end of the FRAP measurement. Depending on the time scale of the recovery, it may be necessary to add delay times between images (step 10 below), particularly if the number of post-bleach frames has been reduced to avoid excessive observational photobleaching (see Note 7).

   Example: 1,100 post-bleach frames at 0 fps for a FRAP recovery that requires ~20 s to equilibrate (Fig. 2c).

10. Delay time between images. The delay time should be the same for both pre-bleach and post-bleach imaging. Use a delay time of 0 s for FRAPs that recover within a few seconds, but for slower FRAPs, introduce delay times to limit observational photobleaching.

    Example: 0 s delay time. The settings in steps 7–10 yield a pre-bleach imaging duration of 2.5 s and a post-bleach imaging duration of 28 s (Fig. 2b).

3.1.2. Set Up the Intentional Photobleach

The first three steps below (steps 1–3) are designed to provide a deep bleach in a short time. This minimizes movement of proteins during the photobleach (see Note 8). The bleach depth, which is the initial value of the normalized FRAP recovery (computed in step 5, Subheading 3.1.4 below), should be from 0.3 to 0.5 (i.e., 30–50% of the initial fluorescence before the bleach). If this bleach depth is not achieved, longer intentional bleaching and/or more bleach iterations should be used.

1. AOTF setting for bleaching. Set the AOTF to 100% for the intentional photobleach. This maximizes the number of fluorescent molecules that can be photobleached with one bleach pulse.

   Example: 100% AOTF of a 488 nm/100 mW diode laser used in DuoScan mode. In this setting, 80% of the total laser power is used for bleaching. Thus the final fraction of the 100 mW laser that is used for bleaching is 100% of 80% or 80%.

2. Scan speed of the intentional bleach. Bleach as fast as possible. If the bleaching scan speed can be set independently, then set it to as high a rate as possible. If the bleaching scan speed is determined by the imaging scan speed, then set the imaging scan speed as high as possible (step 7, Subheading 3.1.1). This usually results in a photobleach of 15–50 ms.

   Example: Bleaching scan speed of 63 fps. Note that this bleaching scan speed is faster than the imaging scan speed of 40 fps (step 7, Subheading 3.1.1). Since the Zeiss LSM 5 Live permits setting these scan speeds independently, we chose an even faster scan rate for the bleach.

3. Number of iterations for the intentional photobleach. Perform one iteration of the photobleach.

   Example: With one iteration of the bleach and the settings in steps 1–2, the bleach depth is ~0.3 (Fig. 2c).

4. Bleach geometry. For homogeneously distributed proteins, use a circular bleach spot whose area does not exceed 10% of the total nuclear area. This will leave a sufficient reservoir of fluorescence to allow for recovery and for the loss of fluorescence from observational photobleaching. Alternate bleach geometries, such as a square, rectangular or strip bleaches, are also possible (see Note 9). For bleaching a specific structure, the bleached region should be drawn to match the shape of the structure. The goal is to bleach the entire structure without bleaching any of its surroundings.

   Example: GR fluorescence is relatively homogeneous. A circular bleach region is selected with a diameter of 2.5 μm (Fig. 2a). (The diameter of the nucleus is about 15 μm). Using the scan speed setting in step 2 above, the bleach time is 16 ms.

5. Positioning of the photobleached region. For FRAP at a specific structure, the photobleach region should overlay it. Otherwise, for FRAP of a generic region in the nucleus, place the bleach region far away from the nuclear membrane and also try to avoid nucleoli.

   Example: GR is relatively homogeneous in its distribution, so the bleach is set up for a generic region of the nucleus. The bleach region is positioned with sufficient nuclear fluorescence surrounding it (as far away as possible from the nuclear membrane and nucleoli) to allow for influx of fluorescence from all sides (Fig. 2a).

3.1.3. Collect the FRAP Data

Configure the microscope as described above in Subheadings 3.1.1 and 3.1.2, and then run the FRAP macro in the confocal software.

3.1.4. Process the FRAP Data

Typically the processing is done either with software available on the confocal microscope or with freely available software such as ImageJ or Fiji. The numbers extracted from this software can be transferred to a spreadsheet program such as Excel, and then the calculations described below can be performed. If the FRAP recovery is slow, cell movement may become a problem and this must be corrected before performing the following measurements (see Note 10).

1. Extraction of time-stamps. The time at which the bleach occurs and at which the images are recorded can usually be extracted from the software that is used to operate the confocal microscope. These times are transferred to a spreadsheet file, and then corrected by subtracting the time value at the first post bleach frame from all the time points. This sets the time at the first post-bleach frame to zero, which is the convention in displaying FRAP data.
2. Measurement of background intensity. Background intensity is produced by the detection system. It corresponds to the fluorescence signal recorded with no fluorescent molecules in the field of view. This is best measured in a region outside of the cell, since even for a nuclear localized protein there may be some residual fluorescence in the cytoplasm.

   The averaged intensity value in the background region can then be calculated by the software at each time point. These values should be copied to the spreadsheet program and averaged. This averaged background value, designated here as BGD, will then be used for background correction in step 5 below. There are several quality checks that should be performed to determine if background is correctly measured (see Note 11).

   Example: Background is measured in the dotted gray circle in Fig. 2a.
3. Measurement of fluorescence loss due to photobleaching. To correct for the losses of fluorescence due to the intentional

photobleach and observational photobleaching, measure the fluorescence loss in a control region far away from the region that was photobleached. This can be done with a region of interest comparable in size to the region used for photobleaching. For consideration of how best to position this region in the nucleus, see Note 12. The region's averaged intensities designated here as *CORR* will then be used to generate the corrected FRAP curve in step 5 below.

Example: The correction curve is measured in the black circle in Fig. 2a.

4. Measurement of fluorescence recovery. Generally, use a region of interest that is the same shape and size of the region that was photobleached, but other sizes are possible. Position this measurement region over the region that was photobleached. Use the image processing software to compute the average intensity within the measurement region at all pre and post bleach time points. Copy the averaged intensities at each time point to the spreadsheet file. These numbers designated here as *BLEACH* will then be used to generate the corrected FRAP curve in step 5.

   Example: Recovery is measured in the gray circle in Fig. 2a.

5. Calculation of the FRAP curve. The spreadsheet file should now contain the following data: (1) the time points (t) for each frame, where $t=0$ s corresponds to the first post-bleach frame; (2) the averaged background intensity BGD; (3) the fluorescence intensity at each time point in the control region CORR(t); (4) the fluorescence intensity at each time point in the bleached zone BLEACH(t).

   Use the spreadsheet program to calculate the corrected FRAP curve using the following formula:

$$\text{FRAP}(t) = (\text{BLEACH}(t) - \text{BGD}) / (\text{CORR}(t) - \text{BGD}). \quad (1)$$

   Then normalize the preceding curve such that the pre-bleach intensity is 1. This is done by first calculating the averaged pre-bleach intensity. Then each time point of the FRAP(t) curve from above is divided by this value, yielding a renormalized FRAP curve.

   Example: Fig. 2b shows BLEACH(t), CORR(t), BGD, and Fig. 2c shows the final corrected curve for the GR FRAP data

3.1.5. Compare FRAP Curves Under Different Biological Conditions

1. Replicating the FRAP experiment. Repeat the procedures in Subheadings 3.1.3 and 3.1.4 on at least ten different cells of comparable brightness and then average these FRAP curves. For dimmer cells or smaller nuclei, up to 50 curves may be required to get a smooth, averaged curve. Calculate standard errors at each time point, and then plot the FRAP curve with these error bars.

2. Comparison with FRAP under different biological conditions. Repeat step 1 above on the mutant, or whatever alternate biological condition is being studied. Plot the two averaged FRAP curves on the same plot.

3. Repeating experiments on different days. Often there are only small differences between FRAP curves, with one curve being slightly faster or slightly slower than the other. These differences could be biologically significant (8). To test for this, repeat steps 1 and 2 above on different days. If the differences between curves are reproducible, then they are probably biologically significant.

   Example: Small differences in FRAP curves are found for the transcription factor GR in the presence and absence of the proteasome inhibitor (Fig. 2d). These are reproducibly seen and also correlate with different transcription levels with and without the proteasome inhibitor (8).

3.2. Protocol for Semi-quantitative FRAP

Use semi-quantitative FRAP for simple numerical comparisons between FRAP curves.

3.2.1. Microscope Set-Up, Data Collection, and Processing

Perform as described above for qualitative FRAP (Subheadings 3.1.1–3.1.5).

3.2.2. Semi-quantitative Measurements

Measure either the mobile and immobile fractions (step 1 below) or a FRAP recovery half-time (step 2 below) or both.

1. Measuring mobile and immobile fractions. Some FRAP curves recover on two very different timescales. The first phase may be complete within a few seconds while the second phase may require tens of minutes or hours or even longer. Data for such FRAP curves are often collected for only a few minutes yielding a second phase of the curve that appears to be flat because the timescale of recovery is very slow. The first fast phase is referred to as the mobile fraction M, while the second slow phase is referred to as the immobile fraction IM. The magnitude of the two fractions can be calculated with the following simple equation:

$$M = \frac{\eta - F_0}{1 - F_0}, \quad \mathrm{IM} = 1 - M, \tag{2}$$

   where η is the final, plateaued value of the FRAP and F_0 is the value of the FRAP at the first post-bleach frame. η can be easily estimated from the FRAP curve by averaging the intensity values once the curve reaches a final plateau.

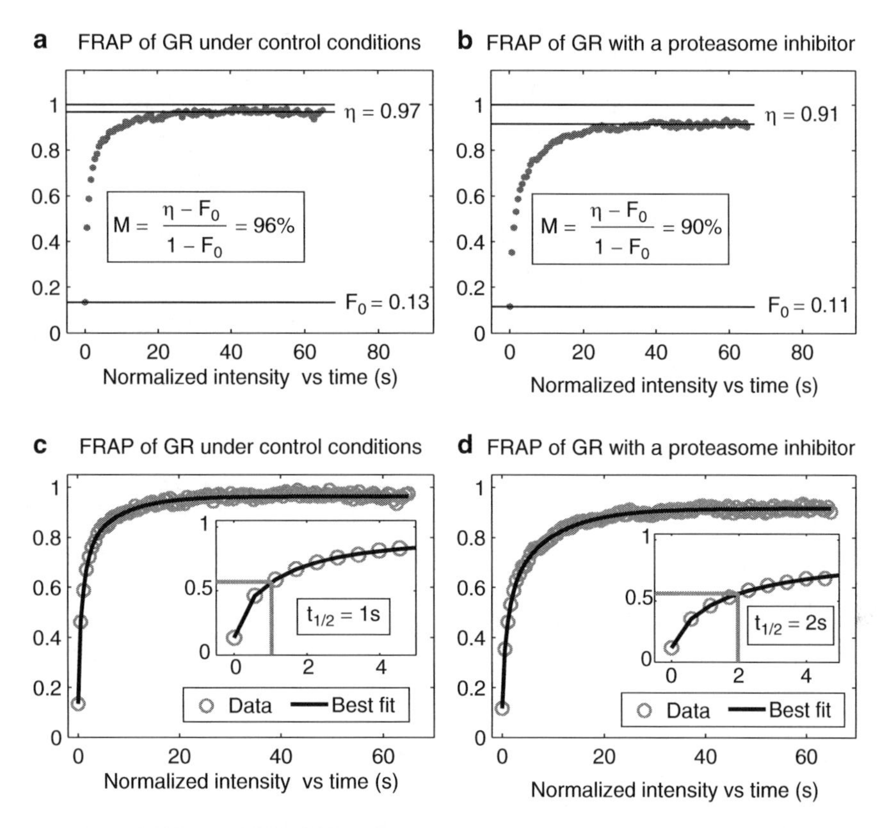

Fig. 3. Semi-quantitative FRAP. (**a**, **c**) FRAP of GR under control conditions (*gray circles*). The mobile fraction of $M=96\%$ is determined from the starting point of the FRAP curve and the averaged intensity once the FRAP shows a plateau. For estimation of $t_{1/2}$, the FRAP is first fit with an exponential function (*black line*). Then $t_{1/2}$ is estimated from this model fit (see inset, $t_{1/2}=1$ s). (**b**, **d**) FRAP of GR after treating cells with a proteasome inhibitor. Treatment leads to an increased immobilization (only 90% mobile fraction) and a slowdown of the FRAP ($t_{1/2}=2$ s). This has been interpreted as an increased GR residence time on the DNA after proteasome treatment (8).

When applying this measurement, be certain that the destruction of fluorescence caused by the intentional photobleach is accounted for in the correction for observational photobleaching. This is the case for the qualitative FRAP protocol described above, but not for the quantitative FRAP protocol described below (see Note 13)

Example: GR exhibits a larger immobile fraction in the presence of the proteasome inhibitor (Fig. 3a, b).

2. Measuring the recovery half-time. The recovery half-time is the time required for the FRAP curve to reach 50% of its final value minus its starting value. It can be measured either by eye, or by fitting a sum of exponentials to the FRAP data. We recommend using the exponential fit since it allows describing a large number of FRAP curves automatically, and it works more reliably for curves with fast recoveries. Furthermore, this method can also be easily adapted to measure other times for

recovery to different final fractions (e.g., 75% instead of 50%), which can be useful to describe FRAP curves with recoveries on different timescales.

Use the following bi-exponential function to fit the FRAP data (which should already be normalized to one):

$$\mathrm{FRAP}(t) = \eta - A\exp(-t/k_1) - B\exp(-t/k_2), \qquad (3)$$

where η is the measured final recovery level from step 1 above, and A, B, k_1 and k_2 are parameters to be estimated from the fit. Once the four free parameters are estimated, then $t_{1/2}$ is obtained by solving the equation:

$$0.5 \times (A + B) = \eta - A\exp(-t_{1/2}/k_1) - B\exp(-t_{1/2}/k_2) \qquad (4)$$

This can be done numerically with the Matlab function fzero.

Note that the estimated half-times are useful numbers for semi-quantitative comparison, but they do not directly reflect the diffusion rate or the binding rate or the residence time of the protein under study.

Example: The GR half-time increases from $t_{1/2} = 1$ s to $t_{1/2} = 2$ s after proteasome inhibition (Fig. 3c, d).

3.3. Protocol for Quantitative FRAP

Quantitative FRAP allows estimation of a protein's diffusion rate (D_{prot}) and its association (k^*_{on}) and dissociation (k_{off}) rates of binding. The binding estimates can be used to calculate several biologically relevant parameters, namely the time to diffuse from one binding site to the next ($1/k^*_{\mathrm{on}}$), the residence time at a binding site ($1/k_{\mathrm{off}}$), the bound fraction $\left(k^*_{\mathrm{on}}/\left(k^*_{\mathrm{on}}+k_{\mathrm{off}}\right)\right)$ and the free fraction $\left(k_{\mathrm{off}}/\left(k^*_{\mathrm{on}}+k_{\mathrm{off}}\right)\right)$.

Binding estimates are obtained by fitting the experimental data with a kinetic model that describes the FRAP recovery (1, 2). In recent years several different kinetic FRAP models have been developed for a variety of different proteins and experimental situations. For a comprehensive list of different modeling approaches see ref. 9.

This section will focus on a description of the approach developed in our group, which is suitable for analyzing FRAP data from nuclear proteins that are binding to chromatin (10). The nuclear protein distribution should be relatively homogeneous and the nucleus should be at least 5 μm in diameter such that a 1–2 μm diameter bleach spot can be used. (With smaller bleach spots, it is difficult to accurately measure the photobleaching profile, which is a critical for performing the quantitative analysis).

It is important to recognize that uncertainties remain about the accuracy of in vivo binding estimates obtained from FRAP kinetic modeling (9). Thus, cross validation with other approaches is important, including by fluorescence correlation spectroscopy (see Chapter 12).

3.3.1. Microscope Set-up and Data Collection

Perform as described above in Subheadings 3.1.1–3.1.3 for qualitative FRAP. However, in Subheading 3.1.1, step 3, it is critical for quantitative FRAP that images of the entire nucleus be recorded, since the entire nuclear fluorescence intensity is used to correct for observational photobleaching, rather than just a sub-region as in qualitative FRAP. It is also critical for the mathematical model described here that the bleach geometry in step 4, Subheading 3.1.2 is circular.

3.3.2. Process and Fit the FRAP Data

The entire workflow below (steps 1–7, 9, 11, 12) has been implemented in Matlab and the scripts are available on request. The basic processing steps (steps 1–6) could also be performed in ImageJ, and the model equations in Subheading 4 could be implemented in other software packages, such as Mathematica, which allow parameter estimation.

1. Extraction of time-stamps. Perform as described in Subheading 3.1.4, step 1 in qualitative FRAP.
2. Measurement and subtraction of background intensity. Measure background as described in Subheading 3.1.4, step 2 in qualitative FRAP. Then for quantitative FRAP, subtract the averaged background value from all recorded images. This is necessary because the intensity profile across the circular bleach spot must be measured in quantitative FRAP.
3. Measurement and correction for observational photobleaching. Measure the total nuclear fluorescence F at each time point before and after the photobleach. To correct in the pre-bleach phase, multiply the image at each time point t in the pre-bleach phase by the scaling factor $F(t)/F(t_{-1})$, where $F(t_{-1})$ is the total nuclear fluorescence in the last pre-bleach image. To correct in the post-bleach phase, multiply the image at each time point t in the post-bleach phase by the scaling factor $F(t)/F(t_0)$, where $F(t_0)$ is the total nuclear fluorescence in the first image after the photobleach.
4. Measurement of fluorescence recovery. Perform as described in Subheading 3.1.4, step 4 in qualitative FRAP. However, a circular region must be used for the measurement and the radius of the circle R_M should be recorded since it is one of the parameters needed for the kinetic model.
5. Generate the FRAP curve for fitting. The measurement of fluorescence recovery in step 4 produces a FRAP curve that has already been corrected for background (step 2) and observational photobleaching (step 3). Normalize this curve to one by following the same normalization procedure described in step 5, Subheading 3.1.4. Note that the final curve will not recover to one since the correction method only accounts for loss of fluorescence due to imaging, but does not account for

the fluorescence destroyed by the intentional photobleach (see Note 13).

Finally, resample the FRAP data such that the slower part of the FRAP recovery will not be overrepresented in the fitting procedure (11). To do so, identify a time point at which the FRAP curve begins to flatten out. For explanatory purposes here, we will assume that this turning point in the FRAP curve occurs at the 25th time point, but in general the turning point depends on the particular protein under study and the temporal sampling selected for the FRAP. Resample the FRAP curve beyond this time point by first computing the difference between the first two resampled time points on a logarithmic time scale: $\Delta = \ln(t_{27}) - \ln(t_{26})$. Then define a new set of time points: $t_i = t_1, t_2, t_3, \ldots, t_{25}, t_{25} + 10^{\Delta}, t_{25} + 10^{2\Delta}, \ldots$. Use these time points to generate a new FRAP data set $(t_i, \mathrm{FRAP}(t_i))$, where the value of FRAP(t_i) is unchanged for t_1–t_{25}, while the value of FRAP(t_i) at the subsequent time points is given by the average of the original FRAP data that lie within the new time point interval. This data set will be used for fitting.

6. Determine the initial conditions. The FRAP model requires the intensity profile of the circular photobleach. This is the average intensity as a function of distance from the center of the photobleach, i.e., the radial intensity profile $I_0(r)$ (Fig. 4a). Use the images obtained in step 3 above. Measure the radial intensity profile from the center of the photobleach using the first image after the photobleach and divide this profile by the radial intensity profile centered at the same location but measured from the last image right before the photobleach. (This division normalizes out any variation in the intensity profile due to cellular structures). Note that these measurements are typically performed over a circle with a larger diameter than the bleach circle such that the fluorescence at the edge of the circle reaches the equilibrium value in the rest of the cell (Fig. 4a). The resultant profile should be fit with Eq. 6 (see Note 14) to yield a functional form for $I_0(r)$ (Fig. 4b). Although Eq. 6 provides a good fit for our circular photobleaches, any functional form can be used to fit $I_0(r)$.
7. Determine the effective size of the nucleus. The effective nuclear size is based on the requirement that the photobleach in the model nucleus destroys the same proportion of fluorescence as in the real nucleus. The effective size can be estimated by using the function $I_0(r)$ determined in step 6 above combined with the final recovery level η of the FRAP curve estimated as described in step 1 of Subheading 3.2.2. The values of η and $I_0(r)$ are substituted into Eq. 7 and then solved for the effective size of the model nucleus R_{Nuc} (see Note 15).

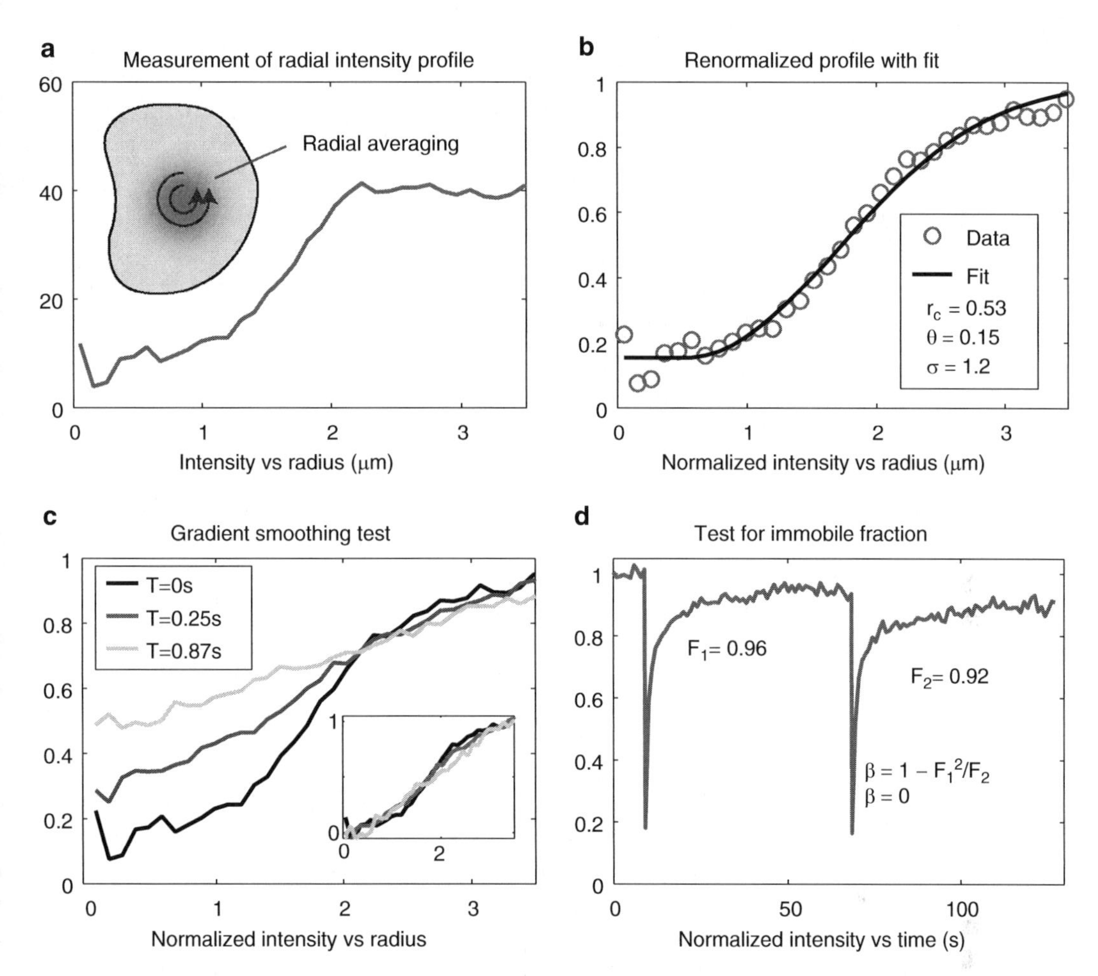

Fig. 4. Quantitative FRAP. (**a**) A plot of the radial profile of the photobleach is shown. The inset shows how the profile is measured by calculating the average intensity in annuli centered on the photobleach. (**b**) The profile is used to supply the initial conditions for the model equations. Shown is the profile from (**a**) normalized to the measured pre-bleach profile, and then fit using the function in Eq. 6. (**c**) To determine which version of the quantitative model to use, it is necessary to assess if diffusion plays a role in the recovery. This is done by comparing radial intensity profiles as a function of time, as shown here for GR. The inset shows the same profiles normalized between 0 and 1. The normalized profiles change their shape, indicating that diffusion should be incorporated into the FRAP model. (**d**) Before applying the quantitative model, it is necessary to test for an immobile fraction. The final recovery levels after the first and second bleach are $F_1 = 0.96$ and $F_1 = 0.92$. This yields an immobile fraction of $\beta = 1 - F_1^2 / F_2 = 0\%$, i.e., GR is fully mobile on the timescale of the FRAP experiment, and so the model in this protocol is applicable.

8. Test for the role of diffusion during the recovery. The FRAP data will be fit with different quantitative models depending on whether or not diffusion plays a role during the recovery. Two different tests for diffusion can be used, the bleach-spot size test (12) or the gradient-smoothing test (13), but we recommend the latter because it is more sensitive (unpublished data).

To perform the gradient smoothing test, determine intensity profiles across the bleach spot at all time points using the same procedure as described in step 6 above. Then renormalize the intensity profiles from 0 to 1 at each time point, and plot them on the same graph. If the profiles systematically change their shape over time, then diffusion plays a role during the FRAP recovery.

Example: The changing profile shapes for a GR FRAP show that diffusion plays a role in this FRAP recovery (Fig. 4c).

If the test for diffusion shows that diffusion does not play a role during the FRAP recovery, then proceed to step 9, otherwise proceed to step 10.

9. If the preceding test in step 8 showed that diffusion does not play a role during the FRAP recovery, then the FRAP curve obtained in step 5 should be fit with Eq. 8 with $N = 1$, which is a model for binding to a single type of binding site (see Note 16). If this yields a good fit, then an estimate of the binding off rate is given by the value of k_{off} obtained in the fit.

   If Eq. 8 with $N = 1$ does not yield a good fit, then fit with $N = 2$, which is a model for binding to two distinct types of binding sites. If this gives a good fit, then the values of $k_{off,1}$ and $k_{off,2}$ obtained from the fit provide estimates of the two distinct binding off rates. If this two-binding state model also fails to fit, then it is possible that there are three binding states present, and a fit with $N = 3$ could be tried. However, caution is advised as a three-binding state model uses six free parameters, and so the predictive power of such a fit becomes questionable.

   In any event, the fitting procedure is now complete and the subsequent steps can be ignored since they apply only to curves where diffusion plays a role in the recovery.

10. Test for an immobile fraction. If the test in step 8 showed that diffusion plays a role in the FRAP recovery, then before applying the kinetic models that account for diffusion, it is also necessary to test if the FRAP recovery has an immobile fraction. This can be done by performing two consecutive FRAP experiments in the same region of the nucleus (14). The first bleach affects both mobile and – if present – immobile proteins, whereas the second bleach affects only mobile proteins. By comparing the final recovery levels of the two FRAP curves, the immobile fraction β can be calculated by

$$\beta = 1 - F_1^2 / F_2, \tag{5}$$

where F_1 and F_2 are the final, plateaued recovery values reached after the first and second photobleach, respectively. If there is an immobile fraction, then β will be greater than zero, and the kinetic models described below should not be used.

To perform the immobile fraction test follow the same protocol as for quantitative FRAP with the following changes. (a) Use multiple bleach iterations to achieve a complete photobleach (five iterations are usually sufficient). (b) Repeat the same FRAP at the same location immediately after the first FRAP experiment is completed.

Example: This test shows that GR has no immobile fraction (Fig. 4d).

11. Fit with a model for diffusion. If there is no immobile fraction (step 10) and diffusion plays a role in the FRAP recovery (step 8), then the FRAP data should be fit with Eq. 9, which is a model that accounts only for diffusion (see Note 17).

    The fit with Eq. 9 requires the initial profile $I_0(r)$ measured in step 6, the effective size of the nucleus R_{Nuc} determined in step 7, the radius of the measurement region R_M noted in step 4, and the FRAP data generated in step 5. If the FRAP data are well fit with Eq. 9, then the fit yields an estimate of the fusion protein's diffusion constant (D_{prot}).

    If this diffusion constant is slow enough, then it likely contains information about binding, allowing an estimate of the ratio of the protein's association and dissociation rates $\left(k_{\mathrm{on}}^{*} / k_{\mathrm{off}}\right)$. To determine if this is possible, compare the fusion protein's diffusion constant with a freely diffusing protein, such as unconjugated GFP. This can be done by performing a quantitative FRAP of unconjugated GFP and fitting with the same diffusion model to estimate the GFP diffusion constant (D_{GFP}) (Fig. 5a). Then use this to estimate the mass of the fusion protein in kD (M_{prot}) according to the following formula: (12) which assumes roughly spherical proteins, and uses 27 kD as the mass of GFP

$$M_{\mathrm{prot}} = 27(D_{\mathrm{GFP}} / D_{\mathrm{prot}})^3 ,$$

    If the estimated mass M_{prot} is much larger than would be expected for the fusion protein (or a molecular complex containing the protein), then the diffusion constant for the fusion protein contains information about DNA binding, which retards the free diffusion of the fusion protein. This process is called effective diffusion (2, 12), and the diffusion constant measured for the fusion protein is called the effective diffusion constant. The following formula can then be used to estimate the ratio of the association to the dissociation rate of binding of the fusion protein: $\left(k_{\mathrm{on}}^{*} / k_{\mathrm{off}}\right) = \left(D_{\mathrm{GFP}} / D_{\mathrm{prot}}\right) - 1$. Please note that when effective diffusion occurs, it is impossible to obtain independent estimates of the association and dissociation rates from the FRAP data. Such independent estimates might be obtained by using FRAP with a smaller bleach-spot size (12) or by using fluorescence correlation spectroscopy (15) or temporal image correlation spectroscopy (16) (see also Chapter 12).

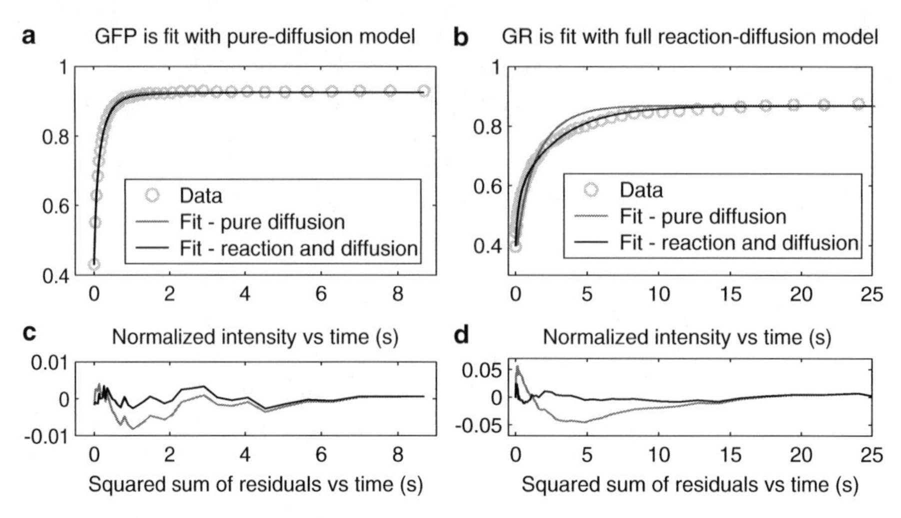

Fig. 5. Fits of FRAP models to unconjugated GFP and GR. (**a**) Quantitative FRAP of unconjugated GFP fused to a nuclear localization signal. The data are well fit by a diffusion model (Eq. 9) yielding an estimate for the diffusion constant of GFP in the nucleus ($D_{GFP} = 18.8\ \mu m^2/s$). When the same data are fit with a full reaction diffusion model (Eq. 10), the fit is only slightly better (compare the squared sum of residual curves below) and the estimates are very similar ($D_{GFP} = 20.0\ \mu m^2/s$ with negligible bound fraction). (**b**) Quantitative FRAP of GR. In this case, the diffusion model (Eq. 9) is a poor fit to the data, while the full reaction diffusion model yields a good fit ($D_{prot} = 6.3\ \mu m^2/s$, $k_{on}^* = 0.44\ s^{-1}$, $k_{off} = 0.38\ s^{-1}$).

12. Fit with a reaction-diffusion model. If the fit in step 11 is not good, then the FRAP data should be fit with a model that explicitly incorporates the binding terms, namely Eq. 10 (see Note 18). As in step 11, the initial profile $I_0(r)$ from step 6 above, the measurement radius R_M and the effective size of the nucleus R_{Nuc} from step 7 above must also be input into Eq. 10. If this equation yields a good fit to the FRAP data (Fig. 5b), then independent estimates for the association k_{on}^* and dissociation rates k_{off} can be obtained. If this fit fails, it is possible that the FRAP data reflect two distinct binding states. It is a straightforward mathematical problem to extend the one-binding state model described in Mueller et al. (10) to account for two distinct binding sites, but this has not been done.

4. Notes

1. Different experiments may require their own set of optimized time-lapse imaging parameters tuned to the protein under study and the microscope in use.
2. The parameters of the objective lens that are relevant for FRAP are the magnification and the NA. Higher magnification corresponds to a smaller field of view. Higher NA provides better light collection efficiency and better bleaching efficiency, but the bleach profile along the optical (z) axis becomes more

conical. This could lead to inaccurate estimates in quantitative FRAP where a cylindrical photobleach is presumed (17).

3. The pixel size is the number of microns per pixel in the image. It is affected by the NA of the objective and the zoom factor set in the confocal software. The smallest pixel size necessary is 0.1 μm, which is half of the best resolution that can be achieved by conventional light microscopy (~0.2 μm using a 1.4 NA objective). The confocal zoom can be adjusted to achieve the 0.1-μm pixel size. Larger pixel sizes can be used if the structure or region of interest is larger.
4. The whole nucleus should be imaged whenever possible because this simplifies the correction procedures. However, if the protein recovers quickly, only a sub-region of the nucleus should be imaged to enable faster data collection. The sub-region should be large enough to include a non-bleached region that can be used for correction of observational photobleaching. On the other hand, if the FRAP recovery is slow and cells move during the recovery process, then a larger imaged area should be used. This allows for correction for cell movement before processing the data (see Note 10).
5. The confocal pinhole sets the thickness of the optical section from which the recovery data are collected. Thicker optical sections (more open pinholes) are desirable for FRAP measurements, since they include more signals and thus improve the signal-to-noise ratio. Note, however, that the confocal pinhole affects only the detection process, but not the pattern of the photobleach, which always extends along the optical (z) axis in a conical profile.
6. On most confocals, the settings for gain and offset can be checked by changing the color palette to a "range indicator." Pixels with intensity of 0 are displayed as one color (blue in the Zeiss LSM 5 Live software) and pixels with intensity of 255 (for 8 bit images) are displayed as another color (red in the Zeiss LSM 5 Live software). Optimal settings are obtained when very few pixels are either red or blue.
7. In selecting the number of frames to collect after the bleach, the goal is to achieve frequent sampling of the FRAP recovery, but at the same time limit observational photobleaching. If a FRAP has a very fast initial phase followed by a very slow second phase, two separate FRAP experiments can be performed: one optimized for the fast initial part and one optimized for the slow second part. However, do not change the temporal sampling rate during the recovery measurement as this introduces artifactual reversible photobleaching (18).
8. Proteins move during the photobleach (19). To minimize this effect, bleach as fast as possible (use the highest scan speed) with as few iterations of the bleach as possible. Increase the

number of bleach iterations only if the initial value of the normalized FRAP recovery (the bleach depth) is larger than 0.5 (see Fig. 1b for an illustration of the bleach depth).

9. There are advantages and disadvantages to different bleach geometries. Circles can be more easily positioned to avoid inhomogeneous distributions of fluorescence, but the noise levels may be higher because of the small circular area (10, 20). Strips increase the measurement area and so reduce noise levels, but are more difficult to position and still avoid inhomogeneous distributions of fluorescence (1, 14, 21). Half-nuclear FRAP reduces the signal-to-noise ratio but takes longer and cannot be used for smaller structures (13, 22). In quantitative FRAP protocols, the bleach geometry is usually predefined. The protocol described here requires a circular bleach, whereas other quantitative FRAP protocols make use of strip bleaches (1, 14, 21) or half-nuclear bleaches (13).
10. Movement of the bleached region due to cell migration will produce an artifactual fluorescence recovery if measurements are made in a fixed location. To track the bleached region, use image registration methods such as the stackreg plugin with rigid body registration in ImageJ.
11. There are two quality checks to determine if background has been correctly measured. First, check if background intensities are constant over time. A background signal that decays over time indicates that some residual fluorescence has been measured (e.g., from the cytoplasm). Second, check if background intensities remain constant on a given day for the same imaging conditions. A background intensity that is much higher than other background intensities measured on the same day indicates again that some residual fluorescence has been measured.
12. Proper correction for observational photobleaching requires correct placement of the correction region. The fluorescence intensity in the correction region should be comparable to the pre-bleach intensity of the bleached region. Further, the correction region should not be positioned too close to the bleached region, i.e., the correction measurement should not show any FRAP.
13. Fluorescence is lost in a FRAP experiment due to both intentional photobleaching and observational photobleaching. When a FRAP curve is corrected for both of these effects, then it will recover to one (provided the protein has no immobile fraction). When a FRAP curve is corrected only for observational photobleaching, then it will never recover to one even if the protein has no immobile fraction. This is because the destruction of fluorescence produced by the intentional photobleach is not accounted for, so the final recovery level is

automatically reduced by the fraction of fluorescence destroyed by the intentional photobleach.

The qualitative FRAP protocol incorporates a correction procedure that corrects for the loss of fluorescence from both intentional photobleaching and observational photobleaching, whereas the quantitative protocol incorporates a correction procedure that accounts only for the loss of fluorescence due to observational photobleaching. In qualitative FRAP the correction is measured in a separate control region away from the FRAP. The loss of fluorescence in this region is due both to observational photobleaching and the influx of bleached fluorophores from the bleached region. Thus, this correction curve accounts for both sources of fluorescence loss, and so the corrected FRAP curve will recover to one for fully mobile proteins. In quantitative FRAP, the correction procedure for the post-bleach phase is designed to keep the total nuclear fluorescence constant relative to the first image after the photobleach. Since the fluorescence loss from the photobleach has already occurred in this first image after the photobleach, this procedure does not account for this loss and so the FRAP curve will not recover to one even for fully mobile proteins.

14. The initial conditions for a circular photobleach are typically well fit by the following equation, which describes a constant function with Gaussian flanks (10). The values for θ, r_C, and σ can be estimated by fitting Eq. 6 to the measured intensity profile $I_0(r)$

$$I_0(r) = \begin{cases} \theta & \text{for } r \le r_c \\ 1 - (1-\theta)\exp\left(-\dfrac{(r-r_c)^2}{2\sigma^2}\right) & \text{for } r > r_c \end{cases}. \quad (6)$$

15. The kinetic model assumes that the nucleus is circular and that its fluorescence is conserved. The real nucleus is a three dimensional ellipsoid and often contains large regions (nucleoli) that are mostly devoid of fluorescence. Nevertheless, the model nucleus can be a reasonable approximation to the real nucleus if its size is set such that the photobleach destroys the same proportion of fluorescence as in the real nucleus.

    The size of the model nucleus can be estimated from two measurable parameters: the profile of the photobleach, $I_0(r)$ (step 6. Subheading 3.3.2) and the final recovery level of the FRAP, η (step 1, Subheading 3.2.2) (10). Since intensities are normalized to one, the amount of fluorescence before the photobleach, F_B, equals the area of the nucleus, whereas the amount of fluorescence after the photobleach, F_A, equals the integral of

$I_0(r)$ over the nuclear area. These parameters are therefore related by

$$\frac{F_A}{F_B} = \eta = \frac{2\pi \int_0^{R_{\mathrm{Nuc}}} I_0(r) r\,dr}{\pi R_{\mathrm{Nuc}}^2}, \tag{7}$$

where R_{Nuc} is the radius of the model nucleus. Substituting the measured values for η and $I_0(r)$ allows solving the equation for R_{Nuc}. If $I_0(r)$ is given by Eq. 6, then Eq. 7 can be solved analytically for R_{Nuc}. For more complex forms of $I_0(r)$, Eq. 7 can be solved numerically for R_{nuc}.

16. When diffusion is very fast compared both to the time to associate with a binding site and to the timescale of the FRAP recovery, then the FRAP behavior is termed reaction dominant (12). The gradient smoothing test will show no change in the shape of the recovery profiles. In this case, diffusion can be neglected in the model and the recovery is solely determined by the binding rates, more specifically by the off-rate(s). The FRAP can be described by a simple equation

$$\mathrm{FRAP}(t) = \eta - \sum_{i=1}^{N} A_i \exp(-t / k_{\mathrm{off},i}), \tag{8}$$

where η is set to the final recovery level of the FRAP (step 1, Subheading 3.2.2) and N, which denotes the number of different binding sites, is set to one or two. The fit yields estimates for the dissociation rates $k_{\mathrm{off},i}$ and the A_i's. The A_i's yield the relative fraction sizes of the two bound fractions, but they do not provide estimates of absolute bound fraction sizes. This requires knowledge of the size of the freely diffusible fraction, which is difficult to estimate because it recovers so rapidly (10).

17. The equation describing the FRAP recovery for a molecule that diffuses with a diffusion constant D_{prot} is (10):

$$\mathrm{FRAP}(t) = U_0 \exp(-D_{\mathrm{prot}} \alpha_0^2 t) + \sum_{k=1}^{\infty} U_k \exp(-D_{\mathrm{prot}} \alpha_k^2 t) \left[2 J_1(\alpha_k R_M) / (\alpha_k R_M)\right] \tag{9}$$

where J_1 is a Bessel function $\alpha_k = \chi_k / R_{\mathrm{Nuc}}$ with χ_k the kth zero of the Bessel function of the first kind, and the constants U_k are given by:

$$U_k = \frac{2}{R_{\mathrm{Nuc}}^2 J_0^2(\chi_k)} \int_0^{R_{\mathrm{Nuc}}} I_0(r) J_0(\alpha_k r) r\,dr,$$

R_{Nuc} is the radius of the nucleus (see Note 15), R_M is the radius of the measurement region of the FRAP curve (noted in step 4, Subheading 3.3.2), and $I_0(r)$ is the initial profile of the photobleach (step 6 in Subheading 3.3.2).

18. The full reaction diffusion solution, incorporating terms for both diffusion and binding, is given by (10):

$$\mathrm{frap}(t) = \left[(U_0 + W_0)\exp(-(w_0 + v_0)t) + (V_0 + X_0)\exp(-(w_0 - v_0)t)\right] + \sum_{k=1}^{\infty}\left[(U_k + W_k)\exp(-(w_k + v_k)t) + (V_k + X_k)\exp(-(w_k - v_k)t)\right] \left[2J_1(\alpha_k R_M)/(\alpha_k R_M)\right] \quad (10)$$

with the constants given by:

$$w_k = \frac{1}{2}\left(D_{\mathrm{prot}}\alpha_k^2 + k_{\mathrm{on}}^* + k_{\mathrm{off}}\right) \text{ and}$$

$$v_k = \sqrt{\frac{1}{4}\left(D_{\mathrm{prot}}\alpha_k^2 + k_{\mathrm{on}}^* + k_{\mathrm{off}}\right)^2 - k_{\mathrm{off}} D_{\mathrm{prot}}\alpha_k^2}$$

$$U_k = \frac{1}{-2k_{\mathrm{off}} v_k}\left[(-w_k - v_k + k_{\mathrm{off}})(w_k - v_k)\right] \frac{2\left(\frac{k_{\mathrm{off}}}{k_{\mathrm{off}} + k_{\mathrm{on}}^*}\right)}{R_{\mathrm{Nuc}}^2 J_0^2(\chi_k)} \int_0^{R_{\mathrm{Nuc}}} I_0(r) J_0(\alpha_k r) r\,dr$$

$$V_k = \frac{1}{2k_{\mathrm{off}} v_k}\left[(-w_k + v_k + k_{\mathrm{off}})(w_k + v_k)\right] \frac{2\left(\frac{k_{\mathrm{off}}}{k_{\mathrm{off}} + k_{\mathrm{on}}^*}\right)}{R_{\mathrm{Nuc}}^2 J_0^2(\chi_k)} \int_0^{R_{\mathrm{Nuc}}} I_0(r) J_0(\alpha_k r) r\,dr$$

$$W_k = U_k \frac{k_{\mathrm{on}}^*}{-(w_k + v_k) + k_{\mathrm{off}}}, \quad X_k = V_k \frac{k_{\mathrm{on}}^*}{-(w_k - v_k) + k_{\mathrm{off}}},$$

The J's are Bessel functions, and $\alpha_k = \chi_k / R_N$ with χ_k the kth zero of the Bessel function of the first kind. R_{Nuc} is the radius of the model nucleus (see Note 15), R_M is the radius of the measurement region of the FRAP curve (step 4, Subheading 3.3.2), and $I_0(r)$ is the photobleach profile (step 6, Subheading 3.3.2). The series expansion can typically be truncated at 500 terms for fitting of most FRAP curves. The fit yields estimates for the unknown parameters, namely the protein's diffusion constant D_{prot} and the association k_{on}^* and dissociation rates k_{off} of binding.

References

1. Houtsmuller AB (2005) Fluorescence recovery after photobleaching: application to nuclear proteins. Adv Biochem Eng Biotechnol 95:177–199
2. Sprague BL, McNally JG (2005) FRAP analysis of binding: proper and fitting. Trends Cell Biol 15:84–91
3. Hager GL, McNally JG, Misteli T (2009) Transcription dynamics. Mol Cell 35:741–753
4. Sprouse RO, Karpova TS, Mueller F, Dasgupta A, McNally JG, Auble DT (2008) Regulation of TATA-binding protein dynamics in living yeast cells. Proc Natl Acad Sci USA 105: 13304–13308
5. Nishiyama A, Mochizuki K, Mueller F, Karpova T, McNally JG, Ozato K (2008) Intracellular delivery of acetyl-histone peptides inhibits native bromodomain-chromatin interactions and impairs mitotic progression. FEBS Lett 582:1501–1507
6. Phair RD, Misteli T (2001) Kinetic modelling approaches to in vivo imaging. Nat Rev Mol Cell Biol 2:898–907
7. Karpova TS, Chen TY, Sprague BL, McNally JG (2004) Dynamic interactions of a transcription factor with DNA are accelerated by a chromatin remodeller. EMBO Rep 5:1064–1070
8. Stavreva DA, Muller WG, Hager GL, Smith CL, McNally JG (2004) Rapid glucocorticoid receptor exchange at a promoter is coupled to transcription and regulated by chaperones and proteasomes. Mol Cell Biol 24:2682–2697
9. Mueller F, Mazza D, Stasevich TJ, McNally JG (2010) FRAP and kinetic modeling in the analysis of nuclear protein dynamics: what do we really know? Curr Opin Cell Biol 22: 403–411
10. Mueller F, Wach P, McNally JG (2008) Evidence for a common mode of transcription factor interaction with chromatin as revealed by improved quantitative fluorescence recovery after photobleaching. Biophys J 94:3323–3339
11. Waharte F, Brown CM, Coscoy S, Coudrier E, Amblard F (2005) A two-photon FRAP analysis of the cytoskeleton dynamics in the microvilli of intestinal cells. Biophys J 88:1467–1478
12. Sprague BL, Pego RL, Stavreva DA, McNally JG (2004) Analysis of binding reactions by fluorescence recovery after photobleaching. Biophys J 86:3473–3495
13. Beaudouin J, Mora-Bermudez F, Klee T, Daigle N, Ellenberg J (2006) Dissecting the contribution of diffusion and interactions to the mobility of nuclear proteins. Biophys J 90:1878–1894
14. Hinow P, Rogers CE, Barbieri CE, Pietenpol JA, Kenworthy AK, DiBenedetto E (2006) The DNA binding activity of p53 displays reaction-diffusion kinetics. Biophys J 91:330–342
15. Michelman-Ribeiro A, Mazza D, Rosales T, Stasevich TJ, Boukari H, Rishi V, Vinson C, Knutson JR, McNally JG (2009) Direct measurement of association and dissociation rates of DNA binding in live cells by fluorescence correlation spectroscopy. Biophys J 97:337–346
16. Stasevich TJ, Mueller F, Michelman-Ribeiro A, Rosales T, Knutson JR, McNally JG (2010) Cross-validating FRAP and FCS to quantify the impact of photobleaching on in vivo binding estimates. Biophys J 99:3093–3101
17. Mazza D, Cella F, Vicidomini G, Krol S, Diaspro A (2007) Role of three-dimensional bleach distribution in confocal and two-photon fluorescence recovery after photobleaching experiments. Appl Opt 46:7401–7411
18. Sinnecker D, Voigt P, Hellwig N, Schaefer M (2005) Reversible photobleaching of enhanced green fluorescent proteins. Biochemistry 44:7085–7094
19. Weiss M (2004) Challenges and artifacts in quantitative photobleaching experiments. Traffic 5:662–671
20. Kang M, Kenworthy AK (2008) A closed-form analytic expression for FRAP formula for the binding diffusion model. Biophys J 95:L13–L15
21. Carrero G, McDonald D, Crawford E, de Vries G, Hendzel MJ (2003) Using FRAP and mathematical modeling to determine the in vivo kinetics of nuclear proteins. Methods 29:14–28
22. Phair RD, Scaffidi P, Elbi C, Vecerova J, Dey A, Ozato K, Brown DT, Hager G, Bustin M, Misteli T (2004) Global nature of dynamic protein-chromatin interactions in vivo: three-dimensional genome scanning and dynamic interaction networks of chromatin proteins. Mol Cell Biol 24:6393–6402

Chapter 12

Monitoring Dynamic Binding of Chromatin Proteins In Vivo by Fluorescence Correlation Spectroscopy and Temporal Image Correlation Spectroscopy

Davide Mazza*, Timothy J. Stasevich*, Tatiana S. Karpova, and James G. McNally

Abstract

Live-cell microscopy has demonstrated that many nuclear proteins bind transiently to target sites in chromatin. These binding interactions can be detected and quantified by two related live-cell imaging techniques, Fluorescence Correlation Spectroscopy (FCS) and Temporal Image Correlation Spectroscopy (TICS). With proper quantitative modeling, it is possible to obtain estimates from FCS and TICS data of the association and dissociation rates of nuclear protein binding to chromatin. These binding rates permit calculating the fractions of free and bound protein in the nucleus, plus the time required to diffuse from one binding site to the next and the dwell time on a chromatin target. In this protocol, we summarize the underlying principles of FCS and TICS, and then describe how these data should be collected and analyzed to extract estimates of in vivo binding.

Key words: Fluorescence correlation spectroscopy, Temporal image correlation spectroscopy, Binding, Nuclear protein, In vivo

1. Introduction

1.1. Mobility of Nuclear Proteins

It is now well established that many nuclear proteins bind only transiently to their chromatin sites (1). Recent work has focused on quantifying these transient interactions to obtain estimates of the association and dissociation rates of chromatin binding in live cells (2). The association rate measures how fast the protein finds a chromatin target site. Its inverse yields the time it takes the protein to move from one binding site to the next. The dissociation rate

*equal contributions

Randall H. Morse (ed.), *Chromatin Remodeling: Methods and Protocols*, Methods in Molecular Biology, vol. 833,
DOI 10.1007/978-1-61779-477-3_12,

measures how fast a protein leaves a target site. Its inverse yields the residence time of the protein at the binding site.

Most of the procedures to extract estimates of association and dissociation rates have come from quantitative analysis of fluorescence recovery after photobleaching (FRAP) data (see Chapter 11). However, the accuracy of the quantitative FRAP procedures remains uncertain because in vivo binding estimates are relatively new and so gold standards do not exist (3). These will come from cross validation with other techniques.

Fluorescence correlation spectroscopy (FCS) and its cousin Temporal image correlation spectroscopy (TICS) provide complementary approaches to FRAP. Until now binding analysis has not been the primary focus of FCS and TICS, which instead have been applied predominantly to measure diffusion in living specimens (for a detailed protocol on FCS measurements of diffusion in the nuclear environment see ref. 4). This has changed, however, with the recent development of FCS and TICS procedures to specifically estimate the association and dissociation rates of proteins bound to chromatin (5, 6). These new approaches are outlined in the protocols below.

1.2. Fluorescence Correlation Spectroscopy and Temporal Image Correlation Spectroscopy in the Nucleus

FCS was developed in the mid 1970s to quantify the local mobility of fluorescent molecules (7). Although the technique was limited to specialists for many years, its popularity has rapidly grown with the ability to tag proteins with GFP. Biologists now regularly perform FCS on a variety of commercial instruments to measure molecular dynamics in living samples.

In conventional FCS, the fluorescence signal collected from a diffraction limited volume is recorded over time (see Fig. 1). As fluorescent molecules move in and out of the volume, the measured fluorescence fluctuates. The more slowly the molecules move, the longer the measured fluorescence will remain correlated with itself at an earlier time. Thus, the probability a molecule remains within the volume for a total time τ can be estimated by calculating the autocorrelation $G(\tau)$ of the fluorescence fluctuations. In general, $G(\tau)$ is maximal at $\tau=0$ and eventually goes to zero because mobile molecules cannot reside within the measurement volume indefinitely (see Fig. 2). By fitting $G(\tau)$ with a kinetic model, it is possible to quantify molecular diffusion and flow rates as well as binding times.

FCS is well suited for measuring live-cell dynamics because it is relatively non-invasive. For one, FCS works best at low concentrations when fluctuations are relatively large. Thus, only a small number of molecules need to be fluorescently labeled. Adequate concentrations can be as low as a few nanomolar. In comparison, FRAP generally requires higher concentrations, on the order of 1 μM. In addition, FCS does not require a strong intentional photobleach like FRAP does. Of the FCS variants, two-photon FCS is generally considered to be the least invasive because longer wavelength

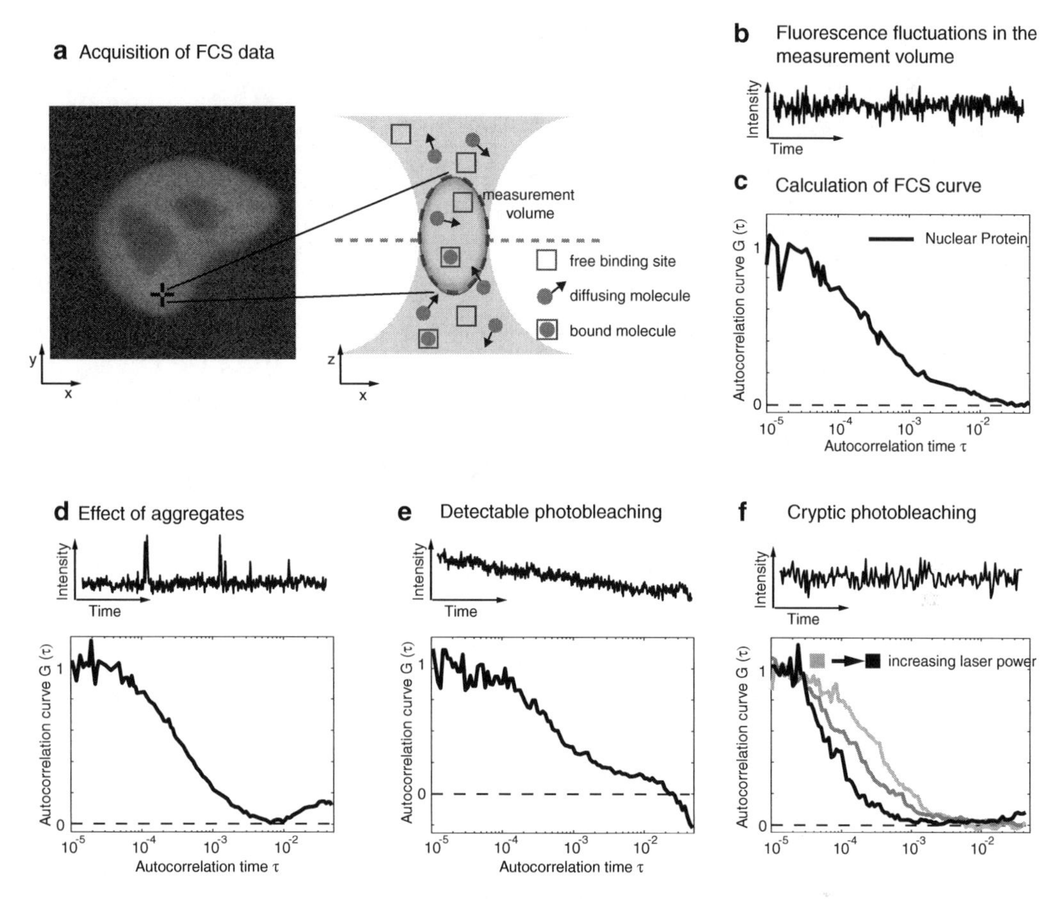

Fig. 1. Principles of FCS. (**a**) FCS measurements are performed by parking a focused laser beam at a selected position within the sample of interest. This generates a femtoliter-sized measurement volume (*dashed ellipse*), which permits monitoring the behavior of a few molecules at a time with a very high temporal resolution. The measurement is then performed by determining the fluorescence intensity within the volume over time (**b**) and by computing the autocorrelation curve of the measured fluctuations (**c**). If large aggregates cross the observation volume, the fluorescence time-course will show abnormally high spikes (**d**) and the autocorrelation curve may not decay to zero for long autocorrelation times. Photobleaching can also corrupt the autocorrelation curve. If the photobleaching is severe, it can cause the raw intensity fluctuations to decay with time (**e**). If the photobleaching is less severe, it may be balanced by an influx of unbleached fluorophores, in which case the intensity flucutations do not decay. Such cryptic photobleaching can only be detected by repeating measurements at lower laser powers to see if the autocorrelation curve decays less quickly (**f**).

excitation light can be used and excitation is localized to the measurement region alone.

TICS is the imaging analog of FCS (8). In TICS, FCS is performed at all pixels within an image sequence, so dynamics at multiple locations within a sample can easily be compared and high resolution maps of molecular mobility can quickly be constructed (see Fig. 3). The main limitation of TICS is a loss of temporal resolution since larger imaged regions take more time to acquire. This loss can partially be compensated for by using microscopes that image at multiple locations in parallel, for example spinning-disk (9) or line-scanning (10) microscopes. In addition, illumination in TICS is more uniform than in FCS, so photobleaching is also more

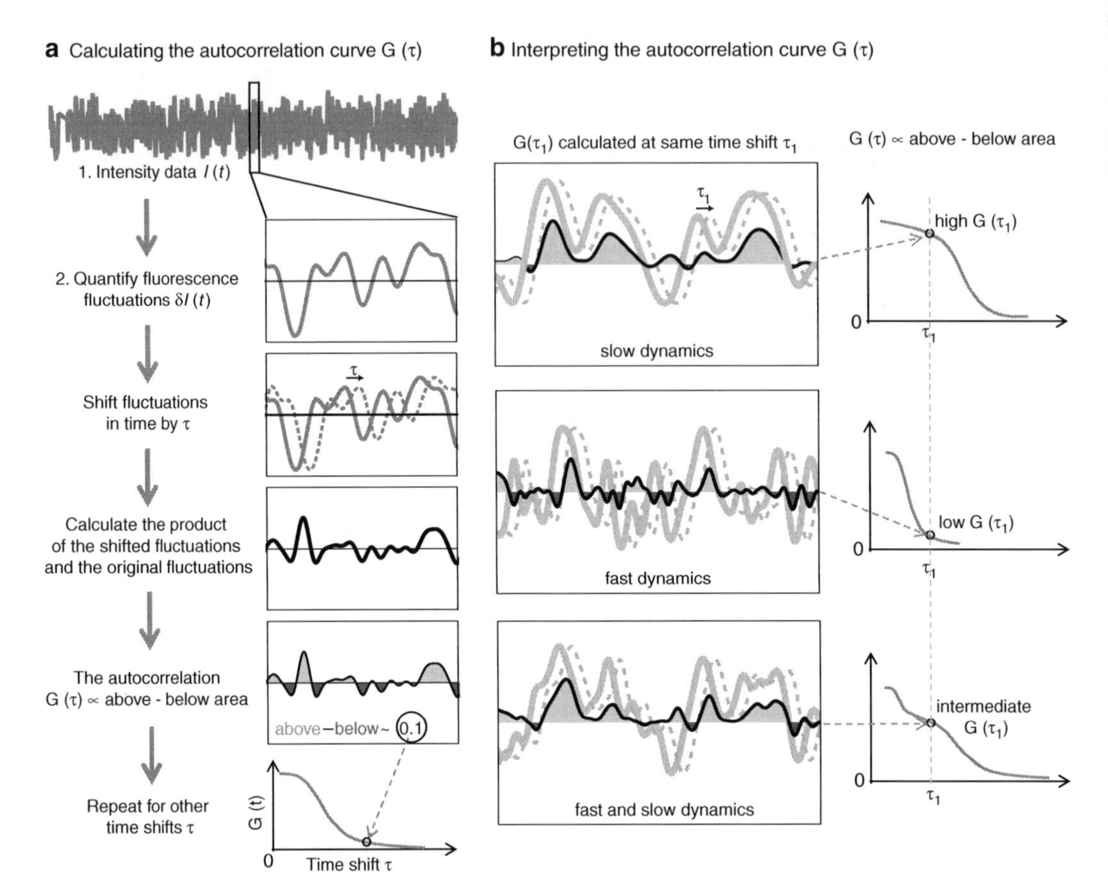

Fig. 2. Calculating the autocorrelation curve. (**a**) The scheme depicts the series of steps that are performed in order to calculate the autocorrelation of the FCS/TICS data. Starting from the measured time-course of the fluorescence intensity $I(t)$, the fluctuations $\delta I(t)$ around the mean value are calculated. The fluctuation time-series is then shifted in time by a factor τ to obtain $\delta I(t-\tau)$ and the product $\delta I(t)\,\delta I(t-\tau)$. The average of this product (which is proportional to the difference between the area above and below the curve) provides the autocorrelation of $\delta I(t)$ for a delay τ. By repeating this calculation for different τ, the entire autocorrelation curve can be constructed. (**b**) Effect of fast and slow dynamics on the autocorrelation curve. When the same time shift τ is applied to different data sets to calculate $G(\tau)$, the product $\delta I(t)\,\delta I(t-\tau)$ is greater than 0 more often for slower dynamics than it is for faster dynamics. Thus, the slower the dynamics, the slower the autocorrelation curve decays. A combination of fast and slow dynamics (for example, fast diffusion and slow binding) can therefore result in a biphasic autocorrelation curve.

uniform, making it easier to detect (i.e., less cryptic, see Fig. 1e, f) and correct for (6). For these reasons, TICS is generally preferable to FCS when measuring slower processes such as the binding of some nuclear proteins to chromatin (where chromatin residence times may be on the order of seconds or longer).

FCS and TICS can be used in a qualitative mode to compare molecular dynamics and assess changes in dynamics after various biological perturbations. The same methods can also be used in a quantitative mode by fitting the data with mathematical models to estimate the underlying molecular kinetics. So far, most of the models that have been applied are for pure or multicomponent diffusion (see Note 1) or flow. These popular models are often

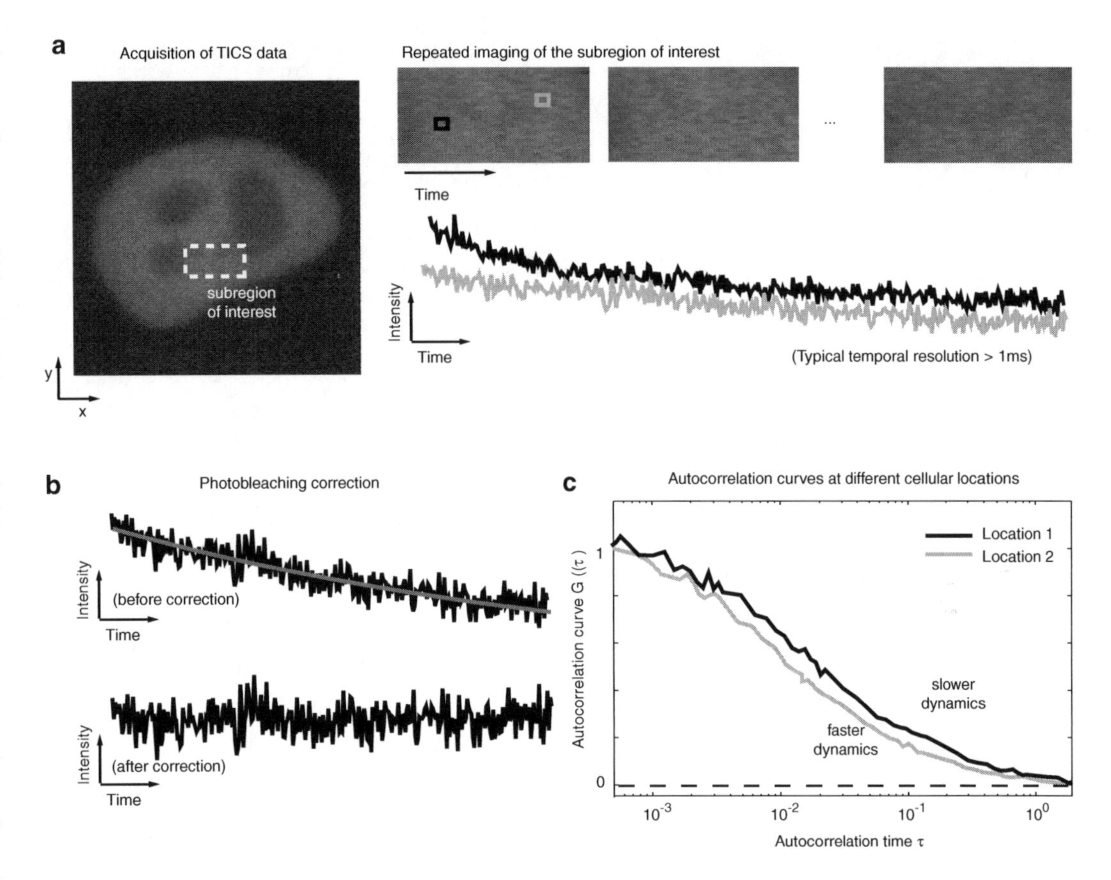

Fig. 3. Principles of TICS. (**a**) In TICS an extended subregion of the sample is imaged repeatedly over time. This makes it possible to compare protein dynamics at different locations by measuring the fluorescence intensity from different pixels of the image (*light and dark gray boxes in the figure*). A second advantage of TICS is that it is less subject to cryptic photobleaching than two-photon FCS and so better suited to measure longer residence times as can occur for binding to chromatin. (**b**) The large imaged area in TICS produces relatively uniform photobleaching that can usually be detected as a decay in the raw intensity fluctuations. This can be partially corrected for by fitting the fluorescence time series from each pixel with a mono- or bi-exponential function, and then re-normalizing the data by dividing it with the fitted exponential function. (**c**) By comparing the autocorrelation curves from different pixels, molecular dynamics can be mapped out in the imaged subregion.

inadequate for describing nuclear dynamics, however, because most nuclear proteins interact strongly with chromatin. To more accurately describe nuclear dynamics, we have recently developed a model that describes the diffusion and binding of nuclear proteins to immobile chromatin substrates (see Fig. 4 and Note 2) (5). The model is applicable to nuclear proteins that are distributed relatively homogeneously either throughout the nucleus or within nuclear substructures that are at least 1–2 μm in diameter.

In this protocol we describe how to perform FCS and TICS measurements in the cell nucleus using commercially available microscopes and describe how to fit the data to our reaction-diffusion model to estimate nuclear binding times.

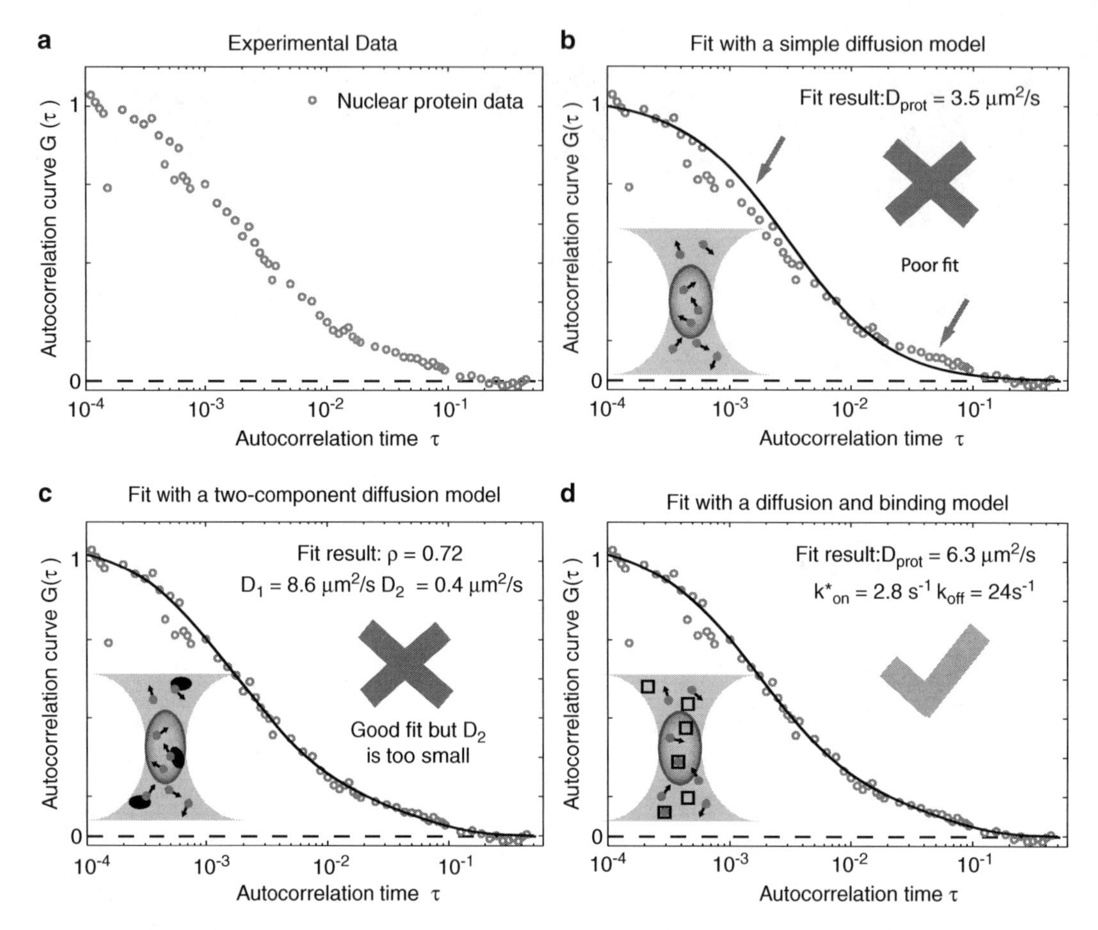

Fig. 4. Quantitative modeling of FCS/TICS. Quantitative analysis of binding by FCS/TICS is realized by fitting experimental FCS curves with a series of models of increasing complexity. Starting from the raw autocorrelation data (**a**), a one-component pure-diffusion model is applied (**b** – inset, full circles reflect molecules that simply diffuse), providing an estimate for the diffusion coefficient D_{prot}. If no satisfactory fit is found with this model, a two-component diffusion model can be fit (**c** – inset, full circles and ellipses reflect molecules that diffuse at fast and slow rates, respectively) to obtain estimates for two diffusion coefficients D_1 and D_2, together with the fraction of molecules ρ with diffusion rate equal to D_1. This may result in a good fit, but the resultant slower diffusion coefficient may be too low to be realistic (see Note 15). In this case, a reaction-diffusion fit is preferred (**d** – inset, full circles reflect molecules that simply diffuse and open squares reflect immobile binding sites for the same molecules). This model results in estimates for the diffusion coefficient and the binding association and dissociation rates k^*_{on} and k^*_{off}.

2. Materials

2.1. Sample Preparation for FCS and TICS

1. Cell chambers with a No. 1.5 coverslip at the bottom (e.g., LabTek Chambers, Nalge Nunc, Naperville, IL, USA) designed for fluorescence imaging on an inverted microscope, hereafter referred to as live-cell imaging chambers.
2. Calibration solutions diluted in distilled water, phosphate buffered saline (PBS: 137 mM NaCl, 2.7 mM KCl, 10 mM Na_2HPO_4O (dibasic, anhydrous), 2 mM KH_2PO_4O (monobasic,

anhydrous)), or glycerol, or mounted with Aquamount (Thermo Scientific).

(a) Purified eGFP.

(b) 100-nm/200-nm diameter Fluorescent Beads (e.g., Fluorospheres, Molecular Probes, Eugene, OR).

(c) Fluorescent dyes such as Alexa 488, rhodamine 6G or rhodamine 110.

3. Cells expressing fluorescently tagged proteins at appropriate concentrations for FCS and TICS (~10–300 nM) (see Note 3).

2.2. Instrumentation

2.2.1. Two-Photon FCS

Here we describe how to perform two-photon FCS on the ISS Alba 2 (ISS Inc. Champaign, IL, USA), although most of our discussion can be generalized to other systems. The ISS Alba 2 has a pair of sensitive avalanche photodiode detectors (APD) and a multitau-autocorrelation board for calculating the FCS autocorrelation curve. The fluorescence signal collected by the objective can be split on the two detectors by a 50/50 beam splitter. This can be used to assess if there are after-pulse effects (see Note 4). This is in general not a problem at the timescales of interest for nuclear proteins, and so the signal from one of the detectors can typically be used for the analysis described below. The following components are also included with the ISS Alba2:

1. An Infrared (IR) femtosecond pulsed laser source for two-photon excitation (Mai Tai, Spectra-Physics-Newport, Santa Barbara, CA) equipped with a neutral density filter wheel (Thorlabs, Newton, NJ, USA) to adjust the laser power.
2. An inverted microscope frame (Zeiss Axiovert 135 M, Zeiss Inc., Jena, Germany).
3. A 100× NA 1.3 oil immersion objective heated to 37°C.
4. An E700 SP 2P dichroic filter to eliminate detection of the exciting light (Chroma Technology, Bellows Falls, VT, USA).
5. A scanning stage to collect images of the specimen before and after the FCS experiment.
6. A piezo-electric device for z-control (Nano-F100, Mad City Labs, Madison, WI).
7. A laser power meter (Thorlabs, Newton, NJ, USA).

2.2.2. TICS

TICS can be performed on any confocal or two-photon microscope equipped with a sensitive detector capable of acquiring data without correlated noise (such noise produces artifacts in the autocorrelation function). Here we describe how to perform TICS on a Zeiss LSM 5 LIVE (Zeiss Inc.) with the following characteristics:

1. A line CCD detector, capable of imaging a line of pixels simultaneously.

2. A 63× NA 1.4 oil objective.
3. A 488 nm/100 mW diode laser with acousto-optical filters (AOTFs) for adjusting laser power.
4. An adjustable slit pinhole.
5. An environment chamber for keeping cells at 37°C with 5% CO_2.

2.3. Software for Analyzing FCS/TICS Data

1. Qualitative analysis: All commercially available FCS setups provide software packages to compute and visualize the autocorrelation of the data. For TICS, however, the raw data must be exported and manually autocorrelated (see Note 5). Our Matlab code for doing this is available upon request.
2. Quantitative Analysis: Most software packages for FCS allow users to fit autocorrelation curves with pure and multicomponent diffusion models, diffusion plus flow models, or anomalous diffusion models. However, no commercial software at the moment includes a reaction-diffusion model needed for quantifying binding. Our Matlab code for doing this is available upon request.
3. Microsoft Excel or another spreadsheet application to store autocorrelation data and analysis results.

3. Methods

The quantitative analysis of FCS and TICS data is quite similar although the data are acquired and preprocessed in different ways. We therefore describe data acquisition and preprocessing for each technique separately (Subheadings 3.1 and 3.2) and conclude with a common section describing the quantitative analysis (Subheading 3.3).

3.1. Data Acquisition for Two-Photon FCS

3.1.1. Calibration of the FCS Measurement Volume

In order to quantitatively analyze FCS data it is necessary to first calibrate the size of the FCS measurement volume (see Fig. 1a). This is critical because processes may appear more/less dynamic if the measurement volume is overestimated/underestimated. For example, when fitting the autocorrelation curve $G(\tau)$, the extracted diffusion coefficient is proportional to the axial area of the measurement volume (see Note 1, Eq. 2), so errors in the volume directly propagate to errors in diffusion estimates.

To good approximation, the measurement volume can be described by a three-dimensional Gaussian with extensions w_x, w_y, and w_z, with $w_x \sim w_y$. To calibrate the dimensions of this volume, FCS is performed on a solution containing a fluorescent dye with a known diffusion coefficient. By fitting the autocorrelation of these data with a pure-diffusion model (see Note 1, Eq. 1), estimates for w_x and w_z are obtained.

1. Select the fluorophore excitation wavelength. For two-photon excitation of Alexa 488 and GFP we use a wavelength of 970 nm. Ensure that the laser is correctly pulsing. (On our two-photon Alba laser this is indicated by the "mode locked" setting on the control panel).
2. Use a laser power meter positioned at the back aperture to measure laser power there. We use powers between 8 and 12 mW, although lower laser powers are preferable whenever possible.
3. Prepare a solution of a bright and photostable fluorescent probe, such as Alexa 488, diluted in sterilized H_2O to approximately 10 nM. Sonicate a 1-ml aliquot for a few minutes to dissociate potential aggregates and pipette into a live-cell imaging chamber. Mount the chamber on the microscope stage using the proper immersion medium (see Note 6).
4. Focus the objective well above the bottom of the imaging chamber to avoid artifacts due to diffusion near a surface. Sufficient fluorescence intensity should be measured by the APD. Photon counts from 10^2 to 10^5 per second are typical.
5. Measure the fluorescence intensity fluctuations for ~20 s using an imaging rate of 100 kHz. At laser powers between 8 and 12 mW, there should be little to no photobleaching.
6. Repeat step 5 five times at the same location. Preprocess and autocorrelate the data (see Subheading 3.1.3, steps 1–4). All autocorrelation curves should be similar and go smoothly to zero.
7. Fit the averaged autocorrelation curves (weighted by the standard deviation of each point) with a single-component pure-diffusion model (see Note 1, Eq. 1). When fitting, fix the diffusion coefficient to the expected value and allow w_x and w_z to change. For Alexa 488 diffusing in water at room temperature $D \sim 435\ \mu m^2/s$ (11). In our two-photon experiments we typically obtain $w_x \sim 0.35\ \mu m$, $w_z \sim 1.5\ \mu m$ using the 100× 1.4 NA oil immersion objective (5).
8. All autocorrelation curves should be well fit, and similar values for w_x and w_z should be obtained. If possible, simultaneous (global) fitting of all calibration curves should be used to determine optimal values for w_x and w_z. Otherwise, calculate the average values obtained from separate fits. The estimated values of w_x and w_z will be used subsequently as fixed parameters in the models used to fit cellular data.

3.1.2. Acquisition of FCS Data on Living Cells

Now that the FCS measurement volume has been properly calibrated, live-cell data can be collected.

1. Prepare cells expressing a fluorescently labeled protein of interest at a suitable concentration (see Note 3) in a live-cell imaging

chamber and mount the chamber on the microscope stage using the same immersion medium used for the calibration in step 3 of Subheading 3.1.1.

2. Match the laser power (8–12 mW), excitation wavelength (970 nm), and acquisition rate (100 kHz) to the ones used in Subheading 3.1.1.
3. Identify a healthy target cell with the fluorescent lamp and center it within the field of view. The cell should have low expression levels of the tagged protein, and so should be barely visible with a standard 100 W mercury arc lamp (see Note 3).
4. Acquire an image of the selected cell by scanning the stage under the excitation laser beam. It might be necessary to collect a series of images at different *z*-positions to identify the most suitable focal plane. The selected plane should lie in the middle of the nucleus (at least 2 μm away from the cell surface). Save the image of the selected plane.
5. Select a measurement spot in the middle of the region of interest, preferably in a fluorescently homogeneous region, far from both other structures and the nuclear envelope to avoid mobility artifacts from border regions.
6. Check the photon counts per second obtained at the selected spot. If the count rate is too high ($>10^5$ counts per second) select a dimmer spot within the cell (or repeat steps 3–5 for a dimmer cell). If the count rate is too low (<100 counts per second), select a brighter spot (or repeat steps 3–5 for a brighter cell). After identifying a suitable spot, annotate its coordinates.
7. Record the raw fluorescence fluctuations at the selected spot for a suitable length of time (see Note 7). Discard data that decay due to photobleaching or do not fluctuate about a constant mean (see Notes 8 and 9).
8. Collect an additional image of the selected plane and compare it with the original image. If the cell moved or changed shape, discard the collected data and select another cell.
9. Collect four additional fluorescence fluctuation time series (steps 6 and 7) at the same spot.
10. Select a different cell and repeat the experiment. The acquisition of data from at least 20 different cells should ensure good statistics to discriminate between cell-to-cell variability and systematic artifacts.
11. To generate data for a subsequent test of cryptic photobleaching (see Subheading 3.1.3, step 5 below and Note 10), repeat the preceding measurements over a range of laser powers. In general, lower laser powers that produce sufficient photon count rates are best.

3.1.3. Preprocessing and Qualitative Analysis of FCS Data

1. For each acquisition, inspect the raw time series. Check for the presence of abnormally high spikes that may indicate unwanted aggregates (see Fig. 1d). If spikes are visible, then the autocorrelation can either be performed selectively on the data preceding and following the sequence of spikes or the data can be entirely discarded.
2. Compute the autocorrelation of the data (see Note 5 and Fig. 2).
3. Visually inspect autocorrelation curves obtained from the same location using the same laser power. Do the curves behave similarly? Do they all converge to zero? If not, discard either the outliers or the whole data set (see Notes 8 and 9).
4. Export the autocorrelation curves to a spreadsheet (such as Excel) and calculate the average and standard deviation of curves originating from the same location in the nucleus. This generates the data for fitting.
5. Compare autocorrelation curves from data taken at different laser powers (from step 11 in Subheading 3.1.2 above). If there is a trend in the autocorrelation curves such that those acquired at higher laser powers go to zero more quickly than those acquired at lower laser powers (see Fig. 1f), then cryptic photobleaching is problematic (see Note 10). In this case, determine if a set of curves at the lowest laser powers go to zero at the same rate. If so, then keep only these curves. If not, then try to reduce the laser power further and collect more autocorrelation data (step 11 in Subheading 3.1.2 above). If the data become too noisy at lower laser powers before a plateau is reached in the rate at which the autocorrelation curve goes to zero, then the data cannot be used for analysis, and an alternative approach such as TICS (Subheading 3.2) should be tried.
6. For qualitative comparisons, repeat the analysis for FCS data collected on another set of samples subjected to a biological perturbation of interest. Normalize each of the autocorrelation curves to one and then plot the normalized autocorrelation functions together on the same plot. The autocorrelation curves of less-mobile samples will decay more slowly (see Fig. 2).

3.2. Data Acquisition and Analysis for TICS

TICS has some advantages over conventional FCS in that measurements can be made at multiple spatial locations simultaneously. In addition, TICS is less subject to cryptic photobleaching (see Note 10) and so is more suitable for measuring binding of nuclear proteins to chromatin, where residence times may be long enough to be artifactually shortened by cryptic photobleaching in FCS.

The shape of the measurement volume in TICS may differ from FCS depending on the microscope used. For the Zeiss LSM 5 LIVE, the measurement volume within the specimen detected by each well in the camera is extended along the line of simultaneous illumination, so $w_x > w_y$ (see Note 11). Because of this complication, we recommend directly measuring the microscope point spread function (PSF) by imaging sub-resolved fluorescent beads and then confirming or refining this measurement volume estimate by performing TICS on a solution containing a fluorescent species whose diffusion constant is known a priori (as described above in Subheading 3.1.1 for FCS).

3.2.1. Direct Measurement of the TICS Measurement Volume

1. Apply a drop of Aquamount onto the bottom of a live-cell imaging chamber and pipette a few microliters of solution containing 100-nm fluorescent beads into the drop. Once dry, the beads will be mounted in a thin layer at the bottom of the imaging chamber.
2. Mount the chamber on the microscope stage and choose an appropriate excitation wavelength. We use 488 nm for TICS on GFP.
3. Adjust the laser AOTF to 5% (see Note 12) and the pinhole to 1 Airy unit (0.94 μm).
4. Focus onto the surface of the imaging chamber and search for an isolated bead (typically found near the edges of the original Aquamount drop).
5. Adjust the pixel size to ~50 nm^2.
6. Acquire a *z*-stack of the bead with *z*-resolution of ~50 nm.
7. Repeat ~20 times, either on the same bead or different beads.
8. Fit each 3D image stack of the beads with a 3D Gaussian to quantify the shape of the measurement volume. We have obtained the following values: $w_x = 0.3 \pm 0.13$ μm, $w_y = 0.25 \pm 0.09$ μm, and $w_z = 0.90 \pm 0.20$ μm on our Zeiss LSM 5 Live microscope using a 63× 1.4 NA oil immersion objective (6). Our Matlab code for doing this fit is available upon request.

3.2.2. Confirming/Refining the TICS Measurement Volume

1. Prepare a 1-ml aliquot of 100-nm and 200-nm diameter fluorescent beads diluted in sterilized H_2O to approximately 100 nM. Sonicate each aliquot for a few minutes to dissociate potential aggregates and pipette into an imaging chamber.
2. Prepare the sample for imaging as described in steps 2 and 3 above in Subheading 3.2.1.
3. Focus well above the bottom of the imaging chamber to avoid artifacts due to diffusion near a surface. Diffusing beads should be visible by eye.
4. Adjust the pixel size to 200 nm^2.
5. Change to "Line" mode and image a single line (512×1 pixels) repeatedly at 1,000–5,000 Hz for 100 s. Imaging at 1,000 Hz for 100 s produces a single spatiotemporal image with dimensions 512 pixels × 1,000 Hz × 100 s = 512×10^5 pixels.

6. Preprocess and autocorrelate the data (see Subheading 3.2.4).
7. Fit the averaged autocorrelation curves (weighted by the standard deviation of each point) with a single-component diffusion model (see Note 1, Eq. 1) assuming the average w_x, w_y, and w_z measured from above. This should yield the theoretical diffusion coefficient of the beads: (~3.28 ± 0.15 μm²/s and 6.56 ± 0.3 μm²/s at 37 ± 2°C for 200-nm and 100-nm beads, respectively (6)). If not, try refitting with small leeway for w_x, w_y, and w_z. If unable to fit the theoretical diffusion coefficient using values for w_x, w_y, and w_z within error of the direct measurements from Subheading 3.2.1, then a 3D Gaussian may not be adequate for describing the measurement volume and more complicated shapes may be required (see Note 11).

3.2.3. Acquisition of TICS Data in Live Cells

Now that the TICS measurement volume has been properly calibrated, live-cell data can be collected.

1. Prepare cells in a live-cell imaging chamber with expression levels ~2–3× higher than for two-photon FCS (see Note 3) and mount the chamber on the microscope stage using the same immersion medium as used for the calibrations in Subheadings 3.2.1 and 3.2.2.
2. Adjust the laser AOTF to 12% and pinhole to 1 Airy unit (0.94 μm).
3. Identify a healthy target cell and center it within the field of view.
4. Acquire an image of the cell with a nuclear region of interest centered within the frame using a pixel dwell time of 0.2–1 ms (to match the TICS imaging rate used later). The cell should be dim but viewable above background. If no signal is apparent, readjust the AOTF to a higher value and repeat or search for a brighter cell.
5. Change to "Line" mode and image a single line (512 × 1 pixels) passing through the region of interest within the nucleus. Image this line repeatedly at 1,000–5,000 Hz for a suitable length of time (see Note 7). Imaging at 1,000 Hz for 100 s produces a single spatiotemporal image with dimensions 512 pixels × 1,000 Hz × 100 s = 512 × 10^5 pixels. Note the *y* coordinate of the line that was imaged such that its location within the imaged region can be found in the future. Save the spatiotemporal line scan data.
6. Reimage the cell. Discard data if notable movement occurred during the previous TICS data acquisition step.
7. Repeat on at least 20 different cells.
8. To generate data for a subsequent test of cryptic photobleaching (see Subheading 3.2.4, step 5 below and Note 10), repeat over

a range of lower laser powers by adjusting the AOTF. In general, lower laser powers that produce sufficient signal above background are the best. For GFP-tagged proteins, we set the AOTF between 5 and 20%.

3.2.4. Preprocessing and Qualitative Analysis of TICS Data

1. For each spatiotemporal image (512×10^5 pixels if imaging was 1,000 Hz for 100 s), separate the data from each of the pixels within the region of interest into individual time series, each having 10^5 time points. Our Matlab code for doing this is available upon request.
2. Correct for observational photobleaching by fitting the decaying average fluorescence from each pixel with a decaying double-exponential. Divide the fit out of the original time series to produce a new time series that fluctuates about a non-decaying mean (see Note 13 and Fig. 3b).
3. Calculate the autocorrelation of each time series (see Note 5 and Fig. 2).
4. Visually inspect the resulting autocorrelation curves acquired at the same laser power. Discard anomalous curves with spurious fluctuations and curves that do not go smoothly to zero (see Note 9).
5. Compare autocorrelation curves from data taken at different laser powers (from step 8 in Subheading 3.2.3 above). If there is a trend in the autocorrelation curves such that those acquired at higher laser powers go to zero more quickly than those acquired at lower laser powers, then cryptic photobleaching is problematic (see Note 10). In this case, determine if a set of curves at the lowest laser powers go to zero at the same rate. If so, then keep only these curves. If not, then try to reduce the laser power further and collect more autocorrelation data (step 8 in Subheading 3.2.3 above). If the data become too noisy at lower laser powers before a plateau is reached in the rate at which the autocorrelation curve goes to zero, then the data cannot be used for subsequent analysis.
6. Store the remaining autocorrelation curves for subsequent analysis. Make sure to annotate both the image source of each curve along with their relative pixel positions for later construction of spatial maps of diffusion and binding parameters.

3.3. Quantitative Analysis of FCS/TICS Data

Here we describe how to fit FCS/TICS data obtained in the nucleus to quantify diffusion and binding (see Fig. 4 and Note 14). Our Matlab code for doing this is available upon request.

1. Begin by fitting the autocorrelation curves with a single-component diffusion model (see Note 1, Eq. 1). If multiple autocorrelation curves have been averaged together, then weight the fits by the standard deviation of each point (see Note 14).

If a good fit is obtained, calculate the estimated molecular weight of the diffusing complex by comparing its measured diffusion coefficient D_{prot} to that of GFP (see Note 15). There are two possibilities:

(a) If the calculation results in a realistic value for the mass of the diffusing component, the dynamics of the molecules can be described by pure diffusion. In this case only a small fraction of molecules (<1%) are bound, so the association and dissociation rates cannot be measured.

(b) If the resulting estimated molecular weight is too big to be realistic, the slowed "effective" diffusion may be indirect evidence for transient binding. This behavior is encountered when many binding and unbinding events occur before a molecule escapes the FCS/TICS measurement volume (see Note 16). In this case, only the ratio of binding on and off rates can be measured (see Note 16, Eq. 3).

2. If the single-component diffusion fit does not provide an accurate description of the experimental FCS/TICS curve, a two-component diffusion fit can be tested (see Note 1, Eq. 3). Such a model will usually provide a better fit to the data since there are more parameters: two diffusion coefficients and the fraction of molecules in each state. As above, calculate the molecular weights of the diffusing complexes by comparing their diffusion coefficients to that of GFP (see Note 15). If either weight is unrealistic, the model is not appropriate.
3. Fit with the full diffusion and binding model (see Note 2, Eqs. 4–7). The fit will result in estimates for the diffusion coefficient D_{prot} and the binding rates k_{on}^{*} and k_{off}. These rates can be inverted to obtain the average time to diffuse from one binding site to the next and the average residence time at a binding site, respectively. Also, the rates can be combined to obtain the fraction of free molecules $F_{eq} = \frac{k_{off}}{k_{on}^{*} + k_{off}}$ and the corresponding fraction of bound molecules $C_{eq} = 1 - F_{eq}$.
4. Choose the best fit: As a rule of thumb, if a simpler model fits the data as well as a more complex model, the simpler model is preferable. More sophisticated statistical tests can also be performed to distinguish models (see Note 17).
5. If the two component diffusion and reaction-diffusion models fit the data equally well, or if diffusion appears to be "effective," then additional experiments may be necessary. For example, we recommend the use of mutant constructs lacking binding domains to distinguish models, as well as cross-validation measurements using complementary techniques such as FRAP (3). Gold standards for in vivo binding are still being developed

(3, 12), so cross validation of binding estimates by alternate approaches is especially important.

4. Notes

1. The simplest kinetic model describes freely diffusing molecules. We refer to this model as the "pure diffusion" model (see Fig. 4b). In this case the autocorrelation curve has the following form:

$$G(\tau) = \frac{1}{2^{3/2} N}\left(1+\frac{\tau}{\tau_D}\right)^{-1/2}\left(1+\frac{\tau}{\omega_{xy}^2 \tau_D}\right)^{-1/2}\left(1+\frac{\tau}{\omega_{xz}^2 \tau_D}\right)^{-1/2}, \quad (1)$$

Here N is the number of molecules within the measurement volume, $\omega_{xy} = w_y / w_x$ and $\omega_{xz} = w_z / w_x$ are the shape factors of the measurement volume and τ_D is the characteristic diffusion time, related to the free diffusion coefficient of the protein D_{prot} by:

$$\tau_D = \frac{w_x^{\,2}}{4\mu D_{\text{prot}}}; , \quad (2)$$

where μ represents the number of photons used for the excitation ($\mu = 1$ for confocal and $\mu = 2$ for two-photon excitation). The pure-diffusion model can also be extended to describe multiple non-interexchanging species with different diffusion constants (see Fig. 4c):

$$G(\tau) = \frac{1}{2^{3/2} N}\left[\rho\left(1+\frac{\tau}{\tau_{D,1}}\right)^{-1/2}\left(1+\frac{\tau}{\omega_{xy}^2 \tau_{D,1}}\right)^{-1/2}\left(1+\frac{\tau}{\omega_{xz}^2 \tau_{D,1}}\right)^{-1/2}\right.$$
$$\left. +(1-\rho)\left(1+\frac{\tau}{\tau_{D,2}}\right)^{-1/2}\left(1+\frac{\tau}{\omega_{xy}^2 \tau_{D,2}}\right)^{-1/2}\left(1+\frac{\tau}{\omega_{xz}^2 \tau_{D,2}}\right)^{-1/2}\right], \quad (3)$$

where the different diffusion constants are given by $D_k = w_x^{\,2} / 4\mu\tau_{D,k}$ $(k = 1, 2)$ and ρ is the fraction of molecules with diffusion coefficient D_1

2. To more accurately describe the dynamics of DNA-binding proteins, we developed a model to quantify both diffusion and

transient binding to immobile substrates (5). The model assumes that free molecules undergo simple diffusion and that their interaction with uniformly distributed binding sites can be described by a simple bimolecular reversible reaction (see Fig. 4d). By fitting this model it is possible to quantify the association and dissociation rates k^*_{on} and k_{off} of the labeled molecules to the binding sites, as well as the diffusion coefficient of the free molecules D_{prot}. In this case the autocorrelation curve has the following form:

$$G(t)=\frac{w_x w_y w_z}{8N(2\pi\mu)^{3/2}}\int\Gamma(\vec{q})\,\Omega(\vec{q},t)\,d^3\vec{q}, \tag{4}$$

where $\vec{q}=(q_x,q_y,q_z)$ is the Fourier transform of the spatial variables x, y (the coordinates within the imaging plane) and z (the coordinate perpendicular to the imaging plane). The term $\Gamma(\vec{q})$ accounts for the measurement volume:

$$\Gamma(\vec{q})=e^{-\frac{w_x^2}{4\mu}q_x-\frac{w_y^2}{4\mu}q_y-\frac{w_z^2}{4\mu}q_z} \tag{5}$$

while the term $\Omega(\vec{q},t)$ accounts for the diffusion and binding processes:

$$\Omega(\vec{q},t)=\frac{1}{2}\left[e^{\lambda_1 t}(1-\phi)+e^{\lambda_2 t}(1+\phi)\right]. \tag{6}$$

Here λ_1, λ_2 and ϕ have the following form:

$$\begin{aligned}\lambda_1&=\frac{1}{2}(-q^2D_{prot}-k_{off}-k^*_{on}-\sqrt{\alpha}),\\ \lambda_2&=\frac{1}{2}(-q^2D_{prot}-k_{off}-k^*_{on}+\sqrt{\alpha}),\\ \phi&=\frac{(k^*_{on}+k_{off})}{\sqrt{\alpha}}+\frac{(k^*_{on}-k_{off})}{(k^*_{on}+k_{off})}\frac{(q^2D_{prot})}{\sqrt{\alpha}},\end{aligned} \tag{7}$$

where $q^2=q_x^2+q_y^2+q_z^2$ and $\alpha=\left(q^2D_{prot}+k^*_{on}+k_{off}\right)^2-4q^2D_{prot}k_{off}$. The integral in Eq. 4 should be numerically evaluated on the whole Fourier space (i.e., $-\infty<q_x,q_y,q_z<+\infty$). However, since $\Gamma(\vec{q})$ rapidly decays to zero for increasing $\vec{q}$ we limit the integration to a box with edges: $|q_i|=4\sqrt{2\mu/w_i^2}$.

In certain cases this general solution takes on a simplified form. In particular, if diffusion and binding occur on different timescales ($\tau_D \ll 1/k^*_{on}$), then diffusing molecules travel through the measurement volume with very little probability of binding. In this case the free and bound populations can be considered independently as little exchange is observed. We refer to this as the "uncoupled reaction and diffusion" model, in which case the autocorrelation function becomes:

$$G(\tau)=\frac{F_{eq}}{2^{3/2}N}\left(1+\frac{\tau}{\tau_D}\right)^{-\frac{1}{2}}\left(1+\frac{\tau}{\omega_{xy}^2\tau_D}\right)^{-\frac{1}{2}}\left(1+\frac{\tau}{\omega_{xz}^2\tau_D}\right)^{-\frac{1}{2}} + \left(1-F_{eq}\right)e^{-k_{off}\tau} \quad (8)$$

where $F_{\mathrm{eq}}=\frac{k_{\mathrm{off}}}{k_{\mathrm{on}}^{*}+k_{\mathrm{off}}}$ represents the fraction of free molecules at equilibrium. When a good fit is obtained by Eq. 8, it is important to verify that the fitted parameters are consistent with the assumptions of the simplified model ($\tau_D \ll 1/k_{\mathrm{on}}^{*}$). Otherwise, the full model fit (Eqs. 4–7) must be used. Also note that the same conditions ($\tau_D \ll 1/k_{\mathrm{on}}^{*}$) lead to a simplified model for FRAP (13). However, the time resolution of FRAP is typically too slow to detect the fast diffusive population (the first term in Eq. 8), so the simplified model for FRAP, known as reaction-dominant, has only one term equivalent to the second term in Eq. 8 (see Chapter 11).

In all models, an additional term describing the photophysics of the fluorescent label can be added. This term reflects fluorophore blinking and takes the form of an exponential decay that only alters the autocorrelation curve at very short times (on the order of μs). If the timescale of diffusion and binding is considerably longer than the blinking timescale (as is the case for most nuclear proteins), the initial part of the acquired autocorrelation curve describing the blinking can simply be dropped when performing fits.

3. FCS and TICS require labeled protein concentrations in the nanomolar to micromolar range. Different methods are available to obtain these levels. Transient transfection with Lipofectamine LTX or with other chemical transfection agents can produce a wide range of expression levels. In this case, dimmer cells can be chosen for FCS/TICS. Stable cell lines expressing the construct under the control of an inducible promoter (such as a tetracycline-regulated promoter) can be used as well. In this case, the concentration and the timing of the inducing/repressing compound should be adjusted to obtain the desired amount of expression. For additional control of the labeling efficiency, the use of more novel fluorophore fusion technologies may

be desirable. For example, HaloTag (14) or Snap/Clip tag (15) systems label proteins of interest post-translationally via a cell-permeable fluorescent ligand whose concentration can easily be tuned.

4. After-pulse effects are due to a short timescale correlation of the detector noise and result in a fast-decreasing slope at the early time points of the autocorrelation function (<1 μs) (16). This can be eliminated by splitting the signal from the sample onto two separate detectors and then calculating the cross-correlation of the two fluctuation time-series. This cross-correlation removes afterpulsing effects since the noise from each detector is uncorrelated. The resultant cross-correlation can then be used as the autocorrelation data.

5. In FCS and TICS the mobility of fluorescently labeled molecules is quantified by measuring fluorescence fluctuations $\delta I(t)$, defined as the deviation of the fluorescence intensity $I(t)$ from its mean value $<I(t)>$ at time t, $\delta I(t) \equiv I(t) - <I(t)>$. The autocorrelation of the fluctuation time series with itself shifted by time τ can then be written as

$$G(\tau) \equiv \frac{<\delta I(t)\delta I(t-\tau)>}{\langle I(t)\rangle^2}, \tag{9}$$

where "<>" indicates an average over t. The amplitude of the autocorrelation function $G(\tau = 0)$ is inversely proportional to the number of molecules within the measurement volume. At later times, $G(\tau)$ eventually decays to zero because mobile molecules have zero probability of remaining within the measurement volume indefinitely (see Fig. 2).

An alternative definition of the autocorrelation is also sometimes used:

$$G_1(\tau) = \frac{\langle I(t)I(t-\tau)\rangle}{\langle I(t)\rangle^2} = 1 + \frac{\langle \delta I(t)\delta I(t-\tau)\rangle}{\langle I(t)\rangle^2}, \tag{10}$$

In this case the curve will start from $1 + G(0)$ and approach 1 (instead of 0) for long times.

There are a number of ways to calculate $G(\tau)$ from the raw intensity time series $I(t)$. Although all commercial FCS systems include software for these purposes, in some cases it is preferable or necessary to independently autocorrelate the data, for example when performing TICS. The straightforward application of Eq. 9 is computationally demanding and therefore not recommended for long time series. More refined methods to calculate $G(\tau)$ involve multi-tau correlation (i.e., binning data before autocorrelating for large τ) or Fourier transforming the data (in which case autocorrelating amounts to a simple multiplication) (17).

6. The microscope objective for FCS needs to be of high numerical aperture (NA > 1) in order to limit the observation to a small volume around the focal plane. Oil immersion objectives are preferable for measurements near the coverslip (<5–10 μm), while water immersion objectives are favorable for measurements deeper in the sample.
7. The duration of experiments should be long enough to observe the complete dynamics of the labeled molecules. As a rule of thumb the experiment should last ~100 times longer than the average time a molecule spends within the measurement volume. Failure to do so will result in autocorrelation curves that do not decay to zero. On the other hand, photobleaching and movement limit the total length of an experiment. As a rule of thumb we suggest starting with a total acquisition time of about 20 s for FCS and 100 s for TICS. This time can subsequently be adjusted depending on the shape of the calculated autocorrelation curve.
8. Conventional FCS requires the sample to be in steady state when the experiment is performed. In other words, the average concentration of molecules should not increase or decrease during the experiment and the temperature should be held constant. Non-equilibrium conditions appear as large and extended fluctuations in the recorded fluorescence intensity time series and these lead to artifacts in the autocorrelation curve. If these artifacts are seen, the data should be discarded and attempts should be made to maintain the temperature, stability, and/or fluorescence expression levels of subsequent samples.
9. If the curves consistently exhibit irregular behavior (for example, negative correlations) at long correlation times, the parameters used for the acquisition might not be suitable for the experimental situation. Different solutions might be found depending on the source of the problem:
   (a) The total observation time might not be sufficient to capture the full dynamics of the observed molecules. In this case, try repeating the experiment, but now with a longer total acquisition time (also see Note 7).
   (b) Photobleaching could be producing negative correlations. Try repeating the experiment using a lower laser power (also see Fig. 1e and Notes 10 and 13).
   (c) The labeled proteins might be aggregating. This could also appear in the raw time-series as anomalously high-intensity spikes (see Fig. 1d). For experiments in solution, further sonication might solve the issue. If this situation is

repeatedly encountered in living cells, oligomerization of the fluorescent tag might be the cause. Moving the fluorescent label to the other end of the protein of interest or changing the fluorescent protein might resolve the issue.

10. Cryptic photobleaching occurs when an influx of non-bleached molecules into the measurement volume counterbalances the rate of photobleaching within the volume. When this occurs, the average fluorescence measured throughout the experiment remains constant in spite of the photobleaching. This tends to shorten the apparent diffusion and binding times of molecules since they bleach before exiting the measurement volume (6, 18). To detect cryptic photobleaching, multiple experiments at different laser powers are required. If the autocorrelation curves acquired at lower laser powers consistently go to zero more slowly than those acquired at higher laser powers, cryptic photobleaching is problematic (Fig. 1f). In this case, continue to lower the laser power until the autocorrelation curves go to zero at the same rate.

11. For line-scanning optics, the measurement volume has small side lobes that generally make $w_x > w_y$ (19). In addition, the TICS measurement volume also depends on the size of pixels. As long as pixels are smaller than or on the order of a diffraction limited spot (~200 nm for 488 nm light passing through a 1.4 N.A. objective), the measurement volume remains Gaussian (9). If the shape of the measurement volume diverges considerably from a Gaussian, it is still possible to describe the volume with a sum of Gaussians, in which case the form of the autocorrelation function must be modified (10).

12. To increase the sensitivity of the laser adjustment, we used the Zeiss LSM 5 LIVE in Duoscan mode with 20% of the laser power used for imaging. Thus, when the AOTF was set to 5%, the laser power was ~1 mW, or 5% of 20% of 100 mW.

13. In TICS an image series is acquired, so photobleaching occurs over a wider region than in standard FCS. This extended photobleaching can cause the fluorescence intensity to decay with time (see Fig. 3). Assuming the photobleaching is fairly uniform within the subregion of interest, then the decay in fluorescence caused by photobleaching can be partially corrected for, either after (20) or before (6, 9, 21) autocorrelation is performed. A straightforward way to correct before autocorrelating is to fit the decay with a mono- or bi-exponential curve and then divide this fit out from the data. However, this method amplifies fluctuations at later time-points and this could potentially induce artifacts in the correlated curve.

14. To quantitatively analyze FCS experiments, theoretical models are fit to the measured autocorrelation data to estimate model

parameters that describe the mobility of the fluorescently tagged molecules. If multiple autocorrelation curves have been averaged together, the fit can be weighted by the standard deviations $\sigma(\tau_i)$ using non-linear minimization of the χ^2, defined as:

$$\chi^2 = \sum_i \left(\frac{G(\tau_i) - G_F(\tau_i)}{\sigma(\tau_i)} \right)^2, \tag{11}$$

where $G_F(\tau_i)$ represents the fitted model at the time point τ_i. If no averaging has been performed before fitting, so that no standard deviations are available, the best fit can be found by minimizing the squared sum of residuals, which is equivalent to Eq. 11, provided that $\sigma(\tau_i)$ is set to 1 for all τ_i.

In principle the correct model to fit experimental data should be chosen based on prior knowledge of the phenomena involved in the movement of the species. Often, however, this information is not available beforehand and different models with increasing complexity must be tested in sequence (see Notes 1 and 2).

15. Unconjugated GFP (27 kDa in size) is reported to diffuse in the nucleus of living cells with a rate of approximately D_{GFP} =20 μm²/s. If the protein of interest is globular in shape, then its diffusion coefficient can be roughly estimated as:

$$D_{\mathrm{prot}} = D_{\mathrm{GFP}} \sqrt[3]{\frac{M_{\mathrm{GFP}}}{M_{\mathrm{prot}}}}, \tag{12}$$

where M_P and M_{GFP} are the respective molecular weights of the fluorescent protein of interest and unconjugated GFP. Thus, a molecule that diffuses 10 times slower than eGFP (~2 μm²/s) would be 1,000 times more massive. Although many nuclear proteins appear to diffuse at ~2 μm²/s, it is unlikely that they represent such massive complexes. In these cases, we favor the interpretation that the slowed diffusion represents binding to chromatin (also see Note 16) or perhaps nuclear crowding producing anomalous diffusion.

16. It may occur that a simple diffusion fit provides a good description of the experimental data, but the fit results in an unrealistically slow diffusion coefficient. This is typically observed when the molecules undergo many binding and unbinding events before escaping the measurement volume, or in other words when the binding association rate k^*_{on} is faster than the rate of diffusion through the measurement volume $1/\tau_D$. Consequently, the molecules appear to diffuse at a slower pace over long times. We refer to this as the "effective diffusion" model, in which case the FCS autocorrelation curve can be

described by Eqs. 1 and 2 provided the free diffusion coefficient D_{prot} is replaced by an effective coefficient D_{eff} defined by:

$$D_{eff} = \frac{D_{prot}}{1 + k_{on}^{*} / k_{off}}, \quad (13)$$

where k_{on}^{*} and k_{off} represent the association and dissociation rates of the molecules from their binding sites, respectively. Thus, if the diffusion coefficient of the free molecules D_{prot} is known, the ratio k_{on}^{*} / k_{off} of the association/dissociation rates of binding to chromatin can be determined.

17. Often a better fit is obtained with a more complicated model because the model has more free parameters. In order to determine if the improved fit is due to the increase in free parameters or actually reflects a more accurate representation of the data, it is useful to compare the resulting χ^2 (Eq. 11) using the F-test. To do so, calculate the reduced sum of residuals for each model: $\chi_{v,i}^2 \equiv \chi_i^2 / v$, where v represents the number of degrees of freedom (which is the number of data points minus the number of free parameters) for each model i. Then compute the ratio $\chi_{v,1}^2 / \chi_{v,2}^2$ (with model 1 the simpler model). If the ratio is higher than the tabulated value of the F-distribution at a 95% confidence threshold, the improvement obtained when fitting with model 2 is statistically significant.

References

1. Hager, G. L., McNally, J. G., and Misteli, T. (2009) Transcription dynamics, Mol. Cell 35, 741–753.
2. van Royen, M.E., Zotter, A., Ibrahim, S.M., Geverts, B. and Houtsmuller AB. (2011) Nuclear proteins: finding and binding target sites in chromatin. Chromosome Res. 19, 83–98.
3. Mueller, F., Mazza, D., Stasevich, T. J., and McNally, J. G. (2010) FRAP and kinetic modeling in the analysis of nuclear protein dynamics: what do we really know?, Curr. Opin. Cell Biol 22, 403–411.
4. Weidtkamp-Peters, S., Weisshart, K., Schmiedeberg, L., and Hemmerich, P. (2009) Fluorescence correlation spectroscopy to assess the mobility of nuclear proteins, Methods Mol. Biol 464, 321–341.
5. Michelman-Ribeiro, A., Mazza, D., Rosales, T., Stasevich, T. J., Boukari, H., Rishi, V., Vinson, C., Knutson, J. R., and McNally, J. G. (2009) Direct measurement of association and dissociation rates of DNA binding in live cells by fluorescence correlation spectroscopy, Biophys. J 97, 337–346.
6. Stasevich, T. J., Mueller, F., Michelman-Ribeiro, A., Rosales, T., Knutson, J. R., and McNally, J. G. (2010) Cross-Validating FRAP and FCS to Quantify the Impact of Photobleaching on In Vivo Binding Estimates, Biophysical Journal 99, 3093–3101.
7. Elson, E. L., Schlessinger, J., Koppel, D. E., Axelrod, D., and Webb, W. W. (1976) Measurement of lateral transport on cell surfaces, Prog. Clin. Biol. Res 9, 137–147.
8. Hebert, B., Costantino, S., and Wiseman, P. W. (2005) Spatiotemporal Image Correlation Spectroscopy (STICS) Theory, Verification, and Application to Protein Velocity Mapping in Living CHO Cells, Biophysical Journal 88, 3601–3614.
9. Sisan, D. R., Arevalo, R., Graves, C., McAllister, R., and Urbach, J. S. (2006) Spatially Resolved Fluorescence Correlation Spectroscopy Using a Spinning Disk Confocal Microscope, Biophysical Journal 91, 4241–4252.

10. Heuvelman, G., Erdel, F., Wachsmuth, M., and Rippe, K. (2009) Analysis of protein mobilities and interactions in living cells by multifocal fluorescence fluctuation microscopy, Eur. Biophys. J 38, 813–828.
11. Petrášek, Z., and Schwille, P. (2008) Precise measurement of diffusion coefficients using scanning fluorescence correlation spectroscopy, Biophysical Journal 94, 1437–1448.
12. Erdel, F., Müller-Ott, K., Baum, M., Wachsmuth, M. and Rippe, K. (2011) Dissecting chromatin interactions in living cells from protein mobility maps. Chromosome Res. 19, 99–115.
13. Sprague, B. L., Pego, R. L., Stavreva, D. A., and McNally, J. G. (2004) Analysis of binding reactions by fluorescence recovery after photobleaching, Biophys. J 86, 3473–3495.
14. Schröder, J., Benink, H., Dyba, M., and Los, G. V. (2009) In vivo labeling method using a genetic construct for nanoscale resolution microscopy, Biophys. J 96, L01–03.
15. Gautier, A., Juillerat, A., Heinis, C., Corrêa, I. R., Kindermann, M., Beaufils, F., and Johnsson, K. (2008) An engineered protein tag for multiprotein labeling in living cells, Chem. Biol 15, 128–136.
16. Zhao, M., Jin, L., Chen, B., Ding, Y., Ma, H., and Chen, D. (2003) Afterpulsing and Its Correction in Fluorescence Correlation Spectroscopy Experiments, Appl. Opt. 42, 4031–4036.
17. Selvin, P. R., and Ha, T. (2008) Single-molecule techniques: a laboratory manual. CSHL Press.
18. Petrášek, Z., and Schwille, P. (2008) Photobleaching in Two-Photon Scanning Fluorescence Correlation Spectroscopy, ChemPhysChem 9, 147–158.
19. Wolleschensky, R., Zimmermann, B., and Kempe, M. (2006) High-speed confocal fluorescence imaging with a novel line scanning microscope, J Biomed Opt 11, 064011.
20. Kolin, D., Costantino, S., and Wiseman, P. (2006) Sampling Effects, Noise, and Photobleaching in Temporal Image Correlation Spectroscopy, Biophysical Journal 90, 628–639.
21. Digman, M., Brown, C., Horwitz, A., Mantulin, W., and Gratton, E. (2008) Paxillin dynamics measured during adhesion assembly and disassembly by correlation spectroscopy, Biophysical Journal 94, 2819–2831.

Chapter 13

Analysis of Chromatin Structure in Plant Cells

Mala Singh, Amol Ranjan, Krishan Mohan Rai, Sunil Kumar Singh, Verandra Kumar, Ila Trivedi, Niraj Lodhi, and Samir V. Sawant

Abstract

A vast body of evidence in the literature indicates that nucleosomes can act as barriers to transcriptional initiation. The nucleosome at the promoter inhibits association of transcription factors disallowing active transcription of the gene. We have found a nucleosome on tobacco pathogenesis-related gene-1a (*PR-1a*) core promoter and mapped its boundaries and extension to find its span. The nucleosome covers the TATA box and Inr region of the core promoter and gets disassembled upon induction. Prior to its removal, modifications (i.e., acetylation and methylation of histones) occur at the nucleosome, proving a role of epigenetic modifications in transcriptional regulation. We summarize here various methodologies to analyze promoter chromatin structure in plants using the *PR-1a core* promoter as an example.

Key words: Nucleosome, Transcriptional initiation, Core promoter, Histone modifications

1. Introduction

The nucleosome is the fundamental unit of chromatin and it is composed of an octamer of the four core histones (H2A, H2B, H3, H4) around which 147 base pairs of DNA are wrapped. Chromatin, the nucleoprotein complex found in the nucleus, has approximately twice the protein mass as DNA (1), and half of this mass is the highly basic histones, H1, H2A, H2B, H3, and H4. At the first level of packaging, the DNA is wrapped around histone octamer to form a beaded chain. Each bead is referred to as a core nucleosome and contains an octamer of two molecules of each of the core histones with two turns of DNA wrapped around the proteins (2, 3). The packaging of DNA into nucleosomes and chromatin positively or negatively affects all nuclear processes in the cell. All core histone contains a conserved C-terminal histone

Randall H. Morse (ed.), *Chromatin Remodeling: Methods and Protocols*, Methods in Molecular Biology, vol. 833,
DOI 10.1007/978-1-61779-477-3_13, © Springer Science+Business Media, LLC 2012

fold domain and unique N-terminal tails. The four core histones interact in pairs via a "hand-shake motif" with two H3/H4 dimers interacting together to form a tetramer while the two H2A/H2B dimers associate with the H3/H4 tetramer in the presence of DNA.

Although nucleosomes have long been viewed as stable entities, there is a large body of evidence indicating that they are highly dynamic (4), capable of being altered in their composition, structure, and location along the DNA. The similarity of nucleosome structures among eukaryotic organisms indicates that the mechanism of chromatin assembly is likely to be highly conserved among all eukaryotes. Eukaryotic gene transcriptional regulation is mediated by binding of transcription factors and epigenetic marks on the genes. Following promoter activation, the nucleosomes are removed from functionally important regions of promoters (5–9). In predifferentiated cells, two general transcription factors, TBP and TFIIB, are bound to the core promoter region. When cell differentiation is induced, RNA polymerase II and other general transcription factors are rapidly recruited to PIC before transcriptional activation. Subsequently, an SWI/SNF-like enzyme and two HATs (CBP and P/CAF) are recruited to the promoter and the nucleosomes are disrupted by histone acetylation, resulting in α1-antitrypsin transcription (10). In phaseolin gene also, expression takes place despite the presence of nucleosomes. In this case, transcription is facilitated by chromatin remodeling in the nucleosome present over the core promoter region that covers the TATA box and Inr region (11).

Mapping of protein–DNA interactions is essential to understand the transcriptional regulation of genes. Recent technological advances have allowed us to investigate such interactions in detail both at gene and genome-wide level. These advances include both in vivo and in vitro experimental methods and the development of new computational analysis tools. The in vivo approach, such as chromatin immunoprecipitation (ChIP) when followed by Q-PCR, allows us to study DNA–protein interactions at gene of interest and, when followed by microarray (ChIP-chip) or by sequencing (ChIP-Seq), determines the location of DNA-binding sites of a particular DNA-binding protein of interest in the genome which can identify candidate genes likely to be regulated by the same and aid in interpretation of regulatory processes (12–14).

2. Materials

2.1. Nucleosome Preparation

2.1.1. Isolation of Nuclei and Mononucleosome Preparation

1. Formaldehyde.
2. Blotting sheets.
3. MilliQ grade water.
4. Liquid nitrogen.
5. Mortar and pestles.
6. Homogenization buffer: 1 M hexylene glycol, 10 mM PIPES/KOH (pH 7.0), 10 mM $MgCl_2$, 0.5% (v/v) Triton X-100, 5 mM β-mercaptoethanol (β-ME), and 0.8 mM phenylmethylsulfonyl fluoride (PMSF) (see Note 1).
7. Homogenizer.
8. Muslin cloth.
9. Nylon mesh, 80, 60, 40, and 20 μ.
10. Nuclei wash buffer: 0.5 M hexylene glycol, 10 mM PIPES/KOH (pH 7.0), 10 mM $MgCl_2$, 0.5% (v/v) Triton X-100, 5 mM β-ME, and 0.8 mM PMSF (see Note 1).
11. Percoll dilution buffer: 0.5 M sucrose, 25 mM Tris–HCl (pH 8), 10 mM $MgCl_2$, 0.5% (v/v) Triton X-100, 5 mM β-ME, and 0.8 mM PMSF. Make up volume with milliQ water (see Note 1).
12. Reconstituted Percoll: 0.5 M sucrose, 25 mM Tris–HCl (pH 8.0), 10 mM $MgCl_2$, 0.5% (v/v) Triton X-100, 5 mM β-ME, and 0.8 mM PMSF. Make up volume with 100% Percoll (see Note 1).
13. Sucrose pad: 2 M sucrose, 25 mM Tris–HCl (pH 8.0), 10 mM $MgCl_2$, 0.5% (v/v) Triton X-100, 5 mM β-ME, and 0.8 mM PMSF (see Note 1).
14. Corex tubes.
15. Ultracentrifuge and swinging bucket rotor (e.g., Beckman ultracentrifuge and SW-28 rotor).
16. Sterile paint brush.
17. 80% glycerol.
18. Micrococcal nuclease buffer: 25 mM KCl, 4 mM $MgCl_2$, 1 mM $CaCl_2$, 50 mM Tris–HCl (pH 7.4), and 12.5% glycerol.
19. Micrococcal nuclease (MNase1).
20. Stop buffer: 2% SDS, 0.2 M NaCl, 10 mM EDTA, 10 mM EGTA, and 50 mM Tris–HCl (pH 8.0).
21. Mineral oil.
22. 3 M sodium acetate (pH 5.2).
23. Proteinase K.
24. Phenol.

25. Phenol:chloroform:isoamyl alchohol (25:24:1).
26. Chloroform:isoamyl alchohol (24:1).
27. Absolute ethanol.
28. 70% ethanol.

2.2. Analysis of Nucleosome

1. Software for designing primers, e.g., Primer Express 2.0 Software (Applied Biosystem, USA).
2. Actin primers:
   AT*ACTIN7*F 5′-AAG TCA TAA CCA TCG GAG CTG-3′.
   AT*ACTIN7*R 5′-ACC AGA TAA GAC AAG ACA CAC-3′.
3. Quant-iT™ PicoGreen® dsDNA Reagent (Invitrogen) or other DNA quantitation reagent.
4. Taq polymerase.
5. 10× PCR buffer: 100 mM Tris–HCl, pH 8.3, 500 mM KCl.
6. 4 mM dNTPs.
7. 25 mM $MgSO_4$.
8. Ethidium bromide (10 mg/ml).
9. T4 polynucleotide kinase.
10. 5× T4 Polynucleotide kinase reaction buffer: 350 mM Tris–HCl (pH 7.6), 50 mM $MgCl_2$, 25 mM dithiothreitol.
11. [γ-^{32}P] ATP.
12. Nucleotide removal kit.
13. 25 mM $MgCl_2$.
14. Liquid scintillation counter.
15. Thermo sequenase kit.
16. PAGE unit (e.g., Bio-Rad).
17. Whatman 3 mm paper.
18. Cling film.
19. Kodak/Fuji film.
20. Phosphoimaging system (e.g., Bio-Rad).

2.3. Deciphering Histone Modifications and Protein Interactions on Chromatin Using Chromatin Immunoprecipitation in Plants

2.3.1. Isolation and Fragmentation of Chromatin

1. Suitable media for growing Arabidopsis.
2. Falcon tubes, 50 and 15 ml.
3. MilliQ grade water.
4. Formaldehyde.
5. Vacuum chamber.
6. Extraction buffer 1: 0.4 M sucrose, 10 mM Tris–HCl (pH 8.0), 5 mM β-ME, 0.1 mM PMSF, and 50 μl plant protease inhibitor cocktail (PPIC)/100 ml of buffer (see Notes 1 and 2).
7. 2 M glycine.
8. Vortexer.

9. Tissue paper.
10. Liquid nitrogen.
11. Mortar and pestle.
12. 20 μ nylon mesh.
13. Extraction buffer 2: 0.25 M sucrose, 10 mM Tris–HCl (pH 8.0), 10 mM $MgCl_2$, 1% Triton X-100, 5 mM β-ME, 0.1 mM PMSF, 5 μl PPIC/10 ml of buffer (see Notes 1 and 2).
14. Extraction buffer 3: 1.7 M sucrose, 10 mM Tris–HCl (pH 8.0), 0.15% Triton X-100, 2 mM $MgCl_2$, 5 mM β-ME, 0.1 mM PMSF, and 5 μl PPIC/10 ml of buffer (see Notes 1 and 2).
15. Buffer N: 15 mM Trizma base, 15 mM NaCl, 60 mM KCl, 250 mM sucrose, 5 mM $MgCl_2$. 6 H_2O, 1 mM $CaCl_2$. 2 H_2O, 5 mM β-ME, 0.1 mM PMSF, 5 μl PPIC/10 ml of buffer (pH 7.5) (see Notes 1 and 2).
16. 5 M NaCl.
17. Chloroform:isoamyl alchohol (24:1).
18. Isopropanol.
19. 70% ethanol.
20. Nanodrop spectrophotometer.
21. MNase1.
22. MNase1 stop solution: 110 mM EGTA, 110 mM EDTA (see Note 2).
23. Hydroxyapetite resin.
24. HAP buffer 1: 5 mM $NaPO_4$ (pH 7.2), 600 mM NaCl, 1 mM EDTA, 5 mM β-ME, 0.1 mM PMSF, 5 μl PPIC/10 ml of buffer (see Note 1).
25. Microspin columns.
26. HAP buffer 2: 5 mM $NaPO_4$ (pH 7.2), 100 mM NaCl, 1 mM EDTA, 5 mM β-ME, 0.1 mM PMSF, 5 μl PPIC/10 ml of buffer (see Note 1).
27. HAP elution buffer: 500 mM $NaPO_4$ (pH 7.2), 100 mM of NaCl, 1 mM EDTA, 5 mM β-ME, 0.1 mM PMSF, 5 μl PPIC/10 ml of buffer (see Note 1).
28. Nuclei lysis buffer: 50 mM Tris–HCl (pH 8.0), 10 mM EDTA, 1% SDS, 5 mM β-ME, PMSF, 5 μl PPIC/10 ml of buffer (see Note 1).
29. Sonicator (e.g., Branson sonifier 3210).
30. ChIP dilution buffer: 1.1% Triton X-100, 1.2 mM EDTA, 16.7 mM Tris–HCl (pH 8.0), 167 mM NaCl, 5 mM β-ME, 0.1 mM PMSF, 5 μl PPIC/10 ml of buffer (see Note 1).

2.3.2. Immunoprecipitation and Collection of Complexes Bound to Antibody and DNA Purification

1. Salmon sperm-sheared DNA/protein A agarose beads.
2. Low-salt wash buffer: 150 mM NaCl, 0.1% SDS, 1% Triton X-100, 2 mM EDTA, 20 mM Tris–HCl (pH 8.0).
3. High-salt wash buffer: 500 mM NaCl, 0.1% SDS, 1% Triton X-100, 2 mM EDTA, 20 mM Tris–HCl (pH 8.0).
4. LiCl wash buffer: 0.25 M LiCl, 1% NP-40, 1% sodium deoxycholate, 1 mM EDTA, 10 mM Tris–HCl (pH 8.0).
5. TE buffer: 10 mM Tris–HCl (pH 8.0), 1 mM EDTA.
6. Elution buffer: 1% SDS, 0.1 M $NaHCO_3$ (see Note 2).
7. Proteinase K.
8. 0.5 M EDTA.
9. 1 M Tris–HCl (pH 6.5).
10. RNase A.
11. 3 M sodium acetate (pH5.2).
12. Absolute ethanol.
13. 70% ethanol.

2.3.3. Analysis of the Precipitated DNA Using QPCR

1. Real-time PCR machine.

3. Methods

3.1. Nucleosome Preparation

This method is based on the principle that approximately 150 bp of DNA in the nucleosome core are protected from digestion by MNase1. To start with, chromatin is cross-linked with formaldehyde and intact nuclei of high purity are extracted on a Percoll gradient. These nuclei are then MNase1 digested to obtain mononucleosomes for analysis. The whole process is to be done at 4°C unless stated otherwise.

3.1.1. Isolation of Nuclei

1. Harvest 120 g of tobacco leaves before flowering and wash with milliQ water, and immerse the leaves in 1% v/v formaldehyde for 30 min.
2. Decant the formaldehyde solution and wash the leaves (two times) with cold milliQ water to remove any traces of formaldehyde.
3. Dry the leaves in between blotting sheets and take out their midribs. Tear the leaves into pieces and grind it properly with liquid nitrogen in a pestle and mortar.
4. Transfer the powder in a beaker and add approximately 5–6 volumes of homogenization buffer and with the help of homogenizer, homogenize it properly (see Notes 3 and 4).

5. After proper homogenization, filter the homogenate through four layers of muslin cloth and add Triton X-100 up to a final concentration of 0.5%.
6. Leave it for 5 min and filter homogenate through Nylon mesh starting from 80, 60, 40, and then 20 μm (see Note 5).
7. After filtration, add filtrate to GSA tubes (or equivalent) and centrifuge at 3,000 × *g* for 10 min at 4°C.
8. Decant the supernatant and suspend the pellet properly in nuclei wash buffer.
9. Dilute it up to ten volumes with nuclei wash buffer and centrifuge in SS-34 tubes or equivalent at 3,000 × *g* for 5 min at 4°C.
10. Prepare different dilutions of Percoll to be used at later steps by diluting reconstituted Percoll with Percoll dilution buffer.
11. Decant the supernatant and dissolve the pellet in 6 ml of 5% Percoll.
12. Meanwhile, set up a gradient of Percoll in six corex tubes having 6 ml each of sucrose pad, 80, 60, and 40% of Percoll, from bottom to the top in every tube.
13. Prespin at 4°C in swinging bucket (SW-55) rotor or equivalent at 3,000 × *g* for 30 min.
14. Overlay 1 ml (from 20 gm of fresh leaves) of pellet suspension from step 11 on each prepared step gradient.
15. Run the gradient at 3,000 × *g* for 30 min at 4°C.
16. Isolate the nuclei layer that is between the sucrose pad and 80% Percoll.
17. Dilute nuclei suspension with at least three volumes of nuclei wash buffer and centrifuge it in SS-34 rotor or equivalent at 3,000 × *g* for 10 min at 4°C.
18. Now, suspend the pellet carefully with paint brush in nuclei wash buffer and add one-third volume of 80% glycerol.
19. Aliquot nuclei preparation into 1.5-ml tube, freeze it in liquid N_2, and store at −80°C indefinitely.

3.1.2. Mononucleosome Preparation (Micrococcal Nuclease Digestion and Purification)

1. Either use freshly prepared nuclei or stored nuclei as described in Subheading 3.1.1. To the stored aliquot nuclei, add equal amount of nuclei wash buffer and centrifuge it at 3,000 × *g* for 10 min at 4°C to pellet down the nuclei. This step eliminates glycerol used for storing nuclei.
2. Add 300 μl of micrococcal nuclease buffer to the pellet and appropriate amount of micrococcal nuclease enzyme (1 U/1 μg of genomic DNA in our hand, but has to be optimized for each condition individually) and incubate at 37°C for 20 min.

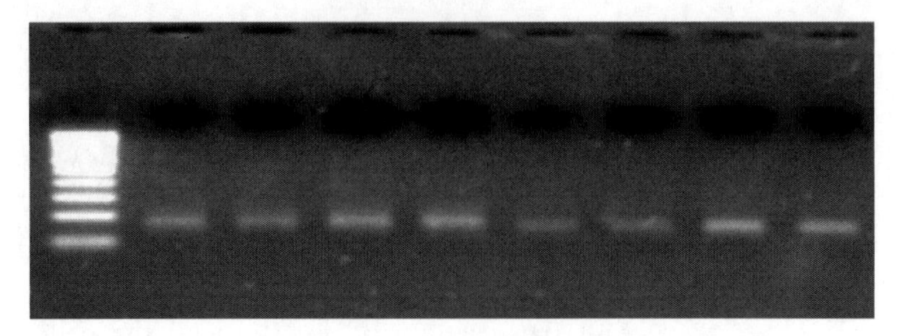

Fig. 1. DNA from purified mononucleosomes isolated from 2-week-old *Arabidopsis thaliana* seedlings in a 2% agarose gel.

3. Stop the reaction by adding equal amount of stop buffer.
4. Add appropriate amount of proteinase K and incubate for 2 h at 37°C or 45 min at 55°C to digest away the proteins.
5. Overlay mineral oil onto the content and incubate at 65°C overnight for reversal of cross-linking.
6. Now, extract the reaction once with equal volume of phenol, then phenol:chloroform:isoamyl alcohol, and finally with chloroform:isoamyl alcohol.
7. To the extracted aqueous layer, add one-tenth volume of 3 M sodium acetate (pH 5.2) and two volumes of ethanol, mix properly, and keep it in −20°C for 1–2 h.
8. Centrifuge at 16,000 × *g* in microfuge for 20 min at 4°C.
9. Decant supernatant and wash the pellet with 70% ethanol.
10. Dry the pellet and dissolve in milliQ water.
11. Check the digested sample on 1.5% agarose gel and excise fragment of an average size of 146 bp.
12. Purify the fragment and store at −20°C. An example of mononucleosomal DNA fragments viewed in 2% agarose gel is shown in Fig. 1.

3.2. Analysis of Nucleosome

Nucleosomes are dynamic building blocks of eukaryotic chromatin. In eukaryotes, transcriptional regulation involves many mechanisms, including posttranslational modifications on N-terminal tail of histones at the nucleosome. Hence, determining the positions and borders of nucleosome on a genes and/or promoter region helps in understanding its regulatory mechanisms. Mononucleosomes isolated as per protocol Subheading 3.1.2 are used to determine their correlation to the regulation of gene using various approaches.

We have previously demonstrated the presence of a distinct nucleosome that masks the TATA box in core promoters of pathogenesis-related gene-1a (*PR-1a*) in the uninduced state (15). The core promoter region in the promoter is transcriptionally active only under salicylic acid (SA)-induced conditions. To establish the correlation of nucleosome remodeling over the core

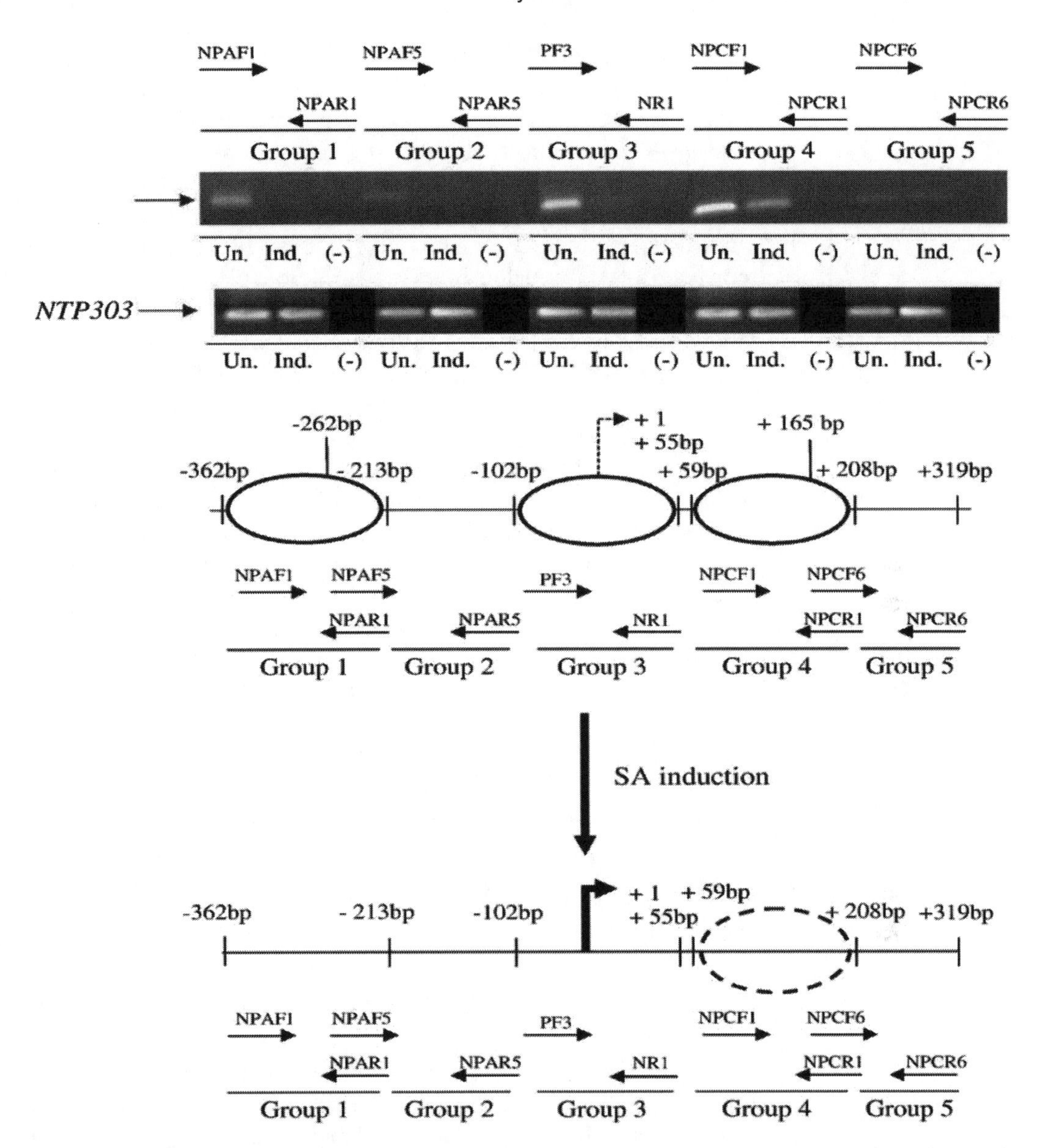

Fig. 2. Semiquantitative PCR was done to detect nucleosome in the core promoter (−102 to +55 bp) and flanking upstream (−103 to −362 bp) and downstream (+59 to +319 bp) regions of the native PR-1a promoter. Negative template controls (−) are shown with each group of PCR. The tobacco NTP303 promoter used as an internal control is shown below each lane. Equal intensity of PCR products with NTP303 primers establishes comparable efficiency of PCR in all reactions. The PCR products are shown below each lane. The models depict the sequential events leading to removal and remodeling of nucleosomes on PR-1a promoter following SA induction.

promoter region with the expression of *PR-1a* gene, we examined nucleosome repositioning using quantitative PCR (QPCR) analysis of core promoter region (group 3, Fig. 2) mononucleosomes prepared from untreated plants (i.e., uninduced) and treated SA (i.e., induced), which allowed us to find that though the promoter is protected by nucleosome in uninduced conditions the nucleosome disappears upon SA induction. This could be either because of nucleosomal sliding or disassembly. In order to understand the fate of the nucleosome, the core promoter region and neighboring regions were arbitrarily divided into five groups (two upstream – group 1 and 2, two downstream – group 4 and 5, and one core region – group 3 each of approximately 150-bp length) and primers were designed from each region of approximate 100-bp amplicon length. Mononucleosomes prepared from plants both untreated and treated with salicylic acid were used for PCR analysis of these above-said regions. Bands in group 1, 3, and 4 under uninduced conditions confirmed the presence of nucleosome in upstream (group 1), core (group 3), and downstream (group 4) regions only in uninduced conditions and not in groups 2 and 5 (Fig. 2). These bands became absent from groups 1 and 3 but not from group 4, when PCR was done from treated sample's mononucleosomes. This clearly demonstrated that the nucleosome is actually removed from the core promoter and is not shifted up/down or remodeled over the locus (15).

3.2.1. Analysis of Nucleosome Using QPCR

QPCR can be used with mononucleosome templates to quantitate the extent of nucleosomal coverage of any specific genomic location.

Designing of primers from region of interest. Both forward and reverse primers are designed from the region of interest, preferably using the primer design software, such as Primer Express 2.0 (Applied Biosystems, USA), at 10 bp resolution to give amplicon length of ≈120 bp from left to right in an overlapping fashion. A schematic diagram is shown in Fig. 3.

1. Optimize the primer concentration. Initial concentration of each primer (forward or reverse) in the mixture is 10 pmol/μl. Use primer sets F_1/R_1, F_2/R_2, F_3/R_3, F_4/R_4, F_5/R_5, F_6/R_6, F_7/R_7, F8/R8, F_9/R_9 to start with.
2. Quantitate the DNA, for example, by Quant-iT™ PicoGreen® dsDNA Reagent according to manufacturer's instructions or alternatively use some other reliable method of accurate quantitation of DNA.

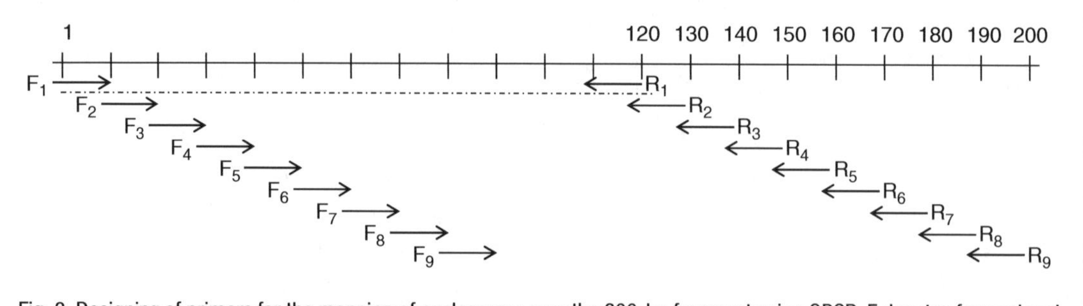

Fig. 3. Designing of primers for the mapping of nucleosome over the 200-bp fragment using QPCR. F denotes forward and R denotes reverse primers. Maximum amplicon length is kept 120 bp.

3. Prepare the PCR reaction in a PCR tube by mixing 2.0 µl 10× buffer, 2.0 µl 4 mM dNTPs, 0.8 µl 25 mM $MgSO_4$, 2.0 µl (≈ 100 ng) mononucleosomal DNA, 1.0 µl AT*ACTIN7*F forward primer, 1.0 µl AT*ACTIN7R* reverse primer, and 0.2 µl Taq-polymerase enzyme (5 U) and make up the volume to 20 µl with 11.0 µl milliQ water. Subject the samples to 28, 30, 32, or 35 cycles of repeated denaturation at 95°C for 1 min, annealing at 60°C (primer-specific temperature) for 30 s, and extension at 72°C for 45 s.
4. Electrophorese the PCR product in a 2.5–3% agarose gel containing 0.5× TBE for 30 min at 50 V.
5. Stain the gel with ethidium bromide and see for the presence of bands of approximate size of 120 bp in a transilluminator at long-range wavelength.
6. Once a particular primer set (say$F_{(N)}$)/$R_{(N)}$ shows the presence of nucleosome, further length of nucleosome can be estimated using primer set $F_{(N-1)}$/$R_{(N)}$. An example of the analysis of mononucleosome with a semiquantitative PCR is shown in Fig. 3.

3.2.2. Nucleosome Border Mapping with PAGE

This method yields high-resolution map of all positioned nucleosomes on the genomic fragment. The primer used is chosen from the results of above-mentioned QPCR analysis.

1. Mix 5 µl of 2 µM oligonucleotide primer (i.e., 10 pM final), 10 µl of 5× T4 polynucleotide kinase reaction buffer, 2 µl of T4 polynucleotide kinase (20 U), 5 µl [γ-^{32}P] ATP (50 µCi), and 28 µl milliQ water to prepare 50 µl reaction mixture for 5′-end labeling of primer.
2. Incubate at 37°C for 10 min. Heat inactivate at 65°C for 10 min.
3. Purify the oligonucleotide using the nucleotide removal kit and elute the oligonucleotide with 200 µl EB buffer (supplied with kit). Count 5 µl by liquid scintillation counting.
4. Sequentially mix 2.0 µl 10× PCR buffer, 2.0 µl 4 mM dNTPs, 2.0 µl radiolabeled forward primer (from step 3), 1.0 µl 25 mM $MgCl_2$, 0.5 µl (≈25 ng) mononucleosome template, and 2.0 µl Taq DNA polymerase (i.e., 6 U final) and make up the volume up to 20 µl with 10.5 µl of milliQ water to prepare forward reaction. Similarly, prepare the reverse reaction using reverse primer instead of forward one. Subject the samples to 35 cycles of repeated denaturation at 95°C for 3 min, annealing at 55°C (primer-specific temperature) for 1 min, and extension at 72°C for 2 min. Heat inactivate the samples at 65°C for 15 min. Cool on ice and then load 5.0 µl on gel.
5. Follow the procedure as given in the protocol supplied with Thermo Sequenase Kit (US Biologicals, USA) for preparing

sequencing ladder reactions for G, A, T, and C. To start with, take four tubes labeled G, A, T, and C and place 4 μl of the ddGTP termination mix in the tube labeled G. Similarly, fill the A, T, and C tubes with 4 μl of the ddATP, ddTTP, and ddCTP termination mixes, respectively. Cap the tubes to prevent evaporation.

6. In a microcentrifuge tube, combine 6.5 μl milliQ water, 2.0 μl (≈100 ng) template DNA, 2.0 μl reaction buffer, 1.0 μl labeled primer from step 3, and 2.0 μl Thermo Sequenase DNA buffer to prepare a final volume of 21.5 μl.
7. Take 4 μl of the above reaction mixture, transfer it to the tube labeled G (from step 5), and mix gently. Similarly, transfer 4 μl of the labeling reaction to the A, T, and C tubes, and mix by pipetting up and down several times. Cap the vials and place them in the thermal cycler.
8. Submit the samples to 40 cycles of repeated denaturation at 95°C for 30 s, annealing at 60°C for 1 min (as per Thermo sequenase protocol), and extension at 72°C for 1 min.
9. Add 4 μl of stop solution (supplied with kit) to each of the termination reactions, mix thoroughly, and centrifuge briefly to separate the oil from the aqueous phase. Alternatively, remove 6 μl from each termination reaction and transfer to a fresh tube containing 3–4 μl of stop solution. Samples may be stored frozen until ready to load the sequencing gel (up to 1 week).
10. Heat the samples to 75°C for 2 min immediately before loading onto the gel. Heating in open vials promotes evaporation of water from the formamide reaction mixture. This increases the signal by concentrating the isotope and promotes more complete denaturation of the DNA. Avoid overheating. Load 2–3 μl in each gel lane.
11. To analyze on polyacrylamide gel, prepare the 8% sequencing (denaturing PAGE) gel (16).
12. Pre-electrophorese the gel in 1× TBE buffer at 30 W for 1 h to raise the temperature up to 50°C. Flush the wells with 1× TBE buffer to remove the gel debris.
13. Incubate the samples of primer extension (from step 4) and sequencing ladder (from step 8) at 65°C for 5 min and load to wells. Load 5 μl of dye (0.3% bromophenol blue and 0.3% xylene cyanol FF) into an empty outer well to track migration of the gel. Electrophorese the gel at 30 W for 3 h. Maintain the temperature of gel at 50°C throughout the run.
14. Transfer the gel after completion of electrophoresis on a 3 mm Whatman sheet and wrap in saran wrap, dry the gel before exposing to Kodak or Fuji films, and keep at room temperature for the required amount of time. Scan the films on the Phospho-Imaging System (Bio-Rad).

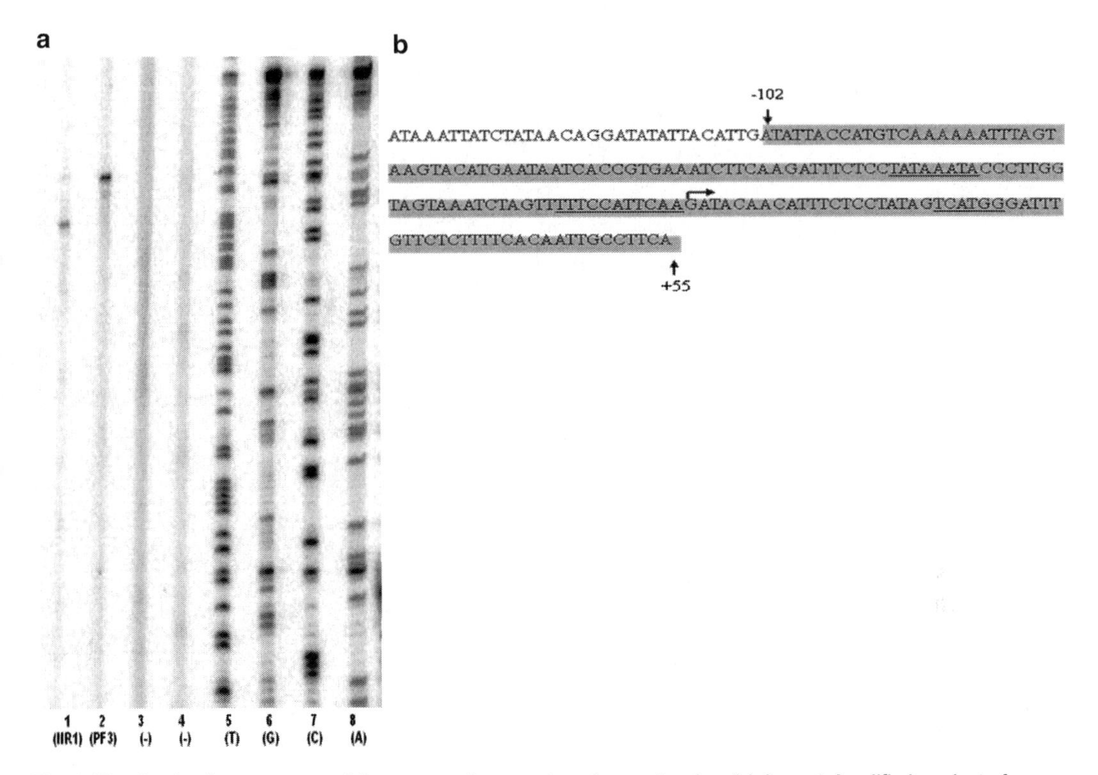

Fig. 4. Mapping border sequences of the core nucleosome by primer extension. (**a**) *Lane 1*: Amplified product of reverse primer (NR1). *Lane 2*: Amplified product of forward primer (PF3). *Lanes 3* and *4*: Nontemplate controls for NR1 and PF3, respectively (negative controls). *Lanes 5–8*: Sequence ladders for T, G, C, and A, respectively. (**b**) Nucleotide sequence of core promoter of tobacco PR-1a promoter showing in dark the −102 to +55 region covered by the nucleosome. The TATA, Inr-like region, and downstream promoter-like sequences are underlined. The transcription start site is shown by *arrow.*

Successful amplification indicates nuclease stability of the primer-bound region due to its protection by a nucleosome. An example of mapping border sequences of the core nucleosome by primer extension is shown in Fig. 4.

3.3. Deciphering Histone Modifications and Protein Interactions on Chromatin Using Chromatin Immunoprecipitation in Plants

ChIP methodology can assess the interplay between the chromatin dynamics, such as DNA methylation, chromatin structure/histone modifications, and the binding of transcription factors (TFs) (17). The ChIP method involves the cross-linking of protein and DNA within the cells, followed by total isolation of protein–DNA mixture/chromatin, its random shearing and immunoprecipitation with the antibodies directed against the protein of interest. Finally, the co-immunoprecipitated DNA is purified and can be analyzed by PCR for the enrichment of specific regions. The relative enrichment or depletion of a particular DNA fragment in the immunoprecipitated fraction reflects its in vivo association with the examined protein (18, 19). ChIP is based on the property of protein A/G (a protein from *Staphylococcus aureus*) to bind F_c portion of immunoglobin IgG. Thus, agarose beads coated with protein A/G precipitate immunocomplexes along with the antibody against protein of interest.

Based on the method of chromatin shearing, the technique is of two types.

(a) X-ChIP, where the DNA and proteins in chromatin are cross-linked (preferably using formaldehyde) and mechanically sheared, is preferred when some weak interactions between DNA and protein are to be investigated, liable to be broken otherwise during long process of ChIP.

(b) N-ChIP, where there is no cross-linking and chromatin is enzymatically sheared with MNase1 to obtain mononucleosomes, is preferred to study histone–DNA interactions.

ChIP is a three-day-long process, but the steps involving overnight incubation with antibody and reverse cross-linking can be optimized for shorter incubations.

The whole process can be divided into four parts (20, 21): (a) isolation and fragmentation of chromatin; (b) incubation of chromatin with antibody, i.e., immunoprecipitation; (c) collection of complexes bound to antibody and DNA purification; and (d) analysis of precipitated DNA using QPCR. The schematic diagram of ChIP is shown in Fig. 5.

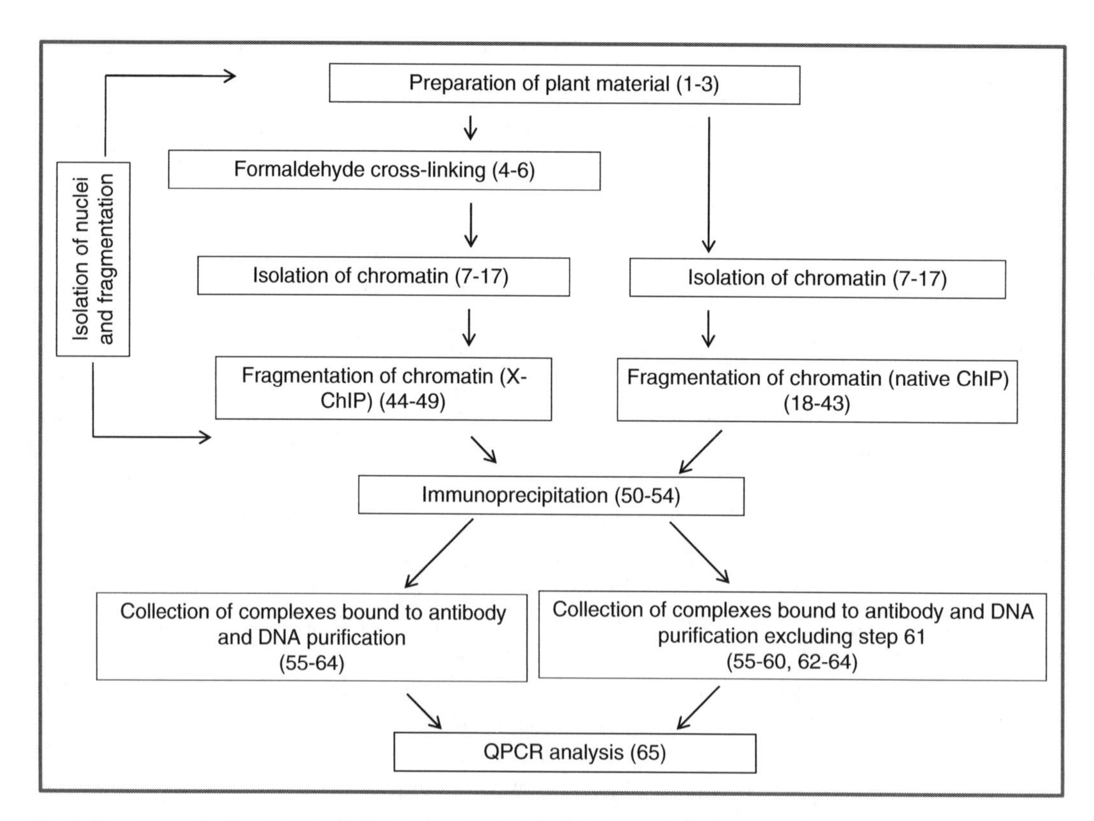

Fig. 5. The schematic flowchart of ChIP is shown to explain the steps to be followed according to the method chosen, i.e., N-ChIP or X-ChIP.

3.3.1. Isolation and Fragmentation of Chromatin

Preparation of plant material

1. Sow *Arabidopsis* seeds on soil or synthetic media as per the requirement of the experiment.
2. After suitable growth period, treat plants as needed for the experiment. Harvest 1.5 g of seedlings per treatment in a 50-ml Falcon tube (see Note 6).
3. Rinse seedlings twice with 40 ml of milliQ water by gently shaking the tube (room temperature). If doing X-ChIP, proceed to section "Formaldehyde cross-linking"; otherwise, move to section "Isolation of chromatin".

Formaldehyde cross-linking

4. After thoroughly removing the water, submerge the seedlings in 37 ml of 1% formaldehyde in extraction buffer 1 (room temperature) and under vacuum for 10 min (see Note 7).
5. Stop the cross-linking by addition of glycine to a final concentration of 0.125 M and apply vacuum for an additional 5 min. At this stage, seedlings should appear translucent.
6. Rinse seedlings twice with 40 ml cold milliQ water.

Isolation of chromatin

7. Remove the water as thoroughly as possible by placing the seedlings on a tissue paper before transferring to a new Falcon tube. At this stage, the cross-linked material can be either frozen in liquid nitrogen and stored at –80°C or can be carried out at 4°C, unless stated otherwise (see Note 8).
8. Precool the mortar and pestle by filling with liquid nitrogen before placing the seedlings in rest of the liquid nitrogen and grind them to a fine powder.
9. Resuspend the powder in 30 ml of extraction buffer 1 in a 50-ml Falcon tube.
10. Filter the solution through 20-μm nylon mesh into a SS-34 tube or equivalent.
11. Spin the filtered solution for 20 min at 2,880 × *g* at 4°C.
12. Gently remove the supernatant and resuspend the pellet in 1 ml of extraction buffer 2.
13. Transfer the solution to 1.5-ml Eppendorf tubes.
14. Centrifuge at 12,000 × *g* for 10 min at 4°C.
15. Remove the supernatant and resuspend the pellet in 300 μl of extraction buffer 3.
16. Overlay the resuspended pellet onto 300 μl of extraction buffer 3 in a fresh Eppendorf tube.
17. Spin for 1 h at 16,000 × *g* at 4°C. If doing native ChIP, proceed to section "Fragmentation of chromatin (native ChIP)";

otherwise, move to section "Fragmentation of chromatin (X- ChIP)".

Fragmentation of chromatin (native ChIP)

18. Remove the supernatant and resuspend the chromatin pellet in 1 ml of buffer N by pipetting up and down (maintaining 4°C temperature). Try to obtain single-nuclei suspensions.
19. Spin nuclei at 3,000 × *g* for 5 min in a microcentrifuge at 4°C. Remove the supernatant by aspiration and discard it.
20. Resuspend the nuclei in 100 μl of buffer N (see Note 9). Place the nuclei on ice. Determine the relative nucleic acid content of the sample by taking a 10-μl aliquot. To this aliquot, add 490 μl of 2 M NaCl and place on ice for 5 min.
21. Add 500 μl of chloroform:isoamyl alcohol, shake vigorously, and spin at 16,000 × *g* for 5–7 min in a microcentrifuge at 4°C.
22. Aspire the upper layer carefully and repeat step 21.
23. Aspire the upper layer and add 0.8 vol–1.0 volume of ice-cold isopropanol.
24. Spin at 16,000 × *g* for 15–20 min at room temperature.
25. Remove the supernatant and wash the pellet with 70% ethanol.
26. Dry and redissolve the pellet in 10 μl milliQ water.
27. Quantitate DNA on Nanodrop spectrophotometer.
28. Estimate the total amount of DNA in rest of the chromatin and add appropriate amount of MNase1 (see Note 10) to 100 μl of single-nuclei suspension.
29. Incubate for 10 min at 37°C with gentle shaking.
30. Stop the reaction by adding 11 μl (one-tenth volume) of MNase1 stop solution, and place on ice.
31. Lyse nuclei by adding 15 μl of 5 M NaCl to the 111 μl of MNase1-digested sample.
32. Add 200 μl of HAP buffer 1 as well as 66 mg of hydroxyapatite resin per 100 μg of DNA, vortex thoroughly, and incubate on a rotator for 10 min at 4°C.
33. Transfer the chromatin/hydroxyapatite slurry to a microspin column (see Notes 11 and 12).
34. Place the microspin column in an Eppendorf tube. Spin in microfuge for 1 min at 6,000 × *g* (4°C). Discard the flow through (see Note 13).
35. Add 200 μl of HAP buffer 1 to a microfuge tube that contains the chromatin/hydroxyapatite slurry to rinse it out, and then load onto the same microspin column.
36. Spin column again in microfuge for 1 min at 6,000 × *g* (4°C). Discard the flow through (see Note 13).

37. Repeat washes (steps 35 and 36) two more times with 200 μl of HAP buffer 1 (four washes in total). Discard the washes.
38. Add 200 μl of HAP buffer 2 to the same column.
39. Spin column again in microfuge for 1 min at 6,000×*g* (4°C). Discard the flow through (see Note 13).
40. Repeat washes (steps **38** and 39) three more times with 200 μl of HAP buffer 2 (four washes in total). Discard the washes (see Note 13).
41. Add 100 μl of HAP elution buffer to the same column. Transfer column to a fresh microfuge tube.
42. Spin column again in microcentrifuge for 1 min at 6,000×*g* (4°C). Keep the elution on ice (see Note 13).
43. Repeat steps 41 and 42 two more times by adding 100 μl of HAP elution buffer each time and using a fresh Eppendorf tube for each elution (to give three elutions in total). Keep the elutions on ice. Most of the chromatin should appear in elutions 1 and 2, with a small amount in elution 3. We generally pool all the three elutions 1, 2, and 3 for proceeding to immunoprecipitation (see Note 13). Add 1,700 μl of ChIP dilution buffer to make 2 ml of chromatin suspension. Move to Subheading 3.3.2.

Fragmentation of chromatin(X-ChIP)

44. Remove the supernatant and resuspend the chromatin pellet in 300 μl of nuclei lysis buffer by vortexing and pipetting up and down (keep solution at 4°C) (see Note 13).
45. Once resuspended, sonicate the chromatin solution for 10 s, four times, to shear the DNA to approximately 0.5–1-kb DNA fragments, with a 45-s pause between each pulse on Branson sonifier or equivalent. The sonicated chromatin solution can be frozen at –80°C or processed to step 20 for immunoprecipitation (see Note 14).
46. Spin the sonicated chromatin suspension for 5 min at 4°C (16,000×*g*) to pellet debris.
47. Remove the supernatant to a new tube. Use an aliquot of 5 μl to check sonication efficiency and electrophoretic determination of the average size of DNA fragments as compared with the aliquot from step 44 (see Note 13).
48. Split the 200 μl into two tubes with 100 μl each.
49. Add 900 μl of ChIP dilution buffer to each tube. This dilutes the SDS to 0.1% SDS.

3.3.2. Immunoprecipitation

50. Preclear each chromatin sample with 40 μl of salmon sperm-sheared DNA/protein A agarose beads per ml of chromatin for 1–2 h at 4°C with gentle agitation. Prior to use, the beads should be rinsed three times and resuspended in ChIP dilution buffer.

51. Spin the chromatin/beads solution at 4°C for 2 min at 16,000 × *g* or at an rpm as directed by manufacturer.
52. Combine the two 1 ml supernatant, and then split into three 2-ml microfuge tubes (666 μl each).
53. Add 4 μg of antibody each to two of the three tubes. The third tube without antibody should be used as mock/negative control (see Notes 15 and 16).
54. Incubate overnight at 4°C with gentle agitation. You can always optimize minimum incubation time sufficient to immunoprecipitate protein–DNA complex so as to decrease the time required for the experiment.

3.3.3. Collection of Complexes Bound to Antibody and DNA Purification

55. Collect the immunoprecipitate with 40 ml of protein A agarose beads (rinsed in ChIP dilution buffer) for at least 1 h at 4°C with gentle agitation.
56. Pellet beads by centrifugation (2 min, 16,000 × *g*) and wash them with gentle agitation for 5–10 min at 4°C each wash, using 1 ml of buffer per wash followed by pelleting the beads. Apply the following washes in the order listed below:
    (a) Low-salt wash buffer (one wash)
    (b) High-salt wash buffer (one wash)
    (c) LiCl wash buffer (one wash)
    (d) TE buffer (two washes)

After the final wash, remove the buffer thoroughly.

57. Prepare elution buffer.
58. Release bead-bound complexes by adding 250 μl of elution buffer to the pelleted beads.
59. Vortex briefly to mix and incubate at 65°C for 15 min with gentle agitation. Alternatively, elution buffer may be prewarmed at 65°C and beads may be incubated at 37°C for 15 min.
60. Spin the beads and carefully transfer the supernatant (elute) to a fresh tube and repeat elution of beads. Combine the two elutes.
61. Add 20 μl 5 M NaCl to elute to reverse the cross-links by an overnight incubation at 65°C. This step is done in case of X-ChIP only.
62. Add 10 μl of 0.5 M EDTA, 20 μl 1 M Tris–HCl (pH 6.5), and 1.5 μl of 14 mg/ml proteinase K to elute and incubate for 1 h at 45°C.
63. Add 3 μl of 10 μg/ml RNase A and incubate at 37°C for ½ h. Extract DNA twice with chloroform:isoamyl alcohol and precipitate with 1/10 V sodium acetate and 2 V absolute

ethanol at –80°C for at least 2 h or overnight. Spin for ½ h at 4°C and wash the pellet with 70% ethanol.

64. Dry and redissolve the pellet in 50 μl of TE (see Note 17).

3.3.4. Analysis of the Precipitated DNA Using QPCR

65. The immunoprecipitated and purified DNA is then used in PCR reactions to amplify examined target sequence, preferably using QPCR in order to quantify the enrichment or depletion of target(s) as compared with the suitable controls. Usually, 2–3 μl is used for a 25 μl PCR reaction. The amount of recovered templates may vary between experiments depending upon the efficiency of immunoprecipitation; thus, PCR conditions may need adaptation.

    ChIP is a long process involving manual handling of samples for approximately 3 days, and hence may suffer variations at different steps among different samples to be compared finally. Hence, there are plenty of controls cited in the literature against which ChIP template amplification data needs to be calibrated before actually comparing experimental values to control. Each of these offers advantage in establishing true correlation of data to hypothesis, but offers some practical problems as well. Hence, all cannot be taken up together in one experiment and it is important to decide upon which normalization factors are to be included in the experiment before actually starting to perform it.

    Below are mentioned those controls.

(a) Percentage of input: Input DNA offers normalization to starting template for immunoprecipitation. After fragmentation of chromatin, some amount (approx. 50 μl) of template is saved before proceeding to step 50. DNA precipitated from this sample is then amplified using some ubiquitous reference gene so as to normalize the differences in the initial amount of DNA from different samples subjected to immunoprecipitation.

(b) No-antibody control: This is also an important control to check that no DNA nonspecifically present in the tube (DNA also adheres nonspecifically to the walls of tube during overnight incubation with antibody) affects final interpretation of the data. Theoretically, there should be no or insignificant, if any, amplification of gene of interest from this sample, but if significant needs to be normalized to correctly interpret the data. Again, because no-antibody control gives very low amplification signal values, small changes in these values can have huge effect on data interpretation. Hence, while representing, the data representation should be made both with and without normalized to no-antibody control.

(c) IgG control: Some DNA can also bind nonspecifically to antibody either directly or indirectly. To avoid such carryover, immunoprecipitation of chromatin with IgG may also be

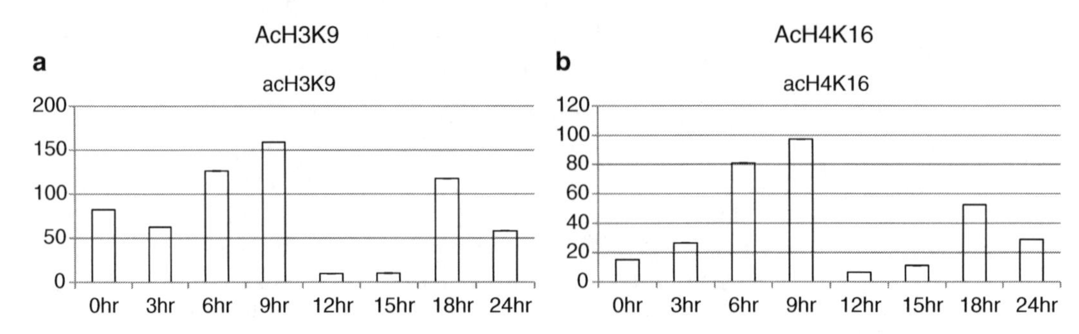

Fig. 6. Time-course analysis of the acetylation status on R1 upon SA induction for different time periods: histone acetylation status on *PR-1a* core promoter region was analyzed by ChIP assay at a given time point post salicylic acid induction using (**a**) anti-acetyl H3K9 and (**b**) anti-acetyl H4K16 antibodies. The immunoprecipitated DNA was analyzed by QPCR. The histogram represents the fold changes w.r.t. 0 h nucleosomal occupancy on the core promoter (measured by anti-H3 ChIP at 0 h) normalized to NoAb control and input signals with SD at each time point.

planned beforehand so that amplifications from this sample may be used to normalize the data (22).

(d) Internal control with some reference gene: The immunoprecipitated DNA is amplified with target region of gene and some internal control reference gene for normalization of initial ChIP template amount. This reference gene region should be physically interactive to the protein of interest but should not be affected by experimental conditions to serve as an internal control. Generally, it is very difficult to find a suitable control gene for internal reference against every protein of interest.

An example of ChIP-QPCR study done to analyze acetylation events over the core promoter region of tobacco *PR-1a* upon induction with salicylic acid is shown in Fig. 6.

4. Notes

1. βME, PMSF, and PPIC must be added and homogenously mixed just ½ h before use.
2. Extraction buffers 1, 2, 3 and buffer N, described in materials of ChIP, must be freshly prepared or a day before, and elution buffer and MNase1 stop solution must be prepared the same day.
3. During nuclei isolation, let the tissue thaw completely after crushing, and then only add homogenization buffer to it.
4. Add Triton only after the tissue is completely homogenized to reduce excess froth.
5. Filtration through nylon mesh of pore size 20 μm should be carefully performed as the size of nuclei ranges between 15 and 20 μm.

6. The ChIP protocol described here is sufficient for studying the interaction of two different proteins with DNA under one set of conditions. If you wish to study the interaction of DNA with more proteins or more set of conditions in one experiment, scale up accordingly.
7. Formaldehyde cross-links interfere with MNaseI activity; one should not combine formaldehyde cross-linking with MNaseI treatment.
8. Although the ChIP experiment can be stalled at many steps, it is preferable to continue the experiment instead of storing the material or nuclei or chromatin as it results in decrease in efficiency of the experiment.
9. It is important to have clump-free nuclei before adding MNaseI; otherwise, access of the MNaseI to chromatin will not be even, which might result in some DNA being of high molecular weight and a loss of resolution in your ChIP studies.
10. The concentration of MNaseI to use in your digestion has to be determined empirically (we usually find good results using 1 U/μg of DNA).
11. When isolating mononucleosomes with HAP, centrifugation speed can be increased if flow through does not pass on time.
12. The microspin column used in ChIP can be either purchased or made by replacing old, used membrane from previously used column with new 40-μm cutoff filter and subjecting the tightly assembled column to sterilization (e.g., multiple washes with milliQ water and TE followed by UV exposure for 10 min).
13. Keep a 5-μl aliquot of all HAP elution steps in ChIP (i.e., steps 34, 36, 39, and 40 first wash, 42, 43, 44, and 47) to visualize nucleosomal arrays or sonication efficiency on an agarose gel after doing DNA precipitation from them (according to steps 21–26) and discard the remaining wash.
14. When performing X-ChIP, sonication needs to be optimized every time when starting material or the instrument changes. Since sonication can cause heating leading to dissociation of proteins from complexes, care must be taken to keep the sample cool while sonicating it.
15. Antibodies to be used in ChIP analysis must be certified ChIP grade. If using for the first time, the antibody should be stringently checked for its specificity and workability during ChIP reaction by additional means.
16. Since immunoprecipitation efficiency of antibody is affected by chromatin constitution, concentration and type of experimental material and it may also vary between batches of antibodies; it is wise to run some pilot experiments in order to determine appropriate titer of chromatin required for successful

immunoprecipitation by each antibody and for every lot before actually performing ChIP (22).

17. The ChIP template should preferably be dissolved in TE and not in milliQ water so that it is stable and in good condition for approximately 4–6 months.

Acknowledgments

The work has been supported by Council of Scientific and Industrial Research, India, under the project OLP0031.

References

1. Butler, P. J. (1983) The folding of chromatin. *CRC Cri. Rev. Biochem.* 15, 57–91.
2. Luger, K. (2003) Structure and dynamic behavior of nucleosomes. *Curr. Opi. Gen. Dev.* 13, 127–135.
3. Khorasanizadeh, S. (2004) The nucleosome: from genomic organization to genomic regulation. *Cell* 116, 259–272.
4. Kamakaka, R.T. (2003) Heterochromatin: proteins in flux lead to stable repression. *Curr. Bio.* 13, R317–R319.
5. Reinke, H., Hörz, W. (2003) Histones are first hyperacetylated and then lose contact with the activated PHO5 promoter. *Mol. Cell* 11, 1599–1607.
6. Boeger, H., Griesenbeck, J., Strattan, J.S., Kornberg, R.D. (2003) Nucleosomes unfold completely at a transcriptionally active promoter. *Mol. Cell* 11, 1587–1598.
7. Lee, C.K., Shibata, Y., Rao, B., Strahl, B.D., Lieb, J.D. (2004) Evidence for nucleosome depletion at active regulatory regions genome-wide. *Nat. Genet.* 36, 900–905.
8. Yuan, G.C., Liu, Y.J., Dion, M.F., Slack, M.D., Wu, L.F., Altschuler, S.J., Rando, O.J. (2005) Genome-scale identification of nucleosome positions in S. cerevisiae. *Science* 309, 626–630.
9. Sekinger, E.A., Moqtaderi, Z., Struhl, K. (2005) Intrinsic histone–DNA interactions and low nucleosome density are important for preferential accessibility of promoter regions in yeast. *Mol. Cell* 18, 735–748.
10. Soutoglou, E., Talianidis, I. (2002) Coordination of PIC assembly and chromatin remodeling during differentiation-induced gene activation. *Science* 295, 1847–1848.
11. Ng, D.W., Chandrasekharan, M.B., Hall, T.C. (2006) Ordered histone modifications are associated with transcriptional poising and activation of the phaseolin promoter. *Plant Cell* 18, 119–132.
12. Johnson, D.S., Mortazavi, A., Myers, R.M., and Wold, B. (2007) Genome-wide mapping of in vivo protein-DNA interactions. *Science* 316, 1497–1502.
13. Kharchenko, P. V., Tolstorukov, M. Y., Park, P. J. (2008) Design and analysis of ChIP-seq experiments for DNA-binding proteins. *Nat Biotechnol* 26, 1351–1359.
14. Park, P. J. (2009) ChIP-seq: advantages and challenges of a maturing technology. *Nat Rev Genet* 10, 669–680.
15. Lodhi, N., Ranjan, A., Singh, M., Srivastava, R., Singh, S. P., Chaturvedi, C. P., Ansari, S. A., Sawant, S. V., Tuli, R. (2008) Interactions between upstream and core promoter sequences determine gene expression and nucleosome positioning in tobacco *PR-1a* promoter. *Biochim Biophys Acta* 1779, 634–644.
16. Sambrook, J., Fritsch, E. F., and Maniatis, T. (1989) *Molecular cloning: A Laboratory Manual. Ed 2.* Cold Spring Harbor Laboratory Press. Cold Spring Harbor. N.Y.
17. Massie, C.E., and Mills, I.G. (2009) Chromatin immunoprecipitation (ChIP) methodology and readouts. *Methods Mol Biol* 505, 123–137.
18. Orlando, V., Strutt, H., and Paro, R. (1997) Analysis of chromatin structure by in vivo formaldehyde cross-linking. *Methods* 11, 205–214.
19. Hecht, A., and Grunstein, M. (1999) Mapping DNA interaction sites of chromosomal pro-

teins using immunoprecipitation and polymerase chain reaction. *Methods Enzymol.* 304, 399–414.

20. Bowler, C., Benvenuto, G., Laflamme, P., Molino, D., Probst, A. V., Tariq, M., Paszkowski, J. (2004) Chromatin techniques for plant cells. *Plant J* 39, 776–789.
21. Brand, M., Rampalli, S., Chaturvedi, C. P., Dilworth, F. J. (2008) Analysis of epigenetic modifications of chromatin at specific gene loci by native chromatin immunoprecipitation of nucleosomes isolated using hydroxyapatite chromatography. *Nat Protoc.* 3, 398–409.
22. Haring, M., Offermann, S., Danker, T., Horst, I., Peterhansel, C., Stam M. (2007) Chromatin immunoprecipitation: optimization, quantitative analysis and data normalization. *Plant Methods* 3:11.

Chapter 14

Analysis of Histones and Histone Variants in Plants

Ila Trivedi, Krishan Mohan Rai, Sunil Kumar Singh, Verandra Kumar, Mala Singh, Amol Ranjan, Niraj Lodhi, and Samir V. Sawant

Abstract

Histone proteins are the major protein components of chromatin – the physiologically relevant form of the genome (or epigenome) in all eukaryotic cells. For many years, histones were considered passive structural components of eukaryotic chromatin. In recent years, it has been demonstrated that dynamic association of histones and their variants to the genome plays a very important role in gene regulation. Histones are extensively modified during posttranslation viz. acetylation, methylation, phosphorylation, ubiquitylation, etc., and the identification of these covalent marks on canonical and variant histones is crucial for the understanding of their biological significance. Different biochemical techniques have been developed to purify and separate histone proteins; here, we describe techniques for analysis of histones from plant tissues.

Key words: Epigenome, Histones, RP-HPLC, Posttranslational modifications

1. Introduction

Eukaryotic gene expression is influenced by changes in both local and long-range chromatin structure (1). Accumulating evidence suggests that the combination of histone modification on nucleosomes and the presence or absence of various histone variants contribute to the encoding of epigenetic information, the ultimate regulation of gene expression. Therefore, the study of histone proteins and the posttranslational modifications that they carry has become increasingly important in relevance to their biological significance.

Randall H. Morse (ed.), *Chromatin Remodeling: Methods and Protocols*, Methods in Molecular Biology, vol. 833,
DOI 10.1007/978-1-61779-477-3_14, © Springer Science+Business Media, LLC 2012

The identification of variant histones is important for the understanding of their biological significance in transcriptional regulation. Variants of histones, namely, H1, H2A, H2B, and H3, have been shown to play important roles in epigenetic regulation of gene expression in eukaryotes. Many histone variants have already been reported in plants, for example His1-1, His1-2, His1-3 in *Arabidopsis thaliana* (2, 3), His1 in *Lycopersicon pennellii* (4), and H1-S in *Lycopersicon esculentum* during drought condition. Other important variants include H3.3, H2A.X, and H2A.Z; the latter has a specific function in transcription in *Saccharomyces cerevisiae* (5).

In addition to the identification of histone variants, posttranslational modifications in the histone N-terminal tails also play an important role in transcriptional regulation and nucleosomal dynamics. These modifications include acetylation, methylation, phosphorylation, ubiquitylation, ADP ribosylation, deimination, proline isomerization, and sumoylation (6). Histone modifications are present at different residues (more than 60 such residues have been reported). Methylation of lysine and arginine residues is associated with both transcriptional activation and repression (7, 8). Earlier, the histone methylation mark was thought to be permanent, but the discovery of LSD1 and the JMJC family of demethylases suggests that methylation is dynamically regulated. LSD1 is a FAD-dependent nuclear amine oxidase, which specifically acts on di- and mono-methylated lysine (9), while JmjC domain containing demethylases are iron-dependent dioxygenases, which demethylate mono-, di-, as well as trimethylated lysine residue (10). In some plants, the relatives of these histone demethylases have already been reported (11) and many groups are working on other plants.

Lys9 and Lys14 on histone H3 and Lys5, Lys8, Lys12, and Lys16 on histone H4 are important positions for acetylation in plants also. Steady-state levels of acetylation of the core histones result from the balance between the opposing activities of histone acetyltransferases (HATs) and histone deacetylases (HDACs) (12). HDACs fall in two broad categories: SIR2 family (NAD-dependent HDAC) and classical HDAC family again subdivided into two subfamilies – class I, RPD3 like, and class II, HDAC II. In addition, plants have one more subfamily – HD2 – which is unique to plant kingdom (13, 14).

Many different biochemical techniques have been developed to purify and separate histone proteins. Here, we present standard protocol for acid extraction of histones from nuclei optimized to work in plants and suitable for their analysis using various standard procedures, such as separation of extracted histones by reversed-phase (RP)-HPLC, analysis of histones by acid urea (AU) gel electrophoresis, and confirmation of histone variants by mass spectrometry-MALDI-TOF.

2. Materials

2.1. Acid Extraction of Histones

1. Milli Q grade water.
2. Blotting sheets.
3. Liquid nitrogen.
4. Mortar and pestle.
5. Homogenization buffer: 1 M hexylene glycol, 10 mM PIPES–KOH, pH 7.0, 10 mM $MgCl_2$, 0.5% (v/v) Triton X-100, 5 mM β-mercaptoethanol (βME), and 0.8 mM phenylmethylsulfonyl fluoride (PMSF) (see Note 1).
6. Homogenizer.
7. Muslin cloth.
8. Nylon mesh (80, 60, 40, 20 μm).
9. Nuclei wash buffer: 0.5 M hexylene glycol, 10 mM PIPES–KOH, pH 7.0, 10 mM $MgCl_2$, 0.5% (v/v] Triton X-100, 5 mM βME, and 0.8 mM PMSF (see Note 1).
10. Corex tubes.
11. Swinging bucket rotor, SW28, or equivalent.
12. Beckmann ultracentrifuge.
13. Percoll dilution buffer: 0.5 M sucrose, 25 mM Tris–HCl, pH 8, 10 mM $MgCl_2$, 0.5% (v/v] Triton X-100, 5 mM βME, and 0.8 mM PMSF. Volume make up with water (see Note 1).
14. Reconstituted Percoll: 0.5 M sucrose, 25 mM Tris–HCl, pH 8.0, 10 mM $MgCl_2$, 0.5% [v/v] Triton X-100, 5 mM βME, and 0.8 mM PMSF. Volume make up with 100% Percoll (see Note 1).
15. Sucrose pad: 2 M sucrose, 25 mM Tris–HCl, pH 8.0, 10 mM $MgCl_2$, 0.5% [v/v] Triton X-100, 5 mM βME, and 0.8 mM PMSF (see Note 1).
16. 0.4 N H_2SO_4.
17. Rotator.
18. Trichloroacetic acid (TCA).
19. Acetone.
20. Lyophilizer.
21. Standard materials for SDS-polyacrylamide gel electrophoresis (PAGE) and subsequent western blot.

2.2. Analysis of Histones

1. Standard instruments for PAGE.
2. Acrylamide:Bis acrylamide solution (29:1).
3. Urea.
4. *N,N,N,'N'*-tetramethylethylene-diamine (TEMED).

5. Ammonium persulfate (APS).
6. 2-Butanol.
7. Potassium acetate.
8. Denaturing sample buffer: 8 M urea, 8% 2-mercaptoethanol, 10% acetic acid, 0.002% methylene blue.
9. AU electrophoresis buffer: 5% glacial acetic acid, deionized water.
10. Staining solution: Coomassie Brilliant Blue R250, 1.75 M glacial acetic acid in methanol, deionized water.
11. Destaining solution: Methanol, 1.75 M glacial acetic acid, deionized water.

2.3. Fractionation of Histones by Reverse-Phase-HPLC

1. Shimadzu 10 AVP HPLC binary gradient system.
2. HPLC sample buffer: 0.1% trifluoroacetic acid (TFA) in HPLC-grade water.
3. RP-HPLC C8-column.
4. Shimadzu Auto Injector (Model SIL-10 ADVP).
5. Solvent A: Solvent A (5% acetonitrile, 0.1% TFA).
6. Solvent B: Solvent B (100% acetonitrile, 0.1% TFA).
7. Shimadzu UV/VIS photodiode-array detector (Model RF-10A XL Fluorescence Detector).
8. Shimadzu Class VP Version 4 software.
9. Lyophilizer.
10. Trypsin.
11. 2× SDS-loading buffer: 0.25% bromophenol blue, 0.5 M DTT.

3. Methods

3.1. Acid Extraction of Histones

Histone proteins are the major protein components of chromatin, the physiologically relevant form of the genome (or epigenome) in all eukaryotic cells. For many years, histones were considered passive structural components of eukaryotic chromatin. But now, it has been proved that histones participate in gene regulation and repression via posttranslational modifications. Since histones are extensively modified during posttranslation, the identification of these covalent marks on canonical and variant histones is crucial for the understanding of their biological significance. Different biochemical techniques have been developed to purify and separate histone proteins. The procedure of acid extraction of histones from plants is as follows.

1. Harvest 120 g of plant material and wash with autoclaved double-distilled water.
2. Dry the leaves in between blotting sheets and take out their midribs. Tear the leaves into pieces and grind it properly with liquid nitrogen in a pestle and mortar.
3. Transfer the powder in a beaker and add approximately five to six volumes of homogenization buffer and with the help of homogenizer, homogenize it properly (see Notes 2 and 3).
4. After proper homogenization, filter the homogenate through four layers of muslin cloth and add Triton X-100 up to a final concentration of 0.5%.
5. Leave it for 5 min and filter homogenate through nylon mesh starting from 80, 60, 40, and then 20 μm (see Note 4).
6. After filtration, add the filtrate to GSA tubes (or other suitable tubes) and centrifuge in RC-5C GSA rotor or equivalent at 3,000 × *g* for 10 min at 4°C.
7. Decant the supernatant and suspend the pellet properly in nuclei wash buffer.
8. Dilute supernatant up to ten volumes with nuclei wash buffer and centrifuge at 3,000 × *g* for 5 min at 4°C.
9. Prepare different dilutions of Percoll to be used at later steps by diluting reconstituted Percoll with Percoll dilution buffer.
10. Decant the supernatant and dissolve the pellet in 6 ml of 5% Percoll.
11. Meanwhile, set up a gradient of Percoll in six corex tubes having 6 ml each of sucrose pad, 80, 60, and 40% of Percoll from bottom to the top in every tube.
12. Prespin at 4°C in swinging bucket (SW-28 or equivalent) rotor at 3,000 × *g* (g_{ave}) for 30 min.
13. Overlay 1 ml (from 20 g of fresh leaves) of pellet suspension from step 11 on each prepared step gradient.
14. Centrifuge the gradient at 3,000 × *g* (g_{ave}) for 30 min at 4°C.
15. Isolate the nuclei layer between the sucrose pad and 80% Percoll in corex tubes.
16. Dilute nuclei suspension with at least three volumes of nuclei wash buffer and centrifuge it in SS-34 rotor or equivalent at 3,000 × *g* for 10 min at 4°C.
17. The pellet containing nuclei is resuspended in 0.4 N H_2SO_4 (use a minimal amount) within corex tubes (see Note 5).
18. Nuclei should be resuspended very well with no clumps left in the solution. If necessary, vortex the solution until clumps are dissolved.
19. Incubate on rotator at 4°C for at least 30 min/overnight.

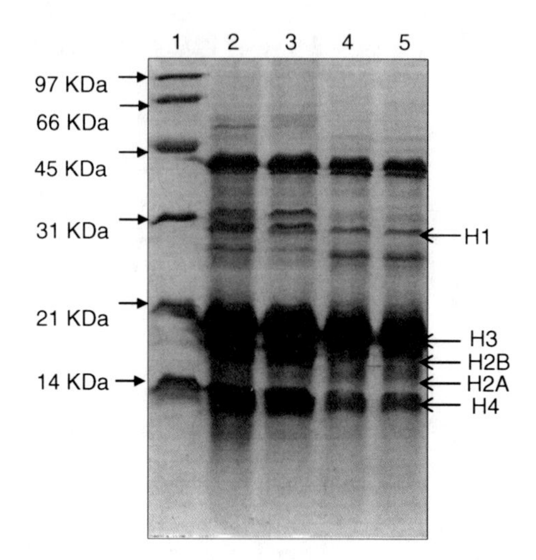

Fig. 1. Lane 1 shows low-range molecular weight marker, and lanes 2–5 show histones from different accessions of cotton plants. Bands corresponding to histones H1 (29 kDa), H2B (16.4 kDa), H3 (17 kDa), H2A (14.3 kDa), and H4 (11 kDa) are indicated.

20. Spin the samples at 16,000 × *g* for 10 min to remove nuclear debris.
21. Transfer the supernatant containing histones into fresh Corex tubes.
22. Add 33% TCA drop by drop to the histone solution and invert the tube several times to mix the solutions.
23. Incubate on ice for 30 min/overnight (see Note 6).
24. Pellet histones by spinning at 16,000 × *g* for 10 min, 4°C.
25. Carefully remove supernatant with pipette and wash the histone pellet with ice-cold acetone without disturbing it (see Note 7).
26. Spin in microcentrifuge at 16,000 × *g* for 5 min, 4°C.
27. Repeat steps 24 and 25 and carefully remove supernatant with pipette. Air dry the histone pellet at room temperature.
28. Dissolve the histone pellet in appropriate volume (typically, 20 μl) of Milli Q.
29. Spin for 10 min at 16,000 × *g*, 4°C.
30. Aliquot the supernatant and then lyophilize completely. Store at −80°C.
31. Run a 15% polyacrylamide-SDS gel to separate the histones on the basis of molecular weights (Fig. 1).
32. The presence of histones can also be confirmed by western blot analysis (Fig. 2).

3.2. Analysis of Histones Using Acid Urea Gel Electrophoresis

AU gel electrophoresis is a method that uses both the molecular size and charge as a base for protein separation at pH 3.0, and thus is routinely used to separate histones according to the individual charges. At pH 3.0, all proteins are likely to be positively charged

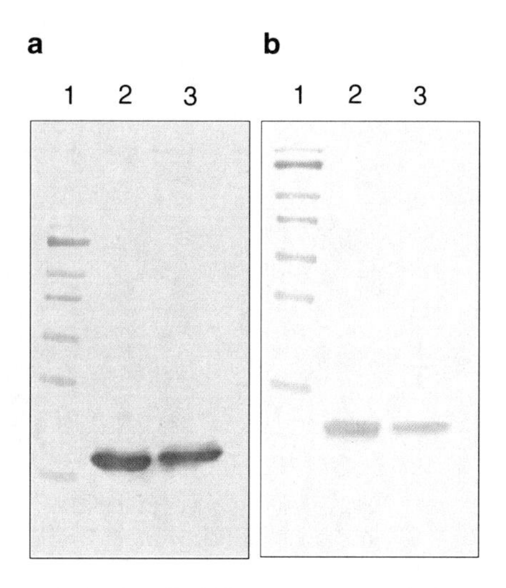

Fig. 2. Western blot analysis of purified histones. (**a**) Histones were blotted with anti-H3 antibody. Lane 1 shows prestained marker, and lanes 2 and 3 show histone H3 of molecular weight 17 kDa. (**b**) Histones were blotted with anti-H4 antibody. Lane 1 shows prestained marker, and lanes 2 and 3 show histone H4 of molecular weight 11 kDa.

(+) and to travel toward the cathode (−) in an electric field. As the proteins are simultaneously denatured by a high concentration of urea without affecting their charge (as opposed to the ionic detergent SDS), the charge of migrating proteins is determined solely by the number of protonated groups under the acidic running conditions. Hence, in an AU-PAGE, two proteins of similar size are separated from each other on the basis of their different charges. Proteins that are studied in the AU-PAGE are minor primary structure variants with slightly different charges or modified forms of the protein.

1. Assemble the clean glass plates and spacers in the form of a chamber, then prepare the separating gel (15% acrylamide:bis acrylamide solution (29:1), 6 M urea, 5% acetic acid, TEMED), and degas the solution (see Note 8).
2. Add 10% APS. Mix gently and pour immediately into the chamber and remove any air bubbles present. (Pour to a level about 0.5 cm below, where the bottom of the well-forming comb comes when it is in position).
3. Carefully overlay the acrylamide solution with 2-butanol without mixing and leave the gel for 1 h at room temperature for polymerization.
4. Now, prepare and degas the acidic stacking gel (6 M urea, 7.5% acrylamide, 0.375 M potassium acetate, TEMED).
5. Add 10% APS to it.
6. Pour off 2-butanol from the polymerized separating gel, wash the gel top with deionized water, and fill the remaining gap in the chamber with the acidic stacking gel mixture.

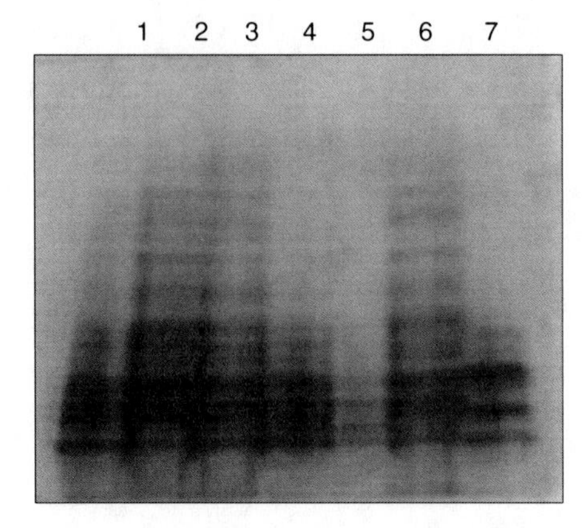

Fig. 3. Acid–urea gel electrophoresis of histones. Lanes 1–7 contain histones from different accessions of *Gossypium* sp.

7. Insert the well-forming comb gently and leave for 1 h for polymerization.
8. When the gel is set, remove the comb without breaking or distorting the sample wells.
9. Install the gel in the PAGE apparatus and fill the reservoirs with AU electrophoresis buffer.
10. Pre-run the gel at 180 V with reverse polarity at room temperature until the current falls to a steady level (3–5 h to overnight).
11. Prepare the samples in denaturing sample buffer in minimum volume (10–30 μL).
12. Electrophorese for 4–7 h at 180 V in fresh running buffer. Decreasing the voltage prolongs the run, whereas an increased voltage generates more heat, which may distort the appearance of the protein bands.
13. After electrophoresis, stain the gel in 500 ml of staining solution for several hours to overnight with gentle agitation.
14. Rinse the unbound dye by washing the gel of destaining solution several times.
15. After destaining, visualize the protein bands to be colored blue to purplish red (Fig. 3). For accurate identification of different bands corresponding to specific histone modifications, this technique can be combined with immunoblotting.

3.3. Fractionation of Histones by Reverse-Phase-HPLC

RP-HPLC separates molecules on the basis of hydrophobicity and is a high-resolution method ideally suited for analysis of histone proteins. Acid-extracted histones can be purified on a standard C-8 or C-18 column (the use of an RP C-8 column helps achieve higher resolution of histones) using an acetonitrile gradient. The method presented is optimized for separation of total histones (see Note 9). These lysine- and arginine-rich proteins require the use of end-capped columns to achieve an effective separation. In addition, the use of ion-pairing agents, such as TFA, is also necessary. The chromatogram shows the elution of acid-extracted histones: the fine peaks correspond to the eluted histones. The additional peaks display contaminating proteins or other chromatin-associated proteins (as analyzed by mass analysis). Purified histones can be further analyzed by mass spectrometry or subjected to enzymatic assays (e.g., methylation/demethylation assays) as needed.

1. We recommend using a Shimadzu (Kyoto, Japan) 10 AVP HPLC binary gradient system consisting of two system pumps (Model LC-10 AR VP) and equipped with a Shimadzu Auto Injector (Model SIL-10 ADVP) for the HPLC analysis. However, this protocol can be adapted to other systems as needed.
2. Based on the protein concentration, resuspend the lyophilized histone aliquots in HPLC sample buffer.
3. Directly inject 20 μL of the sample onto an RP-HPLC C-8 column (4.6 mm × 250 mm × 5 μm).
4. On an average, load 40 μg of proteins on the column for separation. Equilibrate the column in solvent A for 15 min with a flow rate of 1 ml/min.
5. Wash the column with solvent B on a short gradient from 0% B to 100% B for 30 min and re-equilibrate in 0% B for 10 min.

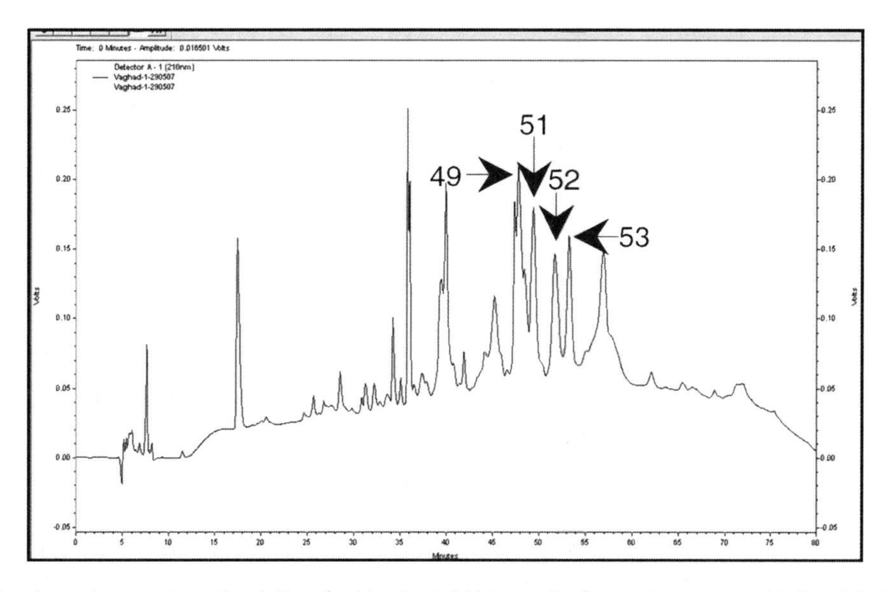

Fig. 4. The chromatogram shows the elution of acid-extracted histones: the fine peaks correspond to the eluted histones. The additional peaks display contaminating proteins or other chromatin-associated proteins.

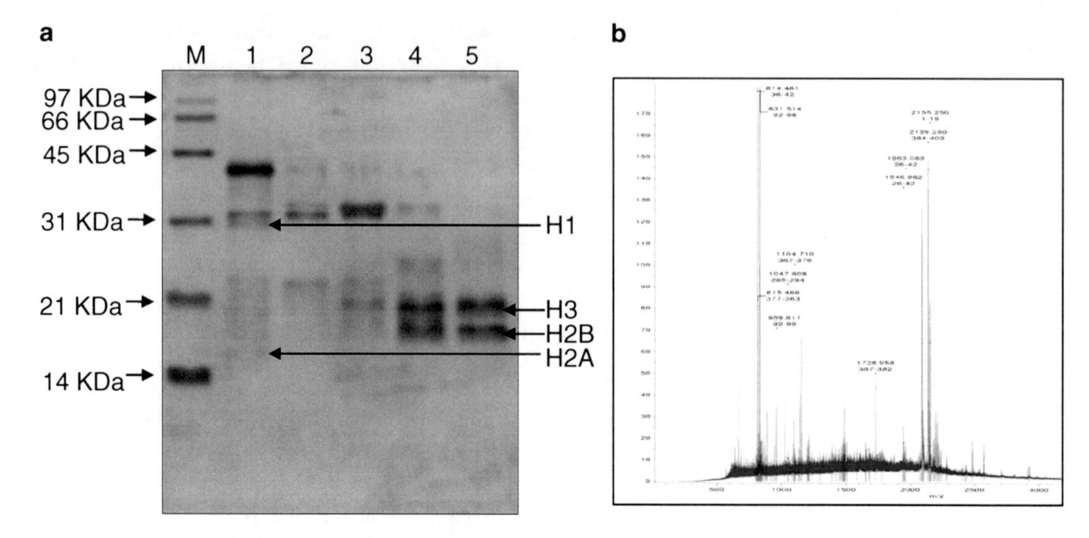

Fig. 5. (**a**) Silver staining after HPLC purification of histones. (**b**) The different peaks in the spectra correspond to different peptides of the same protein. The mass of the total protein is calculated by mass-to-charge (*m*/*z*) ratio.

6. Set a method in which the concentration of solvent A is increased linearly from 0% solvent B to 80% solvent B for 80-min run time, with a flow rate of 0.8 ml/min.
7. Detect histones with a UV/VIS detector (e.g., Shimadzu Model RF-10A XL Fluorescence Detector) set at 214 and 254 nm.
8. Perform the instrument control, data collection, and integration with Shimadzu Class VP Version 4 or other appropriate software.
9. Collect the recovered peaks and lyophilize completely in a speed vacuum (Fig. 4).
10. Dissolve the lyophilized powder in 2× SDS-gel loading buffer and load on a 15% SDS-PAGE (Fig. 5).
11. For analysis by mass spectrometry, excise the protein bands from the gel and partially digest with trypsin.

4. Notes

1. βME and PMSF must be added and homogenously mixed just 30 min before use.
2. During nuclei isolation, let the tissue thaw completely after crushing, and then only add homogenization buffer to it.
3. Add Triton only after the tissue is completely homogenized to reduce excess froth.

4. Filtration through nylon mesh of pore size 20 μm should be carefully performed as the size of nuclei ranges between 15 and 20 μm.
5. Histones are acid-soluble proteins, and most other proteins and nucleic acids precipitate in acid; therefore, the pellet containing nuclei should be properly suspended either in dilute H_2SO_4 or HCl during acid extraction of histones.
6. It is suggested to leave the protein in TCA for overnight precipitation at 4°C to improve the histone yield.
7. Washing the protein pellet with acetone should be done three times to completely remove TCA.
8. Urea, the hydrogen-bond breaking agent, is added to the AU-PAGE system in amounts traversing its entire range of solubility.
9. If further separation between individual histone proteins is required, the fractions can be reloaded onto the column and separated using an RP-HPLC method with a shallower gradient, typically improving the separation.

Acknowledgments

The work has been supported by Council of Scientific and Industrial Research, India, under the project OLP0031.

References

1. Higgs, D. R., Vernimmen, D., Hughes, J., Gibbons, R. (2007) Using genomics to study how chromatin influences gene expression. *Annual Reviews of Genomics and Human Genetics* 8, 299–325.
2. Ascenzi, R., Gantt, S. (1999) Subnuclear distribution of the entire complement of linker histone variants in *Arabidopsis thaliana. Chromosoma* 108, 345–355.
3. Ascenzi, R., Gantt, S. (1999) Molecular genetic analysis of the drought-inducible linker histone variant in *Arabidopsis thaliana. Plant Mol Biol* 41, 159–169.
4. Kahn, T. L., Fender, S. E., Bray, E. A., O' Connell, M. A. (1993) Characterization of expression of drought- and abscisic acid regulated tomato genes in the drought-resistant species *Lycopersicon pennellii. Plant Physiology* 103, 597–605.
5. Larochelle, M., Gaudreau, L. (2003) H2A.Z has a function reminiscent of an activator required for preferential binding to intergenic DNA. *EMBO J.* 22, 4512–4522.
6. Kouzarides, T. (2007) Chromatin modifications and their functions. *Cell* 128, 693–705.
7. Rice, J.C., and Allis, C.D. (2001) Histone methylation versus histone acetylation: new insights into epigenetic regulation. *Curr. Opin. Cell Biol.* 13, 263–273.
8. Zhang, Y., and Reinberg, D. (2001) Transcription regulation by histone methylation: interplay between different covalent modifications of the core histone tails. *Genes Dev.* 15, 2343–2360.
9. Shi, Y., Lan, F., Matson, C., Mulligan, P., Whetstine, J.R., Cole, P. A., Casero, R. A., Shi, Y. (2004) Histone demethylation mediated by the nuclear amine oxidase homolog LSD1. *Cell* 119, 941–953.
10. Tsukada, Y., Fang, J., Erdjument-Bromage, H., Warren, M. E., Borchers, C. H., Tempst, P., Zhang, Y. (2006) Histone demethylation by a

family of JmjC domain-containing proteins, *Nature* 439, 811–816.

11. Jiang, D., Yang, W., He, Y., Amasino, R. M. (2007) Arabidopsis relatives of the human lysine-specific demethylase1 repress the expression of FWA and FLOWERING LOCUS C and thus promote the floral transition. *The Plant Cell* 19, 2975–2987.
12. Wade, P. A. (2001) Transcriptional control at regulatory checkpoints by histone deacetylases: molecular connections between cancer and chromatin. *Hum. Mol.Genet.* 10, 693–698.
13. Pandey, R., Muller, A., Napoli, C.A., Selinger, D. A., Pikaard, C. S., Richards, E. J., Bender, J., Mount, D. W., Jorgensen, R. A. (2002) Analysis of histone acetyltransferase and histone deacetylase families of Arabidopsis thaliana suggests functional diversification of chromatin modification among multicellular eukaryotes. *Nucleic Acids Research* 30, 5036–5055.
14. Ruijter, A. J. M., Gennip, A. H. V., Caron, H. N., Kemp S., Kuilenburg, A. B. P. V. (2003) Histone deacetylases (HDACs): characterisation of the classical HDAC family. *Biochem. J.* 370, 737–749.

Chapter 15

Reconstitution of Modified Chromatin Templates for In Vitro Functional Assays

Miyong Yun, Chun Ruan, Jae-Wan Huh, and Bing Li

Abstract

To study the functions of histone modifications in the context of chromatin, it is necessary to be able to prepare nucleosomal templates that carry specific posttranslational modifications in a defined biochemical system. Here, we describe two sets of protocols for reconstituting designer nucleosomes that contain specifically modified histones. The resulting nucleosomes are suitable for electromobility shift assays, chromatin remodeling assays, and other functional and structural studies.

Key words: Reconstitution, In vitro, Chromatin, Histone modification, Nucleosome

1. Introduction

Posttranslational modifications (PTMs) of histones include acetylation, methylation, ubiquitination, phosphorylation, ADP ribosylation, and sumoylation, which play important roles in regulating transcription, chromatin assembly, DNA repair, recombination, and DNA replication (1, 2). Histone modifications also serve as epigenetic marks that can be inherited through cell division to maintain lineage specificity (3). Therefore, determination of PTM functions has been a central focus of the chromatin field for the past two decades (4).

Chemically modified histone peptides are commonly utilized to identify PTM recognition modules or used as substrates for various enzymatic reactions in vitro. Although this type of assay has led to many major discoveries in the field, some intrinsic shortcomings limit their broad applications. First, short histone peptides may only cover a partial functional surface of histones; second, without DNA, free histone peptides may not recapitulate the native

Randall H. Morse (ed.), *Chromatin Remodeling: Methods and Protocols,* Methods in Molecular Biology, vol. 833, DOI 10.1007/978-1-61779-477-3_15, © Springer Science+Business Media, LLC 2012

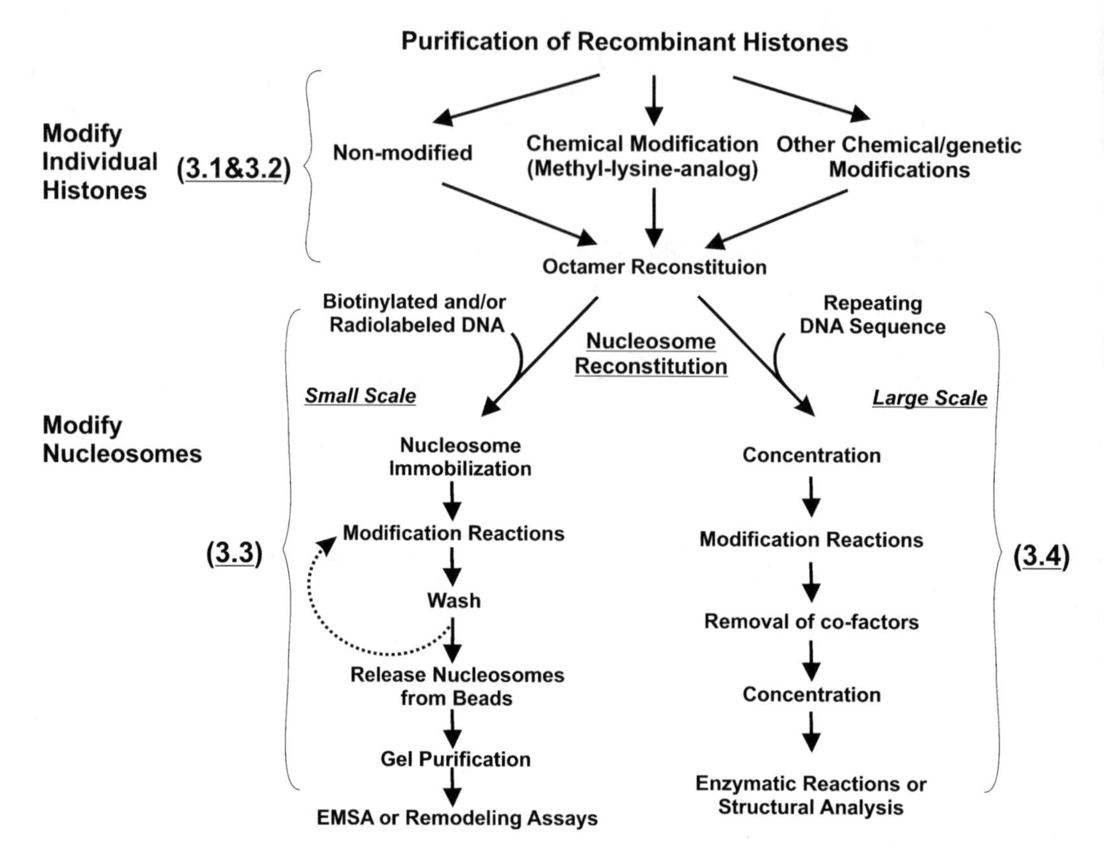

Fig. 1. A schematic diagram of our strategies to modify nucleosomes for studying PTM cross talk. Topics covered in each subtitle are summarized with a bracket.

conformation of intact nucleosomes; third, technical limitations of peptide synthesis may prevent the desired combination of PTM when multiple histones are involved or a great distance between modifications is needed. Therefore, using chromatin templates that carry specific PTM becomes increasingly desirable for biochemical analysis of histone modifications. In this chapter, we describe two sets of protocols for reconstituting designer nucleosomes that contain specifically modified histones. We first present a small-scale reconstitution method in which radiolabeled DNA templates are used and resulting nucleosomes are suitable for electromobility shift assays (EMSAs) or chromatin-remodeling reactions (Fig. 1 and Subheadings 3.1–3.3). We then discuss a generic method to prepare modified nucleosomes on a large scale for functional and structural studies. Histone modifications can be introduced either at the level of individual histones through chemical approaches (Fig. 1, Subheading 3.2) (5–7) or at the level of nucleosomes via a broad range of site-specific histone-modifying enzymes (Fig. 1, Subheadings 3.3 and 3.4) (8, 9). With proper pairings of these two strategies, one can expect to generate a single nucleosome containing various combinations of histone modifications for studying cross talk between PTMs. Due to space limitation, in the cases, where procedures described here are adapted from previously established protocols (5, 10–12), we primarily emphasize the modifications which we have made and the critical parameters that are important for successful subsequent steps.

2. Materials

2.1. Preparation of Recombinant Histone Octamers

1. Recombinant histone H3, H4, H2A, and H2B.
2. Spectra/Por Dialysis membrane (3,500 MWCO with a flat width of 18 mm, Spectrum).
3. Amicon ultra concentrator (5,000 MWCO, Millipore).
4. Unfolding buffer: 6 M guanidinium chloride, 20 mM Tris–HCl (pH 7.5), with freshly added 5 mM DTT.
5. Refolding buffer: 2 M NaCl, 10 mM Tris–HCl (pH 7.5), 1 mM EDTA, 5 mM 2-mercaptoethanol.
6. Superdex 200 column (GE Healthcare).
7. SDS running buffer: 25 mM Tris, 250 mM glycine, pH 8.3, 0.1% SDS.
8. 3× SDS loading buffer: 40% glycerol, 3% SDS, 0.8 M 2-Mercaptoethanol, 15 μM bromophenol blue.
9. Coomassie Blue staining buffer: 3 mM Coomassie Blue R250, 50% methanol, 10% acetic acid.
10. Destaining buffer: 50% methanol, 10% acetic acid.

2.2. Installing Methyl-Lysine Analog onto Histones

1. Recombinant histones.
2. 50°C incubator.
3. Alkylating buffer (made freshly): 1 M HEPES, pH 7.8, 4 M guanidine-HCl, 10 mM methionine.
4. 1 M DTT (*all DTT buffers in this session are made freshly from powder*).
5. Alkylating agents: *Agent1* (2-*N*-Methylaminoethyl chloride HCl for monomethylation); *Agent2* (2-Chloro-*N*,*N*-dimethylethylamine hydrochloride) for dimethylation; *Agent3* ((2-Bromoethyl) trimethylammonium bromide) for trimethylation.
6. 5 mM 2-Mercaptoethanol/H_2O: 5 mM 2-Mercaptoethanol in H_2O.
7. PD-10 Column (GE Healthcare).

2.3. Preparation of Radioactive-Labeled Nucleosomes for EMSA and Sliding

1. Primers for PCR amplification of probe DNA:
   5′ biotinylated Primer 774 (Bio-601 L5′): 5′-Bio -cgAGGC-CTcagctgGATATCacaggatgtatatatctgacacgtgcc. (*Note: The underlined sequence contains multiple restriction enzyme digestion sites which can be used to release the DNA/nucleosomes from the beads. EcoRV is our primary choice.*)
   Primer 773 (601L-216-3′): 5′-TGACCAAGGAAAGCA TGATTCTTCACAC.
2. The 601 positioning sequence containing plasmid (pBL386-601R).
3. γ-^{32}P-ATP, 1 mCi/7 μl.
4. T4 polynucleotide kinase (T4 PNK).
5. LA-Taq™ DNA polymerase (TaKaRa).
6. ProbeQuant G-50 mini column (GE Healthcare).
7. DNA size standards.
8. Gel extraction kit (e.g., QIAquick® gel extraction kit, Qiagen).
9. Recombinant histone octamers (modified and unmodified) as prepared via Subheading 3.1.
10. Initial buffer: 50 mM HEPES (pH 7.5), 1 mM EDTA, 0.1% Igepal CA-630 (Nonidet P-40 substitute), 20% glycerol, 5 mM DTT, 0.5 mM PMSF, 100 μg/ml BSA.
11. Final buffer: 10 mM Tris–HCl (pH 7.5), 1 mM EDTA, 0.1% Igepal CA-630, 20% glycerol, 5 mM DTT, 0.5 mM PMSF, 100 μg/ml BSA.
12. 10×TBE buffer: 890 mM Tris base, 890 mM boric acid, 20 mM EDTA.
13. IF100 buffer: A mixture of 1 volume of initial buffer, 1 volume of final buffer and 0.04 volume of 5 M NaCl.
14. Dynabeads® M-280 Streptavidin coated magnetic beads (Invitrogen).
15. Buffer H600: 25 mM HEPES, pH 7.6, 0.5 mM EDTA, 0.1 mM EGTA, 2.5 mM MgCl2, 10% glycerol, 0.02% Igepal CA-630, 1 mM DTT, 0.1 mg/ml BSA, and 600 mM KCl.
16. Histone-modifying complexes that are active on nucleosome templates (see Note 7).
17. 100 μM Acetyl-CoA.
18. 200 μM S-Adenosyl methionine (SAM).
19. 5×HAT buffer: 250 mM Tris–HCl, pH 8.0, 250 mM KCl, 5 mM DTT, 25% glycerol, 0.03% Igepal CA-630.
20. 5×HMT buffer: 250 mM Tris–HCl, pH 8.0, 250 mM NaCl, 5 mM $MgCl_2$, 10 mM DTT, 25% glycerol, 0.03% Igepal CA-630.

21. RE digestion buffer: 50 mM HEPES, pH 7.9, 100 mM NaCl, 2 mM $MgCl_2$, 0.04% Igepal CA-630, 0.1 mg/ml BSA, and 5 mM DTT.
22. Gel elution buffer: 10 mM Tris-HCl, pH 7.4, 100 mM NaCl, 1 mM EDTA, 5 mM DTT, 0.5 mM PMSF, 0.1 mg/ml BSA.
23. 5× gel final buffer: 10 mM Tris-HCl, pH 7.4, 100 mM NaCl, 1 mM EDTA, 5 mM DTT, 0.5 mM PMSF, 0.1 mg/ml BSA, 50% glycerol, 0.25% Igepal CA-630.

2.4. Preparation of Modified Nucleosome in a Large Scale

1. EcoR V (100,000 U/ml).
2. Ice-cold 100% ethanol.
3. Isopropanol.
4. 5 M NaCl, autoclaved.
5. 3 M sodium acetate (pH 5.2), autoclaved.
6. TAE buffer; 0.04 M Tris–acetate, 1 mM EDTA (pH 8.0), 3.3 mM glacial acetic acid.
7. 10× DNA loading dye.
8. 40% PEG8000, autoclaved.
9. Spectra/Por Dialysis membrane (3,500 MWCO with a flat width of 18 mm, Spectrum).
10. GelRed™ Nucleic Acid Gel Stain; 10,000× in water (Biotium).
11. 10×TEB: 100 mM Tris–HCl, pH 7.5, 10 mM EDTA, 10 mM 2-Mercaptoethanol.
12. TEBS (dialysis buffer): 1× TEB supplemented with different concentrations of NaCl as follows: TEBS2.0, TEBS1.2, TEBS1.0, TEBS0.8, and TEBS0.6 contain 2, 1.2, 1, 0.8, and 0.6 M NaCl, respectively.
13. Native PAGE gel and electrophoresis equipment.
14. 10×TBE buffer: 890 mM Tris base, 890 mM boric acid, 20 mM EDTA.
15. Acetyl Coenzyme A, (Acetyl-^{3}H) and nonlabeled Acetyl-CoA.
16. Sephadex G-50 purification column.
17. P81 Phosphocellulose membrane.
18. 1× HAT wash buffer: 50 mM $NaHCO_3/Na_2CO_3$ pH 9.2.
19. Acetone.
20. Scintillation fluid for filters.
21. TE 10/0.1: 10 mM Tris–HCl, 8.0; 0.1 mM EDTA.

2.5. Histone Deacetylase Assays

1. 1× histone deacetylase (HDAC) buffer: 20 mM Tris–HCl, pH 8.0, 5 mM 2-Mercaptoethanol, 5% glycerol.

2. 1 M HCl/0.4 M acetic acid.
3. Ethyl acetate.
4. Scintillation cocktail (e.g., ScintiSafe 30% Cocktail (Scintanalyzed*, Fisher)).

3. Methods

3.1. Preparation of Recombinant Histone Octamers

1. Lyophilize 0.15 μmol of recombinant histone H3, H4, H2A, and H2B (6 h to overnight) (see Note 1).
2. Dissolve each histone pellet in 1 ml of unfolding buffer for 1 h at room temperature (RT) on an orbital mixer. Determine the concentration of histones by measuring OD_{276} and using the calculated extinction coefficients (see Note 2).
3. Mix the four histones at an equimolar ratio and adjust the final concentration to 1 mg/ml with unfolding buffer. Place the mixture at RT for 30 min and then dialyze (3,500 MWCO) against prechilled refolding buffer at 4°C as follows: 800 ml for 1 h; another 800 ml for 1 h; and then overnight in remaining 2.4 L.
4. Centrifuge at 27,000 × *g* (SS-34 rotor or equivalent) for 10 min at 4°C to remove any aggregates, and then concentrate the supernatant to a final volume of 1 ml using an Amicon-Ultra concentrator (5,000 MWCO). Centrifuge the sample at 21,000 × *g* for 5 min at 4°C to eliminate precipitates and then load it onto a Superdex 200 column (CV = 120 ml) pre-equilibrated with refolding buffer. The flow rate is set at 1 ml/min and 1-ml fractions are collected after 30 ml.
5. Run peak fractions on a 16% SDS-PAGE gel to check the stoichiometry of histone octamers. Wild-type histone octamers elute around 65–68 ml, and tailless histone octamers come out at 70–74 ml.
6. Pool all fractions containing histone octamers and concentrate to about 1 ml (producing solution of about 3–8 mg/ml). To increase the consistency of nucleosome reconstitution, we determine the concentration of histone octamers through a gel quantification method, in which a standard curve is created using a sample of histone at known concentration. When Xenopus, Drosophila, or human histones are used, the yield of octamers is about 80% of total histone proteins. Recombinant yeast histones are reconstituted less efficiently, with an average yield of about 20%.

3.2. Installing Methyl-Lysine Analog to Histones

The protocol in this session is based on a previous method with minor modifications (5).

1. Lyophilize 0.45 μmol (three 0.15-μmol aliquots) of genetically engineered Kc (the lysine of interest replaced by a cysteine) mutant histones (see Note 3).
2. Resuspend entire pellets in 1 ml of freshly prepared alkylating buffer supplemented with 20 μl of 1 M DTT. Place the tube on a roller mixer at RT for 30 min until histones are completely dissolved, and then transfer the tube to 37°C for 1 h to ensure that cysteine residues are completely reduced by DTT. Wrap the tubes with foil to avoid direct light exposure in subsequent steps.
3. Chemical reactions:

   (a) Monomethylation:
   - Add 100 μl of 1 M freshly made *Agent1* (2-*N*-Methylaminoethyl chloride HCl) and incubate at RT for 4 h.
   - Add 10 μl of 1 M DTT and incubate at RT for another 10 h

   (b) Dimethylation:
   - Add 50 μl of 1 M freshly made *Agent2* (2-Chloro-*N*,*N*-dimethylethylamine hydrochloride) and incubate at RT for 2 h.
   - Add 10 μl of 1 M DTT and incubate at RT for 30 min
   - Replenish 50 μl of 1 M *Agent2* and incubate at RT for 2 h.

   (c) Trimethylation:
   - Directly add 100 mg of *Agent3* (2-Bromoethyl trimethylammonium bromide) powder into the tube and incubate at 50°C for 2.5 h with occasional stirring at the beginning to dissolve the powder.
   - Add 10 μl of 1 M DTT and incubate at 50°C for another 2.5 h.
4. Chemical reactions are stopped by addition of 50 μl of 2-Mercaptoethanol (14 M).
5. Discard the upper storage buffer in PD-10 columns and pre-equilibrate the column four times with 5 ml of 5 mM 2-Mercaptoethanol/H_2O.
6. Load the reaction mixtures onto the PD-10 columns. Once samples are completely absorbed, add 1.5 ml of 5 mM 2-Mercaptoethanol/H_2O, but do NOT collect the flow through.
7. Transfer the PD-10 column to a fresh 50-ml conical tube. Add 3.5 ml of 5 mM 2-Mercaptoethanol/H_2O and collect eluates, which contain purified chemically modified histones.

8. Determine the concentration of histones by measuring OD_{276} and using deducted extinction coefficients. Aliquot 0.15 μmol of modified histones in each tube and store them at –80°C.
9. To reconstitute these MLA histones into octamers, proceed to step 1 in Subheading 3.1.

3.3. Preparation of Radioactive Nucleosomes for EMSA and Sliding Assays

End label biotinylated DNA probe containing the 601 positioning sequence (see Note 4)

1. End label one primer by setting up the following reaction: 1 μl of 20 μM Primer 773 (601L-216-3′), 1 μl of 10× kinase buffer (T4 PNK NEB), 0.5 μl of γ-^{32}P-ATP, 6.5 μl of H_2O, 1 μl of T4 PNK. Incubate at 37°C for 60 min and then heat inactivate PNK at 70°C for 10 min.
2. PCR amplify the biotinylated DNA probe: Mix 10 μl of 10×LA Taq buffer, 8 μl of 2.5 mM dNTPs, 10 μl of the entire kinase reaction from step 1, 1 μl of 20 μM Primer 774 (Bio-601-5′), 1 μl of pBL386-601R plasmid (10 ng/μl), 67 μl of H_2O, and 0.5 μl of LA Taq. Twenty-eight to thirty cycles of standard PCR conditions should result in adequate amount of probes.
3. Pass the entire PCR reaction above through a ProbeQuant G-50 mini-column to remove unincorporated γ-^{32}P ATP as suggested by manufacturers. The eluates are then directly loaded onto a 2.0% agarose gel in 1.5× TAE buffer. The DNA band with the correct size is excised and purified using standard Qiagen Gel extraction purification Kits. We elute DNA with 20 μl of prewarmed (42°C) 0.1× TE. Another 10 μl elution is performed to increase yield. Eluates are combined before the final quantification (see Note 5).

Nucleosome reconstitution via a serial salt dilution method

4. Nucleosome reconstitution is started by sequentially adding 5 M NaCl, 2 mg/ml BSA, H_2O, 1 pmol of radio-labeled DNA, and the proper amount of histone octamers (see Note 6) into a 10 μl reaction so that the final concentrations of NaCl and BSA are 2 M and 0.1 mg/ml, respectively.
5. The salt concentration is gradually reduced by adding 3.3, 6.7, 5, 3.6, 4.7, 6.7, 10, 30, and 20 μl of initial buffer with a 15-min interval between additions while samples are incubated at 30°C.
6. The reaction is brought to 0.1 M NaCl by adding 100 μl of final buffer and incubated for another 15 min at 30°C. The efficiency of nucleosome reconstitution is examined by running 2 μl of each sample on a 5% native PAGE gel in 0.3× TBE buffer.

Immobilization of radio-labeled nucleosomes on magnetic beads

7. Transfer 20 μl of streptavidin-coated magnetic beads slurry (10 mg/ml) to a screw-cap tube, and wash beads twice with

200 μl of IF100 Buffer using a magnetic stand. Withdraw the supernatant after the last wash.

8. Add the reconstitution mixture from step 6 to the tube containing prewashed magnetic beads and incubate overnight at 4°C on a rotating mixer to allow biotinylated nucleosomes' binding to the streptavidin-coated magnetic beads.
9. Nucleosome-bound beads are washed twice with 1 ml of Buffer H600 to eliminate nonspecific binding proteins. The beads are finally resuspended in 200 μl of IF100 buffer and store in 4°C or directly proceed to the next step. This method typically results in a 90–95% of immobilization efficiency as estimated by radioactivity (cpm) recovery.

Histone modification reaction catalyzed by chromatin-modifying enzymes

10. Spin briefly the screw-cap tube containing immobilized nucleosomes from step 9, and then let the beads settle on a magnetic stand for 2 min. Withdraw the supernatant carefully with a pipet, and then wash the beads twice using 300 μl corresponding reaction buffer supplemented with 0.1 mg/ml BSA (in our case here, either 1× HAT (BSA) or 1× HMT (BSA) is used).
11. At this point, the beads can be directly subjected to various histone modification reactions (see Note 7). We typically set up the reactions in a separate tube. The last wash of 1x reaction buffer in step 10 is removed right before the reaction mixtures are applied to the beads.
    (a) HAT reaction (100 μl):
    Mix 20 μl of 5× HAT buffer, 1 μl of 10 mg/ml BSA, 2.5 μl of 100 μM Acetyl-CoA, and the proper amount of histone acetylases complexes (see Note 8) and bring the final volume to 100 μl with H_2O.
    (b) HMT reaction (100 μl):
    Mix 20 μl of 5× HMT buffer, 1 μl of 10 mg/ml BSA, 2.5 μl of 200 μM SAM, and the proper amount of histone methyltransferase and bring the final volume to 100 μl with H_2O.
12. Gently invert the tubes for a few times, and then place them on a rotating mixer at 30°C for more than 4 h.
13. Collect beads on a magnetic stand. Wash them three times with 500 μl of Buffer H600 to remove cofactors and enzymes.
14. (Optional) Repeat steps 10 and 13 for additional modification reactions.

Restriction enzyme digestion to release nucleosomes from magnetic beads

15. The beads from the above procedures (steps 13 or 14, containing ~1 pmol DNA) are washed with 200 μl of RE digestion buffer twice.
16. Add 50 units of EcoRV to 50 μl of RE digestion buffer and then apply the mixture to prewashed beads from step 15. Incubate the tubes on a rotating mixer at 37°C for 4–8 h. Transfer the supernatant to another tube; monitor and compare the radioactivity of supernatant with the beads using a Geiger survey meter. A typical recovery rate is about 70–80%. The resulting nucleosomes can be stored at 4°C for up to 2 months.

Gel purification of all nucleosomes at the final stage

17. If the volume of the final elution is well above 50 μl, use Amicon ultra concentrator (10,000 MWCO) to bring it down to about 50 μl.
18. Load samples on a 5% PAGE gel in 0.3×TBE. Run at 11 V/cm at 4°C for 5 h.
19. Cover the wet gel with plastic wrap and expose directly to an X-ray film for 30 min. Use a black sharpie to mark down the gel position so that the bands corresponding to the desired nucleosomes can be precisely excised when the developed film is used as a reference.
20. Smash the excised gel slice into small pieces. Add about 1.5 gel volume of gel elution buffer (~200–400 μl), and elute the nucleosomes overnight at 4°C on a roller mixer.
21. Centrifuge at 21,000×*g* for 5 min at 4°C. Use barrel tips to withdraw the liquid from the gel and then transfer to a 0.5-ml Eppendorf tube. Add ¼ volume of 5× gel final buffer to the eluates and store them at 4°C.

    These nucleosomes can be directly used as substrates in EMSA assays or chromatin-remodeling assays.

3.4. A Large-Scale Preparation of Modified Nucleosomes

1. Construct vectors that contain multiple copies of certain nucleosome-positioning sequences (such as the 601 sequence and 5 S) or other designer sequences. Our strategy to construct repeating 601 sequences in pBlueScript vector is illustrated in Fig. 2, which is adopted from a previously established method (10) (see Note 9).
2. Prepare milligram quantities of plasmids that contain the repeating sequence (see Note 10).
3. Digest 5 mg of plasmids using 500 units of EcoRV (16 copies of 216 bp fragments) in a 10 ml reaction at 37°C for more

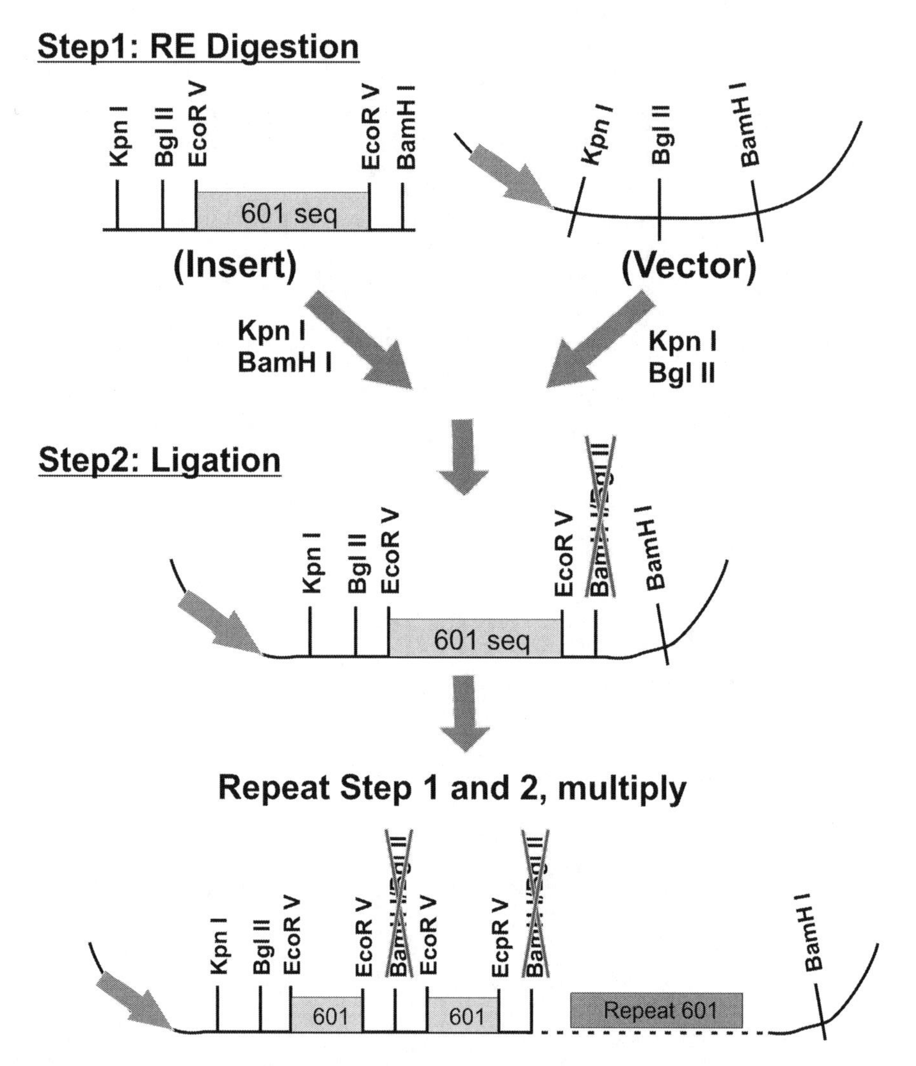

Fig. 2. Construction of vectors that contain multiple copies of nucleosome-positioning sequence. A plasmid containing a single copy of 601 sequence was digested with two sets of restriction enzymes (Step 1) and ligated to produce 2× copies of 601 (Step 2).

than 16 h. Check the completeness of digestion by running 1 μl of the sample on a 1.0% agarose gel. If undigested products remain, add more EcoRV and incubate for another 4 h.

Optimize the conditions for PEG purification of small DNA fragments

4. Set up five testing conditions each containing 1 μg of digested plasmid (2 μl of the reaction mixture in step 3) and 0.5 M NaCl. A different amount of 40% PEG8000 is then added to each condition so that the final concentration of PEG is 6, 7, 8, 9, or 10%, respectively. Bring the final volume of each reaction to 100 μl using 1× Buffer 3 (NEB) and mix well.

5. Incubate these tubes on ice for 1 h, and then centrifuge at 21,000×g at 4°C for 20 min. Carefully transfer the supernatant, which contains the EcoRV-digested small fragments, to a new Eppendorf tube and resuspend the pellets (vector backbone) in 20 μl TE 10/0.1.
6. Add 250 μl (or 2.5 volumes) of prechilled ethanol to each tube and incubate at –20°C for 30 min. Spin down the fragments at 21,000×*g* at 4°C for 20 min. Dissolve the pellets in 50 μl of TE 10/0.1.
7. Load 2 μl of samples from steps 5 and 6 on a 1% agarose gel (Fig. 3). Select the condition which results in the highest yield of small fragment without any backbone contamination.

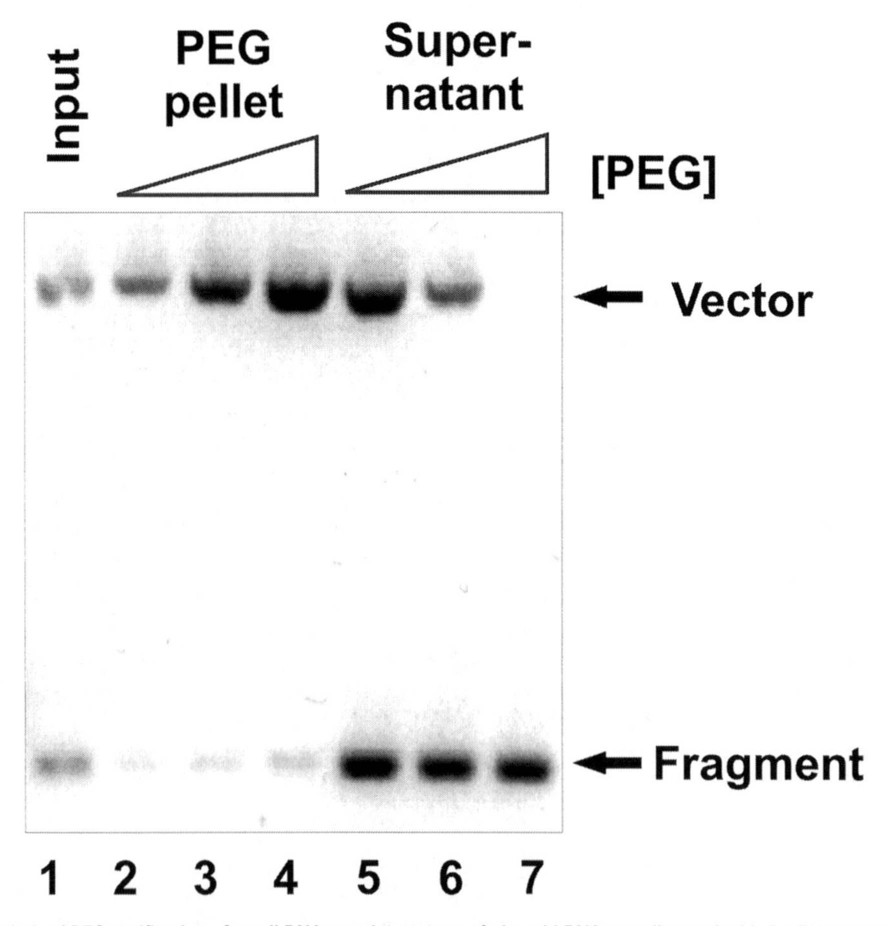

Fig. 3. Optimized PEG purification of small DNA templates. 1 μg of plasmid DNA was digested with EcoRV and precipitated with 7% (lanes 2 and 5), 8% (lanes 3 and 6), and 9% of PEG (lanes 4 and 7). DNA purified from both pellet (lanes 2–4) and supernatant (lanes 5–7) after PEG precipitation was run on 1% agarose gel and the ratio of small fragment (fragments) and backbone (vector) was determined. 0.5 μg of DNA directly purified after digestion was loaded in lane 1 (Input).

Purification of small DNA fragments for nucleosome restitution

8. Add 2 ml of 5 M NaCl to the 10 ml RE digestion mixture (step 3) and the proper amount of 40% PEG8000 as determined above (steps 4–7). Bring the final volume to 20 ml using 1× Buffer 3 before mixing the solution and incubate on ice for 1 h.
9. Centrifuge at 27,000 × g at 4°C for 20 min. Transfer the supernatant to another Oak Ridge tube. Add 20 ml of isopropanol and incubate at RT for 15 min.
10. Centrifuge at 27,000 × g at 4°C for 20 min. Rinse the pellets with 10 ml of 70% EtOH. Dissolve the air-dried pellets in 1 ml of TE 10/0.1, and determine the concentration of DNA fragments by comparing to known concentration DNA marker on a 2% agarose gel.

Reconstitution of nucleosome core particles

11. Empirically determine the optimal ratio of histone/DNA for the large-scale nucleosome reconstitution. Set up nucleosome reconstitution reactions as described in Subheading 3.3 (steps 4–7). Use 2 μg of DNA fragment for each reaction and titrate histone octamers in 1.25-fold increments. Load 20 μl of the final mixtures on a 4.5% native PAGE gel and run it at 11 V/cm for 1.5 h at RT. Stain the gel with the GelRed dye for 20 min, and visualize DNA (nucleosomes) under UV light (Fig. 4).

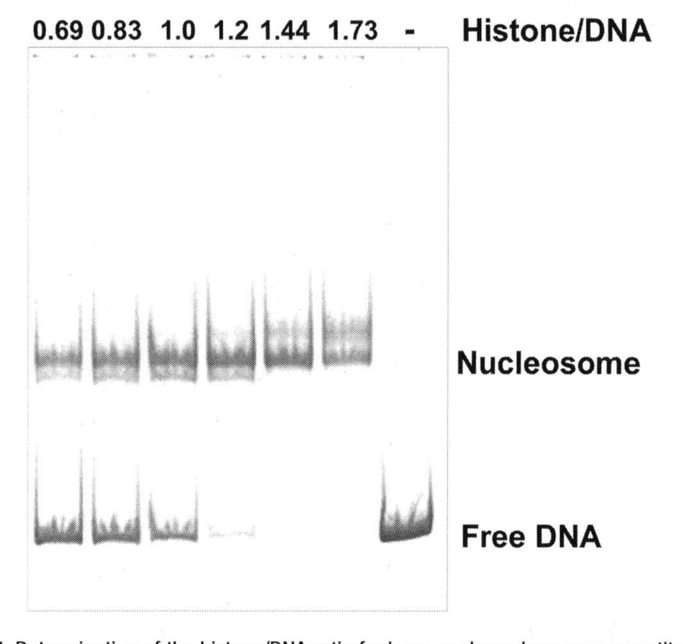

Fig. 4. Determination of the histone/DNA ratio for large-scale nucleosome reconstitution. Nucleosome core particles were reconstituted with 2 μg of DNA templates and histone octamers with 1.25-fold increments. 10% of resulting nucleosomes were run on a 4.5% native PAGE gel and visualized by GelRed staining under UV light. Numbers on top of each lane indicate the ratio of histone to DNA.

12. Mix 500 μg of DNA fragments (from step 10), 4 ml of 5 M NaCl, and 100 μl of 10× TEB buffer. Use H_2O to bring the final reconstitution mixture to 1 ml (including histone octamers). Gently invert the tube several times and then add the proper amount of histone octamers as determined above. Incubate the tube at 30°C for 30 min, and then transfer the mixture to a dialysis bag (3,500 MWCO).
13. Dialyze against 800 ml of each buffer listed below at RT with indicated duration:

TEBS 2.0 (2 M NaCl)	1 h
TEBS 1.0	2 h
TEBS 0.8	2 h
TEBS 0.6	2 h
1×TEB	Overnight (at 4°C)

14. Transfer the reconstitution mixture to an Eppendorf tube and incubate at 37°C for 30 min. Remove aggregates by centrifuging at 21,000 × *g* for 10 min at 4°C. Adjust the final concentration of nucleosomes to 1–2 mg/ml using Amicon ultra concentrators.

Histone modification reactions (acetylation)

15. Mix 20 μg of reconstituted nucleosomes, 30 μl of 5× HAT buffer, 4 μl of 20 μM 3H-acetyl CoA, and the proper amount of histone acetyltransferases and adjust the final volume to 150 μl with H_2O. Incubate at 30°C for 20 h.
16. Carefully load the entire reaction onto a G-50 mini column prewashed with 1× HAT buffer. Centrifuge at 700 × *g* for 2 min to collect modified nucleosomes. Unincorporated cofactors (acetyl-CoA in this case) should remain in the column (see Note 11).
17. (Optional) Repeat steps 15 and 16 to install other modifications enzymatically.
18. The overall acetylation level can be determined by the standard filter-binding assay (see Note 12) using a scintillation counter. To check the acetylation specificity, 0.5 μg of purified acetylated nucleosomes is subject to western blotting using antibodies against specifically acetylated histones.

3.5. Histone Deacetylase Assays Using Nucleosomal Templates

1. Mix 0.5 μg of acetylated nucleosome substrates (3H labeled) with a proper amount of HDAC. Bring the final volume to 36 μl using 1× HDAC buffer.
2. Incubate at 30°C for 1 h.

3. Add 36 μl of 1 M HCl/0.4 M acetic acid, and briefly vortex the tubes to stop the reactions.
4. Add 800 μl of ethyl acetate and vigorously vortex for 5 s.
5. Centrifuge at 12,000 × *g* for 10 min at 4°C.
6. Transfer supernatant (~720 μl) to a scintillation vial containing 4 ml of scintillation fluid for liquid counting.
7. Vortex each vial vigorously for 5 s, and then measure the radioactivity using a scintillation counter.

4. Notes

1. We prepare each individual recombinant histone based on the general protocol described by the Luger lab (10). We typically use 3 L culture for wild-type histones and 1.5 L for mutants, which yield about 100–500 mg of histones. We found that applying the solubilized inclusion bodies through a single ion-exchange column (5 ml Hi-Trap SP (GE)) results in adequate purity for the procedures described here.
2. The concentrations of histones are determined by OD_{276} using the theoretical extinction coefficients calculated on the ExPASy Proteomics Server. For mutant histones, concentrations are further confirmed by running on an SDS-PAGE gel and comparing to a standard curve of a histone with known concentration.
3. For histones containing native cysteines, these residues should first be mutated to alanine. The resulting "wild-type" histones should be tested in available in vitro and/or in vivo functional assays to ensure that the C-to-A mutation itself introduces minimal interference within the intended experiments. The commonly used histone H3C110A does not result in any noticeable phenotype in all tested assays thus far. Your favorite lysine(s) can then be substituted with cysteine(s) in this "wild-type" background. Multiple cysteines can be introduced in one histone (such as H3Kc9 and H3Kc36); however, only one type of methylation analog can be installed at these residues using the current protocol. We have tested various combinations of double-lysine mutant histones, and found that the recipes described here are sufficient for completion of such double reactions.
4. If chemical approach can generate all intended histone modification combinations, nonbiotinylated DNA templates can be used. Proceed to nucleosome reconstitution (steps 4–6). In this case, these nucleosomes can be directly gel purified starting from step 17 in Subheading 3.3. ^{32}P also can be incorporated into DNA using a body-labeling method in which PCR

reactions using the same primer sets are carried out in the presence of ^{32}P-labeled nucleotides.

5. We measure radioactivity incorporation (cpm) by counting 1 μl of purified DNA on a scintillation counter. We then determine the DNA concentration (ng/μl) by running 1 μl of DNA on a 2% agarose gel with known concentration DNA markers. This allows us to deduce the ratio of cpm/ng for the labeled DNA and estimate the amount of DNA during subsequent steps by monitoring the radioactivity (cpm).
6. Theoretically, histone and DNA should be in a 1:1 molar ratio of nucleosomes binding sites to octamers or a 1:1.3 mass ratio of histone octamers to DNA. We typically perform a titration test with 1.25-fold increments. Once the optimal histone/DNA ratio for any given histone octamer is determined, it can be consistently used for the same preparation with different DNA probes.
7. The histone-modifying enzymes that are suitable for the assays described in this chapter should be highly active and possess nucleosomal activity. Preliminary tests are normally performed to determine the necessary amount of enzymes and the completeness of the modification reactions. If 80–90% of a target population is modified, these nucleosomal templates are generally satisfactory for most other assays.
8. The generic histone acetyltransferase complexes used here are purified from yeast using the TAP method (13). We typically use Ada2-TAP, use of which yields a mixture of three HAT complexes: SAGA, SLIK, and ADA, to provide the histone H3-specific HAT, and the NuA4 (Epl1-TAP) complex, to provide the histone H4 HAT.
9. A pBlueScript vector carrying a single-copy 601 sequence flanked by several restriction sites (Fig. 2, top-left panel) is digested with KpnI and BamHI; the small fragment is ligated to the same vector digested with KpnI and BglII to produce the 2× 601 plasmid. Since the junction site of BamHI and BglII can no longer be cut by either enzyme, the two steps in Fig. 2 can be multiplied several times to generate more repeats. We found that for the single sequence that is less than 300 bp in length, a 16× version of construct can be routinely obtained.
10. A previous established procedure for a large-scale plasmid preparation (10) is adopted with minor modifications. We normally start with 2 L of 2×TY culture. After 18–22-h incubation at 37°C, about 20 g of cells can be harvested, which yield about 10 mg of plasmid DNA.
11. The simple procedure described here only removes free cofactors. Most modifying enzymes remain in the final nucleosome fraction, which should be factored into the data interpretation as appropriate.

12. The following procedures are based on a previous protocol (12). Spot 5 μl of acetylated nucleosomes on a piece of negatively charged P81 phosphocellulose filter. Air dry filters for 5 min. Wash filters with 50 ml of HAT wash buffer at RT for 5 min and then repeat two more times. Filters are briefly rinsed with 30 ml acetone and then air dried for 5 min. Place the filters into scintillation vials containing 4 ml of scintillation fluid (for filters) and measure the radioactivity using a scintillation counter.

References

1. Kouzarides, T. (2007) Chromatin modifications and their function. *Cell* **128,** 693–705
2. Li, B., Carey, M., Workman, J. L. (2007) The role of chromatin during transcription. *Cell* **128,** 707–719
3. Henikoff, S., McKittrick, E., Ahmad, K. (2004) Epigenetics, histone H3 variants, and the inheritance of chromatin states. *Cold Spring Harb Symp Quant Biol* **69,** 235–243
4. Jenuwein, T., Allis, C. D. (2001) Translating the histone code. *Science* **293,** 1074–1080
5. Simon, M. D., Chu, F., Racki, L. R., de la Cruz, C. C., Burlingame, A. L., Panning, B., Narlikar, G. J., Shokat, K. M. (2007) The site-specific installation of methyl-lysine analogs into recombinant histones. *Cell* **128,** 1003–1012
6. Neumann, H., Hancock, S. M., Buning, R., Routh, A., Chapman, L., Somers, J., Owen-Hughes, T., van Noort, J., Rhodes, D., Chin, J. W. (2009) A method for genetically installing site-specific acetylation in recombinant histones defines the effects of H3 K56 acetylation. *Mol Cell* **36,** 153–163
7. Shogren-Knaak, M., Ishii, H., Sun, J. M., Pazin, M. J., Davie, J. R., Peterson, C. L. (2006) Histone H4-K16 acetylation controls chromatin structure and protein interactions. *Science* **311,** 844–847
8. Li, B., Gogol, M., Carey, M., Lee, D., Seidel, C., Workman, J. L. (2007) Combined action of PHD and chromo domains directs the Rpd3S HDAC to transcribed chromatin. *Science* **316,** 1050–1054
9. Carrozza, M. J., Florens, L., Swanson, S. K., Shia, W. J., Anderson, S., Yates, J., Washburn, M. P., Workman, J. L. (2005) Stable incorporation of sequence specific repressors Ash1 and Ume6 into the Rpd3L complex. *Biochim Biophys Acta* **1731,** 77–87; discussion 75–76
10. Dyer, P. N., Edayathumangalam, R. S., White, C. L., Bao, Y., Chakravarthy, S., Muthurajan, U. M., Luger, K. (2004) Reconstitution of nucleosome core particles from recombinant histones and DNA. *Methods Enzymol* **375,** 23–44
11. Owen-Hughes, T., Utley, R. T., Steger, D. J., West, J. M., John, S., Cote, J., Havas, K. M., Workman, J. L. (1999) Analysis of nucleosome disruption by ATP-driven chromatin remodeling complexes. *Methods Mol Biol* **119,** 319–331
12. Grant, P. A., Berger, S. L., Workman, J. L. (1999) Identification and analysis of native nucleosomal histone acetyltransferase complexes. *Methods Mol Biol* **119,** 311–317
13. Puig, O., Caspary, F., Rigaut, G., Rutz, B., Bouveret, E., Bragado-Nilsson, E., Wilm, M., Seraphin, B. (2001) The tandem affinity purification (TAP) method: a general procedure of protein complex purification. *Methods* **24,** 218–229

Chapter 16

A Defined In Vitro System to Study ATP-Dependent Remodeling of Short Chromatin Fibers

Verena K. Maier and Peter B. Becker

Abstract

ATP-dependent remodeling factors regulate chromatin structure by catalyzing processes such as nucleosome repositioning or conformational changes of nucleosomes. Predominantly, their enzymatic properties have been investigated using mononucleosomal substrates. However, short nucleosomal arrays represent a much better mimic of the physiological chromatin context. Here, we provide a protocol for the enzyme-free reconstitution of regularly spaced nucleosomal arrays. We then explain how these arrays can serve as substrates to monitor ATP-dependent nucleosome movements and changes in the accessibility of nucleosomal DNA.

Key words: ATP-dependent chromatin remodeling, Nucleosome sliding, Chromatin reconstitution, Nucleosomal arrays, ACF, ATPase, Linker histone, Chromatin accessibility, Nucleosome mapping

1. Introduction

The discovery of ATP-dependent chromatin remodeling factors has replaced the former static concept of nucleosomes with a much more dynamic view of these smallest chromatin units. Chromatin remodelers use the energy of ATP to reposition, evict or introduce conformational changes into nucleosomes or replace individual histones by specialized variants (1, 2). These processes are easily conceivable on a mononucleosomal level, but it is more difficult to envision entire 10- and 30-nm fibers in constant motion and rearrangement. Likewise, mononucleosomes represent convenient and therefore widely used model substrates for chromatin remodelers, whereas it is more challenging to observe the action of these enzymes on chromatin fibers. Studies using mononucleosomes yielded valuable insights into the mechanisms of remodeling. However,

Randall H. Morse (ed.), *Chromatin Remodeling: Methods and Protocols*, Methods in Molecular Biology, vol. 833,
DOI 10.1007/978-1-61779-477-3_16, © Springer Science+Business Media, LLC 2012

access to nucleosomes within the nucleus is limited by their neighbors and by chromatin fiber folding. Assaying remodeling of linear nucleosomal arrays rather than mononucleosomes takes these limitations into account. This is crucial when investigating how factors that affect chromatin compaction influence the remodeling reaction.

Arrays of regularly spaced nucleosomes can be reconstituted by salt dialysis from purified components on DNA containing repetitive nucleosome positioning sequences. Such nucleosomal arrays are very homogeneous and convenient to analyze because the same DNA sequence is wrapped around every histone octamer. In this chapter, we describe the reconstitution of short nucleosomal fibers by salt dialysis on linear arrays of the 601 nucleosome positioning sequence. The system has been pioneered by Rhodes and colleagues (3–6) and recently applied by us to study the remodeling of arrays containing 12 positioned nucleosomes or chromatosomes (7). In contrast to enzymatic assembly systems (8–12), this method requires no enzymes or cell extracts that could otherwise affect enzymatic assays conducted with the arrays. For an easy-to-interpret and sensitive detection later on, we also provide a protocol to selectively radiolabel one end of the arrays. To exemplify the incorporation of a factor promoting chromatin compaction, the assembly of nucleosomal arrays containing stoichiometric amounts of linker histones is described. We then explain how to monitor the stoichiometric binding of core and linker histones to the DNA arrays. Finally, we present three assays to investigate chromatin remodeling with a focus on nucleosome repositioning: a chromatin accessibility assay to determine the extent of chromatin remodeling, a chromatin footprinting assay to observe nucleosome repositioning throughout the nucleosome array, and a primer extension assay, which allows to compare the exact positions of nucleosomes before and after the remodeling reaction.

2. Materials

2.1. Reconstitution of Regularly Spaced Nucleosomal Arrays

2.1.1. Preparation of DNA

1. Plasmid containing repeats of the 601 nucleosome positioning sequence, i.e., pUC18 12×601 (3).
2. Restriction enzymes EcoRI, HindIII, DraI (20 U/μl each) and NEB buffer 2: 50 mM NaCl, 10 mM Tris–HCl pH 7.9, 10 mM $MgCl_2$, 1 mM dithiothreitol.
3. 0.8% agarose gels containing 0.5 μg/ml EtBr (with regular and with preparative, 1 cm wide wells) in TAE (40 mM Tris–acetate, pH 8.0, 1 mM EDTA).
4. 6× DNA loading buffer (50% v/v glycerol, 0.25% w/v Orange G).
5. 100% and 70% ethanol.

6. 3 M NaOAc pH 5.2.
7. TE buffer: 10 mM Tris–HCl, 1 mM EDTA pH 7.5.
8. Low melting agarose.
9. Tris-buffered phenol pH 7.5 (equilibrate 2× each with 1 volume 0.5 M Tris–HCl pH 7.5 and 0.1 M Tris–HCl pH 7.5).
10. Butanol.

2.1.2. Radioactive End-Labeling of the HindIII-End

1. DNA Polymerase I, Large (Klenow) Fragment or Klenow Fragment (3′®5′ exo-) (5 U/μl).
2. [α-^{32}P]-dCTP (10 mCi/ml, 3,000 Ci/mmol).
3. Glycogen 1 mg/ml.

2.1.3. Chromatin Reconstitution

1. DB0, DB50, DB500, and DB2000 (DB*x*: 10 mM Tris–HCl pH 7.5, 1 mM EDTA pH 8.0, 1 mM β-mercaptoethanol, *x* in M NaCl).
2. High salt buffer: 2 M NaCl, 10 mM Tris–HCl pH 7.6 and 0.12 μg/μl bovine serum albumin (BSA).
3. BSA 98% pure.
4. Dialysis chambers (MWCO 6–8 kDa) (see Note 1).
5. Floating rack to hold the dialysis chambers.
6. Low retention 1.5-ml reaction tubes.
7. Histone octamers.
8. If applicable: linker histones.
9. Peristaltic pump or FPLC.

2.1.4. Chromatin Purification

1. 10 mM $MgCl_2$.

2.2. Quality Controls of Nucleosomal Arrays

2.2.1. Electrophoretic Mobility Shift Assays

1. Highly pure agarose, e.g., Biozym LE *GP* agarose.
2. 0.2× TB: 9 mM Tris–borate pH 8.0.
3. Ethidium bromide (EtBr).

2.2.2. Digestion to Mononucleosomes/ Chromatosomes

1. RB50: 10 mM Hepes pH 7.6, 50 mM KCl, 1.5 mM $MgCl_2$, 0.5 mM EGTA pH 8.0.
2. AvaI (10 U/μl).
3. Highly pure agarose, e.g., Biozym LE *GP* agarose.

2.2.3. Control of Histone Stoichiometry

1. 15% SDS-polyacrylamide gel and corresponding running buffer.
2. Coomassie brilliant blue staining and destaining solutions.
3. LI-COR Odyssey machine.

2.3. Remodeling Assays

2.3.1. Chromatin Accessibility Assay

1. 50 mM ATP stock solution (see Note 2).
2. AluI (10 U/μl).
3. Chromatin remodeling enzyme.
4. Nonspecific free DNA (i.e., plasmid DNA).
5. 10 mg/ml Proteinase K dissolved in 10 mM Tris–HCl pH 7.5, 3 mM $CaCl_2$ and stored at −20°C.
6. 1.3% agarose gel in Tris–glycine buffer: 50 mM Tris base, 384 mM glycine.
7. Whatman DE81 paper.
8. PhosphoImager and PhosphoImager screen.
9. Image analysis software, e.g., AIDA Image Analyzer software.

2.3.2. Chromatin Footprint Assay Using Micrococcal Nuclease

1. Micrococcal nuclease (MNase).
2. 50 mM $CaCl_2$.
3. 50 mM EDTA pH 8.0.

2.3.3. Nucleosome Repositioning Assay Using Primer Extension

1. SYBR gold (10,000× concentration, Invitrogen).
2. Gel Extraction Kit.
3. Primers A-fw (5′-A TCTGACACGTGCCTGGA-3′), A-rv (5′-TCCAGGCACGTGTCAGA T-3′), B-fw (5′-CGTACG-TGCGTTTAAGC-3′), and B-rv (5′-GCTTAAACGCACGT-ACG-3′).
4. 10-bp DNA ladder (1 μg/μl).
5. T4 Polynucleotide Kinase (10 U/μl) and supplied reaction buffer.
6. [γ-^{32}P]-ATP (10 mCi/ml, 6,000 Ci/mmol).
7. G-25 spin columns.
8. Taq polymerase (5 U/μl) and supplied reaction buffer.
9. 2 mM dNTP mix.
10. 80% formamide containing 0.25% w/v bromophenol blue.
11. 7 M urea, 7% polyacrylamide gel (17×40 cm, 0.4 mm thick) in TBE (45 mM Tris–borate, 1 mM EDTA), corresponding sequencing gel unit.

3. Methods

3.1. Reconstitution of Regularly Spaced Nucleosomal Arrays

3.1.1. Preparation of DNA

1. Digest 50 μg pUC18 12×601 per tube with EcoRI, HindIII and DraI (50 U each) in 100 μl NEB buffer 2 at 37°C for 3 h (see Note 3).
2. Check digest by running 250 ng DNA on an agarose gel. If necessary, add more enzyme until the digest is complete.

3. Precipitate DNA by adding 1/10 volume of 3 M NaOAc pH 5.2 and 2 volumes of ethanol, incubate on ice or at –20°C for 1 h to overnight and wash with 70% ethanol. Let the pellet dry for 5 min and resuspend in TE pH 7.5.
4. For the reconstitution of end-labeled chromatin arrays, at least 20% of the 601 repeats should be radioactively labeled (see Subheading 3.1.2). This fraction has to be purified from the vector DNA, to avoid labeling the carrier DNA. For this purpose, apply the digested plasmid (this fraction does not have to be precipitated after the digestion) on an about 8 cm long 0.8% agarose gel in TAE containing 0.5 μg/ml EtBr with 1 cm preparative wells (about 10 μg DNA per well) (see Note 4). Separate the DNA fragments by electrophoresis and use a low-energy handheld UV lamp to visualize the DNA. Cut out a rectangle just below the 2.4-kb band containing the 601 repeats.
5. Fill the empty rectangle with warm 0.7% low melting agarose dissolved in TAE without EtBr and let it solidify at 4°C.
6. Resume electrophoresis until the band has moved into the low melting agarose gel.
7. Excise gel slices containing 601 repeats and incubate them at 65°C until fully melted (see Note 5).
8. Add 1 volume of Tris-buffered phenol, mix, and centrifuge in a table-top microfuge for 10 min, 13,000 rpm at room temperature.
9. Collect the aqueous phase and re-extract the organic phase with 300 μl TE pH 7.5.
10. Combine the aqueous phases from several gel slices in a 15-ml conical tube. Fill the tube with butanol, mix, centrifuge for a few seconds and aspirate the upper phase containing butanol, water, and EtBr. Repeat until the volume of the lower aqueous phase is reduced to about 400 μl, but does not disappear entirely.
11. Precipitate DNA with ethanol (see step 3) and resuspend in TE pH 7.5.

3.1.2. Radioactive End-Labeling of the HindIII-End

1. Shortly before the chromatin reconstitution, isotope-label HindIII 3′-overhangs of the required amount of gel-purified DNA. Use 2 U DNA Polymerase I, Large (Klenow) Fragment or Klenow Fragment (3′→5′ exo-) for 2 μg of DNA in 20 μl NEB buffer 2 in the presence of dGTP, dATP, dTTP (33 μM each) and 1 μl of [α-^{32}P]-dCTP (10 mCi/ml, 3,000 Ci/mmol). Incubate at 37°C for 15 min (see Note 6).
2. Stop reaction by adding 1 μl 0.5 M EDTA and inactivate the enzyme for 20 min at 70°C.
3. Add 1 μl glycogen, perform ethanol precipitation (see Subheading 3.1.1), and resuspend in TE pH 7.5 at a concentration of 100 ng/μl.

3.1.3. Chromatin Reconstitution

1. Pre-adsorb dialysis chambers by filling each of them with 0.5 ml 2 mg/ml BSA in DB500. Place them in a floating rack in DB500 for at least 2 h at 4°C (see Note 7).
2. Mix digested plasmid containing 2 μg 601 repeats (Subheading 3.1.1, step 2), 500 ng isotope-labeled 601 repeats (Subheading 3.1.1, step 14) (for assays Subheadings 3.3.1 and 3.3.2) or digested plasmid containing 2.5 μg 601 repeats (for assay Subheading 3.3.3) with core histones and, if desired, linker histones in a final volume of 50 μl high salt buffer in low retention tubes (see Note 8).
3. Transfer mixture into dialysis chambers, which have been rinsed several times with DB500.
4. Put floating rack with dialysis chambers in 200 ml cold DB2000, making sure that the membrane touches the buffer (see Note 9).
5. Dilute DB2000 1:10 by continuous addition of cold DB0 over a 13–15 h period at 4°C. Ensure even mixing of the buffers by stirring on a magnetic stirrer (see Note 10).
6. Place floating rack in cold DB50, make sure that the membrane touches the buffer and keep on stirring for 1 h at 4°C.
7. Transfer assembled chromatin to low retention tubes and store at 4°C. Do not freeze.

3.1.4. Chromatin Purification

1. Add 1 volume of 10 mM $MgCl_2$ to reconstituted chromatin arrays, mix gently by tapping.
2. Incubate on ice for 15 min.
3. Centrifuge in a table-top microfuge for 15 min at 13,000 rpm, 4°C (see Note 11).
4. Carefully remove supernatant, do not allow the pellet to dry (see Note 12).
5. Redissolve in TE pH 7.5 and store at 4°C (see Fig. 1b).

3.2. Quality Controls of Nucleosomal Arrays

The following three quality controls allow determination of whether the 601 arrays are saturated with core and linker histones, whether all 601 positioning sequences are occupied by a histone octamer and a linker histone, and whether the stoichiometry of linker to core histones is correct. EMSA assays (Subheading 3.2.1) are preferentially performed before $MgCl_2$ precipitation in order to monitor binding of histones to both 601 arrays and carrier DNA. Digestion (Subheading 3.2.2) can be carried out either before or after precipitation, whereas determination of histone stoichiometry (Subheading 3.2.3) must be performed after precipitation.

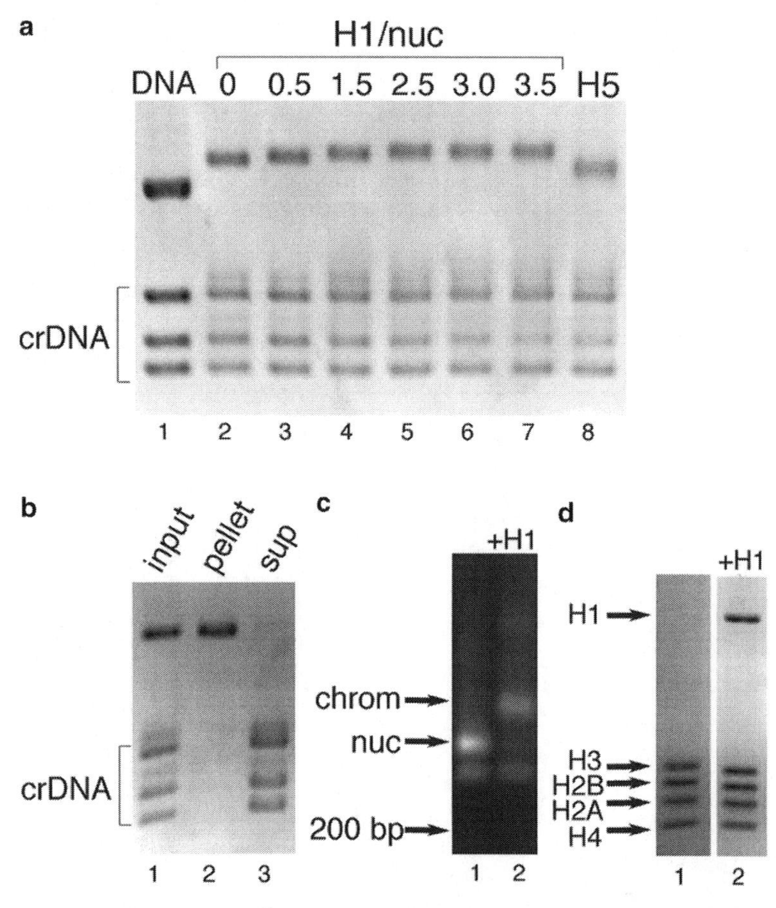

Fig. 1. Quality controls of nucleosomal arrays. (**a**) Native agarose gel of 12mer nucleosomal arrays assembled in the presence of carrier DNA (crDNA) and with increasing molar ratios of linker histones per 601 positioning sequence (H1/nuc) or saturating amounts of H5. For comparison, DNA not bound by histones is shown as well (DNA) (see Subheading 3.2.1 and Note 16). (**b**) $MgCl_2$ precipitation of 12mer nucleosomal arrays. Corresponding amounts of input, pellet, and supernatant (sup) were run on a native agarose gel (see Subheading 3.1.4 and Note 11). (**c**). Digestion of 12mer nucleosomal arrays without or with H1 to nucleosomes (nuc) or chromatosomes (chrom), respectively. The absence of free 200-bp DNA fragments (200 bp) indicates occupation of all positioning sequences by histone octamers (*see* Subheading 3.2.2 and Note 17). (**d**) Histone content of nucleosomal or chromatosomal (+H1) arrays after $MgCl_2$ precipitation visualized by SDS-PAGE (see Subheading 3.2.3 and Note 18). (Data originally published in ref. 7).

3.2.1. Electrophoretic Mobility Shift Assays

1. Apply reconstituted chromatin arrays (100 ng of 601 repeat DNA) with DNA loading buffer on an agarose minigel in 0.2× TB (see Note 13).
2. Run gel at 10 V/cm until Orange G has migrated to the bottom of the gel (see Note 14).
3. Stain the gel in a solution of 5 μg/ml EtBr in 0.2× TB for 1 h (see Note 15) and destain in 0.2× TB for 30 min.
4. Visualize arrays using a gel documentation unit (see Note 16 and Fig. 1a).

3.2.2. Digestion to Mononucleosomes/ Chromatosomes

1. Digest reconstituted chromatin arrays (100 ng of 601 repeat DNA) with 15 U AvaI in 15 μl final volume of RB50 at 26°C for 1 h.
2. Add DNA loading buffer and apply on a 20 cm agarose gel in 0.2× TB (see Note 13), run gel at 10 V/cm until Orange G reaches almost the bottom of the gel.
3. Stain with 5 μg/ml EtBr and destain as in Subheading 3.2.1.
4. Visualize mononucleosomes/chromatosomes using a gel documentation unit (see Note 17 and Fig. 1c).

3.2.3. Control of Histone Stoichiometry

1. Apply reconstituted and purified chromatin arrays on a 15% SDS polyacrylamide gel and electrophorese until the bromophenol blue reaches the bottom of the gel.
2. Visualize proteins with Coomassie brilliant blue (see Fig. 1d).
3. Measure relative amounts of histones by densitometry on a LI-COR Odyssey machine (see Note 18).

3.3. Remodeling Assays

The described assays monitor different aspects of the remodeling reaction. The chromatin accessibility assay (Subheading 3.3.1) measures ATP-dependent changes in the accessibility of nucleosomal DNA. It does not distinguish between DNA bulging, nucleosome repositioning, or histone eviction, but is very useful for quantitative comparison of the extent of remodeling. The chromatin footprint assay (Subheading 3.3.2) monitors global repositioning of nucleosomes. The nucleosome repositioning assay (Subheading 3.3.3) allows determination of the positions of nucleosomes before and after remodeling. In addition, with this assay, it is possible to specifically study remodeled chromatosomes that are still bound by a linker histone.

3.3.1. Chromatin Accessibility Assay

1. Dilute 6 fmol chromatin arrays (corresponds to 10 ng DNA for 12mer 200 bp arrays) in 8 μl RB50 final volume containing 25 μM ATP (will be 20 μM in the final reaction). Prepare negative controls without ATP and without remodeling enzyme (see Note 19).
2. Add 5 U AluI along with the remodeling enzyme in 2 μl RB50 (see Notes 20 and 21).
3. Incubate at 26°C for 1 h.
4. Stop the reaction by adding 200 ng free DNA (i.e., plasmid DNA) and putting it on ice (see Note 22).
5. Add 1 μl 10 mg/ml Proteinase K and incubate at 37°C for 1 h.
6. Precipitate the DNA with ethanol and resuspend in 5 μl TE. Add 1 μl 6× DNA loading buffer.
7. Load DNA on a 20-cm long, thin (about 3 mm) 1.3% agarose gel in Tris–glycine buffer and electrophorese until Orange G has migrated almost to the bottom of the gel.

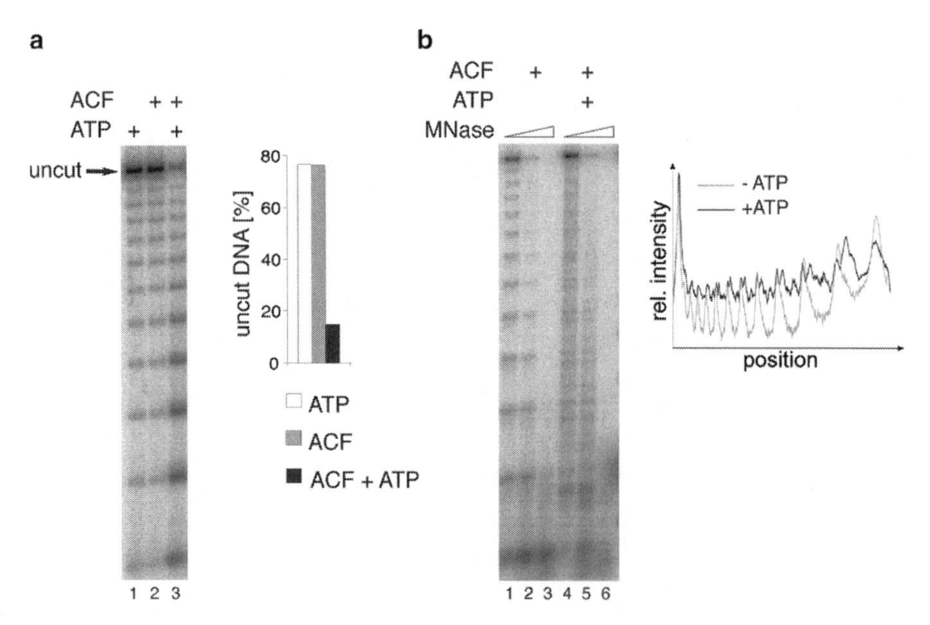

Fig. 2. (**a**) Chromatin accessibility assay with ACF (see Subheading 3.3.1 and Note 23). (**b**) Chromatin footprint assay with ACF. Densitometry profiles of *lanes 1* and *4* are compared (see Subheading 3.3.2 and Note 25). (Data originally published in ref. 7).

8. Dry the gel in a vacuum gel dryer at 50°C for 1–2 h onto Whatman DE81 paper.
9. Expose dried gel overnight on a PhosphoImager screen.
10. Scan the gel in a PhosphoImager.
11. Determine percentage of full-length (2.4 kb) versus cut DNA fragments present in one lane, e.g., using AIDA Image Analyzer software (see Note 23 and Fig. 2a).

3.3.2. Chromatin Footprint Assay Using Micrococcal Nuclease

1. Dilute 18 fmol chromatin arrays (corresponds to 30 ng DNA of 12mer 200 bp arrays) in 24 μl RB50 final volume containing 25 μM ATP.
2. Add remodeling enzyme in 6 μl RB50 (see Note 20).
3. Incubate at 26°C for 1 h.
4. Stop the reaction by adding 600 ng free DNA (i.e., plasmid DNA) and putting it on ice.
5. Add 15 μl MNase mix (containing about 4×10^{-3} U of MNase (Sigma unit definition) and 10 mM $CaCl_2$) and incubate reaction at 26°C (see Note 24).
6. After 1, 3, and 5 min, remove 15 μl of the reaction and place into 1.5-ml tubes containing 5 μl 50 mM EDTA.
7. Process samples, run, dry, and expose 1.3% agarose gel in Tris–glycine buffer as in Subheading 3.3.1, steps 5–10.
8. Compare digestion patterns of samples incubated in the absence or presence of ATP (see Note 25 and Fig. 2b).

3.3.3. Nucleosome Repositioning Assay Using Primer Extension

1. Incubate 30 fmol chromatin arrays (corresponds to 50 ng DNA for 12mer arrays of 200 bp) with the remodeling enzyme in 15 μl RB50 final volume containing 20 μM ATP.
2. Incubate at 26°C for 1 h.
3. Add 10^{-3} U MNase (Sigma unit definition; see Note 24) and $CaCl_2$ to 1 mM and incubate at 26°C for 20 min (see Notes 26 and 27).
4. Add 3 μl 6× DNA loading buffer and apply mixture on a 20-cm long agarose gel in 0.2× TB (see Note 13).
5. Run gel at 10 V/cm until Orange G migrates close to the bottom of the gel and stain with SYBRgold according to the manufacturer's instructions.
6. Excise gel slices containing mononucleosomes or monochromatosomes (see Note 28 and Fig. 3a).
7. Extract DNA using a gel extraction kit (e.g., Qiagen Gel Extraction Kit).
8. Label primers A-fw, A-rv, B-fw, and B-rv and 1 μl of 10-bp DNA ladder in individual 20-μl reactions in polynucleotide kinase buffer with 10 U T4 polynucleotide kinase and 1 μl [γ-^{32}P]-ATP for 20 min at 37°C (see Fig. 3b for the positions of the primers with respect to the 601 positioning sequence).
9. Remove excess ATP using G-25 spin columns according to the manufacturer's instructions.
10. In 50 μl final volume of 1× Taq polymerase buffer, mix 10% of the recovered DNA from step 7 with one of the labeled primers from step 8 (final concentration 5 μM), dNTPs (0.2 mM each) and 3 U Taq polymerase. Prepare four primer extension reactions – one with each of the primers – per remodeling reaction.
11. Subject these mixtures to 12 cycles of primer extension reactions (30 s at 94°C, 30 s at 60°C, and 60 s at 68°C).
12. Precipitate samples with glycogen and ethanol (see Subheading 3.1.2, for convenient precipitation, unlabeled DNA may also be added) and resuspend in 5 μl 80% formamide containing bromophenol blue.
13. Load samples onto a 0.4-mm thick, 17×40-cm 7% polyacrylamide, 20% urea TBE gel poured in a sequencing gel unit, which has been prerun for 1 h. Apply radiolabeled 10-bp DNA ladder, choose an amount per lane with similar activity as the samples.
14. Run gel at 40 W until the bromophenol blue reaches the bottom of the gel.
15. Carefully disassemble the gel cassette and transfer the gel onto DE81 paper (see Note 29).

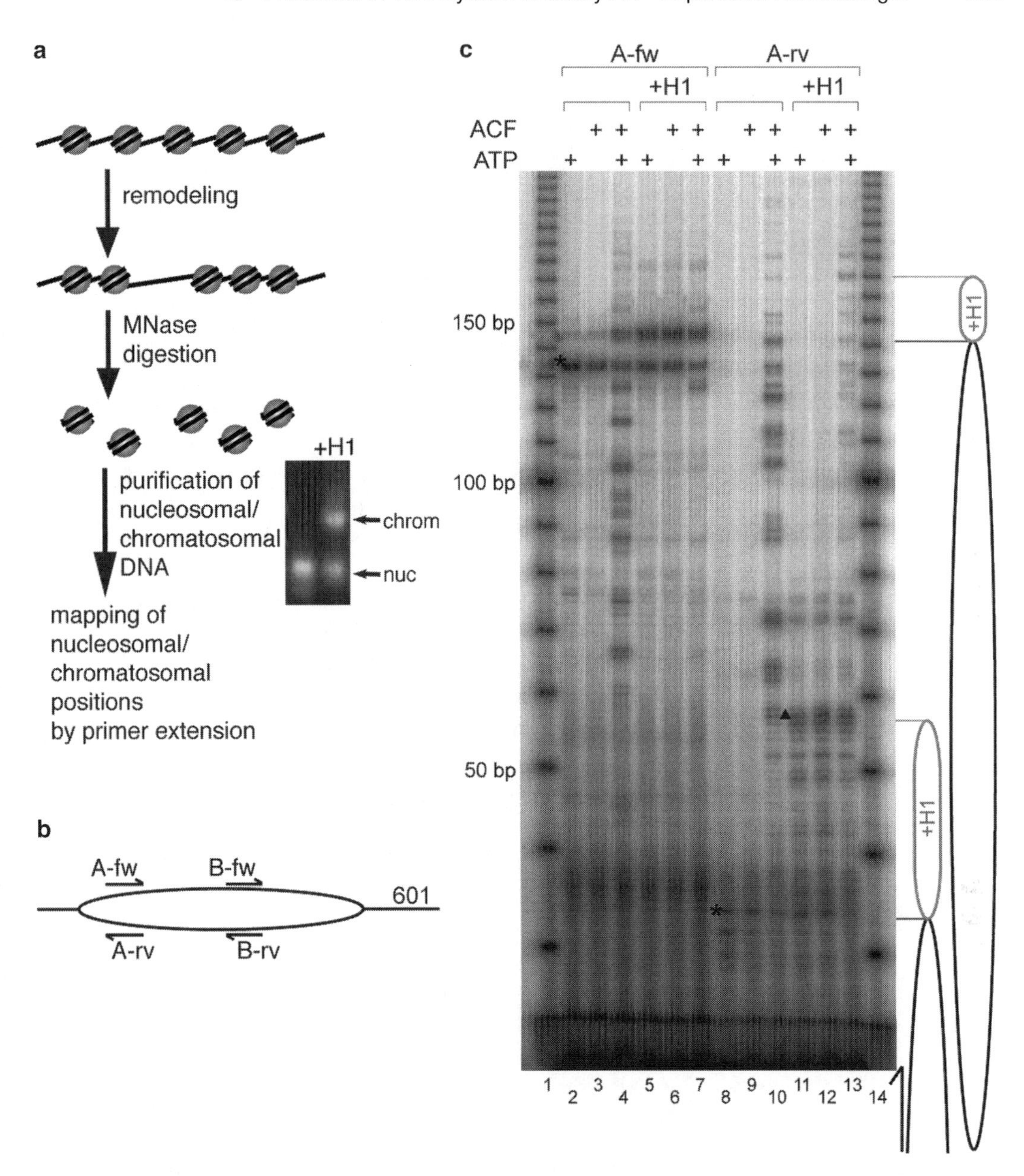

Fig. 3. (**a**) Overview of the nucleosome repositioning assay using primer extension. Nucleosomal and chromatosomal (+H1) arrays digested by MNase were electrophoresed on a native agarose gel (see Subheading 3.3.3 and Note 28). (**b**) Positions of primers A-fw, A-rv, B-fw, and B-rv with respect to the 601 positioning sequence (see Subheading 2.3.3). (**c**) Denaturing polyacrylamide gel of primer extension reactions with primers A-fw and A-rv on nucleosomal and chromatosomal DNA. DNA was obtained from 12mer arrays incubated with or without ACF and ATP as indicated. Bands corresponding to unremodeled nucleosomes or chromatosomes are marked by an *asterisk* or a *triangle*, respectively. Positions of the nucleosome and H1 before remodeling, and one prominent position after remodeling as revealed by extension of A-rv are represented on the *right* (see Note 30). (Data originally published in ref. 7).

16. Dry gel and expose to a PhosphoImager screen overnight.
17. Scan screen in a PhosphoImager (see Note 30 and Fig. 3c).

4. Notes

1. Commercial (e.g., Slide-A-Lyzer MINI dialysis units, Thermo Scientific) or self-made dialysis chambers work equally well. For preparing a dialysis chamber, cut a circular hole in the lid of a low retention 1.5-ml tube, e.g., using a hole puncher. Remove the bottom of the tube – e.g., by using a scalpel that has been heated in the flame of a Bunsen burner – and use the bottom upside down as a lid of the resulting dialysis chambers. Pin the dialysis membrane between the lid and the tube. These tubes are then placed upside down on a floating rack (for a more detailed description including an explanatory figure see ref. 13).
2. To prepare an ATP stock solution, dissolve the appropriate amount of ATP in water, adjust the pH to 7.0 with 0.1 M NaOH and add water to reach the final volume. Acidic conditions would result in hydrolysis of ATP.
3. EcoRI and HindIII excise the insert containing 601 positioning sequences. DraI digests the pUC18 vector backbone into three fragments (692, 811, and 1,113 bp). The vector DNA, which does not contain nucleosome positioning sequences, serves as carrier DNA binding excess histone octamers and linker histones. The fragments are easier to remove than the full length vector both when isolating 601 repeat DNA for radioactive labeling (see Subheading 3.1.1, steps 4–11) and after chromatin assembly by $MgCl_2$ precipitation (see Subheading 3.1.4).
4. Alternatively, 601 repeat DNA can be isolated by selective polyethylene glycol (PEG) precipitation as described in ref. 3, but it has to be ensured that remaining PEG does not affect the enzymatic activity of the remodeling enzyme.
5. Gel slices may be frozen at this point until further use.
6. dCTP labels 5′ overhangs generated by HindIII, but not by EcoRI, since the latter ones do not contain a complementary dGTP.
7. Both core and linker histones readily stick to tubes and tips. Pre-adsorbing of dialysis chambers can be omitted, but more histone octamers will be necessary to saturate the chromatin arrays. Blocking is highly recommended when adding additional factors such as linker histones, since the amount of available histone octamers will be more consistent. In addition, tips can be blocked by briefly pipetting a solution of 20 mg/ml BSA up and down.

8. The amounts of histone octamers and linker histones have to be titrated (see Subheading 3.2 for assessing the extent of occupancy of positioning sequences by histone octamers and linker histones). We observed saturation of positioning sequences – meaning that each positioning sequence was bound by a histone octamer – at a molar ratio of histone octamers to positioning sequences of about two. We attribute this requirement of a slight excess of histone octamers to the loss of proteins on tips and tubes and the fact that a moderate excess of histones is needed to achieve occupancy of all positioning sequences. Similarly, an excess of linker histones is required to achieve saturation. Surplus histones are absorbed by carrier DNA.
9. If using self-made dialysis chambers, air bubbles underneath the membrane have to be removed with the help of a Pasteur pipette, which has been shaped into a flat hook.
10. We found that this simple dilution method results in high quality nucleosomal arrays, although the salt concentration decreases much faster in the beginning than in the end. However, a steady decrease in salt concentration can be achieved by pumping DB0 into the beaker and at the same time pumping buffer out of the beaker into a waste container. Alternatively, a FPLC can be programmed to add DB0 at an exponentially increasing rate.
11. $MgCl_2$ results in the reversible compaction and precipitation of chromatin arrays. Unbound DNA and proteins as well as DNA with only few bound histones remain in the supernatant (see Fig. 1b) (14).
12. The chromatin pellet has a gel-like and colorless appearance and should not be touched. We recommend to carefully remove the supernatant with a Pasteur pipette that has been heated and drawn out into a capillary.
13. Highly pure agarose should be used for a good resolution on native agarose gels. We use Biozym Seakem LE *GP* agarose, which is easier to handle than low melting agarose. The optimal percentage of the agarose gel has to be determined empirically. It will depend on the length of the array and on which component's incorporation is being monitored. For 12mer arrays, the incorporation of histone octamers and linker histone H1 is best resolved on a 0.7% agarose gel, whereas we recommend 1.4% agarose to monitor binding of linker histone H5 (see also Note 16). Mononucleosomes and monochromatosomes should be visualized on a 1.1% agarose gel.
14. Due to the low salt concentration of the running buffer, the applied voltage can be higher than for regular agarose gels.

15. This high concentration of EtBr strips histones off the DNA. Consequently, the DNA is more accessible for the dye. Still, nucleosomal arrays will stain less well than free DNA.
16. The incorporation of histone octamers results in a markedly slower migration of 601 arrays on native agarose gels (see Fig. 1a, compare lanes 1 and 2). Incompletely assembled arrays run as a broad, fuzzy band due to the inhomogeneity of the chromatin, whereas arrays saturated with histone octamers appear as a sharp band. Once saturation has been reached, the addition of more histone octamers does not result in a further shift of the migration of the 601 arrays. Instead, surplus histones bind to and shift the carrier DNA. Binding of linker histone H1 to the arrays results in an even slower migration on native agarose gels (see Fig. 1a, lanes 3–7). Again, upon saturation of the arrays, excess proteins bind to the carrier DNA. Binding of linker histone H5 results in faster migrating arrays, presumably due to the high compaction induced by this specialized linker histone, which seems to outweigh the increase in molecular weight (see Fig. 1a, lane 8). It is also advisable to control the completeness of the $MgCl_2$ precipitation by applying corresponding amounts of input, pellet and supernatant on native agarose gels. Arrays saturated with histone octamers should be completely precipitated, carrier DNA should remain in the supernatant (see Fig. 1b).
17. If all positioning sequences are occupied by histone octamers, no free 200-bp fragments are obtained. Instead, the digestion results in much slower migrating mononucleosomes (see Fig. 1c, lane 1). Digestion of arrays containing the linker histone H1 yields even slower migrating monochromatosomes (see Fig. 1c, lane 2).
18. If no LI-COR Odyssey machine is available, the intensity of the bands can be estimated by eye. Upon saturation of the arrays, adding more linker histones to the chromatin assembly should not result in a significant increase of the linker histone's band intensity, since surplus proteins bind to the carrier DNA and are therefore removed during the $MgCl_2$ precipitation.
19. At 1.5 mM $MgCl_2$, the chromatin fiber is not compacted and therefore accessible for remodeling enzymes.
20. The amount of remodeling enzyme to be used depends on the enzyme, the activity of the enzyme preparation and on the desired extent of remodeling. We generally use one ACF for every three nucleosomes present in the reaction. If comparing different enzymes, they can be normalized based on their ATPase activity.
21. Alternatively, other restriction enzymes with corresponding sites on the 601 positioning sequence like BanI or RsaI may be used to monitor remodeling in different parts of the nucleosome.

22. Other quenchers like EDTA or apyrase may also be used. Adding unlabeled DNA is still recommended to facilitate the subsequent ethanol precipitation.
23. Some digestion may already occur in the negative control reactions (absence of ATP or of remodeling enzyme). An enzyme- and ATP-dependent decrease in the percentage of uncut DNA indicates chromatin remodeling (see Fig. 2a).
24. The appropriate amount of MNase has to be titrated for every enzyme batch. The extent of MNase digestion may be adjusted via the concentration of MNase or $CaCl_2$ or the time of digestion. Note that the definition of MNase units differs greatly between companies.
25. Upon nucleosome repositioning, the MNase pattern is expected to be significantly altered. For a better comparison of digestion patterns, densitometry profiles of corresponding lanes can be generated, e.g., using AIDA Image Analyzer software (see Fig. 2b).
26. Again, the amount of MNase in the reaction has to be titrated in order to yield only mononucleosomes and monochro-matosomes.
27. We found that using only 1 mM rather than 3 mM $CaCl_2$ results in less unwanted DNA nicking and therefore a cleaner profile after primer extension (meaning only one predominant band from unremodeled nucleosomes).
28. Arrays not containing linker histones should give rise to one mononucleosomal band. When linker histone H1 containing arrays are digested, a mixture of mononucleosomes and slower migrating monochromatosomes is obtained (see Fig. 3a), since MNase treatment leads to displacement of a fraction of H1 from the chromatosomes (15). To facilitate interpretation, only the chromatosomal band should be excised after remodeling of linker histone containing arrays. We were not able to observe a "chromatosome stop" band when digesting linker histone H5-containing arrays.
29. The gel is very thin and easily tears apart. To make sure it remains on one plate of the gel cassette, one glass plate may be siliconized before pouring the gel. To transfer the gel onto DE81 paper, the paper is placed on the gel and supported by a layer of regular Whatman paper. The glass plate along with the gel and the paper is then turned upside down. The paper with the gel sticking to it is removed carefully.
30. The positions of nucleosomes and chromatosomes can be deduced from the position of the primer with respect to the 601 sequence and the length of the labeled DNA fragments. Binding of linker histone H1 partially protects additional 20 bp on one side of the nucleosome from MNase digestion (see Fig. 3c, compare lanes 8 and 11). If repositioning occurred, additional bands will be detected (see Fig. 3c, lanes 4, 7, 10, and 13).

Acknowledgments

Work in the laboratory of P. Becker on nucleosome remodeling is supported by the Deutsche Forschungsgemeinschaft (SFB594). We thank Mariacristina Chioda, Henrike Klinker, and Felix Müller-Planitz for helpful discussion.

References

1. Chioda, M., and Becker, P. B. (2010) Soft skills turned into hard facts: nucleosome remodelling at developmental switches, Heredity 105, 71–79.
2. Gangaraju, V. K., and Bartholomew, B. (2007) Mechanisms of ATP dependent chromatin remodeling, Mutat Res 618, 3–17.
3. Huynh, V. A., Robinson, P. J., and Rhodes, D. (2005) A method for the in vitro reconstitution of a defined "30 nm" chromatin fibre containing stoichiometric amounts of the linker histone, J Mol Biol 345, 957–968.
4. Robinson, P. J., An, W., Routh, A., Martino, F., Chapman, L., Roeder, R. G., and Rhodes, D. (2008) 30 nm chromatin fibre decompaction requires both H4-K16 acetylation and linker histone eviction, J Mol Biol 381, 816–825.
5. Robinson, P. J., Fairall, L., Huynh, V. A., and Rhodes, D. (2006) EM measurements define the dimensions of the "30-nm" chromatin fiber: evidence for a compact, interdigitated structure, Proc Natl Acad Sci USA 103, 6506–6511.
6. Routh, A., Sandin, S., and Rhodes, D. (2008) Nucleosome repeat length and linker histone stoichiometry determine chromatin fiber structure, Proc Natl Acad Sci USA 105, 8872–8877.
7. Maier, V. K., Chioda, M., Rhodes, D., and Becker, P. B. (2008) ACF catalyses chromatosome movements in chromatin fibres, EMBO J 27, 817–826.
8. Becker, P. B., Tsukiyama, T., and Wu, C. (1994) Chromatin assembly extracts from Drosophila embryos, Methods Cell Biol 44, 207–223.
9. Bulger, M., Ito, T., Kamakaka, R. T., and Kadonaga, J. T. (1995) Assembly of regularly spaced nucleosome arrays by Drosophila chromatin assembly factor 1 and a 56-kDa histone-binding protein, Proc Natl Acad Sci USA 92, 11726–11730.
10. Fyodorov, D. V., and Kadonaga, J. T. (2003) Chromatin assembly in vitro with purified recombinant ACF and NAP-1, Methods Enzymol 371, 499–515.
11. Glikin, G. C., Ruberti, I., and Worcel, A. (1984) Chromatin assembly in Xenopus oocytes: in vitro studies, Cell 37, 33–41.
12. Loyola, A., and Reinberg, D. (2003) Histone deposition and chromatin assembly by RSF, Methods 31, 96–103.
13. Sandaltzopoulos, R., and Becker, P. B. (2009) Analysis of reconstituted chromatin using a solid-phase approach, Methods Mol Biol 523, 11–25.
14. Schwarz, P. M., Felthauser, A., Fletcher, T. M., and Hansen, J. C. (1996) Reversible oligonucleosome self-association: dependence on divalent cations and core histone tail domains, Biochemistry 35, 4009–4015.
15. Nightingale, K., Dimitrov, S., Reeves, R., and Wolffe, A. P. (1996) Evidence for a shared structural role for HMG1 and linker histones B4 and H1 in organizing chromatin, EMBO J 15, 548–561.

Chapter 17

In Vitro Reconstitution of In Vivo-Like Nucleosome Positioning on Yeast DNA

Christian J. Wippo and Philipp Korber

Abstract

Genome-wide nucleosome mapping in vivo highlighted the extensive degree of well-defined nucleosome positioning. Such positioned nucleosomes, especially in promoter regions, control access to DNA and constitute an important level of genome regulation. However, the molecular mechanisms that lead to nucleosome positioning are far from understood. In order to dissect this mechanism in detail with biochemical tools, an in vitro system is necessary that can generate proper nucleosome positioning de novo. We present a protocol that allows the assembly of nucleosomes with very much in vivo-like positioning on budding yeast DNA, either of single loci or of the whole-genome. Our method combines salt gradient dialysis and incubation with yeast extract in the presence of ATP. It provides an invaluable tool for the study of nucleosome positioning mechanisms, and can be used to assess the relative stability of properly positioned nucleosomes. It may also generate more physiological templates for in vitro studies of, e.g., nucleosome remodeling or transcription through chromatin.

Key words: Nucleosome positioning, In vitro reconstitution, *Saccharomyces cerevisiae*, Yeast extract, Salt gradient dialysis

1. Introduction

The majority of nucleosomes are nonrandomly but rather well positioned in vivo, which regulates the access to functional DNA sites in eukaryotic genomes (1–3). This prominent level of genome regulation was recently underscored by genome-wide nucleosome maps for many organisms (4–9). However, we know rather little about the molecular determinants for this primary order of chromatin. In order to understand the molecular mechanisms of nucleosome positioning, a cell-free in vitro system is necessary that allows generating in vivo-like nucleosome positioning de novo.

Randall H. Morse (ed.), *Chromatin Remodeling: Methods and Protocols*, Methods in Molecular Biology, vol. 833,
DOI 10.1007/978-1-61779-477-3_17, © Springer Science+Business Media, LLC 2012

Classically, nucleosomes are reconstituted in vitro via salt gradient dialysis (10–12), where histones and DNA are initially mixed at high salt concentration, which is step-wise or gradually dialyzed away such that nucleosomes form on DNA. This technique has been used extensively to probe intrinsic DNA-sequence preferences for nucleosome formation (13, 14), and thus generated nucleosome occupancy (= probability of a given base pair to be in any nucleosome (2)) distribution may in some cases correlate reasonably well with in vivo distributions (15, 16). However, in vitro reconstitution of sheared genomic yeast DNA by salt gradient dialysis could not recapitulate the majority of in vivo nucleosome positions (= defined position of a particular nucleosome relative to a given base pair) (15–18). Therefore, it is common practice to resort to special "nucleosome positioning sequences," e.g., Sea Urchin 5S rDNA (19), satellite DNA (20), or the in vitro selected "601" sequence (21), as templates for salt gradient dialysis assembly if in vitro assays require well-positioned nucleosomes.

Several chromatin assembly systems based on extracts or purified histone chaperones with or without ATP-dependent remodeling enzymes are available (22). In the presence of ATP, they are especially powerful in generating extensive nucleosomal arrays with physiological spacing, but they usually do not achieve nucleosome positioning, apart from some cases where sequence-specific DNA-binding factors were added (23, 24).

Based on the pioneering work in the group of Michael Schultz who used yeast extracts for chromatin assembly in vitro (25–27), we established an in vitro chromatin reconstitution system that is able to generate in vivo-like nucleosome positioning on yeast DNA sequences (28, 29). In a first step, nucleosomes are preassembled onto plasmid DNA by classical salt gradient dialysis. In a second step, these chromatin templates are incubated with a yeast whole cell extract (WCE) in the presence of ATP to induce proper positioning (Fig. 1). Very recently, we showed that our method can be applied to yeast whole-genome plasmid libraries and combined with high-throughput sequencing (30). This allows to study genome-wide nucleosome positioning mechanisms in vitro.

If different nucleosome positions, either on the same template or on different templates present in the same reconstitution reaction, are compared between conditions of saturating and limiting histone concentrations, it is possible to assess the relative stability of these positioned nucleosomes (28, 29). As defined by this assay, a nucleosome that remains properly positioned at subsaturating histone concentrations is more stably positioned than a nucleosome that requires high assembly degrees for proper positioning.

Chromatin with in vivo-like nucleosome positioning as assembled by the here described method may be used as template for in vitro nucleosome remodeling (31) or transcription assays. However, if purified chromatin templates are required, it is not trivial and

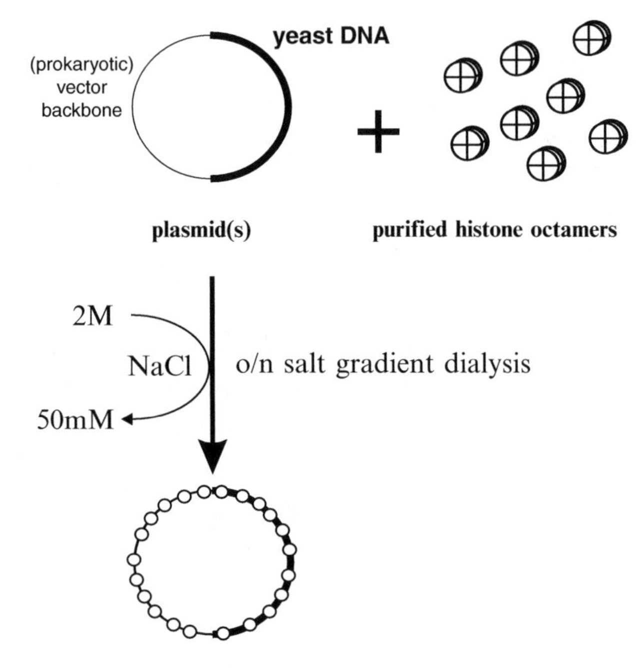

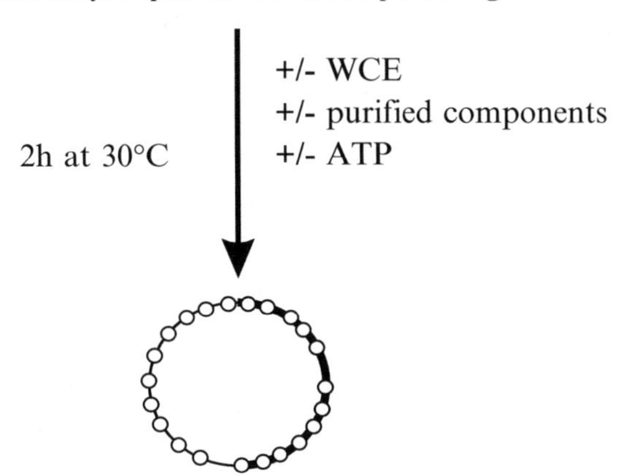

Fig. 1. Schematic overview of the method. Plasmid DNA and purified histones are preassembled into chromatin by salt gradient dialysis. This yields nucleosome positions that are specified by the intrinsic DNA preferences. Subsequent incubation with *S. cerevisiae* whole cell extract (WCE) and ATP will generate in vivo-like nucleosome positioning on *S. cerevisiae* DNA sequences. WCE fractions instead of WCE may be used (36) and additional components, like purified factors or acetyl CoA (31), can be added at any step.

remains to be established how the factors from the WCE can be removed without perturbation of nucleosome positioning. Nonetheless, WCEs prepared from mutants or that are immunodepleted may be used in order to assess the role of factors of interest in downstream assays.

2. Materials

2.1. Preparation of Yeast WCE

1. Yeast strain, e.g., BY4741, or any mutant of interest. For WCE from wild type cells, we successfully used household baker's yeast from a convenience store. Huge amounts of extract can be prepared this way at very low cost.
2. Extraction buffer without protease inhibitors: 200 mM HEPES-KOH, pH 7.5, 10 mM $MgSO_4$, 20% (v/v) glycerol, 1 mM EDTA, 390 mM $(NH_4)_2SO_4$, 1 mM freshly added DTT.
3. Extraction buffer with protease inhibitors: As above, with 1× Complete™ (Roche Applied Science) or equivalent protease inhibitors.
4. Cold spatula.
5. 5 or 10 ml plastic syringe with cut off nozzle.
6. 100 and 400 ml plastic beaker.
7. 250 and 50 ml conical tube.
8. Liquid nitrogen.
9. Porcelain mortar and pestle.
10. Clear ultracentrifuge tubes (Beckman Ultra-Clear™ tubes or equivalent).
11. Cold mineral oil.
12. Beckman Optima LE-80k ultracentrifuge with SW55Ti rotor, or equivalent.
13. 5 ml plastic syringe with rubber seal plunger; syringe needle, e.g., 1.1 × 40 mm.
14. Beckman Optima MAX-E ultracentrifuge with TLA55 rotor and Microfuge® Polyallomer tubes, or equivalent.
15. Ammonium sulfate (solid).
16. Disposable inoculation loops.
17. Rotating wheel in coldroom.
18. Dialysis buffer: 20 mM HEPES-KOH, pH 7.5, 80 mM KCl, 10 or 20% (v/v) glycerol, 1 mM EGTA and freshly added 5 mM DTT, 0.1 mM phenylmethylsulfonyl fluoride (PMSF) and 1 mM sodium metabisulfite.
19. Dialysis tubing (MWCO 3.5 kDa).

2.2. Chromatin Assembly by Salt Gradient Dialysis

1. Template DNA. The DNA of your region of interest, usually in the context of a plasmid backbone. Several plasmids or even a whole-genome plasmid library may be used for a single reconstitution reaction. Prepare by using a Qiagen (or similar) plasmid preparation kit according to manufacturer's directions. Store DNA preparation in TE buffer (10 mM Tris–HCl, pH 8.0, 1 mM

EDTA). We do not have much experience with linear templates, but it is usually more difficult to achieve high nucleosome assembly degrees on linear than on supercoiled templates (32).

2. Histones. *Drosophila* embryo histones (33) or recombinant *Drosophila*, *Xenopus*, or yeast histones (12) (see Note 1). Histones are typically stored in 1 M NaCl, 50% (v/v) glycerol, 5 mM DTT, 1× Complete™ (Roche Applied Science), or equivalent protease inhibitors at –20°C.
3. β-mercaptoethanol.
4. Low salt buffer: 10 mM Tris–HCl, pH 7.6, 50 mM NaCl, 1 mM EDTA, 0.05% (w/v) Igepal CA630, prepare as 20× stock.
5. High salt buffer: 10 mM Tris–HCl, pH 7.6, 2 M NaCl, 1 mM EDTA, 0.05% (w/v) Igepal CA630. Prepare also a small amount as 2× stock without Igepal CA630.
6. Magnetic stirrer and large (e.g., 4 cm) stir bar.
7. Peristaltic pump.
8. Siliconized 1.5 ml tubes.
9. Dialysis tubing (MWCO 3.5 kDa).
10. Two 3 l plastic beakers; small, e.g., 100 ml, beaker.
11. Drawn out Pasteur pipets.
12. Floater for 1.5 ml tubes.
13. Bovine serum albumin (BSA) 5 mg/ml in water.

2.3. Proper Nucleosome Positioning Upon Incubation with WCE and ATP

1. Block solution: 2 mg/ml BSA, 0.1% (w/v) Igepal CA630, 20 mM HEPES-KOH, pH 7.5.
2. Creatine kinase (CK): Dissolve the lyophilized CK powder in 0.1 M imidazole-HCl, pH 6.6 at 20 mg/ml, and flash freeze in liquid nitrogen as 20 μl aliquots. Store at –80°C (see Note 2).
3. 0.1 M imidazole-HCl, pH 6.6.
4. 0.25 M ammonium sulfate.
5. 4× reconstitution mix: 80 mM HEPES-KOH, pH 7.5, 320 mM KCl, 12 mM $MgCl_2$, 2 mM EGTA, 10 mM DTT, 48% (v/v) glycerol, 12 mM ATP, 120 mM creatine phosphate, can be stored at –20°C, for conditions without energy, omit ATP, $MgCl_2$, and creatine phosphate (see Note 3).

2.4. Chromatin Digestion with DNaseI, MNase, or Restriction Enzymes

1. Proteinase K: 20 mg/ml in ddH_2O.
2. Glycogen: 20 mg/ml in ddH_2O.
3. 50 U/ml apyrase in (New England Biolabs).
4. DNaseI-solution (e.g., Roche Applied Science).
5. MNase: Resuspend lyophilized MNase in Ex50 buffer (10 mM HEPES-KOH pH 7.6, 50 mM KCl, 1.5 mM $MgCl_2$, 0.5 mM

EGTA, 10% (v/v) glycerol, 1 mM DTT, and 0.2 mM PMSF) for example at 1 U/μl.

6. Appropriate restriction enzymes can be obtained from any manufacturer. It is usually advantageous to use the highest available concentrations.
7. DNaseI digestion buffer: 20 mM HEPES-KOH, pH 7.5, 80 mM NaCl, 12% (v/v) glycerol, 5.5 mM $MgCl_2$, 5.5 mM $CaCl_2$, 2.5 mM DTT, 0.1 mg/ml BSA.
8. DNaseI solutions: Dilute DNaseI with DNaseI digestion buffer to concentrations in the range of 0.005–0.02 U/ml (free DNA), 0.02–0.1 U/ml (salt gradient dialysis chromatin), or 2–10 U/ml (salt gradient dialysis chromatin with WCE). These DNaseI solutions are freshly prepared on ice and not stored.
9. Stop buffer: 100 mM EDTA, 4% (w/v) sodium dodecyl sulfate.
10. Restriction enzyme digestion buffer: 20 mM HEPES-KOH, pH 7.5, 4.5 mM $MgCl_2$, 2.5 mM DTT, 80 mM NaCl, 0.5 mM EGTA.
11. Sheared salmon sperm DNA: Salmon sperm DNA is dissolved in TE buffer (10 mM Tris–HCl, pH 8.0, 1 mM EDTA) at 2.5 mg/ml and sheared by sonication such that a mixture of DNA fragments in the range of 300 bp to several kb is generated. Store at –20°C.
12. Orange G or bromophenol blue loading dye: 40% (w/v) sucrose, 10 mM Tris–HCl, pH 8.0, 0.25% (w/v) Orange G or bromophenol blue, respectively.

3. Methods

3.1. Preparation of Yeast WCE

1. Grow 2–6 l of mid log phase yeast culture, harvest (30 min, 6,000 × *g*, 4°C, 1 l centrifuge bottles), wash (use 50 ml of ice-cold water per 1 l of yeast culture to combine pellets in 250 ml conical bottom centrifuge tube), and centrifuge again (15 min, 6,000 × *g*, 4°C).
2. For buffer exchange, resuspend the washed cell pellet in 20 ml extraction buffer without protease inhibitor per 1 l of culture, collect cells (10 min, 6,000 × *g*, 4°C), and resuspend in 10 ml extraction buffer with protease inhibitor per 1 l of culture, collect cells (5 min, 6,000 × *g*, 4°C).
3. Determine wet cell weight (usually 1–2 g per 1 l of culture).
4. Scrape cell pellet with cold spatula into cold 5 or 10 ml syringe with cut off nozzle.
5. Cover a 400 ml plastic beaker with aluminum foil and poke a 50 ml conical tube through the foil such that it can stand

upright in the beaker. Fill about 200 ml liquid nitrogen into the beaker and about 20 ml into the conical tube (see Note 4).

6. Extrude cell pellet with syringe into the liquid nitrogen in the conical tube, such that it looks like "frozen spaghetti." Carefully (see Note 4) pour off the liquid nitrogen without losing the cell pellet material, or let the nitrogen evaporate away. The frozen spaghetti may be stored at −80°C.
7. Fill porcelain mortar (see Note 5) repeatedly with liquid nitrogen until the mortar is cooled down enough to keep the liquid nitrogen for a while. Have plenty of liquid nitrogen in a Dewar at hand for repletion during grinding.
8. Add the frozen spaghetti into the liquid nitrogen in the mortar and carefully start to crush them into small pieces with the pestle. Grind the frozen cell material carefully (avoid spills), but forcefully, as this is the only cell lysis step, for 45 min. Always replenish liquid nitrogen shortly after it is all evaporated. This is somewhat hard work, but can be interrupted at any moment by storing mortar with pestel and cell powder at −80°C. After about 20 min of grinding, add 0.4 ml extraction buffer with protease inhibitors per gram wet cell mass. The resultant ice particles are very crunchy and help with the lysis during the grinding. In the end, this generates a very fine powder.
9. Let all liquid nitrogen evaporate and scrape powder into 100 ml beaker. Let warm quickly at room temperature under continuous stirring with a metal spatula until the powder turns into a thick paste (see Note 6). Place on ice immediately to avoid warming beyond 0°C.
10. Scrape paste into precooled Ultra-Clear™ or equivalent tubes. If necessary, top off with cold mineral oil in order to fill the tube sufficiently and avoid tube collapse during ultracentrifugation.
11. Spin in SW55Ti rotor or equivalent for 2 h at 29,500 rpm (82,500 × *g* average) and 4°C with brake on.
12. Preform a hole into the ice of your ice bucket using an empty SW55Ti centrifuge tube and put there the sample tube after the ultracentrifugation. Be careful not to disturb the phase separation in the tube. There are four different phases now: (a) the compact pellet of cell debris at the bottom, (b) a cloudy yellowish layer on top, which fades into a (c) clear supernatant, and finally a (d) whitish lipid rich top phase at the meniscus.
13. Using a precooled 5 ml syringe with needle (a rubber seal instead of all plastic plunger facilitates gentle suction) carefully remove the middle part of the clear supernatant (see (c) in step 12) by poking the needle through the lipid top layer. Avoid as much of the yellowish cloudy layer (see (b) in step 12) as possible, but usually it is not possible to avoid all of it (see Note 7).

Transfer the withdrawn lysate into Microfuge® Polyallomer TLA55 or equivalent tubes on ice.

14. Determine the volume of the lysate with a pipet.
15. Grind ammonium sulfate into a fine powder and add in small portions 337 mg per ml of lysate; it may help to use a small funnel folded from a piece of paper. After each addition, mix with innoculation loop and place on rotating wheel at 4°C. Avoid foam generation. After all the ammonium sulfate has dissolved (check if you still see tiny crystals sinking to the tube bottom when holding it against the light) rotate tubes for an additional 30 min.
16. Spin the solution in TLA55 rotor for 20 min at 26,000 rpm (30,300 × *g* average) and 4°C.
17. Carefully withdraw the supernatant with a cold 5 ml syringe and needle and discard it.
18. Redissolve the pellet in 0.2–0.5 ml of dialysis buffer per gram wet cell mass, depending on how well it dissolves and how concentrated the final extract shall be. Again, twirling with an inoculation loop helps.
19. Transfer the solution into a dialysis tube and dialyze twice for 1.5 h against 40- to 50-fold excess volume of dialysis buffer.
20. Remove dialyzed extract, flash freeze 50–1,000 μl aliquots in liquid nitrogen, and store at −80°C. The nucleosome positioning activity tolerates at least two freeze-thaw cycles. Such extracts usually retain their nucleosome positioning activity for at least 2 years (see Note 8).

3.2. Chromatin Assembly by Salt Gradient Dialysis

3.2.1. Pump and Beakers

1. Set up the salt dialysis apparatus in a hood as high concentrations of β-mercaptoethanol are used.
2. Fill one 3 l beaker with 3 l of 1× low salt buffer and another 3 l beaker with 300 ml 1× high salt buffer. Add 300 μl β-mercaptoethanol to the beaker with low salt buffer and mix well.
3. Place the beaker with high salt buffer on a magnetic stirrer and add a large stir bar.
4. Set up the peristaltic pump and place into each of the 3 l beakers one end of the tube. Fix the tube at each 3 l beaker with tape such that the tube cannot slide off. Make sure that the tube end in the beaker with the low salt buffer is situated at the bottom of the beaker such that all buffer can be pumped out.

3.2.2. Dialysis Mini Chamber

1. Cut off the end of a siliconized 1.5 ml tube, just above the 0.5 ml mark.
2. Using pointed scissors, puncture the thin center part of the tube lid that is circumscribed by the elevated edge that fits into the tube upon closing the lid, and scrape out the plastic up to

the elevated edge. Basically, you generate a lid with a hole of about 0.8 cm diameter. Make sure not to generate sharp edges that could puncture or rip the dialysis membrane later on.

3. Cut off the thus perforated lid from the previously truncated tube.
4. Cut off about 1.5–2 cm of dialysis tubing and place in a small beaker filled with ddH_2O for about 10 min. Cut the tubing open at one side so that the dialysis membrane can be folded open as a single layer.
5. Place the perforated lid top down onto a sheet of cling film, which serves as a convenient and clean surface to prepare the dialysis mini chamber. Place the dialysis membrane centered on top of the lid. Press the truncated siliconized tube with its top over the dialysis membrane onto the lid such that the membrane becomes wedged in between lid and tube like a drumhead (see Note 9). Cut away most of the excess membrane sticking outward from the tube.
6. Use the floater to let the dialysis mini chamber float on top of the high salt buffer in the 3 l beaker with lid and membrane facing downward and the truncated tube facing upward. Airbubbles right underneath the membrane have to be removed. It is convenient to suck away the bubbles with a drawn out Pasteur pipet that has been bent twice into a U-shape over a Bunsen burner flame.

3.2.3. Samples

Combine 10 μg plasmid DNA (see Note 10), 20 μg BSA and variable amounts of histones (see Note 11), 50 μl 2× high salt buffer without Igepal CA630, 5 μl 1% (w/v) Igepal CA630, and ddH_2O to make up 100 μl. Mix thoroughly by pipetting and avoid foam generation.

3.2.4. Salt Gradient Dialysis

1. Pipet samples through the open end of the floating dialysis mini chamber onto the membrane. Be careful not to damage the membrane with the pipet tip!
2. Adjust magnetic stirrer underneath the high salt buffer beaker such that slow mixing is achieved without compromising easy floating of the dialysis mini chambers.
3. Add 300 μl β-mercaptoethanol to the high salt buffer beaker and cover the beaker with cling film. Make sure, e.g., by using tape or placing a heavy glass plate on top, that the beaker is properly sealed (see Note 12).
4. Set speed of peristaltic pump such that all of the 3 l low salt buffer will be pumped into the high salt buffer over the course of at least 15 h. A trial run with water and without samples is advisable to determine the right pump speed.
5. After complete transfer of the low salt buffer, transfer the floater with the dialysis mini chambers to a jug with 1 l fresh low salt buffer plus 300 μl β-mercaptoethanol. Remove again air bubbles from underneath the membranes.

6. Dialyze for 1–2 h with slow stirring to ensure complete buffer exchange.
7. Transfer the samples with a pipet from the dialysis mini chambers into fresh siliconized 1.5 ml tubes and determine the volume with the pipet. The volume sometimes increases from 120 to 130 μl. The salt gradient dialyzed chromatin samples can be stored at 4°C for several weeks up to a few months.

3.3. Proper Nucleosome Positioning Upon Incubation with WCE and ATP

1. Block siliconized 1.5 ml tubes by pipetting 1 ml block solution into and out of the tubes. The block solution can be reused many times. Collect remaining solution in the tubes by short centrifugation in table top centrifuge, remove the last droplet with yellow tip, and let the tubes air dry. Such blocked tubes can be prepared in large quantities beforehand and stored indefinitely.
2. Prepare a fresh 1:20 dilution of CK by adding 380 μl 0.1 M imidazole buffer to a freshly thawed 20 μl CK aliquot (see Note 2). Mix by pipetting and keep on ice.
3. Spin down salt gradient dialysis chromatin and thawed WCE for 3 min at full speed in a cooled table top centrifuge to avoid carryover of aggregates. Especially, the WCE usually shows a visible pellet. In this case, avoid disturbance of the pellet when taking out aliquots.
4. Combine 25 μl 4× reconstitution mix, 4 μl 0.25 M ammonium sulfate, 2 μl CK 1:20 dilution, salt gradient dialysis chromatin corresponding to 0.5–1 μg of preassembled DNA, 10 μl of WCE (if the protein content is about 20 mg/ml, see Note 8) and ddH_2O to make up a volume of 100 μl (see Note 13). Start with water, 4× reconstitution mix, and ammonium sulfate, all three of which can be combined to a master mix if several reactions are done in parallel. If called for, any purified component, e.g., the transcription factor Pho4 or a remodeling enzyme (31), may be added.
5. Incubate for 2 h at 30°C. 1 h can be sufficient (28) and the incubation can even be extended overnight.
6. Analyze chromatin by your favorite assay. As an example we describe briefly the digestion with DNaseI or restriction enzymes for indirect end-labeling analysis and the generation of MNase ladders.

3.4. Chromatin Digestion with DNaseI, MNase, or Restriction Enzymes

(see also ref. (34) for a detailed description of these methods).

1. For DNaseI digestion, add 25 μl aliquots of a 100 μl reconstitution reaction to 25 μl of DNaseI solutions with appropriate DNaseI concentrations (see Note 14), incubate for exactly 5 min at room temperature, and stop the digest with 10 μl stop buffer.

2. For restriction enzyme digestion, ATP must be removed by the addition of 0.2 U apyrase per 100 μl reconstitution reaction and incubation for 30 min at 30°C (see Notes 3 and 14). 1–2 μl aliquots of such an ATP-depleted reconstitution reaction are mixed with 30 μl of restriction enzyme digestion buffer and treated with various amounts of selected restriction enzymes (see Note 15) for 2 h at 37°C. Stop digest with 7.5 μl stop buffer.
3. For both types of nuclease digestion, the DNA is deproteinized by addition of 0.06 μl proteinase K per microliter digestion reaction together with 1 μl glycogen (as carrier for precipitation) and incubation at 37°C over night, precipitated with ethanol, resuspended in TE buffer, digested with an restriction enzyme appropriate for secondary cleavage (see Note 16), again ethanol precipitated, and resuspended in TE buffer.
4. Southern blot and hybridization of the DNA is described elsewhere (35). For examples of plasmid-borne yeast loci reconstituted by the here described method, see Fig. 2.
5. For the generation of MNase ladders, MNase is used instead of DNaseI in the above protocol of step 1. Higher degrees of digestion are chosen, and the secondary cleavage step is omitted. The resulting purified DNA samples are electrophoresed in 1.3% agarose gels with Orange G as loading dye (Bromophenol blue migrates close to the dinucleosomal band and may confound the pattern). As MNase may cut at several sites within the linker DNA, there will not be clearly defined fragment sizes but rather fuzzy bands. Include ethidium bromide in the gel for sharper appearance of the bands. Note that MNase will "trim" toward the nucleosome cores, i.e., the fragment sizes will get somewhat shorter with increasing MNase concentrations. Include a suitable size marker, e.g., the 123 bp ladder (Invitrogen) or the 100 bp ladder (NEB).

4. Notes

1. Histones are very sticky proteins. Use siliconized (and maybe even blocked, see Subheading 3.3, step 1) tubes. As many others, we noted that recombinant yeast histones are more difficult to work with, i.e., it is more difficult to achieve high assembly degrees and proper positioning of tricky loci like at the yeast *PHO5* promoter (31).
2. Prepare CK-dilution freshly before use and always use a fresh aliquot! Do not refreeze!
3. The concentration of ATP may be determined, especially after apyrase treatment, using a luciferase-based essay, e.g., Enliten,

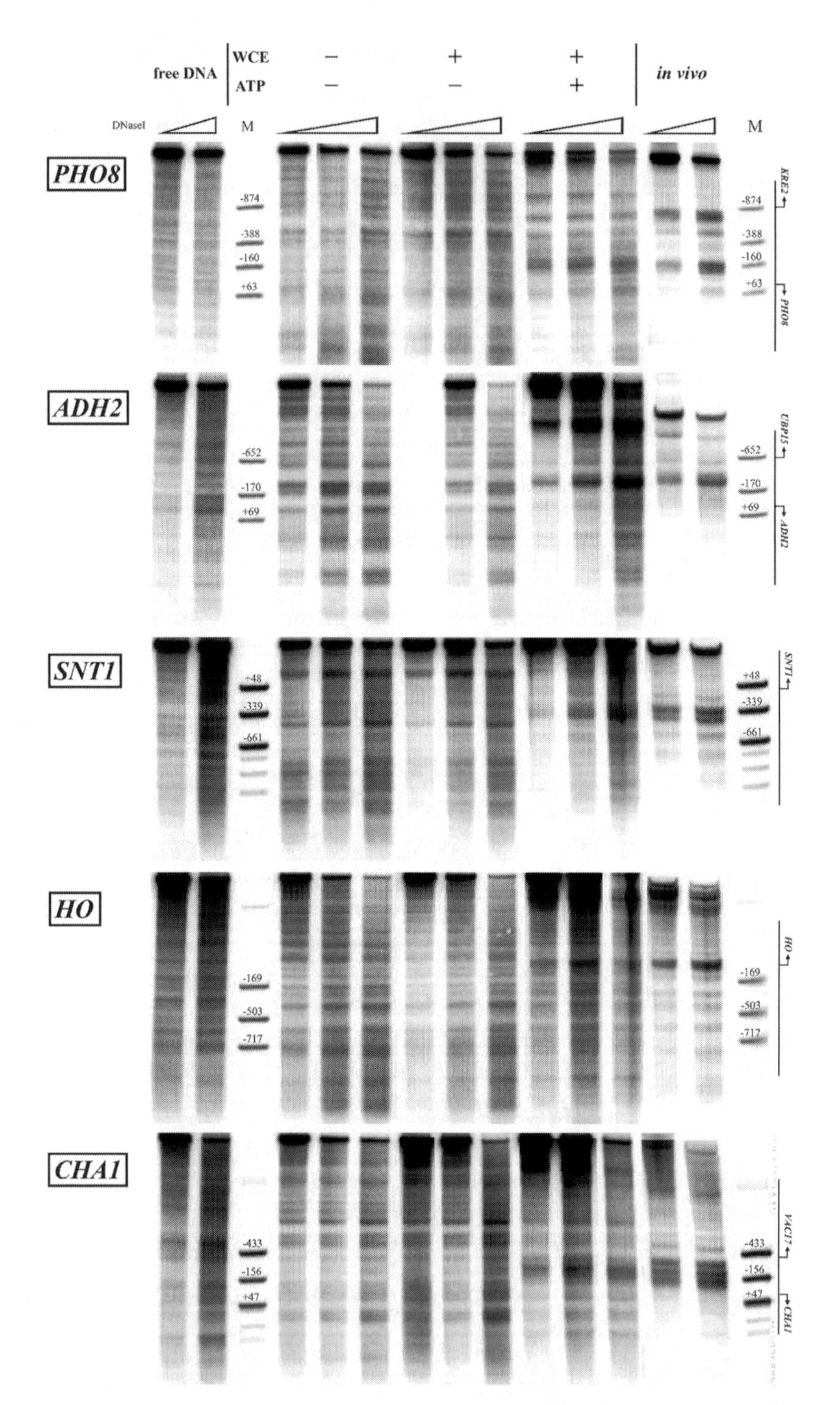

Fig. 2. Examples of yeast loci with in vitro reconstituted in vivo-like nucleosome positioning. In vitro reconstitution of nucleosome positioning at the *PHO8*, *ADH2*, *SNT1*, *HO*, and *CHA1* loci. pUC19 plasmids containing ~3.5 kb of the indicated locus were assembled into chromatin by salt gradient dialysis and incubated in the presence or absence of WCE and ATP. *Lanes 4–6* show the pattern of chromatin reconstituted by salt gradient dialysis. Nucleosome positioning was analyzed by DNaseI indirect endlabeling. Free DNA samples were generated from nonassembled plasmids in the absence of WCE and ATP but under otherwise identical conditions. The in vivo samples were prepared from nuclei isolated from wildtype strain BY4741 grown logarithmically at 30°C. Ramps indicate increasing DNaseI concentrations. The numbers above the marker bands refer to the position (in base pairs) relative to the start of the corresponding ORF. The approximate start of the indicated ORFs is indicated in the schematics on the *right*.

Promega, FF2021, in connection with a Berthold Lumat luminometer. Attention: This assay is very sensitive and therefore easily saturated. Measure serial tenfold dilutions (up to 10^{-6}) in water in order to find the actual working range of the assay. The high dilutions will also slow down ATPases from further depleting ATP if you are interested in the ATP concentration at a particular point in your procedure.

4. Be careful with liquid nitrogen! Wear safety glasses and insulating gloves.
5. Alternatively, we use an electric mortar (Retsch RM100). Fill the electric mortar with liquid nitrogen and close lid with pestle. After most of the liquid nitrogen has evaporated, open the lid and refill the mortar with liquid nitrogen. Immediately add the "frozen spaghetti," close lid and start grinding at pestel setting of one ("1"). After all of the spaghetti fragments have been ground into a powder add the appropriate amount of extraction buffer. Subsequently, increase pestle setting to ~5.5. Keep grinding at this setting for ~8–10 min (assuming 10 g of wet weight material as input, shorter grinding for less material). Refill mortar with nitrogen through the small window at the top of the mortar each time shortly after its evaporation.
6. In our view, it is a common misconception that sensitive biological samples should be flash frozen, but slowly thawed on ice. As thawing is the reversal of freezing, it should also be fast, e.g., at room temperature, but "to point," i.e., do not let the sample get warmer than 0°C.
7. It is possible to recentrifuge after this step to allow better phase separation. But this is usually not necessary.
8. Our yeast extracts usually contain 10–30 mg/ml protein as assayed by Bradford assay with BSA as standard. Of these, we usually take 5–15 μl per nucleosome positioning reaction. In contrast to the histone:DNA mass ratio (see Note 11), the amount of extract per nucleosome positioning reaction is much less critical, i.e., variation by a factor of 2 or 0.5 usually has hardly any effect. Nonetheless, too much extract leads to chromatin aggregation. We routinely adjust our extracts with dialysis buffer to 50 mg "protein" per milliliter according to nanodrop reading at 280 nm and use 10 μl per nucleosome positioning reaction. This usually corresponds to a protein concentration of ~20 mg/ml as measured by Bradford assay. The nanodrop reading is somewhat confounded by varying amounts of nucleic acids, especially RNA. Nonetheless, this procedure works just fine as a quick measure for how much extract to use per nucleosome positioning reaction. Very recently, we showed that WCE fractions may be used as well in order to identify involved factors (36).

9. It is important that the membrane is tightly sealed between lid and tube and that the membrane surface is tense and smooth. Otherwise, the dialysis mini chamber may be leaky. If several dialysis mini chambers are prepared at the same time, make sure that the membranes do not dry out at any point. You can make a small puddle of ddH_2O onto the cling film and store there the dialysis mini chambers lid-down, which will keep the membranes wet. Do not allow any water into the dialysis chamber as this will dilute your sample.
10. Mixtures of plasmids and even a whole-genome library (30) are also possible, but requires increasingly more material or more sensitive methods to analyze the chromatin structure at loci of interest. As formation of a nucleosome corresponds to about one negative supercoil (37), nucleosome reconstitution is more efficient on supercoiled plasmids (32, 38).
11. The histone:DNA ratio is probably the most crucial parameter for the reconstitution of in vivo-like nucleosome positioning by this method as well as for other chromatin reconstitution protocols (22, 39). Ideally, a physiological mass ratio of 1:1 should be achieved. In practice, one should aim at as high an assembly degree as possible without aggregation of the chromatin and without packing the nucleosomes too tightly such that they will be refractory to ATP-dependent remodeling. Aggregation can be tested by MNase ladders analysis as it will result in an increasing amount of undigestible material and less signal within the lane. The assembly degree can be estimated also via topology assay (30) if the template is a plasmid that is not too large for separation of topoisomers in agarose gels. Fully assembled chromatin usually has a similar degree of superhelicity as the plasmid prepared from *E. coli* (39). A more direct and functional read out with regard to nucleosome positioning is indirect end-labeling of a locus of interest. Both too low (28) and too high (Längst, G., Wippo, C.J., Ertel, F. and Korber, P., unpublished observation) degrees of assembly can interfere with the proper repositioning of nucleosomes upon incubation of salt gradient dialysis chromatin with WCE and ATP. In summary, the optimal assembly degree is difficult to be calculated from measured concentrations of DNA and histones, but usually found by careful and repeated titration using the mentioned assays as read out. The estimation of DNA and histone concentration, e.g., spectrophotometrically or by comparing band intensities on gels to standard samples, serves as an initial reference point to set up assembly reactions with histone:DNA mass ratios in the range of 0.5–2.0. We keep the DNA concentration constant and vary the histone concentration. Titrate in histones until you see overassembly by the assays mentioned above, then go back again to lower mass ratios and perform more assemblies with more and more finely varied

mass ratios until the best ratio for the desired application is found. Importantly, this kind of titration has to be repeated for each new preparation of histones.

Prokaryotic DNA has an intrinsically lower propensity to be incorporated into nucleosomes (16). Therefore, including prokaryotic DNA, either in *cis* as part of the vector backbone or in *trans*, may serve as buffer for excess histones regarding the eukaryotic DNA fraction. The assembly of prokaryotic competitor DNA as monitored in native agarose gel electrophoresis can be used as indicator for full assembly of eukaryotic or other high affinity target DNA as described elsewhere (40).

12. As this is in the hood and runs overnight, the sample volume decreases substantially due to evaporation if the beaker is not covered.
13. Addition of protease inhibitors is usually not necessary. We compared reconstitution reactions with and without inhibitors several times and never saw a difference in our assays. Nonetheless, depending on the application and readout adding protease inhibitors may become advisable.
14. Indirect end-labeling requires that each template has on average only one double strand cut in the region of interest. Typical DNaseI concentrations are given in Subheading 2. One should always do several (typically three) different concentrations in parallel in order to catch a proper degree of digestion. Due to the single-cut limit digestion regime, DNaseI indirect end-labeling corresponds to a snap shot of the time and population average chromatin structure. This is why ATP – and concomittantly remodeling activity – need not be removed prior to digestion. In contrast, if remodeling enzymes are active during the exhaustive restriction enzyme digest, they will continuously generate windows of opportunity for DNA cleavage and the irreversibly cut DNA templates will accumulate over time resulting in apparent high accessibility, even though on time and population average the respective cutting site may be covered by a nucleosome (41).
15. The easiest way to ensure that the restriction enzyme digest was complete is to compare two different, e.g., fourfold, restriction enzyme concentrations, which should yield roughly the same accessibility value (34).
16. The gel will have to resolve the fragments resulting from the nuclease cuts in your region of interest in combination with the secondary cleavage (34). So the secondary cleavage site has to be chosen such that resulting fragments are within the resolution of the gel, and, of course, the restriction enzyme for secondary cleavage must not cut within the region of interest. Typically, for 1.5% TAE agarose gels, the secondary cleavage site is about 0.7–1.2 kb up- or downstream of the region of interest.

Acknowledgments

Work in our laboratory is funded by the German Research Community (DFG, grant within the SFB/Transregio 5) and through the 6th Framework Programme of the European Community (NET grant within the Network of Excellence The Epigenome). We thank Nils Krietenstein for critical reading of the manuscript. This paper is dedicated to the memory of Eduard Buchner, who founded biochemistry by demonstrating the power of yeast extracts.

References

1. Radman-Livaja, M. and Rando, O. J. (2010) Nucleosome positioning: how is it established, and why does it matter? *Dev. Biol. 339*, 258–266.
2. Segal, E. and Widom, J. (2009) What controls nucleosome positions? *Trends Genet. 25*, 335–343.
3. Jiang, C. and Pugh, B. F. (2009) Nucleosome positioning and gene regulation: advances through genomics *Nat. Rev. Genet. 10*, 161–172.
4. Yuan, G. C., Liu, Y. J., Dion, M. F., Slack, M. D., Wu, L. F., Altschuler, S. J., and Rando, O. J. (2005) Genome-scale identification of nucleosome positions in S. cerevisiae *Science 309*, 626–630.
5. Mavrich, T. N., Jiang, C., Ioshikhes, I. P., Li, X., Venters, B. J., Zanton, S. J., Tomsho, L. P., Qi, J., Glaser, R. L., Schuster, S. C., Gilmour, D. S., Albert, I., and Pugh, B. F. (2008) Nucleosome organization in the Drosophila genome *Nature 453*, 358–362.
6. Valouev, A., Ichikawa, J., Tonthat, T., Stuart, J., Ranade, S., Peckham, H., Zeng, K., Malek, J. A., Costa, G., McKernan, K., Sidow, A., Fire, A., and Johnson, S. M. (2008) A high-resolution, nucleosome position map of C. elegans reveals a lack of universal sequence-dictated positioning *Genome Res. 18*, 1051–1063.
7. Schones, D. E., Cui, K., Cuddapah, S., Roh, T. Y., Barski, A., Wang, Z., Wei, G., and Zhao, K. (2008) Dynamic regulation of nucleosome positioning in the human genome *Cell 132*, 887–898.
8. Lee, W., Tillo, D., Bray, N., Morse, R. H., Davis, R. W., Hughes, T. R., and Nislow, C. (2007) A high-resolution atlas of nucleosome occupancy in yeast *Nat. Genet. 39*, 1235–1244.
9. Lantermann, A. B., Straub, T., Stralfors, A., Yuan, G. C., Ekwall, K., and Korber, P. (2010) Schizosaccharomyces pombe genome-wide nucleosome mapping reveals positioning mechanisms distinct from those of Saccharomyces cerevisiae *Nat. Struct. Mol. Biol. 17*, 251–257.
10. Stein, A. (1989) Reconstitution of chromatin from purified components *Methods Enzymol. 170*, 585–603.
11. Rhodes, D. and Laskey, R. A. (1989) Assembly of nucleosomes and chromatin in vitro *Methods Enzymol. 170*, 575–585.
12. Luger, K., Rechsteiner, T. J., and Richmond, T. J. (1999) Expression and purification of recombinant histones and nucleosome reconstitution *Methods Mol. Biol. 119*, 1–16.
13. Widom, J. (2001) Role of DNA sequence in nucleosome stability and dynamics *Q. Rev. Biophys. 34*, 269–324.
14. Schnitzler, G. R. (2008) Control of nucleosome positions by DNA sequence and remodeling machines *Cell Biochem. Biophys. 51*, 67–80.
15. Kaplan, N., Moore, I. K., Fondufe-Mittendorf, Y., Gossett, A. J., Tillo, D., Field, Y., LeProust, E. M., Hughes, T. R., Lieb, J. D., Widom, J., and Segal, E. (2009) The DNA-encoded nucleosome organization of a eukaryotic genome *Nature 458*, 362–366.
16. Zhang, Y., Moqtaderi, Z., Rattner, B. P., Euskirchen, G., Snyder, M., Kadonaga, J. T., Liu, X. S., and Struhl, K. (2009) Intrinsic histone-DNA interactions are not the major determinant of nucleosome positions in vivo *Nat. Struct. Mol. Biol. 16*, 847–852.
17. Pugh, B. F. (2010) A preoccupied position on nucleosomes *Nat. Struct. Mol. Biol. 17*, 923.
18. Kaplan, N., Hughes, T. R., Lieb, J. D., Widom, J., and Segal, E. (2010) Contribution of histone sequence preferences to nucleosome organization: proposed definitions and methodology *Genome Biol. 11*, 140.

19. Simpson, R. T. and Stafford, D. W. (1983) Structural features of a phased nucleosome core particle *Proc. Natl. Acad. Sci. USA 80*, 51–55.
20. Neubauer, B., Linxweiler, W., and Hörz, W. (1986) DNA engineering shows that nucleosome phasing on the African green monkey alpha-satellite is the result of multiple additive histone-DNA interactions *J. Mol. Biol. 190*, 639–645.
21. Lowary, P. T. and Widom, J. (1998) New DNA sequence rules for high affinity binding to histone octamer and sequence-directed nucleosome positioning *J. Mol. Biol. 276*, 19–42.
22. Lusser, A. and Kadonaga, J. T. (2004) Strategies for the reconstitution of chromatin *Nat. Methods 1*, 19–26.
23. Pazin, M. J., Bhargava, P., Geiduschek, E. P., and Kadonaga, J. T. (1997) Nucleosome mobility and the maintenance of nucleosome positioning. *Science 276*, 809–812.
24. Langst, G., Becker, P. B., and Grummt, I. (1998) TTF-I determines the chromatin architecture of the active rDNA promoter *EMBO J. 17*, 3135–3145.
25. Robinson, K. M. and Schultz, M. C. (2003) Replication-independent assembly of nucleosome arrays in a novel yeast chromatin reconstitution system involves antisilencing factor Asf1p and chromodomain protein Chd1p *Mol. Cell. Biol. 23*, 7937–7946.
26. Schultz, M. C. (1999) Chromatin assembly in yeast cell-free extracts *Methods 17*, 161–172.
27. Schultz, M. C., Hockman, D. J., Harkness, T. A., Garinther, W. I., and Altheim, B. A. (1997) Chromatin assembly in a yeast whole-cell extract *Proc. Natl. Acad. Sci. USA 94*, 9034–9039.
28. Hertel, C. B., Längst, G., Hörz, W., and Korber, P. (2005) Nucleosome stability at the yeast PHO5 and PHO8 promoters correlates with differential cofactor requirements for chromatin opening *Mol. Cell. Biol. 25*, 10755–10767.
29. Wippo, C. J., Krstulovic, B. S., Ertel, F., Musladin, S., Blaschke, D., Sturzl, S., Yuan, G. C., Hörz, W., Korber, P., and Barbaric, S. (2009) Differential cofactor requirements for histone eviction from two nucleosomes at the yeast PHO84 promoter are determined by intrinsic nucleosome stability *Mol. Cell. Biol. 29*, 2960–2981.
30. Zhang, Z., Wippo, C. J., Wal, M., Ward, E., Korber, P., and Pugh, B. F. (2011) A packing mechanism for nucleosome organization reconstituted across a eukaryotic genome. *Science 332*, 977–980.
31. Ertel, F., Dirac-Svejstrup, A. B., Hertel, C. B., Blaschke, D., Svejstrup, J. Q., and Korber, P. (2010) In vitro reconstitution of PHO5 promoter chromatin remodeling points to a role for activator-nucleosome competition in vivo *Mol. Cell. Biol. 30*, 4060–4076.
32. Patterton, H. G. and von Holt, C. (1993) Negative supercoiling and nucleosome cores. I. The effect of negative supercoiling on the efficiency of nucleosome core formation in vitro *J. Mol. Biol. 229*, 623–636.
33. Simon, R. H. and Felsenfeld, G. (1979) A new procedure for purifying histone pairs H2A + H2B and H3 + H4 from chromatin using hydroxylapatite *Nucleic Acids Res. 6*, 689–696.
34. Svaren, J., Venter, U., and Hörz, W. (1995) *In vivo* analysis of nucleosome structure and transcription factor binding in *Saccharomyces cerevisiae Methods in Mol. Genet. 6*, 153–167.
35. Sambrook, J., Fritsch, E. F., and Maniatis, T. (1989) *Molecular Cloning: A Laboratory Manual (Edition)* Cold Spring Harbor Laboratory Press, Cold Spring Harbor, NY.
36. Wippo, C. J., Israel, L., Watanabe, S., Hochheimer, A., Peterson, C. L., and Korber, P. (2011) The RSC chromatin remodelling enzyme has a unique role in directing the accurate positioning of nucleosomes *EMBO J. 30*, 1277–1288.
37. Germond, J. E., Hirt, B., Oudet, P., Gross-Bellark, M., and Chambon, P. (1975) Folding of the DNA double helix in chromatin-like structures from simian virus 40 *Proc. Natl. Acad. Sci. USA 72*, 1843–1847.
38. Pfaffle, P. and Jackson, V. (1990) Studies on rates of nucleosome formation with DNA under stress *J. Biol. Chem. 265*, 16821–16829.
39. Nightingale, K. P. and Becker, P. B. (1998) Structural and functional analysis of chromatin assembled from defined histones *Methods A Companion To Methods In Enzymology 15*, 343–353.
40. Huynh, V. A., Robinson, P. J., and Rhodes, D. (2005) A method for the in vitro reconstitution of a defined "30 nm" chromatin fibre containing stoichiometric amounts of the linker histone *J. Mol. Biol. 345*, 957–968.
41. Korber, P. and Hörz, W. (2004) In vitro assembly of the characteristic chromatin organization at the yeast PHO5 promoter by a replication-independent extract system *J. Biol. Chem. 279*, 35113–35120.

Chapter 18

Activator-Dependent Acetylation of Chromatin Model Systems

Heather J. Szerlong and Jeffrey C. Hansen

Abstract

Regulatory mechanisms underlying eukaryotic gene expression, and many other DNA metabolic pathways, are tightly coupled to dynamic changes in chromatin architecture in the nucleus. Activation of gene expression generally requires the recruitment of histone acetyltransferases (HATs) to gene promoters by sequence-specific DNA-binding transcriptional activators. HATs often target specific lysines in the core histone amino-terminal "tail" domains (NTDs), which have the potential ability to alter higher order chromatin structure. In order to better characterize the impact targeted histone acetylation has on chromatin structure and function, we have characterized a novel model system derived from the human T-cell lymphoma virus type 1 promoter. Using this system as an example, here we describe the use of a combination of biochemical and biophysical methods to investigate the effect of activator-dependent acetylation on higher order chromatin structure and transcription by RNA polymerase II.

Key words: Transcription, Nucleosomal array, Nucleosome, Histone acetyltransferase, HTLV-1, Higher order chromatin structure, Acetylation

1. Introduction

In eukaryotes, a hierarchical packaging system organizes an entire genome into a single nucleus (1). Histone and nonhistone proteins assemble with DNA to form a macromolecular complex called chromatin, which is the fundamental component of the packaging system. The subunit of chromatin is the nucleosome, which consists of 147 base pairs (bp) of DNA wrapped around an octamer of core histone proteins (H2A, H2B, H3, and H4). Nucleosomes spaced at 20–60 bp along a DNA molecule are called nucleosomal arrays, whereas nucleosomal arrays bound to other chromosomal

Randall H. Morse (ed.), *Chromatin Remodeling: Methods and Protocols*, Methods in Molecular Biology, vol. 833, DOI 10.1007/978-1-61779-477-3_18, © Springer Science+Business Media, LLC 2012

proteins are called chromatin fibers or just chromatin. In low salt, nucleosomal arrays and chromatin fibers exist in a beads-on-a string primary structure. Under physiological salt conditions, nucleosomal arrays and chromatin fibers condense to form higher order chromatin structures at both the secondary and tertiary levels (1). Secondary chromatin structure involves short-range internucleosomal interactions that lead to localized folding of the array or fiber. Tertiary chromatin structure involves long-range fiber–fiber interactions in vivo and is characterized by oligomerization of model nucleosomal arrays and chromatin fibers in vitro (2). Transitions among primary, secondary, and tertiary structural states in vivo are mediated by a diverse group of chromatin regulatory factors that restructure and remodel chromatin, including histone chaperones (3, 4), linker histones (5, 6) and other chromatin architectural proteins (7), ATP-dependent chromatin remodelers (8, 9), histone acetyltransferases (HATs) (10, 11), histone deacetylases (HDACs), and many other histone-modifying enzymes (12). The location and activity of such chromatin regulatory factors correlate with modulation of numerous nuclear functions, including transcription by RNA polymerase II (RNAPII). However, due to the difficulty involved in simultaneously assaying transcription and higher order chromatin structure in vitro, the biochemical properties and structural features of active and repressed chromatin are not well-understood.

To better define the structure/function relationships that link higher order chromatin structure, histone acetylation, and transcription, a novel chromatin model system has been developed that allows one to biochemically characterize transcriptionally active and repressed chromatin states, and determine how the transitions between the two states are regulated by specific p300-mediated histone acetylation (13). In our approach, the transcriptional activators, cAMP response element-binding protein (CREB) and Tax, recruit p300 HAT to a human T-cell leukemia virus type-1 (HTLV-1)-based promoter that is centered in an array of positioned nucleosomes. p300 subsequently targets promoter-proximal nucleosomes and acetylates the N-terminal "tail" domains (NTDs) of the core histones. We further find that activator-dependent p300 histone acetylation is directly correlated with both enhanced RNAPII transcription and disruption of secondary and tertiary chromatin structures, and describe the corresponding assays here (13). Importantly, the HTLV-1 chromatin template and the methods described here are general and can be modified to accommodate alternative gene promoters, activators, HATs, recombinant histones (such as mutants or variants), or histone modifications (such as methylation, phosphorylation, or ubiquitylation) for future chromatin structure/function studies.

2. Materials

2.1. The HTLV-1 DNA Template

1. A human T-cell lymphoma virus type 1 (HTLV-1) DNA fragment containing the HTLV-1 promoter and G-less cassette (used as a reporter for RNA synthesis) centered in an array consisting of eight 5S rDNA nucleosome-positioning sequences has previously been cloned into pUC19 (Fig. 1) (13). The HTLV-1 template is similar in size and sequence to the prototypical 208-12 5S array used for chromatin structural studies (14).

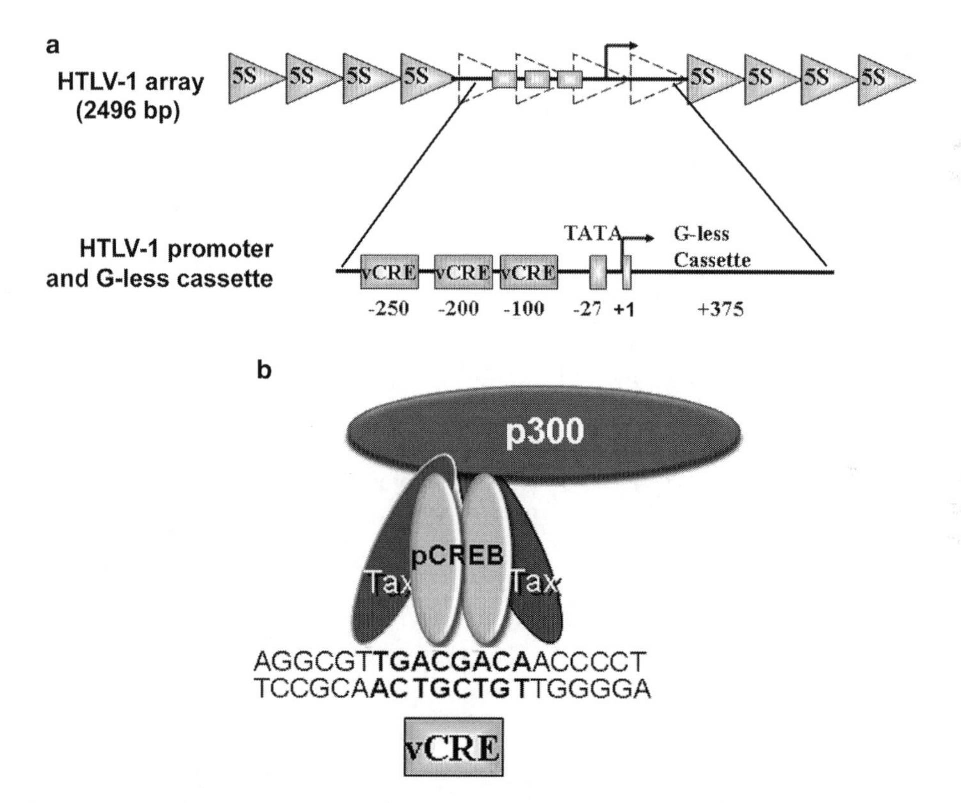

Fig. 1. Chromatin model system used to study activator-dependent p300 acetylation. (**a**) Schematic depicting the HTLV-1 chromatin model system. The HTLV-1 model system is based on a 2,496 bp DNA fragment consisting of an HTLV-1 promoter, G-less cassette, and ~150 bp of pUC13 polylinker DNA (totaling 832 bp), flanked on both sides by four repeats of the 208 bp 5S rDNA sequence. The positions of the HTLV-1 promoter (three viral cyclic AMP response elements (vCRE)) and TATA box are indicated relative to the transcription start site (+1) linked to the G-less cassette. After reconstitution with core histones, *shaded triangles* indicate the approximate positions of the 5S rDNA nucleosomes while approximate locations of promoter nucleosomes (in a saturated array) are indicated by *dashed triangles*. (**b**) Schematic depicting p300 recruitment to the HTLV-1 promoter by activators (pCREB/Tax). S133-phosphorylated CREB, Tax, and p300 form a ternary complex with a single vCRE consisting of a core CRE element (*bold*) flanked by 5′ G-rich and 3′ C-rich sequences within the HTLV-1 promoter (33, 34). (This research was originally published in Szerlong, H. J., Prenni, J. E., Nyborg, J. K., and Hansen, J. C. (2010) Activator-dependent p300 acetylation of chromatin in vitro: enhancement of transcription by disruption of repressive nucleosome–nucleosome interactions. *J Biol Chem* ***285***, 31954–64 © the American Society for Biochemistry and Molecular Biology).

2. TAE buffer: Prepare 10× stock with 24.2 g Tris base, 5.71 mL glacial acetic acid, 10 mL 0.5 M EDTA (pH 8.0), and water to 500 mL. A working solution is prepared by dilution of 1 part buffer with 9 parts water (see Note 1).
3. 6× DNA sample buffer: 0.25% bromophenol blue, 0.25% xylene cyanol, 30% glycerol in water. A working solution is prepared by dilution of 1 part buffer with 5 parts water or sample.
4. 1% agarose gel: Dissolve 1% (w/v) agarose in 1× TAE and heat until dissolved. Once agarose solution is cooled to ~65°C, pour into a gel electrophoresis mold, insert comb, and let set for 1 h at room temperature.
5. Ethidium bromide stain: 1 μg/mL ethidium bromide in 1× TAE (ethidium bromide is a carcinogen and care should be taken not to receive exposure).
6. Ethidium bromide destain: Water.
7. Competent DH5-alpha *E. coli.*
8. Ampicillin sodium salt is dissolved in water at 50 mg/mL, sterilized with a 0.2-μm syringe filter system, aliquoted, and stored at –20°C. Working solutions are prepared by dilution in liquid medium to 50 μg/mL.
9. Luria-Bertani (LB) medium and agar plates supplemented with 50 μg/mL ampicillin (LB Amp) (15).
10. Thiamine hydrochloride is dissolved in water at 1% (w/v), sterilized with a 0.2-μm syringe filter system, aliquoted, and stored at –20°C. A working solution is prepared by dilution of 1 part thiamine stock to 1,000 parts total liquid medium volume.
11. Tryptic Soy Broth (TSB)/Ampicillin/Thiamine (A/T): 30.0 g TSB is dissolved in 1 L water, autoclaved, and cooled to room temperature. TSB media is then supplemented with 1 mL 1% thiamine and 1 mL 50 mg/mL ampicillin.
12. Ethyl alcohol (200-proof) is flammable and should be stored in a fume hood at room temperature.
13. Chloramphenicol is dissolved in ethyl alcohol at 34 mg/mL. Working solutions are prepared by dilution in liquid medium to 10 μg/mL.
14. Phosphate-buffered saline (PBS) (pH 7.4): Prepare a 10× stock solution with 80 g NaCl, 2.0 g KCl, 14.4 g Na_2HPO_4, and 2.4 g KH_2PO_4 in 1 L of water and autoclave. A working solution is prepared by dilution of 1 part buffer to 9 parts water.
15. Qiagen Plasmid Giga Kit (Qiagen) or equivalent.
16. *Cfo*I restriction endonuclease 10 U/μL.
17. SuRE/Cut buffer L: 10 mM Tris–HCl, 10 mM $MgCl_2$, 1 mM dithioerythritol, pH 7.5, at 37°C.

18. Ethylenediamine tetraacetic acid (EDTA) is dissolved in water to 0.5 M, pH is adjusted to 8.0, and autoclaved (see Note 2). 0.5 M EDTA stock solution is used in preparation of all subsequent buffers containing EDTA.
19. 3.0 M sodium acetate (NaOAc) (pH 5.2).
20. Isopropyl alcohol (99.5%) is flammable and should be stored in a fume hood at room temperature.
21. Sephacryl™ S-1000 Superfine size-exclusion resin (GE Healthcare).
22. Sephacyl size exclusion column (125 cm long × 2.5 cm in diameter, ~600 mL volume).
23. Peristaltic pump.
24. Tris(hydroxymethyl) aminomethane (Tris) base is dissolved in water to 1.0 M, pH adjusted to 7.9 or 7.5 with concentrated HCl at room temperature, and autoclaved. 1.0 M Tris stock solutions are used in all subsequent buffers containing Tris.
25. TEN50 Buffer: 10 mM Tris–Cl (pH 7.9), 0.25 mM EDTA, 50 mM NaCl.
26. TE buffer: 10 mM Tris–Cl (pH 7.9), 0.25 mM EDTA.

2.2. Nucleosomal Array Reconstitution and Biophysical Characterization

1. Purified HTLV-1 DNA template (see Subheadings 2.1 and 3.1).
2. Recombinant *Xenopus laevis* core histones expressed, purified, and refolded to octamers as described (16, 17). Purified octamers are stored in refolding buffer (10 mM Tris (pH 7.5), 1.0 mM EDTA, 2.0 M NaCl, 5.0 mM β-mercaptoethanol (BME)) at 4°C and are quantified by A_{276} using a UV spectrophotometer. The micromolar concentration of octamer is equal to $A_{276} \times 10^6/39{,}120$ molar extinction coefficient (ε). BME is toxic and decomposes under the influence of moisture, water, and acids, forming toxic and combustible gas (hydrogen sulfide), so care should be taken for proper storage and to avoid exposure.
3. Dithiothreitol (DTT) is dissolved in water to 1.0 M, aliquoted, and stored at −20°C.
4. Phenylmethylsulfonyl fluoride (PMSF) is dissolved in ethyl alcohol to 0.2 M, aliquoted, and stored at −20°C (PMSF is toxic, so care should be taken not to receive exposure). PMSF solution is stable for ~1 month at −20°C.
5. Sodium chloride (NaCl) is dissolved in water to 5.0 M and filtered through a 0.2-μm hydrophilic polypropylene membrane filter using a Buchner or Hirsch vacuum filtration apparatus. NaCl stock solution is used in preparing all solutions containing <500 mM NaCl.
6. B0 Buffer: 10 mM Tris–HCl (pH 7.9), 0.25 mM EDTA, 2.0 M NaCl, 1.0 mM DTT.

7. B1 Buffer: 10 mM Tris–HCl (pH 7.9), 0.25 mM EDTA, 1.0 M NaCl, 1.0 mM DTT.
8. B2 Buffer: 10 mM Tris–HCl (pH 7.9), 0.25 mM EDTA, 0.75 M NaCl, 1.0 mM DTT.
9. B3 Buffer: 10 mM Tris–HCl (pH 7.9), 0.25 mM EDTA, 2.5 mM NaCl, 1.0 mM DTT.
10. B4 Buffer: 10 mM Tris–HCl (pH 7.9), 0.25 mM EDTA, 2.5 mM NaCl, 0.2 mM PMSF.
11. Spectra/Por molecular porous membrane tubing or equivalent (flat width 10 mm, diameter 6.4 mm, 12–14,000 Dalton (Da) molecular weight cutoff) (see Note 3).
12. Dialysis tubing clamps.
13. Dialysis buoy (e.g., from Pierce Applied Science).

2.3. Activator-Dependent Acetylation of HTLV-1 Nucleosomal Arrays

1. Recombinant human CREB protein is expressed in *E. coli*, purified to homogeneity (18), and subsequently phosphorylated at Serine 133 (pCREB) using the catalytic subunit of PKA (19) (see Note 4).
2. Recombinant Tax-His_6 is expressed (20) and purified (21) as described (see Note 4).
3. Human His_6-tagged p300 is expressed in Sf9 cells from a baculovirus-based plasmid and purified as described (22) (see Note 4).
4. Reconstituted HTLV-1 nucleosomal arrays (see Subheadings 2.2, 3.2 and Note 5).
5. Protein assay kit (e.g., Bio-Rad Protein Assay Kit (Bio-Rad Laboratories)).
6. HAT buffer: Prepare a single-use 5× stock with 100 mM Tris (pH 7.5), 250 mM NaCl, 25% glycerol, 20 mM BME.
7. Acetyl coenzyme A sodium salt (acetyl-CoA) is dissolved in water to 3.0 mM to prepare a 10× stock, aliquoted, and stored at –70°C. Working solutions are prepared by dilution of 1 part acetyl-CoA to 10 parts total reaction volume.
8. Acetyl coenzyme A [acetyl-1-^{14}C] (60 mCi/mmol) (radiolabeled acetyl-CoA is a radiation hazard and should be handled accordingly).

2.4. HTLV-1 Nucleosomal Array Folding and Oligomerization Assays

1. 1.0 M $MgCl_2$.
2. Beckman XL-A or XL-I analytical ultracentrifuge equipped with scanner optics (Beckman Coulter).
3. 12-mm double sector cells and a four-hole analytical rotor (Beckman Coulter).

4. UltraScan software version 9.9 (Borries Demeler (http://www.ultrascan.uthscsa.edu/)).
5. Transcription buffer: Prepare a single-use 2× stock solution with 2.0 mM DTT, 50 mM Tris–HCl (pH 7.9), 100 mM KCl, 12.5 mM $MgCl_2$, 1.0 mM EDTA, and 20% glycerol. Working solutions are prepared by dilution of 1 part buffer to make 2 parts total reaction volume.
6. Software for data analysis (e.g., Microsoft Excel).
7. UV spectrophotometer and quartz cuvette.

2.5. Preparation of Nuclear Extract from CEM Cells

1. 1–10 L of CEM T-lymphocytes are grown to 3×10^6 cells/mL (see Subheading 3.5, step 11, for typical protein yield). CEM cells have been used to prepare nuclear extracts to be used for in vitro transcription for the HTLV type I promoter used in this example, but other cells may be used as appropriate for the system being used.
2. 1× PBS, pH 7.4, plus 1 mM $MgCl_2$: 137 mM NaCl, 2.7 mM KCl, 10 mM Na_2HPO_4O (dibasic, anhydrous), 2 mM KH_2PO_4O (monobasic, anhydrous), 1 mM $MgCl_2$.
3. Hypotonic lysis buffer: 10 mM Tris–HCl, pH 7.9, 10 mM KCl, 1.5 mM $MgCl_2$, 1.0 mM DTT, 0.5 mM PMSF, 1 mM benzamidine, 20 mM AEBSF, 0.16 μM aprotinin, 10 μM bestatin, 3.0 μM E-64, 4.0 μM leupeptin, 4.0 μM pepstatin.
4. Dounce homogenizer with B pestle.
5. Ammonium sulfate, finely ground.
6. Extraction buffer: 50 mM Tris–HCl, pH 7.9, 420 mM KCl, 5.0 mM $MgCl_2$, 0.1 mM EDTA, 2.0 mM DTT, 20% glycerol, 10% sucrose, 0.5 mM PMSF, 1 mM benzamidine, 20 mM AEBSF, 0.16 μM aprotinin, 10 μM bestatin, 3.0 μM E-64, 4.0 μM leupeptin, 4.0 μM pepstatin.
7. Transcription buffer: Prepare a single-use 2× stock solution with 2.0 mM DTT, 50 mM Tris–HCl (pH 7.9), 100 mM KCl, 12.5 mM $MgCl_2$, 1.0 mM EDTA, and 20% glycerol. Working solutions are prepared by dilution of 1 part buffer to make 2 parts total reaction volume.

2.6. In Vitro Transcription

1. CEM nuclear extract is prepared as described for mammalian cells (23) (see Subheadings 2.5, 3.5 and Note 4).
2. Safe-Lock Tubes 1.5 mL (Eppendorf); screw-cap microfuge tubes may be substituted.
3. Ribonucleotide tri-phosphate (rNTPs) mix: Prepare a 10× stock solution with 2.5 mM ATP, 2.5 mM CTP, and 0.12 mM UTP in TE buffer. Working solutions are prepared by dilution of 1 part mix with 10 parts total reaction volume (see Note 6).

4. Uridine 5′-triphosphate [α-^{32}P] UTP (3,000 Ci/mmol) (radiolabeled [α-^{32}P] UTP is a radiation hazard and should be treated accordingly).
5. Ribonuclease T1 (T1 RNase) (1,000 U/μL). Working solutions are prepared by dilution of 1 part T1 Rnase with 4 parts 50 mM Tris–HCl (pH 7.5) and 2.0 mM EDTA.
6. Proteinase K is dissolved in 50 mM Tris–HCl (pH 8.0), 3.0 mM $CaCl_2$, and 50% glycerol at 10 mg/mL, aliquoted, and stored at –20°C.
7. Transcription stop buffer: 250 mM NaCl, 1% SDS (w/v), 20 mM Tris–HCl (pH 7.5), 5 mM EDTA (pH 8.0). Store at room temperature. Transcription stop buffer is supplemented with proteinase K to 0.75 mg/mL and recovery standard (RS) (see Subheading 2.5, step 14, and Subheading 3.5, step 7) for same day use.
8. Precipitation solution: 175 mM NaOAc (pH 5.2) and 40 μg/mL glycogen in ethyl alcohol for same day use (see Note 7).
9. 70% ethyl alcohol. Store at –20°C.
10. Formamide loading dye (10 mL): 5.0 mg xylene cyanol, 5.0 mg bromophenol blue, 0.4 mL 0.5 M EDTA (pH 8.0), 0.1 mL water, 9.5 mL ultrapure formamide. The solution is aliquoted and stored at –20°C.
11. 5× TBE: 445 mM Trizma base, 445 mM boric acid, 2 mM EDTA.
12. 6.5% polyacrylamide 6.0 M urea gel solution: Prepare 300 mL gel solution with 108.1 g urea, 60 mL 5× TBE and add water up to 251 mL. The solution is heated until all urea has dissolved, and then cooled to room temperature. Then, 48.75 mL of 40% acrylamide/bis (19:1) (Amresco) is added to the urea solution. The gel solution is stored in an amber glass bottle at 4°C (acrylamide is a neurotoxin when unpolymerized and so care should be taken not to expose skin).
13. Whatman 3MM Chromatography Paper or equivalent.
14. Plastic wrap.
15. Radiolabeled molecular weight marker (MWM). The MWM used here is the *Hpa*II endonuclease (New England Biolab) digestion product of pBR322 (New England Biolab). The recovery standard is a single DNA fragment (~622 bp) from pBR322 digested with *Hpa*II and purified using a QIAquick Gel Extraction Kit (Qiagen) or an equivalent. Both the MWM and RS (100 ng each) are dephosphorylated with Antarctic Phosphatase (New England Biolab) or an equivalent, and 5′ end labeled with adenine 5′-triphosphate [γ-^{32}P] ATP (6,000 Ci/mmol) using T4 polynucleotide kinase (PNK). Unincorporated nucleotide is removed using a BioSpin 6

(Bio-Rad Laboratories) buffer exchange column according to the manufacturer's instructions or an equivalent procedure. Store at −20°C. Radiolabeled [γ-^{32}P] ATP is a radiation hazard, and should be treated accordingly.

16. Software for quantitation of phosphorimager data, such as ImageQuant version 5.1 (Molecular Dynamics).

3. Methods

The HTLV-1 DNA template is first reconstituted into nucleosomal arrays using the stepwise salt dialysis protocol developed previously for the 208-12 5S rRNA model system (14, 24). After assembly, the number of nucleosomes per DNA template (i.e., nucleosome saturation) is determined from the sedimentation coefficient (S, corrected to 20°C and water ($s_{20,w}$)) derived from analytical ultracentrifugation experiments. For reference, a fully saturated array containing 12 nucleosomes/template will sediment at 29S in low-salt buffer. Importantly, both in vitro transcription and chromatin structural studies are sensitive to the degree of nucleosome saturation of the DNA template. Therefore, to maintain consistency between array preparations, the $s_{20,w}$ distributions of all array reconstitutions must closely match to document a similar degree of nucleosome saturation.

Chromatin in vitro is in salt-dependent equilibrium between secondary and tertiary higher order conformational states (1). As measured in the analytical ultracentrifuge, intrafiber nucleosome–nucleosome interactions within a 12-mer nucleosomal array/chromatin fiber result in moderately folded (40S) and maximally folded (55S) secondary structures, the latter of which equal the canonical 30-nm fiber in their extent of compaction (1). Tertiary interfiber nucleosome–nucleosome interactions are measured by an oligomerization assay that uses differential centrifugation in a microfuge to quantitate the fraction of the sample that has formed very large multimers as a function of salt concentration. For HAT and in vitro transcription assays, the methods described below for HTLV-1 chromatin are modeled after those defined previously for the HTLV-1-306/G-less-derived construct (25).

In our experimental approach, unacetylated and p300-acetylated HTLV-1 chromatin model systems are subjected to both structural and transcription assays under identical salt concentrations. By directly comparing the results obtained by analytical ultracentrifugation, differential centrifugation, and in vitro transcription methodologies, we find that that acetylation of promoter-proximal nucleosomes by p300 simultaneously leads to significant decondensation of nucleosomal arrays and concurrent enhancement of RNA synthesis.

Of note, the HTLV-1 chromatin template and the methods described here are general and can be modified to include alternative gene promoters, activators, HATs, mutants or variant recombinant histones, or other histone modifications for future chromatin structure/function studies.

3.1. Purification of the HTLV-1 DNA Template

1. The HTLV-1 DNA fragment (13) is purified by a method similar to the 208-12 DNA template used for previous chromatin structural studies (24, 26). The HTLV-1 construct cloned into pUC19 is transformed into competent DH5α *E. coli* and selected on LB Amp agar plates. For 1 L culture: A well-isolated cell colony is used to inoculate a 25 mL TSB/A/T starter culture and incubated at 37°C shaking overnight for ~18 h.
2. The 25 mL overnight culture is added to 1 L of TSB/A/T and incubated at 37°C for 6–8 h while shaking. 0.5 mL of chloramphenicol (10 μg/mL) is added to each liter of culture and incubated for an additional ~18 h at 37°C while shaking.
3. Cells are centrifuged at 6,300 × *g* for 15 min and resuspended in 35 mL 1× PBS. Cells are transferred to 50-mL falcon tubes and centrifuged at 3,200 × *g* for 30 min, the supernatant is discarded, and pellets are stored at −20°C (if necessary).
4. Plasmid DNA is purified from the cell pellets using a Qiagen Plasmid Giga Kit (Qiagen) or the equivalent (*6 L culture yields 30–50 mg DNA*) according to the manufacturer's instructions.
5. Plasmid DNA is digested with *Cfo*I (0.5 U/μg DNA) in a 250-mL Erlenmeyer flask in SuRE/Cut buffer L and incubated at 37°C for ~48 h while shaking (see Note 8).
6. The reaction is stopped by the addition of 5.0 mM EDTA (pH 8.0). DNA is precipitated by the addition of 0.3 M NaOAC (pH 5.2) and 80% total volume isopropanol. DNA is transferred to 33-mL Oakridge centrifuge tubes. The DNA is centrifuged at 27,000 × *g* for 30 min at 4°C and the supernatant is discarded. DNA pellets are resuspended in 70% ethanol and centrifuged at 27,000 × *g* for 30 min at 4°C. The supernatant is discarded and the pellet is air dried.
7. The DNA pellet is resuspended in TE overnight at 4°C while shaking and the DNA concentration adjusted to ~2 mg/mL (A_{260} 40.0) using a UV spectrophotometer in TE. DNA is stored at either 4°C or −20°C.
8. *Cfo*I-digested DNA (4 mg or 2 mL) is loaded on a pre-equilibrated sephacyl size-exclusion column (125 cm long × 2.5 cm in diameter, ~ 600 mL volume) in TEN50 buffer. The flow rate is maintained at 1 mL/min using a peristaltic pump. After 1 h, 4-mL fractions are collected (~ 79 fractions for ~5 h). 10 μL of each fraction is mixed with DNA electrophoresis sample loading dye and loaded onto a 1% agarose gel

(see Subheading 2.1, step 4). The gel is run at ~13 V/cm in TAE buffer, transferred to ethidium bromide stain solution, and submerged for 20 min at room temperature while rocking. The gel is then transferred to ethidium bromide destain solution, submerged for 20 min at room temperature while rocking, and analyzed on a UV transilluminator. Fractions containing a single (~2,500 bp) DNA fragment are pooled, precipitated (see Subheading 3.1, step 6), and resuspended in TE buffer to a DNA concentration of ~1 mg/mL (A_{260} 20.0). Purified HTLV-1 template DNA is aliquoted and stored at −20°C.

3.2. Reconstitution of HTLV-1 Nucleosomal Arrays and Determination of Nucleosome Saturation

1. HTLV-1 nucleosomal arrays are reconstituted by salt dialysis as described (26). 150 μg (97 pmol) of HTLV-1 DNA template (see Subheadings 2.1 and 3.1) is incubated with 125 μg (1.16 nmol) purified histone octamer in a 400 μL reaction in B0 Buffer (see Note 9). For saturated arrays (i.e., 12 nucleosomes/HTLV-1 DNA template), a molar ratio of octamer to DNA is near 12:1 (see Note 10).
2. The assembly reaction is carried out in dialysis tubing (10–14,000 MWCO) in a stepwise NaCl concentration gradient over a 48-h time span. First, assembly reactions are incubated on ice for 30 min in B0 buffer and placed into hydrated dialysis tubing (clamped closed on one end). Once transferred, the remaining opening is clamped closed, secured to a buoy and suspended in 1.0 L B1 buffer in a 1-L beaker with a magnetic stir bar, and stirred for 6–7 h at 4°C.
3. B1 buffer is replaced with B2 buffer in the beaker containing the dialysis tubing and stirred for ~16 h at 4°C.
4. B2 buffer is replaced with B3 buffer in the beaker containing the dialysis tubing and stirred for ~6–7 h at 4°C.
5. B3 buffer is replaced with B4 buffer in the beaker containing the dialysis tubing and stirred for ~16 h at 4°C.
6. The dialysis tubing is removed from the beaker. While held upright, the top clamp is removed and the sample is removed with a pipetteman and transferred to a 1.5-mL Eppendorf tube. The sample is centrifuged at 22,000 × *g* for 10 min at 4°C and the supernatant is saved in a clean 1.5-mL Eppendorf tube. The A_{260} is measured using a UV spectrophotometer to determine DNA concentration (typical concentrations range between 200 and 270 ng/μL).
7. The degree of nucleosome saturation (number of nucleosomes/HTLV-1 DNA template) in a reconstitution reaction is measured by sedimentation velocity using either a Beckman XL-A or XL-I analytical ultracentrifuge equipped with scanner optics as described (27). Reconstitution reactions are diluted to a final A_{260} ranging from 0.4 to 0.6 in 400 μL of B4 buffer and

loaded into 12-mm double-sector cells. 420 μL of B4 buffer is loaded in the reference sector. Samples are spun at $54,000 \times g$ in a four-hole analytical rotor (An 60 Ti). The temperature is held at 22 ± 0.1°C for the length of the run. Digital scans at 0.001-cm radial increments (20–50 total scans for each sample) are analyzed by the method of Demeler and van Holde (28) and van Holde and Weischet (29) using the Ultrascan data analysis software program (Dr. B. Demeler, University of Texas Health Science Center, San Antonio, TX). The buffer and solute settings for the van Holde–Weishet analysis are the following: density (0.998700), viscosity (1.000100), vbar (0.6500), divisions (30), data smoothing (1–3), % of boundary (90), boundary pos.(%) (5).

8. Data are plotted as boundary fraction versus $s_{20,w}$ to yield the diffusion corrected integral distribution of $s_{20,w}$, $G(s)$ and are shown in Fig. 2.

9. The degree of saturation of reconstituted HTLV-1 nucleosomal arrays is determined by the sedimentation coefficient at the 50% boundary (30). Nucleosomal arrays in the range of

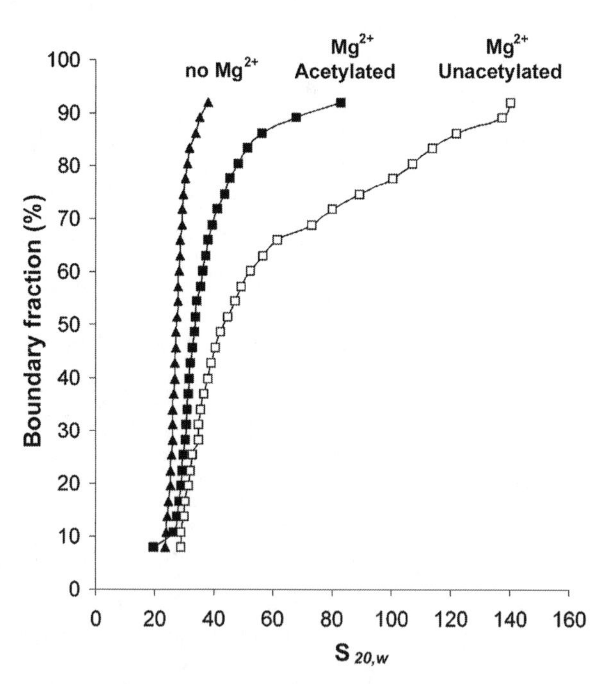

Fig. 2. Activator-dependent p300 acetylation disrupts folding of HTLV-1 nucleosomal arrays. HTLV-1 nucleosomal arrays in B4 buffer (*filled triangle*) were subjected to sedimentation velocity in 1.4 mM $MgCl_2$ and compared to unacetylated (*open square*) and acetylated (*filled square*) HTLV-1 chromatin. Shown are the sedimentation coefficient distributions, $G(s)$. (This research was originally published in Szerlong, H. J., Prenni, J. E., Nyborg, J. K., and Hansen, J. C. (2010) Activator-dependent p300 acetylation of chromatin in vitro: enhancement of transcription by disruption of repressive nucleosome–nucleosome interactions. *J Biol Chem* ***285***, 31954–64 © the American Society for Biochemistry and Molecular Biology).

27S–29S contain 11–12 nucleosomes per DNA template and are subsequently used for acetylation, in vitro transcription, folding, and oligomerization assays described here.

3.3. Activator-Dependent Acetylation of HTLV-1 Chromatin

1. Protein concentrations for pCREB, Tax, and p300 are determined by Bio-Rad Protein Assay (Bio-Rad Laboratories) relative to a BSA standard (according to the manufacturer's instructions). Other standard methods for establishing protein concentrations may be substituted here.
2. A standard HAT assay contains HTLV-1 nucleosomal arrays, pCREB, and Tax mixed at molar ratios of 1:12:12 in 1× HAT buffer for 20 min at 4°C. p300 (at a molar ratio of array to p300 of 1:2) is then added, followed by the addition of 100 μM acetyl-CoA, and the mixture then incubated at 30°C for 60 min. For array folding and oligomerization assays (see Subheadings 3.4 and 3.5), 110 μL reactions contained HTLV-1 nucleosomal arrays (80 nM) (see Subheading 3.2, step 9) pCREB (0.9 μM), Tax (0.9 μM), p300 (160 nM), and acetyl-CoA (100 μM) in lx HAT buffer (see Note 4).
3. HAT buffer (containing acetyl-CoA) is exchanged for 10 mM Tris–HCl (pH 8.0) by passage through a BioSpin 6 (Bio-Rad Laboratories) column (according to the manufacturer's instructions); equivalent exchange methods could be substituted. Buffer exchange is used to stop the HAT reaction and reduce the concentration of unincorporated acetyl-CoA (due to its high absorbance at 260 nm), NaCl, glycerol, and BME for subsequent analysis.

3.4. Chromatin Folding and Oligomerization Assays

1. Mg^{2+}-dependent folding is assayed by sedimentation velocity analysis using either a Beckman XL-A or XL-I ultracentrifuge as described (27). Samples are diluted to a final A_{260} ranging from 0.4 to 0.6 in B4 buffer (supplemented with 0–2 mM $MgCl_2$) to a final volume of 400 μL total and loaded into 12-mm double-sector cells. 420 μL of B4 buffer (with equivalent $MgCl_2$ concentration) is loaded in the reference sector.
2. Samples are centrifuged at 21,000 × *g* in a four-hole analytical rotor (An 60 Ti). The temperature is held at 22 ± 0.1°C for the length of the run. Digital scans at 0.001-cm radial increments (20–50 total scans for each sample) are analyzed by the method of Demeler and van Holde (28) and van Holde and Weischet (29) using the Ultrascan data analysis software program (Dr. B. Demeler, University of Texas Health Science Center, San Antonio, TX). The buffer and solute settings for the van Holde–Weishet analysis for transcription buffer are the following: density (1.030500), viscosity (1.375900), vbar (0.6500), divisions (30), data smoothing (1–3), % of boundary (90), boundary Pos.(%) (5). The buffer and solute settings for the

van Holde–Weishet analysis for TEN buffer supplemented with 1.4 mM $MgCl_2$ are the following: density (0.998800), viscosity (0.998700), vbar (0.6500), divisions (30), data smoothing (1–3), % of boundary (90), boundary Pos.(%) (5).

3. Data are plotted as boundary fraction versus $s_{20,w}$ to yield the diffusion-corrected integral distribution of $s_{20,w}$, $G(s)$ and are shown in Fig. 2.
4. Data are plotted as A_{260} versus radius (cm) to yield the corrected A_{260} of acetylated HTLV-1 nucleosomal arrays. The corrected A_{260} of acetylated arrays subtracts the contribution of unincorporated acetyl-CoA from the total absorbance. The absorbance from acetyl-CoA is the fraction of sample that fails to sediment under the conditions described in Subheading 3.4, step 2. The corrected A_{260} is equal to A_{260} (total) − A_{260} (acetyl-CoA) and is used to calculate the percent (%) of sample in the supernatant after differential centrifugation of acetylated arrays in oligomerization assays (see Subheading 3.4, step 7).
5. Acetylated HTLV-1 nucleosomal arrays or unacetylated arrays (see Subheading 3.3) are diluted in B4 buffer to a total A_{260} 1.0–1.5 in 60 μL total. 60 μL of 2× transcription buffer is added to 60 μL of diluted arrays for a final A_{260} ~0.5–0.75 (see Notes 11 and 12). Reactions are mixed well and incubated at room temperature for 5 min. 60 μL of each sample (or ½ total volume) is transferred to a clean Micro Cell (8 mm high) (Beckman Coulter) cuvette. Air bubbles are avoided by gently tapping of the cuvette prior to obtaining the A_{260} using a UV spectrophotometer. This initial A_{260} reading corresponds to 100% of the array *plus* any unincorporated acetyl-CoA in the sample (see Subheading 3.4, step 4). The remaining 60 μL of the sample is centrifuged at 22,000 × *g* in a tabletop centrifuge at room temperature for 5 min. 55 μL of the supernatant is carefully removed without disturbing the pellet, transferred to a cuvette, and the A_{260} is measured using a UV spectrophotometer. The A_{260} of the supernatant corresponds to the fraction of the array that does not sediment *plus* any unincorporated acetyl-CoA in the sample.
6. The percent (%) of sample in supernatant equals (A_{260} of the supernatant − A_{260} from unincorporated acetyl-CoA)/(the initial (or total) A_{260} − A_{260} from unincorporated acetyl-CoA) × 100. The standard deviation for each condition is determined from at least three independent reactions. An example of an oligomerization assay with acetylated and unacetylated HTLV-1 nucleosomal arrays is shown in Fig. 3.

3.5. Preparation of Nuclear Extract from CEM Cells

CEM cells have been used to prepare nuclear extracts to be used for in vitro transcription for the HTLV type I promoter used in this example, but other cells may be used as appropriate for the system being used.

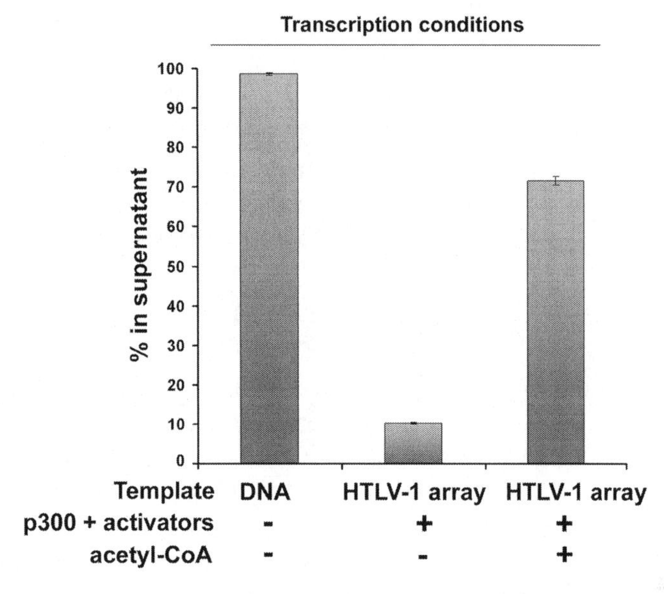

Fig. 3. Activator-dependent p300 acetylation disrupts oligomerization of HTLV-1 nucleosomal arrays under transcription conditions. HTLV-1 nucleosomal arrays were incubated with activators (pCREB/Tax) and p300 with (+) and without (–) acetyl-CoA in HAT buffer, prior to buffer exchange and differential centrifugation in transcription buffer. Naked HTLV-1 DNA template (free DNA) in transcription buffer was used as a control. Total A_{260} was corrected for the contribution from unincorporated acetyl-CoA. Data are expressed as the percentage of initial absorbance (260 nm) that remained in the supernatant after centrifugation (% in supernatant). *Error bars* indicate the standard deviation of at least three independent oligomerization assays. (This research was originally published in Szerlong, H. J., Prenni, J. E., Nyborg, J. K., and Hansen, J. C. (2010) Activator-dependent p300 acetylation of chromatin in vitro: enhancement of transcription by disruption of repressive nucleosome-nucleosome interactions. *J Biol Chem* ***285***, 31954–64 © the American Society for Biochemistry and Molecular Biology).

1. 1–10 L of CEM cells suspended in spinner cell culture flasks are collected by centrifugation at 300×*g* for 10 min and the supernatant is discarded.
2. Cells are washed by suspension in 1× PBS supplemented with 1.0 mM $MgCl_2$ and pelleted by centrifugation at 300×*g* for 10 min. The supernatant is discarded and the packed cell volume (PCV) is determined.
3. The cell pellet is resuspended in hypotonic lysis buffer corresponding to four PCV and incubated on ice for 10 min. The cell suspension is homogenized with ten strokes in a Dounce homogenizer using a “B” pestle.
4. Nuclei are pelleted by centrifugation at 1,100×*g* for 5 min. The supernatant is discarded and the pellet is resuspended in two PCV hypotonic lysis buffer.
5. The suspension is centrifuged at 2,000×*g* for 5 min. The supernatant is discarded and the nuclear pellet is resuspended in four PCV extraction buffer and rocked at 4°C for 30 min.

6. The nuclear suspension is centrifuged at 35,000 × *g* for 30 min. The supernatant is transferred to a beaker and the high-speed supernatant volume (HSSV) is determined.
7. Ammonium sulfate (finely ground) is added to the suspension while stirring at 4°C (0.33 g of ammonium sulfate per milliliter HSSV). The ammonium sulfate is allowed to dissolve for 1 h while stirring at 4°C.
8. The nuclear lysate is centrifuged at 30,000 × *g* for 10 min and the supernatant is discarded. The ammonium sulfate precipitant is resuspended in 2× transcription buffer (0.05 times HSSV).
9. The nuclear lysate is dialyzed overnight against 2 L 2× transcription buffer using 6–8,000 MWCO dialysis tubing. The buffer is then replaced with 2 L of fresh 2× transcription buffer and dialyzed for an additional 4–6 h.
10. The nuclear lysate is centrifuged at 16,000 × *g* for 20 min and the supernatant is flash frozen in 100-μL aliquots in liquid nitrogen and stored at −70°C.
11. The total protein concentration is measured via Bradford assay. A typical value is 10–20 mg/mL.

3.6. In Vitro Transcription

1. In vitro transcription reactions were performed as described (32), except that 27S–29S HTLV-1 reconstituted nucleosomal arrays were used in place of immobilized chromatin templates (see Subheading 3.2, step 9).
2. In Safe-Lock Eppendorf tubes, 30 μL reaction mixtures containing HTLV-1 nucleosomal arrays (2.5 nM), pCREB (130 nM), Tax (130 nM), p300 (5 nM), and acetyl-CoA (100 μM) in 1× transcription buffer are incubated at 4°C for 20 min while rotating (see Notes 13 and 14).
3. ~40 μg of CEM nuclear extract is added to each reaction and incubated at 30°C for 30 min. The optimal concentration of CEM nuclear extract is determined empirically by titration (see Notes 13 and 14).
4. RNA synthesis is initiated by the addition of 3 μl of 10× rNTPs (ATP (2.50 mM), CTP (2.50 mM)), and 1.0 μL of [α-^{32}P] UTP to each reaction and incubated at 30°C for 45 min in a water bath. Mix reactions by flicking the tubes every 15 min (see Note 15).
5. Reactions are centrifuged (3-s pulse at 2,000 × *g*) and 2.0 μl T1 RNase (200 U/μl) is added to each reaction, mixed, and incubated at 37°C for 20 min.
6. Reactions are centrifuged (3-s pulse at 2,000 × *g*) and 110 μl of transcription stop solution (supplemented with proteinase K and recovery standard) is added, mixed, and incubated at 45°C for 20 min.

7. Reactions are centrifuged (3-s pulse at 2,000 × *g*) and 655 μL of transcription precipitation solution (supplemented with glycogen (40 μg/mL)) is added, mixed, and incubated at −20°C for at least 1 h. Reactions are centrifuged at 22,000 × *g* at 4°C for at least 30 min. The supernatant (containing unincorporated NTPs and [α-^{32}P] UTP) is properly disposed and the pellet is washed with 100 μL of 70% ethyl alcohol and centrifuged at 22,000 × *g* at 4°C for 30 min. The supernatant (containing unincorporated NTPs and [α-^{32}P] UTP) is properly disposed, the cap is left open, and the RNA pellet is air dried at room temperature.
8. The pellet is resuspended in 6 μL of formamide loading dye, heated to 95°C for 5 min in a heat block, vortexed, centrifuged (3-s pulse at 2,000 × *g*), and load onto a 6.0 M urea, 6.5% acrylamide/bis (19:1) sequencing gel.
9. The 6.0 M urea, 6.5% acrylamide/bis (19:1) sequencing gel is prepared using a gel notch plate and rear plate (~20 cm wide × ~20 cm long), separated by 0.4-mm spacers that are stacked and laid flat on a raised support (such as an Eppendorf rack). The notched plate faces up. The long edges of the plates are secured together using gel clamps. 25 mL solution of 6.5% polyacrylamide 6.0 M urea, 250 μL of 10% APS, and 25 μL of TEMED is mixed and dispensed between the notch plate and the rear plate using a 25-mL pipette. The gel mix enters the space by capillary action. Gentle tapping on the surface of the plates with a spatula while dispensing the gel mix removes unwanted air bubbles. Once the gel mix is dispensed, a 0.4-mm comb is inserted between the plates and secured with gel clamps. Allow the gel to set for at least 1 h at room temperature. The clamps and comb are removed and excess acrylamide is washed from the top of gel and secured to a vertical gel electrophoresis apparatus. A metal plate is clamped to the gel plates to evenly distribute heat during the run. The top and bottom gel wells are filled with 1× TBE, air bubbles are removed from the bottom of the gel, and wells are rinsed using a 22-gauge syringe filled with 1× TBE. The gel is pre-run at 35 W (1.8 W/cm) for at least 30 min.
10. 6.0 μL of sample are loaded and run at 35 W (1.8 W/cm) for ~45 min or until the dye front has run off the bottom of the gel. *The dye front contains a significant amount of unincorporated [α-^{32}P] UTP and should be handled and disposed with care*. The power source is turned off and the gel is removed from the electrophoresis apparatus. The gel is laid flat on a lab bench (preferably near a sink reserved for radiation). The notch plate is separated from both the gel and rear plate using a razor blade or spatula between the two glass plates for leverage. The spacers are carefully removed and a piece of Whatman paper

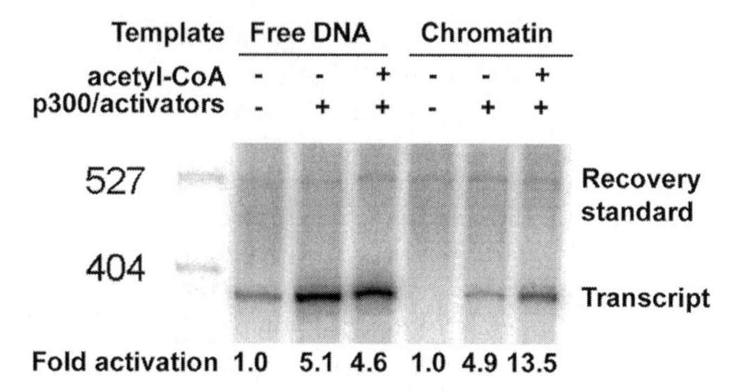

Fig. 4. In vitro transcription of HTLV-1 chromatin and free DNA. Transcription was assayed in the presence (+) and absence (–) of purified activators (pCREB/Tax), p300, and acetyl-CoA. Transcript levels were normalized to basal transcription for free DNA or chromatin in the absence of activators and acetyl-CoA (set at 1.0). The positions of RNA transcript (*right*), recovery standard (^{32}P-labeled DNA) (*right*), and DNA size markers (*left*) are indicated. (This research was originally published in Szerlong, H. J., Prenni, J. E., Nyborg, J. K., and Hansen, J. C. (2010) Activator-dependent p300 acetylation of chromatin in vitro: enhancement of transcription by disruption of repressive nucleosome-nucleosome interactions. *J Biol Chem* ***285***, 31954–64 © the American Society for Biochemistry and Molecular Biology).

(cut to the dimensions of the gel) is placed on top of the gel. Using both (gloved) hands, the Whatman paper is pressed firmly to the gel and slowly peeled away, starting from the bottom of the gel and working toward the top until the gel is completely separated from the rear plate and fully adhered to the Whatman paper (see Note 16). The gel is placed face up on three to four stacked pieces of Whatman paper and covered with plastic wrap. The entire gel assembly is placed on a slab gel dryer, covered, and set at 80°C for at least 1 h with constant vacuum.

11. The cover of the gel drier is lifted to release vacuum, the vacuum pump is turned off, and the gel is placed in an exposure cassette, secured with tape, and set in a dark area (such as a drawer) for 1–2 days. The screen is scanned using a phosphorimager. Data are analyzed with Imagequant software. Figure 4 shows RNA products from an in vitro transcription assay resolved on a 6.5% polyacrylamide, 6 M urea sequencing gel.

4. Notes

1. All solutions should be prepared in water that has a resistivity of 18.2 MΩ-cm from a standard Milli-Q water purification system. This standard is referred to as “water” in this text.
2. EDTA is dissolved more readily in water at pH 8–10. Therefore, the pH is adjusted to ~9.0 with NaOH until dissolved, and then lowered to a final pH 8.0 with HCl.

3. Spectra/Por molecular porous membrane tubing is stored dry at 4°C and hydrated in B1 buffer for 30 min immediately before use by submersion in a 500-mL beaker.
4. pCREB, Tax, p300-purified proteins, and CEM nuclear extract are aliquoted (20–30 μL each), flash frozen in liquid nitrogen, and stored at −70°C. Thaw aliquots on ice and centrifuge at 22,000 × *g* for 10 min at 4°C prior to use. Samples are again flash frozen in liquid nitrogen and stored at −70°C for a second use.
5. Reconstituted HTLV-1 nucleosomal arrays are stored in B4 buffer at 4°C and are stable for ~1 month.
6. The 10× rNTPs mix can be prepared from 100 mM NTP stock solutions that have been aliquoted (20 μl each) and stored for single use at −20°C. It is critical that NTPs are diluted to 2.5 mM for same day use only.
7. Glycogen has low solubility in ethyl alcohol. Therefore, mix NaOAc and glycogen together prior to the addition of ethyl alcohol in a 50-mL plastic screw-top tube. Solution appears cloudy due to glycogen insolubility. Mix well between samples in transcription assays.
8. A complete digestion of pUC19 HTLV-1 plasmid DNA by *Cfo*I (0.5 U/μg DNA) is monitored every 48 h. A 20 μL sample is resolved on a 1% agarose gel, stained with ethidium bromide, destained, and digestion products analyzed on a UV transilluminator. An incomplete digestion may require more *Cfo*I and longer incubation at 37°C. If *Cfo*I is unable to digest DNA completely, a *Cfo*I site may have been mutated during the culture growth and plasmid DNA should be prepared again from a new transformant.
9. Histone octamer stability is sensitive to NaCl concentration; octamers should, therefore, be diluted into buffers containing 2.0 M NaCl. In addition, octamers are typically stored in buffer containing 2.0 M NaCl; therefore, the final concentration of NaCl must account for the contribution of NaCl from the octamer. Assuming the use of a 5.0 M NaCl stock, the volume of NaCl needed in a 400 μL reaction is equal to (400 μL total reaction vol − vol histone octamer) × 2.0 M NaCl final/5.0 M NaCl stock.
10. For optimizing nucleosome saturation, maintain a single concentration of DNA and titrate the amount of histone octamer in separate assembly reactions.
11. The A_{260} should be within the linear range of the UV spectrophotometer (typically, 0.1–1.0). An initial A_{260} in the range of 0.5 to 1.0 allows accurate readings of reactions that sediment 80–90% of the sample.

12. The buffer conditions for the oligomerization assays are the same as in vitro transcription for consistency between experiments. However, oligomerization assays have also included buffer conditions with varying combinations of mono- or divalent salts (e.g., KCl with $MgCl_2$) (31).
13. The amount of 2× transcription buffer in each in vitro transcription reaction is calculated by (total reaction volume/2) − (volume of pCREB + volume of Tax + volume p300 + volume nuclear extract) since protein components are diluted in 2× transcription buffer prior to use. Optimal transcription conditions contain 50 mM KCl and 6.25 mM $MgCl_2$.
14. In vitro transcription is sensitive to the concentrations of template, pCREB, Tax, p300, and nuclear extract and may vary between different preparations. To optimize in vitro transcription assays, each component should be singly titrated in 30 μL reactions.
15. A master mix containing 10× rNTPs (ATP (2.50 mM), CTP (2.50 mM)), and [α-^{32}P] UTP is made by adding [α-^{32}P] UTP (1.0 μL × (# of reactions + 1)) to 10× rNTPs mix (3 μL × (# of reactions + 1)) in a Safe-Lock Tube. Mix components well and transfer 4 μL to each reaction. [α-^{32}P] UTP produces ionizing radiation and so proper lab attire (including safety glasses, gloves, lab coat, closed-toe shoes, and pants) should be worn and samples should be placed behind proper shielding at all times to reduce exposure.
16. Separation of the gel from the rear plate is greatly improved when the rear plate is treated with Rain-X Original Glass Treatment (typically used for windshield protection) prior to use. The notch plate is not treated.

Acknowledgments

We would like to thank Jennifer Nyborg, Teri Mclain, and members of their laboratories for helpful discussions and troubleshooting. This work was supported by National Institutes of Health Grants GM45916 and GM66834 (to J. C. H.) and by an American Heart Association postdoctoral fellowship (to H. J. S.).

References

1. Hansen, J. C. (2002) Conformational dynamics of the chromatin fiber in solution: determinants, mechanisms, and functions. *Annu Rev Biophys Biomol Struct* ***31***, 361–92.
2. Woodcock, C. L., and Dimitrov, S. (2001) Higher-order structure of chromatin and chromosomes. *Curr Opin Genet Dev* ***11***, 130–5.
3. Park, Y. J., and Luger, K. (2008) Histone chaperones in nucleosome eviction and histone exchange. *Curr Opin Struct Biol* ***18***, 282–9.
4. Hansen, J. C., Nyborg, J. K., Luger, K., and Stargell, L. A. (2010) Histone chaperones, histone acetylation, and the fluidity of the chromogenome. *J Cell Physiol* 224(2): 289–99.

5. Happel, N., and Doenecke, D. (2009) Histone H1 and its isoforms: contribution to chromatin structure and function. *Gene 431*, 1–12.
6. Woodcock, C. L., Skoultchi, A. I., and Fan, Y. (2006) Role of linker histone in chromatin structure and function: H1 stoichiometry and nucleosome repeat length. *Chromosome Res 14*, 17–25.
7. McBryant, S. J., Adams, V. H., and Hansen, J. C. (2006) Chromatin architectural proteins. *Chromosome Res 14*, 39–51.
8. Cairns, B. R. (1998) Chromatin remodeling machines: similar motors, ulterior motives. *Trends Biochem Sci 23*, 20–5.
9. Clapier, C. R., and Cairns, B. R. (2009) The biology of chromatin remodeling complexes. *Annu Rev Biochem 78*, 273–304.
10. Roth, S. Y., Denu, J. M., and Allis, C. D. (2001) Histone acetyltransferases. *Annu Rev Biochem 70*, 81–120.
11. Choi, J. K., and Howe, L. J. (2009) Histone acetylation: truth of consequences? *Biochem Cell Biol 87*, 139–50.
12. Kouzarides, T. (2007) Chromatin modifications and their function. *Cell 128*, 693–705.
13. Szerlong, H. J., Prenni, J. E., Nyborg, J. K., and Hansen, J. C. (2010) Activator-dependent p300 acetylation of chromatin in vitro: enhancement of transcription by disruption of repressive nucleosome-nucleosome interactions. *J Biol Chem 285*, 31954–64.
14. Simpson, R. T., Thoma, F., and Brubaker, J. M. (1985) Chromatin reconstituted from tandemly repeated cloned DNA fragments and core histones: a model system for study of higher order structure. *Cell 42*, 799–808.
15. Sambrook J, F. E., Maniatis T (1989) *Molecular Cloning--a laboratory manual*, second ed., Cold Spring Harbor Laboratory Press, New York.
16. Luger, K., Rechsteiner, T. J., and Richmond, T. J. (1999) Expression and purification of recombinant histones and nucleosome reconstitution. *Methods Mol Biol 119*, 1–16.
17. Luger, K., Rechsteiner, T. J., and Richmond, T. J. (1999) Preparation of nucleosome core particle from recombinant histones. *Methods Enzymol 304*, 3–19.
18. Lopez, D. I., Mick, J. E., and Nyborg, J. K. (2007) Purification of CREB to apparent homogeneity: removal of truncation products and contaminating nucleic acid. *Protein Expr Purif 55*, 406–18.
19. Kim, Y. M., Ramirez, J. A., Mick, J. E., Giebler, H. A., Yan, J. P., and Nyborg, J. K. (2007) Molecular characterization of the Tax-containing HTLV-1 enhancer complex reveals a prominent role for CREB phosphorylation in Tax transactivation. *J Biol Chem 282*, 18750–7.
20. Zhao, L. J., and Giam, C. Z. (1991) Interaction of the human T-cell lymphotrophic virus type I (HTLV-I) transcriptional activator Tax with cellular factors that bind specifically to the 21-base-pair repeats in the HTLV-I enhancer. *Proc Natl Acad Sci USA 88*, 11445–9.
21. Giebler, H. A., Loring, J. E., van Orden, K., Colgin, M. A., Garrus, J. E., Escudero, K. W., Brauweiler, A., and Nyborg, J. K. (1997) Anchoring of CREB binding protein to the human T-cell leukemia virus type 1 promoter: a molecular mechanism of Tax transactivation. *Mol Cell Biol 17*, 5156–64.
22. Kraus, W. L., and Kadonaga, J. T. (1998) p300 and estrogen receptor cooperatively activate transcription via differential enhancement of initiation and reinitiation. *Genes Dev 12*, 331–42.
23. Dynan, W. S. (1987) DNase I footprinting as an assay for mammalian gene regulatory proteins. *Genet. Eng. 9*, 75–87.
24. Hansen, J. C., Ausio, J., Stanik, V. H., and van Holde, K. E. (1989) Homogeneous reconstituted oligonucleosomes, evidence for salt-dependent folding in the absence of histone H1. *Biochemistry 28*, 9129–36.
25. Anderson, M. G., Scoggin, K. E., Simbulan-Rosenthal, C. M., and Steadman, J. A. (2000) Identification of poly(ADP-ribose) polymerase as a transcriptional coactivator of the human T-cell leukemia virus type 1 Tax protein. *J Virol 74*, 2169–77.
26. Gordon, F., Luger, K., and Hansen, J. C. (2005) The core histone N-terminal tail domains function independently and additively during salt-dependent oligomerization of nucleosomal arrays. *J Biol Chem 280*, 33701–6.
27. Schwarz, P. M., and Hansen, J. C. (1994) Formation and stability of higher order chromatin structures. Contributions of the histone octamer. *J Biol Chem 269*, 16284–9.
28. Demeler, B., and van Holde, K. E. (2004) Sedimentation velocity analysis of highly heterogeneous systems. *Anal Biochem 335*, 279–88.
29. van Holde, K. E., Weischet, W. O. (1978) Boundary analysis of sedimentation velocity experiments with monodisperse and paucidisperse solutes. *Biopolymers 17*, 1387–1403.
30. Hansen, J. C., and Lohr, D. (1993) Assembly and structural properties of subsaturated chromatin arrays. *J Biol Chem 268*, 5840–8.
31. Schwarz, P. M., Felthauser, A., Fletcher, T. M., and Hansen, J. C. (1996) Reversible oligonucleosome self-association: dependence on divalent cations and core histone tail domains. *Biochemistry 35*, 4009–15.

32. Konesky, K. L., Nyborg, J. K., and Laybourn, P. J. (2006) Tax abolishes histone H1 repression of p300 acetyltransferase activity at the human T-cell leukemia virus type 1 promoter. *J Virol 80*, 10542–53.
33. Harrod, R., Tang, Y., Nicot, C., Lu, H. S., Vassilev, A., Nakatani, Y., and Giam, C. Z. (1998) An exposed KID-like domain in human T-cell lymphotropic virus type 1 Tax is responsible for the recruitment of coactivators CBP/p300. *Mol Cell Biol 18*, 5052–61.
34. Nyborg, J. K., Egan, D., and Sharma, N. (2010) The HTLV-1 Tax protein: revealing mechanisms of transcriptional activation through histone acetylation and nucleosome disassembly. *Biochim Biophys Acta. 1799*, 266–74.

Chapter 19

Mapping Assembly Favored and Remodeled Nucleosome Positions on Polynucleosomal Templates

Hillel I. Sims, Chuong D. Pham, and Gavin R. Schnitzler

Abstract

Positioning of nucleosomes regulates the access of DNA binding factors to their consensus sequences. Nucleosome positions are determined, at least in part by the effects of DNA sequence during nucleosome assembly. Nucleosomes can also be repositioned (moved in *cis*) by ATP-dependent nucleosome remodeling complexes. Most studies of repositioning have used short DNA fragments containing a single nucleosome. It is difficult to use this type of template to analyze the role of DNA sequence in repositioning, however, because the many remodeling complexes are strongly influenced by nearby DNA ends. Mononucleosomal templates also cannot provide information about how repositioning occurs in the context of chromatin, where the presence of flanking nucleosomes could limit repositioning options. This protocol describes a newly developed method that allows the mapping of nucleosome positions (with and without remodeling) on any chosen region of a plasmid polynucleosomal template in vitro. The approach uses MNase digestion to release nucleosome-protected DNA fragments, followed by restriction enzyme digestion to locally unique sites, and Southern blotting, to provide a comprehensive map of nucleosome positions within a probe region. It was developed as part of studies which showed that human remodeling enzymes tended to move nucleosomes away from high affinity nucleosome positioning sequences, and also that there were differences in repositioning specificity between different remodeling complexes.

Key words: ATP-dependent remodeling complex, Nucleosome position, Chromatin, Mapping, Micrococcal nuclease, Restriction enzyme, SWI/SNF, Southern blotting

1. Introduction

Nucleosomes block access of most DNA binding factors to their consensus sequences, while linker DNA between nucleosomes is much more accessible. Several recent studies have shown that nucleosomes assume specific positions on genomic chromatin with a much higher frequency than might have been previously expected. While this observation is most striking in yeast, where the majority

Randall H. Morse (ed.), *Chromatin Remodeling: Methods and Protocols*, Methods in Molecular Biology, vol. 833,
DOI 10.1007/978-1-61779-477-3_19, © Springer Science+Business Media, LLC 2012

of nucleosomes are well positioned over most of the genome, it is also clear that nucleosomes tend to be well positioned over promoters and genomewide in other organisms from nematodes to man ((1–4), and for review see ref. 5). These recent studies showed that "functional" transcription factor binding sites, as identified by being bound by a specific factor in vivo (as assayed by chromatin immunoprecipitation) and/or evolutionarily conserved, are much more frequently found in linker regions than in DNA covered by nucleosomes. They also indicated that nucleosome positions were encoded, at least in part, by the underlying DNA sequence. Taken together, these studies suggest that transcription may frequently be controlled by promoter nucleosome positions which are at least partly controlled by the presence of DNA sequence elements that promote histone octamer occupancy (nucleosome positioning sequences or NPSes) or that exclude histone octamers.

All tested ATP-dependent remodeling complexes have the ability to alter nucleosome positions on DNA, indicating that the "default" positions of nucleosomes may be altered by remodeler action and suggesting that nucleosome repositioning will be a central aspect of remodeling complex regulatory function (5–7). Thus, it is essential to understand the combined effects of DNA sequence and remodeling complexes on nucleosome positions. Unfortunately, remodeling complexes appear to have strong tendencies to move nucleosomes either toward or away from DNA ends, and this "end effect" is likely to obscure any sequence dependence that may characterize the repositioning reaction. Here, we present a recently developed "MNase footprint/restriction enzyme" method which allows the nucleosome positions before and after remodeling to be measured at any desired location on a polynucleosomal plasmid template. The first step in the procedure is to identify a plasmid template and a region within this template to examine (Subheading 1). Next, this plasmid template is assembled into polynucleosomes by salt dialysis (Subheading 2). The plasmid template is then treated with or without remodeler, digested with micrococcal nuclease (MNase), and ~146 bp MNase footprint products corresponding to well-separated nucleosomes are isolated by PAGE. These products are then digested with diagnostic, locally unique restriction enzymes, followed by PAGE and Southern blotting with a probe to the region of interest (Subheading 3). The results from the Southern blots are then analyzed to provide quantitative information about the positions and abundance of nucleosomes in the probe region (Subheading 4).

This approach eliminates end effects, and has an estimated resolution of ±5 bp and an estimated accuracy of ±0.4% for the fractional occupancy of each mapped position. It also allows the effects of adjacent nucleosomes on remodeled and unremodeled positions to be determined. Finally, this approach allows the mapping of structurally distinct unremodeled and remodeled nucleosomal structures, such as closely abutting pairs of nucleosomes (normal dinucleosomes) or altered structures generated by some remodeling complexes (such as the altered dinucleosomes formed by the human SWI/SNF complex (8, 9)). We have applied these techniques using templates containing the promoter sequences of several hSWI/SNF target genes, and have also used them to compare the effects of SWI/SNF and ISWI class human remodeling complexes (9, 10). Our results indicate that nucleosome repositioning by remodeling complexes is influenced by DNA sequence, that significant differences between remodeling effects can be found, and that these effects have the potential to contribute to the regulation of target genes. Note that, while we describe the use of this approach for in vitro mapping studies, the same general technique could also be applied to mapping nucleosome positions in vivo, so long as a strong enough Southern blotting signal can be achieved.

2. Materials

2.1. For Preparation of Template and Probe Fragment

1. Software for restriction enzyme site identification (must allow analysis of 4-cutters).
2. Plasmid template at 0.5 mg/ml concentration or higher (see Subheading 3.1 for the choice of plasmid). Purify the plasmid from *Escherichia coli* using standard techniques (Cesium gradient and Qiagen-kit purified plasmid DNA both work well).
3. Oligonucleotide primers to chosen probe region (see Subheading 3.1), PCR machine, Taq or other thermostable polymerase and manufacturer-supplied buffer.
4. Agarose (normal grade).
5. 10× TBE: 890 mM Tris–HCl, 890 mM boric acid, 20 mM EDTA.
6. Agarose gel box and power supply.
7. 5× or 10× Gel loading buffer (containing bromophenol blue, xylene cyanol, and glycerol, such that 1× buffer is 0.05% bromophenol blue, 0.05% xylene cyanol, and 5% glycerol).
8. DNA molecular weight standards.
9. Ethidium bromide (EtBr), a 10 mg/ml stock can be kept at room temperature covered in foil.
10. UV light box and camera or gel documentation station.

2.2. For Nucleosome Assembly

1. Core histones in a 2 M NaCl-containing buffer and at a concentration of 1 mg/ml or higher. We used core histones purified from HeLa cell chromatin by hydroxylapatite column chromatography (11). Histones should be quickly frozen in

liquid nitrogen and stored at –80°C. They are stable for several years, and aliquots can be frozen and thawed several times. Other sources of histones are also acceptable (with the caveat that it is formally possible that nucleosome positions before or after remodeling may differ using histones from different sources or with different posttranslational modifications).

2. Dilution Buffer: 0.1 mg/ml BSA, 1 mM DTT, 0.2 mM PMSF, 10 mM Tris–HCl, pH 7.5, 1 mM EDTA. Prepare fresh, ~300 μl per assembly, from room temperature stocks of Tris–HCl, EDTA, and 0.1 M PMSF in isopropanol, and from –20°C stocks of BSA and DTT.
3. TE: 10 mM Tris–HCl, pH 7.5, 1 mM EDTA.
4. Wheat germ topoisomerase I. Store in aliquots at –70°C.
5. 2× Topo I reaction mix: 10 mM $MgCl_2$, 10% glycerol, and 0.2 U/μl Wheat Germ Topoisomerase I. Prepare 10 μl per reaction.
6. 10% SDS.
7. Buffer neutralized phenol (pH 7.0). Phenol–chloroform mixes do not do as good a job of removing histones from chromatin samples.
8. Glycogen, 10 mg/ml. Must be DNAse free.
9. 100% Ethanol (EtOH).
10. 3 M Sodium acetate (NaOAc), pH 5.2.
11. Agarose gel electrophoresis materials (as per Subheading 2.1, steps 4–10).

2.3. For Remodeling Reactions and Nucleosome Mapping

1. Assembled polynucleosomes and purified probe-region fragment (from Subheadings 3.1 and 3.2).
2. Chosen remodeling complex, purified by any method that gives nearly pure complex and removes contamination of other remodeling complexes. Concentration and activity need be sufficient to allow full remodeling (see Note 4). The default protocol describes the use of human SWI/SNF, purified from HeLa Ini1 cells by immunoaffinity chromatography against the FLAG-tagged Ini1 subunit as described in ref. 12 (2.4 μg per reaction).
3. BC100 Buffer: 100 mM KCl, 0.2 mM EDTA, 20 mM Tris–HCl, pH 7.9, 0.5 mM DTT, 0.2 mM PMSF, 10% glycerol (or appropriate storage buffer for chosen remodeler).
4. Remodeling Adjustment Buffer: 0.4 mg/ml BSA, 3 mM $MgCl_2$, 0.83 U/μl Topoisomerase I. If your remodeler prefers reaction conditions that are different from hSWI/SNF, you will need to adjust this buffer accordingly (see text for further details). For a discussion of the inclusion of Topo I in this buffer, (see Note 1).

5. 25 mM ATP/$MgCl_2$: 25 mM in ATP and 25 mM in $MgCl_2$ (stored at −20°C). If you prepare ATP from powder, be sure the pH is not far below neutral, or it will be unstable (use Tris–HCl or some other buffer to neutralize it). ATP chelates magnesium, and the $MgCl_2$ in this mix ensures that the addition of ATP does not alter the free Mg^{2+} concentration of your buffer.
6. 100 mM ADP (stored at −20°C).
7. MNase Adjustment Buffer: 0.4 μg/μl BSA, 65 mM KCl, 0.5 mM $MgCl_2$, 3.75 mM $CaCl_2$. For the hSWI/SNF reactions described, this brings the buffer in each tube to 1 mM $MgCl_2$, 3 mM $CaCl_2$, 65 mM KCl/NaCl, 0.4 μg/μl BSA, 0.067 mM ATP, and 2 mM ADP. Prepare fresh or freeze in aliquots. Note that if your remodeling buffer differs significantly from the hSWI/SNF remodeling buffer, you may need to adjust this buffer to come up with the same or similar BSA and cation concentrations.
8. MNase stop solution (1 ml): 1% SDS, 75 mM EDTA. Store at room temperature. Remake if precipitate is visible.
9. PAGE components: 30% acrylamide (30:1 acryl:bis), 10% ammonium persulfate (APS), and TEMED, all stored in the dark at 4°C. 10× TBE, power supply and medium-format plates, spacers, combs, and gel boxes for PAGE, for ~16 × 16 cm gels with 1.5 mm spacers and ~5 mm wells (the ~8 cm plates typically used for SDS-PAGE are too small). Molecular weight markers with bands in the 100–300 bp range. Gel loading buffer, EtBr, and UV light box (as per Subheading 2.1, step 4).
10. Restriction enzymes and buffers as determined by your template and probe (see Subheading 3.1).
11. Phenol:chloroform (1:1) or phenol:chloroform:isoamyl alcohol (~24:24:1).
12. 40% Acrylamide (20:1 acryl:bis, stored in the dark at 4°C) and other PAGE components, as per Subheading 2.3, step 9.
13. A medium-format electrotransfer apparatus, of the same general design as small Western blotting chambers, but capable of accommodating an ~14 × 14+ cm gel.
14. 3MM Paper.
15. Plastic wrap (e.g., Saran wrap).
16. An electrophoresis power supply capable of 550 mA output.
17. Screw-cap microfuge tube(s).
18. Geiger counters and shielding for the use of moderate levels of ^{32}P, and a scintillation counter capable of reading Cerenkov counts.
19. Random deoxynucleotide hexamer oligos, 1 mg/ml, stored at −20°C.

20. α^{32}P-dATP, 6,000 Ci/mmol, 20 mCi/ml (3,000 Ci/mmol may be acceptable, so long the probe has as a sufficient specific activity).
21. Klenow enzyme (DNA Pol fragment), manufacturer-supplied buffer.
22. A nucleotide mix containing 0.6 mM each of dTTP, dCTP, and dGTP, stored at −20°C.
23. Microcentrifuge desalting columns (BioRad Microbiospin 30 columns work well).
24. 6× SSC: 0.9 M NaCl, 90 mM trisodium citrate–HCl, pH 7.0.
25. APH solution: 5× SSC, 1% SDS, 5× Denhardt's solution. 50× Denhardt's solution is composed of 10 mg/ml Ficoll 400, 10 mg/ml BSA, and 10 mg/ml polyvinylpyrolidone, and can be purchased from various suppliers.
26. Sheared salmon sperm DNA (~5 mg/ml), stored at −20°C.
27. Positively charged nylon membrane (such as GeneScreen plus). Beware, the effectiveness of these membranes can vary greatly between manufactures and even between different lots from the same manufacturer. If you find that you get very weak signal, poor DNA adhesion leading to streaking, or high background, try membranes from different sources.
28. UV cross-linking apparatus (e.g., Stratalinker from Stratagene).
29. A hybridization oven (rotating, with hybridization bottles preferred), although you can probably get by with sealed bags and a rotating platform in an incubator.
30. Wash Buffers: 2× SSC, 0.1% SDS and 0.2× SSC, 0.1% SDS.
31. A Molecular Dynamics PhosphorImager screen and access to a PhosphorImager machine (or similar quantitative radiation detection method).

2.4. For Analysis of Mapping Results

1. ImageQuant (IQ) software by Molecular Dynamics (or similar software allowing line scans of PhosphorImager signal).

3. Methods

3.1. Choice of Region to Analyze, and Preparation of Template and Probe DNAs

1. You will first want to select or create a plasmid template which contains the DNA sequence you are interested in mapping nucleosomes on. The region for which you wish to map nucleosome positions should be ~200–400 bp in length and flanked, on each side, by more than ~200 bp (>1 nucleosome) of endogenous sequence to provide the correct local chromatin neighborhood. This will help to ensure that the nucleosome

positions in the region of interest represent positions preferred both by immediate sequence characteristics and also by the sequence-directed positions of neighboring nucleosomes.

2. Before making a final choice of templates, create an extensive restriction site map for the region of interest and surrounding sequences. Make sure to include sites for enzymes with short recognition sequences. From this map, you will need to identify restriction sites that are unique to the region of interest and at least 146 bp (one nucleosome length) on either side (see Note 2). These unique sites should be spaced such that each 146 bp segment contains three unique restriction sites (see Note 3). Do not be surprised if you end up using restriction enzymes you have never heard of before.
3. Next, you should identify the DNA sequence that you will use for your Southern probe. It is often useful to have the probe start and end at two of your chosen restriction sites, since this can sometimes simplify the analysis. Shorter probes (~250 bp) allow for fast unambiguous identification of positions, while longer probes (~300–400 bp) can allow more information per probing, but will necessitate a more complex band assignment process.
4. Design PCR primers that match the left and right edges of the probe region. Use any thermostable polymerase (Taq is fine) to amplify the fragment, and purify it from an agarose gel using an appropriate kit. Quantitate concentration and recovery by any convenient method.

3.2. Polynucleosome Assembly

1. Set up the polynucleosome assembly reaction in a microfuge tube at room temperature as follows:
   - 18 μl of 5 M NaCl (to 1.2 M final).
   - 30 μg of core histones.
   - TE to bring the reaction to 75 μl less the volume of DNA to be added.

   Gently mix to stir components.
2. Add 20 μg of supercoiled plasmid DNA (bringing the reaction to 75 μl), mix gently, and incubate at room temperature for 15 min.
3. Every 10 min, add Dilution Buffer in the following volumes, and carefully mix by flicking the tube: 21, 21, 21, 21, 21, 45, and 75 μl (for a final volume of 300 μl).
4. During dilutions, prepare a mock assembly reaction containing: 6 μl 5 M NaCl, 6.7 μg of plasmid DNA, TE to 25 μl, and then 75 μl of Dilution Buffer.

5. After the final 10 min incubation, the assembled polynucleosomes can be stored at 4°C for up to 2 months.
6. To confirm assembly, mix 7.5 μl of assembly reaction, 2.5 μl of TE, and 10 μl of 2× Topo I reaction mix. Also prepare a bare DNA control containing 7.5 μl of the mock assembly reaction. Prepare a no-Topo control by adding 12.5 μl of TE to 7.5 μl of the mock assembly reaction.
7. Incubate reactions for 30 min at 30°C, and stop by addition of 2 μl of 10% SDS.
8. Add an equal volume of neutralized phenol, vortex, microcentrifuge at max speed for 2 min, remove aqueous (top) solution to new tube.
9. Add 2 μl (20 μg) of glycogen carrier, 3 μl of 3 M NaOAc, pH 5.2, 70 μl of 100% EtOH, mix and precipitate by microcentrifugation at top speed for 10 min at 4°C.
10. Dry the pellet by leaving tube open to air, then resuspend in 10–20 μl of TE.
11. Add gel loading dye to 1× concentration, and resolve samples on a 1/2× TBE 0.8% agarose gel (load half of each sample). Also load a molecular weight marker with bands in the same range as the plasmid length. Choose a voltage that does not cause the gel to get hotter than slightly warm to the touch. Run the bromophenol blue dye to the end of the gel. Do not include EtBr in your gel or running buffer.
12. After running, stain the gel in a bath of 1/2× TBE with 0.3 μg/ml EtBr.
13. Photograph the gel on a UV light box.
14. Analyze your assembly results, and troubleshoot as necessary: The bare DNA control should show slow-moving relaxed topoisomers, while the no-Topo control and the assembled plasmid will run fast. Over-assembly is evidenced by a sizeable fraction of the DNA being in the wells. Under-assembly is evidenced by greater than 25% of the signal for the assembly lane resolving as slower than maximal migrating topoisomers. To correct these problems, adjust the histone:DNA ratio in the assembly reaction. You can also head off this potential issue by initially assembling at several ratios: e.g., 1.25:1, 1.5:1, and 1.8:1. Additional information on the assembly can be obtained by running a chloroquine gel (see Note 4). You will likely also see some nicked DNA (completely relaxed, and thus slow migrating) and a small amount of linear DNA (running at the expected size relative to molecular weight markers, between relaxed and supercoiled). You may also see linked plasmid multimers (a weak, slow-moving reflection of the topo-isomer pattern). None of these things are problematic, unless they constitute the majority of your products. An example of assembly results is shown in Fig. 1.

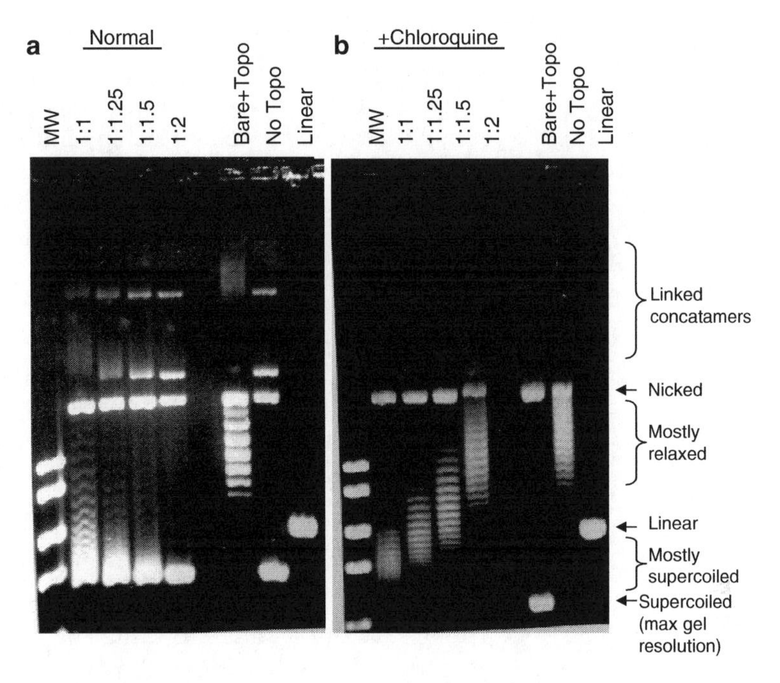

Fig. 1. Analysis of assembly products by agarose gel electrophoresis. Chromatin was assembled onto an ~5 kb c-myc promoter plasmid assembled as per Subheading 2, before 0.8% agarose electrophoresis without chloroquine (**a**) or with 2 μg/ml chloroquine (**b**).

3.3. MNase Footprint/ Restriction Enzyme Mapping, and Southern Blotting

Important note regarding reaction conditions. Reactions should be carried out in an ideal buffer to give good repositioning activity for the remodeling complex being tested. The protocol below details the buffer conditions we used for analysis of the human SWI/SNF complex. You may need to adjust the reaction buffers to best fit the remodeling complex tested. In doing so, remember that the final assembly buffer is 300 mM NaCl, which will contribute 28 mM NaCl to the final reaction. If this poses a problem, the remaining salt can be removed by dialysis into Dilution Buffer. Also, make sure to consider the composition of the buffer that your remodeling complex is in. It is very important that you confirm that your remodeling complex has full repositioning activity in your chosen buffer conditions and using the amount of plasmid chromatin needed for the full analysis. Several ways to assay for sufficient repositioning activity are described in (see Note 5).

Note regarding timing. This section takes several days, which can be most conveniently broken up into: (day 1) remodeling reactions, MNase, gel isolation of fragments, and elution of fragments overnight, (day 2) restriction digestion, (day 3) PAGE, electrotransfer and probe labeling, followed by overnight Southern hybridization, (day 4) washing of blots and overnight PhosphorImager screen exposure. There are also several other reasonable stopping points. Samples can be stored indefinitely at −20°C during any EtOH precipitation step, just after resuspension from an EtOH precipitation or after restriction digestion (steps 9, 10, and 20). You may also stop after transblotting and cross-linking, but before hybridization (step 32). If you do so, wrap the membrane in plastic wrap with enough 2× SCC to keep it damp, and store at 4°C for no more than a few days. Gels can be prepared up to several days in advance, wrapped in plastic wrap and stored at 4°C. The probe can also be labeled in advance and stored at –20°C for up to a week. If you let gels sit after electrophoresis and before fragment isolation or transblotting, keep them at 4°C and for no more than ~12 h.

Day 1

1. On ice, prepare 60 μl reactions containing:
   - 5.7 μl of polynucleosomal assembly (380 ng of plasmid chromatin by DNA weight).
   - 30 μl of Remodeling Adjustment Buffer.
   - 1.2 μl of 25 mM ATP/$MgCl_2$ (to 0.5 mM), or equal volume of water, for no ATP controls.
   - 7.1 μl of water (to bring the final reaction volume, after hSWI/SNF addition, to 60 μl).
   - 16 μl of hSWI/SNF (2.4 μg, to give a hSWI/SNF:nucleosome ratio of 0.4:1) in BC100 Buffer (added last). For control reactions, add 16 μl of BC100.

   Note that control reactions should either be reactions lacking remodeler, containing remodeler but lacking ATP, or containing ATP but with the addition of ADP before the addition of remodeler. Note also that the KCl from BC100 and the NaCl from the assembly bring the reaction to 55 mM monovalent cations, which is near-ideal for hSWI/SNF.
2. Incubate these reactions for 2.5 h at 30°C.
3. While the reactions incubate, pour a 4% polyacrylamide (30:1 acryl:bis) 1/2× TBE gel, with ~1.5 mm spacers and combs. 15 × 15 or 20 cm gel plates, and ~3 or 4 mm wells work well. Pour gel at least 2 h before use, and be sure that it is fully polymerized before use.
4. Stop the remodeling reactions by the addition of 6 μl 100 mM ADP (for discussion of stop conditions, see Note 6).
5. Add 240 μl of MNase Adjustment Buffer (bringing total reaction volume to ~300 μl) and pre-warm the reactions for several minutes at 30°C.
6. Add micrococcal nuclease (MNase) to 0.4 U/μl (Roche unit definition), mix, and incubate for 5 min. MNase can vary by lot and manufacturer. If you do not have an MNase stock that you have characterized before, (see Note 7) for how to optimize MNase concentrations.

7. Stop the reactions by addition of 6 μl of 10% SDS (to ~0.2%) and 9 μl of 500 mM EDTA (to ~15 mM).
8. Add 300 μl of neutralized phenol and extract, as per Subheading 3.2, step 8.
9. Add 50 μg of glycogen, 30 μl of 3 M NaOAc, and 750 μl of 100% EtOH, and precipitate, as per Subheading 3.2, step 9.
10. Resuspend each reaction in 25 μl of TE, and add 1× loading dye/glycerol buffer.
11. Remove the gel comb, flush wells, and pre-run the gel at 100 V for 20 min. Load all but ~4 μl of each sample, and ~0.5 μg of a DNA ladder that gives several bands in the ~300–100 bp range (PhiX HaeIII or 50 bp ladders work well). Leave one lane between the reaction lanes and the molecular weight ladder lane to prevent cross-contamination. Electrophorese at ~100–150 V until bromophenol blue is about 1/3 from the bottom.
12. Remove the gel from the plates and incubate for 30 min with gentle agitation in 1/2× TBE 30 ng/ml ethidium bromide.
13. Place the gel between two layers of UV-transparent plastic wrap (or, alternatively, use a very clean light box). Examine the gel under UV light (long wave preferred, but short wave is acceptable so long as exposure times are kept to a minimum). MNase products should be mostly mononucleosomal (~146 bp) with a weaker dinucleosomal band (~300 bp). If they are not, (see Note 7). An example of an MNase product isolation gel is shown in Fig. 2b.
14. Mark band positions on the plastic wrap, move the gel to a clean surface, and using a clean razor blade, remove the ~146 bp mononucleosomal band (as well as any other products of interest, such as ~290 bp dinucleosomes or the ~220 bp altosomal products of hSWI/SNF). Restrict the height of the gel slice so as to get only the major band (~140–150 bp for mononucleosomes), and try to keep gel slice volumes below ~100 mg (equivalent to an 8×8 mm slice) and relatively constant. If your gel slice volumes are greater than this, (see Note 8).
15. Weigh gel slices, and elute the MNase fragments by continuous shaking at 37°C in 3 gel slice volumes (assuming 1 g = 1 ml) of TE overnight. Recovery can be assumed to be ~75% (or the products from ~285 ng of plasmid).
16. Products can be stored at this stage at 4°C for weeks or –20°C indefinitely.

Day 2

17. For each restriction enzyme, prepare an enzyme mix containing a 4× concentration of the supplier-recommended buffer and between 2 and 4 U/μl of enzyme.

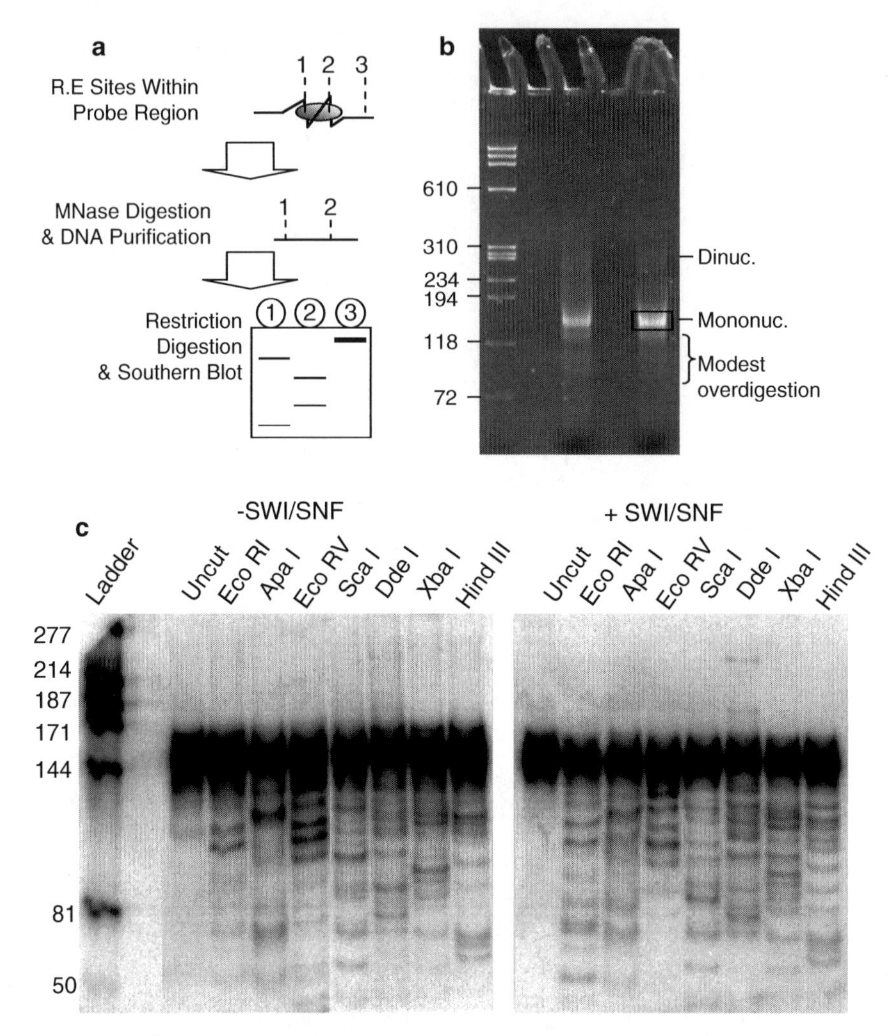

Fig. 2. MNase footprint/Restriction Digestion and Southern blotting. (**a**) Flow chart of mapping procedure, using the hypothetical case of a single, positioned mononucleosome in the probe region. *Numbers* and *circled numbers* denote restriction sites and restriction enzymes, respectively. (**b**) Example gel for isolation of MNase products. Positions of PhiX HaeIII bands in bp are marked on the *left*, and positions of mononucleosome, dinucleosome, and subnucleosomal bands are marked on the *right*. The *box* indicates a typical gel slice for isolation of mononucleosomes. (**c**) Southern blots of restriction enzyme digestion products for the 146 bp MNase mononucleosome fragments in a 359 bp region of the c-myc promoter, with and without (+ or –) SWI/SNF remodeling. Parts of this figure are adapted from ref. 9.

18. To 5 μl of each enzyme mix, add MNase fragments equivalent to 13 ng of the original plasmid. For instance, if your elution volume was 300 μl, the MNase product concentration is 285 ng/300 μl, or 0.95 ng/μl, and to get the products from 13 ng of plasmid, use 13.7 μl. Bring each reaction to 20 μl with TE.
19. As a control for complete digestion, also perform each digestion reaction on ~10 ng of your unlabeled probe DNA.
20. Incubate each reaction at the supplier-recommended temperature for 4 h to ensure complete digestion.

21. Prepare an undigested mononucleosomal control reaction that receives no restriction enzyme and is kept on ice.
22. After digestion, for the bare DNA digestion controls only, add 80 μl of water, 100 μl of phenol:chloroform, and extract. Add 20 μg of glycogen, 10 μl of 3 M NaOAc, and 250 μl of EtOH, ethanol precipitate, and resuspend in 25 μl of TE.
23. Based on the expected sizes of the resulting products, make two mixtures of these restriction reactions such that all restriction reactions are represented in one or the other mix, and each mix has several sizes of fragments between ~30 and 150 bp. These will serve as both digestion controls and as molecular weight ladders for the Southern blots. Store these ladder mixes and all other digestion products at –20°C.

Day 3

24. To all digested and control nucleosomal samples, add glycerol/xylene cyanol/bromophenol blue loading dye to 1× concentration.
25. Create an aliquot of each ladder mix that has a 1/4 molar amount of each restriction digested DNA. For instance, 13 ng is about 5 fmol of a 4 kb plasmid. If your probe is 200 bp long, 5 fmol is 0.65 ng, and you should use ~0.16 ng of each control restriction digest in your ladder mix. Bring each ladder aliquot to 20 μl with TE and add loading dye to 1× concentration.
26. Separate the restriction digested samples by 8% PAGE (20:1 Acryl:Bis, 0.5× TBE). Pour the gel at least an hour in advance, and be sure that polymerization is complete. Pre-run the gel at 100 V for 20 min. If you are using very high salt or otherwise unusual restriction digestion buffers, (see Note 9). We typically use gels of ~15×15 cm with 1.5 mm combs. Electrophorese at ~100 V until bromophenol blue is 1/3 from the bottom of the gel (approximately 5.5 hrs (550 V hours) with 15×15 cm plates).
27. Following PAGE separation, the DNA should be transferred to a positively charged nylon membrane by liquid transfer. Use the same type of electrotransfer apparatus typically used for Western blotting. Assemble the transblot sandwich as follows:
    - Black cassette edge, which will go nearest the negative (black) electrode in the transfer apparatus.
    - Spacer sponge soaked in 1/2× TBE and just barely submerged in 1/2× TBE in a wide flat dish.
    - Two layers of Whatman 3MM filter paper (or equivalent) soaked in 1/2× TBE.
    - The gel (trimmed of wells and notched on one corner to indicate orientation).
    - The nylon membrane (soaked for ~5 min in 1/2× TBE).

- Two more layers of 1/2× TBE-soaked 3MM paper.
- 1/2× TBE-soaked space sponge.

Close the cassette, and place in transblot tank 2/3 full of 4°C 1/2× TBE. The tank should contain a stir bar and is sitting on a stir plate set at medium speed, at 4°C. Fill the tank the rest of the way with 1/2× TBE.

28. Electrophorese at a constant 550 mA for 2.5 h at 4°C.
29. Rinse the blot 2× with 2× SSC.
30. Denature each blot, separately, by gentle submersion in 0.4 mM NaOH for 5 min on each side (do not agitate).
31. Wash three times for 5 min with agitation in 2× SSC.
32. Crosslink the DNA to the membrane by UV treatment, either surrounded by plastic wrap, DNA side down on a short to medium wave UV light box for 45 s, or on plastic wrap face up using a dedicated transblot crosslinker such as Stratagene's Stratalinker.
33. While transfer is in process, label your PCR amplified, purified probe fragment as follows (see Note 10 regarding use of ^{32}P). Mix in screw-cap microfuge tube:
    - 3 μl (3 μg) of random hexamer oligos.
    - Between 30 and 80 ng of purified probe fragment.
    - TE to bring reaction to 20 μl.
34. Boil for 3 min, quickly spin solution to the bottom of tube, and put on ice.
35. Add the following:
    - 3.75 μl of 10× Klenow buffer.
    - 2.8 μl of a mix containing 0.6 mM of dTTP, dCTP, and dGTP.
    - 7.5 μl of 6,000 Ci/mmol, 20 mCi/ml alpha dATP.
    - 1.5 μl (15 U) of Klenow polymerase fragment (add last).
36. Incubate at room temperature for 2 h.
37. Remove free nucleotides by microcentrifugation through a desalting column (BioRad Microbiospin 30 columns work well).
38. Determine labeling efficiency by Cerenkov counting in a scintillation counter. A good labeling should be $>5 \times 10^8$ cpm/μg (but as low as 2×10^8 may be acceptable).
39. While the Klenow reaction is incubating, put each blot in a hybridization tube, and wash 1× in 6× SSC.
40. Replace the SSC with 20 ml APH solution. In a rotating hybridization oven, incubate tube as well as an additional 10 ml of APH solution at 60°C for at least 15 min.
41. Prepare 30 μg of salmon sperm DNA in 30 μl of water.
42. Boil salmon sperm DNA for 10 min, spin briefly, add to the 10 ml of 60°C APH and mix.

43. Add this to the APH in the hybridization tube, mix gently, and incubate each blot with rotation at 60°C for 1 h.
44. For each hybridization bottle, add from 1/3 to all of the probe to 200 μl of prehybridization mix, and mix.
45. Boil this for 10 min and add directly to the prehybridization mix in each bottle.
46. Mix gently and incubate for 16 h at 60°C with rotation.

Day 4

47. Wash each blot 4× in the bottle with 2× SSC 0.1% SDS at room temperature for ~5 min with gentle mixing. If you like, you can save the hybridization mix at –20°C and re-use it up to twice (boiling the mix for 15 min, followed by cooling to 60°C before use).
48. Remove filters from the bottle and wash, with agitation 3× more with 0.2× SSC 0.1% SDS, for 10 min each wash. If background is present after this medium stringency wash, you can increase the stringency by washing 2× for 20 min with 0.2× SSC 0.1% SDS at 37°C, or even 42°C.
49. Lay filters on plastic wrap and remove excess moisture with edge of tissue.
50. Carefully wrap, still slightly damp, in plastic wrap, and expose to a PhosphorImager screen overnight at room temperature (or longer if signal is low). Be sure to orient the blot so that lanes will be nearly vertical or horizontal relative to the screen, and also that no liquid will leak out (since this can damage the screen). Two examples of the images from this Southern blotting procedure are shown in Fig. 2c.

3.4. Mapping and Quantitation of Nucleosome Positions

Initial notes. Below, we describe analysis of mononucleosome positions, but the same approaches can be applied to analysis of dinucleosomes (footprint size ~292) or hSWI/SNF-generated altered dinucleosomes (footprint size ~210 bp) by simply adjusting the calculations for these uncut MNase fragment sizes. Note that while we will describe the quantitation method using ImageQuant (IQ), this program is not absolutely necessary for this analysis – and other programs would be acceptable so long as they allow the quantitation of line scans from 2D intensity data. One free program that at least some basic line scan functionality is NIH Image, which can be downloaded from http://rsbweb.nih.gov/nih-image/.

1. Scan and visualize each blot using a PhosphorImager and ImageQuant software (or appropriate software for your phosphor detection machine).
2. In IQ, first draw a segmented line that follows the ~146 bp uncut MNase product bands across all lanes, and shift this

upwards (toward the top of the gel) by an inch or so on the image (maintaining the position of this line relative to the lanes). This reference line serves to control for gel "smiling" effects as well as any slight angle to the image.

3. Draw a line from this as an upper reference down the center of a central lane. After drawing this first line, click Object > Object Attributes and change the line width so that it spans the width of each lane without overlapping adjacent lanes.
4. Duplicate and move this line until lines cover each lane of the gel (including the marker lanes) and touch the upper reference line. This procedure ensures that all lines are in the same length, with the uncut mononucleosome band in the same position in each.
5. Select the lanes you wish to quantitate, and choose Analysis > Create Graph.
6. For all line graphs, choose Analysis > Peak Finder, select "auto detect" and "lowest point" background, leaving other parameters at default.
7. Use Tools > Define Peaks to create a single wide peak that erases all of the auto-detected peaks.
8. Adjust background for each line upwards from the true lowest point so that the line fits to the approximate average value of the noise in background regions. Do this by using the arrow button to click and drag the red handle boxes on the background line. If you are analyzing multiple lines at once, use the zoom out button to examine each line individually. If necessary, background can also be adjusted to account for evident Southern artifacts (spots and smears) and also to deal with trailing signal from the uncut mononucleosome band (see Note 11).
9. Select each distinguishable peak less than or equal to the uncut size of 146 bp. The best way to do this is to select the whole region (from just above the 146 bp uncut band to the bottom of the gel), and subdivide this peak using Tools > Split Peaks. Create a "peak" region even for places where no peaks are evident (which, if your background correction was done well should have very few counts). This is needed, since later quantitation requires a value for total counts per lane.
10. Choose Analysis > Area Report and export the results to Excel (when the report appears, click to close it, then click "yes" to activate Excel – alternatively, cut and paste the report data into Excel).
11. In Excel, first look at the output for the marker lanes, determining the apex pixel location of each peak.
12. Create a small table in Excel with two columns: pixel location and BP (where BP is the expected length of each control

fragment). Plot this and determine exponential best fit line to calculate length from pixel location (exponential because gel migration is closely related to the inverse log of DNA fragment length).

13. Now use this equation to calculate BP length from the apex pixel locations of each peak in your mononucleosome restriction digestion lanes. Note that you may sometimes need to manually input something other than the apex pixel if a visual inspection indicates that the center of the band is not the same as the apex (e.g., due to a background speck). Ideally, you will have some bands on the gel that are clearly paired (e.g., strong ~100 and ~46 bp bands in a lane where the other bands are much weaker). Moreover, you should be able to see symmetric patterns of strong and weak bands surrounding the 73 bp half-nucleosome position. This should allow you to confirm the accuracy of your length assignments. If the assignments are off, you may need to adjust your equation (the assumption that migration is proportional to log(length) works better for some gels than for others), or even dispense with a single equation and calculate length from separate linear equations established between every two ladder bands.

14. Take the background corrected signal from the larger of each pair of bands in a lane (≥73 bp for mononucleosomes) and multiply by the length of the nucleosome species being probed, and divide by the length of this larger fragment. This calculation corrects for the intensity of the smaller band, simplifying the analysis, and allowing quantitation even when very small restriction fragments (~30 bp or smaller) have run off the gel. As a sanity check, when you apply this calculation to the smaller band in a pair, it should give similar values. We find that this is usually true, since signal is nearly linearly related to length of overlap with the probe. If you find that this is not the case, even when applied to bands in your ladder, you can create an equation that corrects for this nonlinear response by plotting signal versus base pair using your ladder bands.

15. To calculate the percentage of nucleosomal species present in the probe region that was cut by each restriction enzyme to give a specific size fragment, divide this corrected intensity by the signal for the entire lane ("raw % cutting").

16. Draw a map of your probe region, to a convenient scale, such as 20 bp per cm, and mark the locations of each restriction site. For each pair of long and short restriction fragments that add up to 146 bp in a given lane, the long fragment must either be on the left or the right on the map. Say you have a pair of 110 and 36 bp fragments in the lane that was cut at site B, and data from adjacent lanes for restriction fragments cut at site A (60 bp to the left of B) and site C (80 bp to the right of B).

If the long fragment (containing the nucleosome center) is on the left, you will find bands of 50 bp (110 bp – 60 bp) and 96 bp (the corresponding band from mononucleosome digestion, 146 bp – 50 bp) in lane A, and these bands will have similar raw % cutting values as the 110 band in lane B. If, instead, the long band in B lies on the right, you will instead find bands of 30 (110 – 80) and 116 bp in lane C. This can be visualized by using a ruler to measure 110 bp to either side of site B. It is often easiest to work in from the ends of your probe region, especially if you have sites corresponding exactly to the edge of your probe. For instance, a site at the left edge of the probe will contain only one band from each nucleosome position (the rightward pointing fragment), allowing unambiguous establishment of that position without absolutely needing to consult other lanes. Indeed, if you have restriction sites on both ends of your probe and your probe is shorter than two nucleosome lengths (<~250 bp), nucleosome position assignments should be easy. For alternative mapping approaches and strategies for disambiguating complex banding patterns, (see Note 12). Note that you will be able to map nucleosomes whose center positions are somewhat outside the probe region. This depends on the smallest fragment size resolved on your gel. Specifically, you can map center positions up to 73 bp (smallest resolved fragment size).

17. Once you know positions, you can determine adjusted, average % cutting, which is directly proportional to nucleosome occupancy. Note that the intensity of any given band is proportional not to its length, but to the length of its overlap with the probe. For example, if you cut with an enzyme whose site is located 60 bp from the left edge of your probe and see a 110 bp product that overlaps the left probe edge, the signal is due to the 60 bp overlap, not the 110 bp overall length. To correct this, identify any such cases and multiply raw % cutting by (fragment length)/(length of overlap with probe). Next, for all bands corresponding to a given position, average the adjusted "raw % cutting" values determined above. This "average raw % cutting" is the initial quantitative measure for occupancy of each nucleosome position. In calculating this average, watch for obvious disconnects in raw % cutting values, which may indicate that bands are assigned to the wrong nucleosome.
18. After analyzing several gels, it should be evident that many of the same nucleosome positions are present in each gel, but may be stronger or weaker under different control or remodeling conditions (see Note 13). However, you may find that the calculated BP position for the same nucleosome position across several gels will differ up to ~±5 bp (the estimated accuracy of the assay). In order to compare the nucleosome occupancies

for these shared positions across multiple gels, create a master list of bands corresponding to each observed nucleosome position (we often find it convenient to use letters to identify nucleosomes). Then, calculate the average position for each lettered nucleosome as the average of its calculated BP positions from all gels.

19. Next, determine the fraction of DNA in the probe region that is covered by mononucleosomes. To do this, measure the signal from any pair of bands in your marker lanes that come from the same restriction digest and have sizes in the ~100–200 bp range. Calculate: (signal in the no restriction enzyme, control mononucleosome lane)/(4× the signal from this pair of bands). This value, multiplied by the length of the probe region in base pairs and divided by 146 (the mononucleosome length), tells the number of mononucleosomes, on average, in the probe region under each condition.
20. Multiply the average % cutting values for each nucleosome position by the number of nucleosomes in the probe region to get "% occupancy," a quantitative measure of the fraction of templates that have a nucleosome in any given position. (See Note 14) for further discussion of this calculation.
21. Plot nucleosome positions and occupancy. In Excel, create a column of "position bin values" with 5 bp between them, and covering the center positions of all mapped nucleosomes. In the next column, sum the % occupancy values for all nucleosome center positions that fall between each pair of values (fall into the bin). Now, create a bar plot with the first column on the *X*-axis. These plots are useful for identifying individual positions that may change upon remodeling. An example (from the data in Fig. 2c) is shown in Fig. 3a.
22. Plot "nucleosome coverage," or the fraction of templates that have a nucleosome covering any given position in the probe region. To do this, you must take into account the fact that each nucleosome with a center mapped to a certain location covers DNA extending 73 bp to the left and 73 bp to the right of this center position. The simplest way to do this is to create a column for each nucleosome position and put its % occupancy in all bins whose centers are within 73 bp of its center position. These columns can then be summed, in a new column, which when used as the Y-axis along with the bin location column as *X*, will yield % coverage graphs (using the Excel scatter plot function (Fig. 3b)). For a way to get excel to do this for you automatically, (see Note 15). This type of analysis will reveal, for instance, whether the tested remodeler greatly changes DNA accessibility due to nucleosome coverage in any part of the probe region. Note that, for % coverage values to be completely accurate, you need to have mapped and quantitated all

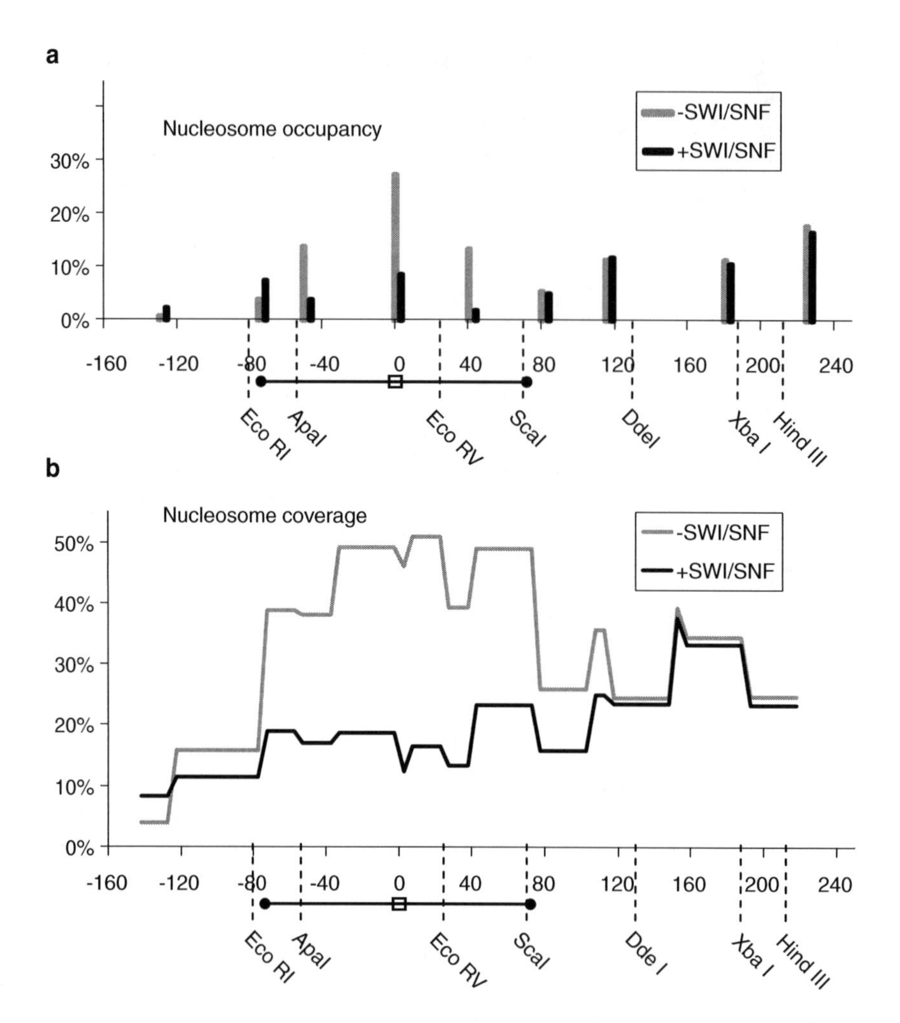

Fig. 3. Nucleosome occupancy and coverage plots. (**a**) Bar graph representing % occupancy (*Y*-axis) and base pair position (*X*-axis) of midpoints for all mapped mononucleosomes, from control (*gray*) and remodeled (*black*) templates. (**b**) Scatter plot representing the summed mononucleosome % coverage considering the area covered by each nucleosome. The core *5S* positioning sequence is indicated by a *barbell* on the *bottom*. Adapted from ref. 9.

nucleosome positions that touch any given position. Because you can map positions whose centers extend 73 bp minus the length of your smallest resolved fragment (typically ~30 bp), this means that nucleosome coverage information will be complete only at this distance from either end of your probe.

23. You can estimate the error in % occupancy for each position by repeating the analysis for one or a few conditions, and calculating the average absolute difference across all positions between the repeat gels. This will provide a basic idea of the accuracy of your mapping results, and can also be used for statistical analyses, as described in (see Note 16).

4. Notes

1. Topoisomerase I was included in remodeling reaction buffer to relax any supercoils that were not constrained by nucleosomes, or any unconstrained supercoils generated by the action of hSWI/SNF, eliminating possible effects of supercoiling on nucleosome positioning. It may not be necessary for remodeling complexes that are known not to alter supercoiling (although some unconstrained negative supercoils are likely to be present after assembly, as evidenced by the slower migration of unrelaxed plasmid than Topo-relaxed polynucleosomes on a chloroquine gel, see Fig. 1b).
2. If you are planning to map positions for species with a larger MNase footprint size, the chosen enzymes should be unique to either side of the chosen region for one full length of that species, and you need three unique enzymes per this length (e.g., 292 bp for dinucleosomes).
3. While we have not found this to be a problem with the promoter sequences we have tested so far, it is possible that you will find >146 bp stretches that do not fit this criterion. If so, your options are to reconsider your template or region of interest, accept this as a stretch where you will get incomplete information, or subdivide the region and use separate Southern probings (if this would allow sites present multiple times to become unique and usable for mapping).
4. Run an 0.8% 1/2× TBE gel containing 2 μg/ml chloroquine in both the gel and running buffer (8). Chloroquine effectively adds positive supercoils, such that initially relaxed DNA (the bare DNA + Topo control) now runs fast, and negatively supercoiled DNA runs slowly (the no-Topo control should now look mostly relaxed, while the assembled chromatin looks almost relaxed). This gel resolves the highly negatively supercoiled DNAs that all pile up as a single fast-migrating band in normal TBE buffer. On this gel, DNA from well-assembled chromatin will run almost, but not quite, as slowly as the supercoiled plasmid DNA.
5. To confirm that your chosen remodeling enzyme has maximal activity, you could titrate the remodeler and perform the complete MNase footprint Restriction Digestion mapping assay, but use only one or a few restriction enzymes. While this will not let you map positions, it will let you determine the amount of remodeler needed for maximal activity. The idea behind this is that as you titrate in your remodeler, you will see changes in the restriction pattern for any enzyme whose site is covered by one or more nucleosomes whose positions are

altered during the repositioning reaction. The remodeler concentration at which these changes level off will indicate the amount of remodeler needed for full activity. Alternatively, you might use another available assay to test for full remodeling. For instance, one common assay for repositioning uses a radiolabeled mononucleosomal template and measures the shift in mobility of the mononucleosome by PAGE and autoradiography (centered octamers migrate more slowly than end-positioned octamers). We used this approach to confirm full nucleosome repositioning in our recent study of human SNF2h complexes. Briefly, we prepared 30 μl reactions with 190 ng of unlabeled polynucleosomes and a tracer amount (less than 1 ng) of ^{32}P-labeled mononucleosome for which the effects of each SNF2h complex was known, and determined the concentration of remodeler necessary to achieve the full repositioning effect on the mononucleosomal template when in the presence of the polynucleosomes (10). In addition to titrating enzyme amount, you might also test whether increased repositioning activity is seen in other buffers (we have found that lowering monovalent or divalent cation concentrations could improve hSWI/SNF activity), at increased temperature or for increased or decreased incubation times. If you find that you cannot achieve full remodeling in a 60 μl reaction with 380 ng template, you might increase the reaction volume or decrease the amount of template (although this latter option will either decrease the number of restriction assays you can perform, or decrease your signal on your Southern blots).

6. Remodeling should be stopped by some method other than chelation of divalent cations (since these will be needed for subsequent MNase digestion). We have successfully used hydrolysis of ATP by apyrase (2.4 U/60 μl reaction) or addition of ADP (as a competitive inhibitor of ATP hydrolysis) to a concentration of 10 mM or both. If apyrase is used, incubate for at least 10 min before MNase digestion to ensure complete hydrolysis.

7. If dinucleosomal and higher multimer bands are not of lower intensity than mononucleosomal bands, you will need to use more MNase. If there are multiple bands of near-equal intensity below the ~146 bp band, or if band signal is much lower than expected (based on marker lanes), you will need to use less. The activity of MNase from different suppliers can vary, so it is recommended to do a test titration of new MNase batches on control samples. Also, we have also noticed that some remodeler preparations appear to have an inhibitory effect on MNase, requiring additional MNase to be added to give complete digestion.

8. If the gel slices are greater than 100 μl, you may need to concentrate the MNase footprint fragments using a 10 kDa Millipore centrifugal filter.

9. We have found that loading directly onto the gel is OK for most restriction digestion buffers, but that some specialized buffers can cause unacceptable salt effects. You can check for this by digesting a sample of plasmid or unlabeled probe DNA with each enzyme and looking for distortions in band migrations after EtBr staining. If such distortions are seen, you can eliminate this effect either by EtOH precipitation (in the presence of 20 μg of glycogen) or by the use of microcentrifugal desalting columns (such as BioRad Microbiospin 30 columns).
10. Use all proper care and follow institutional requirements when dealing with radiation. ^{32}P can be readily detected by Geiger counter. We anticipate that nonradioactive Southern detection methods, such as the use of fluorescent labels and fluorescent blot readers, would work equally well, but have not tested this in our studies.
11. If your signal is high and background is low, a simple correction to what appears to be the lowest point in the actual gel lane may be appropriate. However, if background clearly varies from top to bottom of the gel or background smears are visible in particular regions, these can be adjusted for by drawing an appropriate adjusted background curve (use the keyboard insert key while clicking with the mouse to insert new handle boxes on the background line). Another common problem is that longer fragment bands often appear as shoulders on tailing edge of the 146 bp mononucleosome band. This can lead to overestimated signal for these bands. The simplest approach to dealing with this is to use only bands from other lanes to determine % occupancy for the corresponding nucleosome position. Alternatively, after using the background settings above to quantitate most bands and total signal per lane, you can correct for this tailing background as follows. In your undigested control mononucleosome lane, measure intensity and pixel position at the apex and at several points along the fast-migrating tail of the mononucleosome band (in the region you wish to correct background for). For each restriction digestion lane, establish background points at each of the pixel positions measured on the uncut band tail by creating a handle box and moving it to a value equal to (mononucleosome apex signal from restriction digested lane)/(mononucleosome apex signal from undigested lane) × (signal at that position for undigested lane). If the apex pixel for your restriction digested lane differs from the undigested control, shift the locations of each box accordingly. This adjusts background for the trailing edge of undigested mononucleosomes according to the strength of this peak remaining after each restriction digestion reaction, and requantitation using this background will give more accurate % cutting values for bands near 146 bp.

12. For long probe regions, especially where you have many nucleosome positions (and hence many bands), disambiguating nucleosome positions to the left or right can sometimes be tricky. When banding patterns are complex (especially far from the edges of your probe region), it is important to pay close attention to both fragment size and intensities in adjacent lanes, and to test different combinations of positions with the goal of not leaving any strong bands unexplained. Most challenging are situations where nucleosomes in two locations give the same size fragments for a given restriction site (e.g., nucleosomes with left edges at 10 and -36 will both give 50 + 96 bp fragments for a restriction site at 60). While looking at adjacent lanes will usually allow these situations to be identified, % cutting values for these bands in the lane where they overlap should not be used in determining % occupancy for those nucleosomes.
13. We find that nucleosomes, both before and after remodeling, do not occupy a continuum of all positions (e.g., every base pair). Rather, nucleosome can form at only a fraction of positions, and remodeling often results in the strengthening or weakening of positions seen after assembly, without creation of entirely new positions.
14. The presence of multiple nucleosomes can decrease the % cutting (and thus apparent occupancy) at any one position. For instance, if a region contained two perfectly positioned nucleosomes, digestion at a site covered by the first nucleosome would only result in 50% cutting, because that restriction enzyme will not cut fragments protected by the second nucleosome. To determine the total number of nucleosomes within each probe region, compare the signal for a known amount of probe DNA that was not digested by MNase to the equivalent amount of probe region DNA after MNase digestion of polynucleosomes. In particular, from the gel loading instructions, each pair of bands from a given restriction digest in the ladder lane correspond to one quarter the number of moles of probe region DNA in each restriction digestion lane. Thus, if MNase had not digested away any DNA, the signal for the polynucleosomal template would be four times the signal for each pair of bands in the ladder. Accordingly, the equation (uncut mono signal)/(4× ladder band pair signal) × (probe length)/146 gives the number of mononucleosomes, on average, present in the probe region. Note that this method assumes a 75% recovery of MNase digestion fragments after PAGE isolation, which may not be precisely accurate. It also assumes that there was no significant under or overdigestion resulting in a large fraction of mononucleosomal products that were not present in the PAGE slice that was isolated. An alternative,

which avoids these assumptions, is to run a sample of the nonisolated MNase digestion reaction (equivalent to 13 ng of undigested plasmid) on your Southern gel and to quantitate the signal for mononucleosomes after MNase using a line scan in IQ to measure the percentage of all signal in the mononucleosome peak. This latter method should definitely be used if you wish to quantitate more than one species (e.g., mono- versus dinucleosomes). Note that the MNase digestion conditions required to map positions remove all linker DNA, but can also result in modest overdigestion of nucleosome-protected DNA, resulting in some ~130 to ~110 bp subnucleosomal DNA below the mononucleosomal band. Accordingly, the number of nucleosomes calculated by either of the above methods is likely to be slightly below the actual number in the probe region. If you find that over- or underdigestion effects are strong, one way to correct for this is to establish an MNase titration curve and use this to extrapolate the actual fraction of the probe region covered by mononucleosomes (as described in ref. 10). Alternatively, if there is restriction site on your template that you know to be near-100% occupied under control conditions (and which does not cut too many times in the rest of the plasmid), accessibility at this site can be measured by a simple agarose gel assay, to provide a correction value for your estimate of nucleosome number (see ref. 9 for details).

15. Analysis of nucleosome coverage can also be done using the special excel SUM(IF()) function. If you have two columns containing all mapped nucleosome center positions (column A, rows 2–40) and % occupancy (column B, rows 2–40), and base pair position values every 5 bp in column C rows 2–101, you can calculate % coverage in cell D2 using the following command: =SUM(IF(ABS(A$2:A$40-C2)<73,B$2:B$40)), followed by CTRL-SHIFT-ENTER (needed for Excel to accept this special command). Filling this formula down column D will give summed % occupancy values that can be graphed with a scatter plot. With the same set-up, a similar command can also be used to determine % occupancy in 5 bp bins: in cell E2 input "=SUM(IF(ABS(A$2:$A40-C2)<2.5, B$2:B$40))." Which adds % occupancy values for all positions within 2.5 bp of the reference value in column C (with the one caveat that positions that are an exact multiple of 5 bp will be missed – which can be dealt with by adding 0.0001 to positions that are an even multiple of 5 bp).

16. If you have analyzed multiple gels under the same condition, you can determine the average absolute error in assignment of nucleosome positions (as described in Subheading 3.4, step 23). The standard deviation, *s*, is equal to this average absolute difference divided by the square root of 2. To determine, statistically,

whether the whole pattern of nucleosome positions over a given region has changed significantly, use a *T*-test to compare average absolute differences in % occupancy between conditions versus average absolute difference in % occupancy for repeats of the same condition: $t = (\text{avg\_diff}_1 - \text{avg\_diff}_2) / \sqrt{(2 \times s^2/n)}$, where n equals the number of nucleosome positions used to determine each average. If more than one comparison is made (e.g., each pairwise comparison between four conditions) you can correct this using the Bonferroni method: multiply the resulting *p*-value by the number of comparisons to get adjusted *p* values (10).

References

1. Miele V, Vaillant C, d'Aubenton-Carafa Y, Thermes C and Grange T (2008) DNA physical properties determine nucleosome occupancy from yeast to fly. Nucleic Acids Res 36:3746–3756.
2. Shivaswamy S, Bhinge A, Zhao Y, Jones S, Hirst M and Iyer VR (2008) Dynamic remodeling of individual nucleosomes across a eukaryotic genome in response to transcriptional perturbation. PLoS Biol 6:e65.
3. Schones DE, Cui K, Cuddapah S, Roh TY, Barski A, Wang Z, Wei G and Zhao K (2008) Dynamic regulation of nucleosome positioning in the human genome. Cell 132:887–898.
4. Tillo D, Kaplan N, Moore IK, Fondufe-Mittendorf Y, Gossett AJ, Field Y, Lieb JD, Widom J, Segal E and Hughes TR (2010) High nucleosome occupancy is encoded at human regulatory sequences. PLoS One 5:e9129.
5. Schnitzler GR (2008) Control of Nucleosome Positions by DNA Sequence and Remodeling Machines. Cell Biochem Biophys 51:67–80.
6. Ramachandran A and Schnitzler G (2004) Regulating transcription one nucleosome at a time: Nature and function of chromatin remodeling complex products. Recent Res Devel Mol Cell Biol 5:149–170.
7. Saha A, Wittmeyer J and Cairns BR (2006) Chromatin remodelling: the industrial revolution of DNA around histones. Nat Rev Mol Cell Biol 7:437–447.
8. Ulyanova NP and Schnitzler GR (2005) Human SWI/SNF generates abundant, structurally altered dinucleosomes on polynucleosomal templates. Mol Cell Biol 25:11156–11170.
9. Sims HI, Baughman CB and Schnitzler GR (2008) Human SWI/SNF directs sequence-specific chromatin changes on promoter polynucleosomes. Nucleic Acids Res 36: 6118–6131.
10. Pham CD, He X and Schnitzler GR (2010) Divergent human remodeling complexes remove nucleosomes from strong positioning sequences. Nucleic Acids Res 38:400–413.
11. Utley RT, Owen-Hughes TA, Juan LJ, Cote J, Adams CC and Workman JL (1996) In vitro analysis of transcription factor binding to nucleosomes and nucleosome disruption/displacement. Methods Enzymol 274: 276–291.
12. Schnitzler G, Sif S and Kingston RE (1998) Human SWI/SNF interconverts a nucleosome between its base state and a stable remodeled state. Cell 94:17–27.

Chapter 20

Analysis of Changes in Nucleosome Conformation Using Fluorescence Resonance Energy Transfer

Tina Shahian and Geeta J. Narlikar

Abstract

ATP-dependent nucleosome-remodeling motors use the energy of ATP to alter the accessibility of the underlying DNA. Understanding how these motors alter nucleosome structure can be aided by following changes in histone–DNA contacts in real time. Here, we describe a fluorescence resonance energy transfer-based approach that enables visualization of such changes.

Key words: Histone, ATP-dependent chromatin remodeling, ACF, SNF2h

1. Introduction

The packaging of eukaryotic DNA into chromatin provides a versatile context for regulating access to DNA. A major mode of regulation is through the action of ATP-dependent chromatin-remodeling complexes (1). These complexes catalyze many different types of reactions ranging from moving the nucleosomes *in cis* to exchanging histone components. Fluorescence resonance energy transfer (FRET) is a powerful tool for monitoring changes in distance between two regions of a nucleosome (2–4). In this technique, a donor and an acceptor fluorescent molecule (fluorophores) are attached to two sites of interest. The emission peak of the donor fluorophore must overlap with the excitation peak of the acceptor fluorophore. Excitation of the donor leads to a transfer of energy to the acceptor by dipole–dipole interactions such that the acceptor emits light at its own emission wavelength (2). The higher the efficiency of transfer, the more quenched the emission of the donor and the greater the emission of the acceptor. The transfer efficiency is very sensitive to the distance between the fluorophores as shown in 1.

Randall H. Morse (ed.), *Chromatin Remodeling: Methods and Protocols*, Methods in Molecular Biology, vol. 833,
DOI 10.1007/978-1-61779-477-3_20,

Table 1
Fluorescence donor and acceptor properties (see Note 10)

Donor	Excitation (nm)	Emission (nm)	Acceptor	Excitation (nm)	Emission (nm)	~R_0 (Å)
Fluorescein	495	520	Rhodamine	559	583	55
Cy3	550	564	Cy5	648	668	56

$$E = \frac{R_0^6}{R_0^6 + r^6} \tag{1}$$

Ro, also known as the Förster radius, is the distance between the donor and acceptor probe at which the energy transfer is 50% efficient and r is the distance between the fluorophores. Thus, increases or decreases in inter- or intramolecular distances can be monitored by a decrease or increase in FRET efficiency, respectively. Ro is unique to every FRET pair (Table 1).

The histones are labeled at single cysteines, which are introduced by site-specific mutagenesis (e.g., QuikChange from Stratagene). The location of the cysteine mutation is chosen based on the crystal structure of the nucleosome core particle such that it is within the Förster radius from the labeled DNA residue (5, 6). Further, the mutation should not disrupt any histone–DNA interactions. A critical aspect of preparing fluorescently labeled nucleosomes is ensuring high quantum efficiency of the labels. Therefore, it is critical to use a fresh batch of fluorescent dyes when labeling histones. Furthermore, it is important to shield all fluorescently labeled reagents (dyes, DNA, histones, octamers, and nucleosomes) from light exposure throughout all protocols.

Over the last several years, FRET has been extensively used in several studies of chromatin dynamics. This technique has been used to study the process of exchange of core histones with histone variants (7). FRET has also been used to measure rates of spontaneous exposure of nucleosomal DNA sites and the rates of remodeling by the ATP-dependent chromatin-remodeling complex, ACF (8, 9). In all these studies, it was demonstrated that the FRET probes do not interfere with nucleosome integrity.

Here, we describe protocols for FRET measurements with nucleosomes containing a Cy3 dye on one 5′-end of the DNA and a Cy5 dye on histone H2A at position 120, which is mutated to a cysteine (Fig. 1a, H2A-120C). These locations were chosen based on the nucleosome crystal structure such that distance between the dyes is within their Förster radius (Ro) (Table 1). We describe how

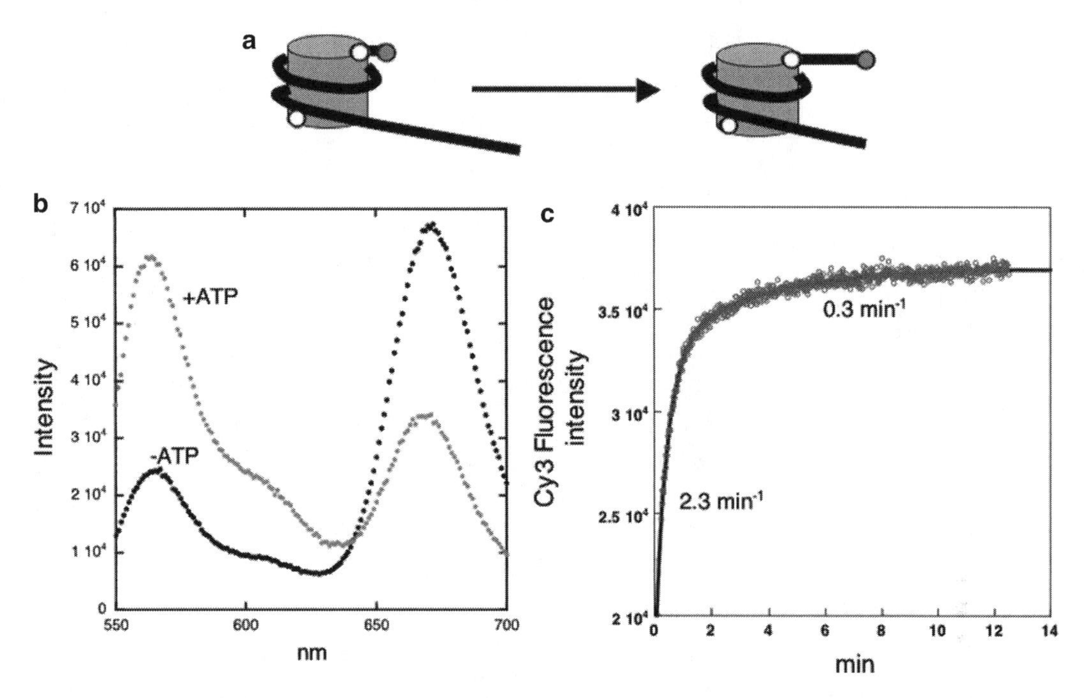

Fig. 1. (**a**) Schematic of a fluorescently labeled nucleosome that is moved toward the center of the DNA upon the action of the ATP-dependent chromatin-remodeling enzyme, SNF2h. The histone octamer is shown as a *cylinder* and the DNA is shown in *black*. The Cy3 dye on the DNA is shown as *filled circle*, and the two Cy5 dyes on each of the two H2A molecules are shown as the *open circles*. Only one of the two Cy5 dyes is close enough to the Cy3 dye to give FRET. (**b**) An emission scan obtained across 500–700 nM after exciting samples at the excitation maximum for Cy3. The samples contain reaction mixtures similar to those described in Subheading 3.4. In the absence of ATP, the Cy3 fluorescence is lower than the Cy5 fluorescence due to FRET (*black line*). After reaction in the presence of ATP, SNF2h moves the nucleosome toward the center as shown in (**a**), which results in an unquenching of Cy3 fluorescence (*grey line*). This increases the distance between the Cy3 and Cy5 dyes and lowers the FRET. As a result, there is unquenching of the Cy3 dye fluorescence and a reduction of the FRET-induced fluorescence of the Cy5 dye. (**c**) The kinetics of remodeling can be measured by following the unquenching of the Cy3 fluorescence as a function of time. In this particular plot, the data is best fit by two exponentials resulting in a fast rate constant (2.3 min^{-1}) and a slower rate constant (0.3 min^{-1}).

to measure the effects of nucleosome remodeling by SNF2h, the ATPase subunit of the human ACF complex. The specific buffer conditions for the remodeling reaction can be altered as needed for other remodeling enzymes.

2. Materials

2.1. Site-Specific Labeling of Histones with Fluorescent Dyes

1. Lyophilized histones with single cysteines (4).
2. Cy5-mono maleimide.
3. Anhydrous dimethylformamide (DMF).
4. Tris(2-carboxyethyl) phosphine hydrochloride (TCEP), 0.5 M, pH 7.
5. 2-mercaptoethanol.

6. Labeling buffer, pH 7.3: 20 mM Tris–HCl, pH 7.0, 7 M guanidinium HCl, 5 mM EDTA.
7. Centrifugal filter device for sample concentration, such as Microcon-10 (Millipore).
8. SYPRO-Red (Molecular Probes).

2.2. Preparation of Histone Octamer from Purified Core Histones

1. Lyophilized histones H2B, H3, and H4.
2. Cy5-labeled histone H2A (prepared in Subheading 3.1).
3. Unfolding buffer, 50 ml: 20 mM Tris–HCl, pH 7.5, 7 M guanidinium HCl, 10 mM DTT (add just prior to use).
4. Refolding buffer, 3.5 l: 10 mM Tris–HCl, pH 7.5, 2 M NaCl, 1 mM EDTA, 5 mM 2-mercaptoethanol (add just prior to use).
5. Dialysis membranes (6–8 kDa cutoff) preboiled in Milli-Q water and cooled to room temperature.
6. Superdex 200 HR 10/30 column (GE, Amersham) or equivalent.
7. Centrifugal filter device for sample concentration, such as Microcon-10 (Millipore).
8. 0.22 μm Millipore Ultrafree centrifugal filter or equivalent.
9. Wash buffer: 50 mM Tris–HCl, pH 7.5, 100 mM NaCl, 1 mM EDTA, 1 mM benzamidine HCl (add just before use).

2.3. Large-Scale PCR Preparation of Fluorescently Labeled DNA

1. 96-well PCR plates.
2. Forward primer labeled with cy3 at 5′ end (HPLC purified).
3. 10× PCR buffer: 100 mM Tris–HCl, pH 8.3, 500 mM KCl, 15 mM $MgCl_2$, 5% Tween 20.
4. 2 mM dNTPs.
5. Taq polymerase.
6. OakRidge centrifuge tubes or equivalent.
7. Large gel electrophoresis apparatus.
8. 3 M sodium acetate, pH 5.0.
9. 200 proof ethanol.
10. Native DNA-loading buffer: 20% glycerol, 20 mM Tris–HCl, pH 7.5.
11. TE buffer: 10 mM Tris–HCl, pH 8.0, 1 mM EDTA.
12. 50-ml Conical tube filter units.
13. 5% acrylamide, 0.5× TBE mini gel.
14. 10-ml syringe without needle.
15. DNA of known concentration (*preferably DNA used to assemble nucleosomes*).

16. Ethidium bromide (EtBr).
17. Scanning device, such as Typhoon Scanner.

2.4. Measurement of Remodeling End Points

1. 15 mM $MgCl_2$.
2. 10 mM ATP–$MgCl_2$.
3. BC100: 20 mM HEPES, pH 7.9, 100 mM KCl, 0.2 mM EDTA, 20% glycerol.
4. Nucleosome buffer: 20 mM Tris–HCl, pH 7.5, 1 mM EDTA, 20% glycerol, 0.1% NP-40.

2.5. Measurement of Remodeling Kinetics

1. BC100 with FLAG peptide: BC100: 20 mM HEPES, pH 7.9, 100 mM KCl, 20% glycerol, 0.2 mM EDTA, 1 mg/ml FLAG peptide.
2. Nucleosome buffer: 20 mM Tris–HCl, pH 7.5, 1 mM EDTA, 20% glycerol, 0.1% NP-40.
3. BC100 without glycerol: 20 mM HEPES, pH 7.9, 100 mM KCl, 0.2 mM EDTA.
4. 100 mM $MgCl_2$.
5. 120 mM Mg.ATP.
6. BC100: 20 mM HEPES, pH 7.9, 100 mM KCl, 0.2 mM EDTA, 20% glycerol.
7. 15 mM $MgCl_2$.

3. Methods

3.1. Site-Specific Labeling of Histones with Fluorescent Dyes

The method described here is for labeling of H2A-120C with Cy5 but can be applied to any histone and maleimide combination. The histones must have single cysteine mutations incorporated in the location of interest, for example by using QuikChange mutagenesis. Location of labeling can be determined based on the crystal structure of the nucleosome (5, 6). Maleimide chemistry is then used to specifically label the cysteines. All steps in this protocol are carried out at room temperature and shielded from light.

1. Dissolve 0.5 mg of *H2A-120C* in 0.4 ml of labeling buffer (~0.1 mM final) at room temperature (see Note 1).
2. Add 1 μl of 0.5 M TCEP to the reaction (~1.25 mM final).
   This step reduces the cysteines, so they can be modified.
3. Incubate at room temperature for 2 h (shield from light by covering the entire rack of tubes with foil and storing in a drawer).
4. Add 12.5 μl of 100 mM Cy5-maleimide (~3 mM final) (see Note 2).

Dissolve a fresh batch of Cy5-mono maleimide *in anhydrous DMF. Usually, all dissolved dye is used up. Leftover dissolved dye is discarded.*

5. Incubate at room temperature for 3 h (shield from light by covering with foil and keeping in a drawer).
6. Stop the reaction by adding beta-mercaptoethanol to 80 mM final concentration.
7. Save a small aliquot to run on a gel. At this stage, the reaction can be stored overnight if needed (shield from light and store at room temperature).
8. To remove most of the free dye from the labeled histone, concentrate the histone in a Microcon-10 or similar filter concentration device at room temperature three times. Each time, concentrate to approximately one-third the original volume, and dilute back down to original volume with labeling buffer. Use a new filter every time. Save the flow through until certain that histones are retained on top of the Microcon (see Note 3).
9. Following the Microcon treatment, run a sample on a 15% polyacrylamide separating gel and stain with Coomassie or SYPRO (for quantification) (see Note 4). Dilute 1 μl of sample into 50 μl of water before adding SDS loading dye and load ~15 μl of the diluted histone on the gel. This dilution step is required because the guanidium in the reaction mixture can cause precipitation with SDS. Load equal volumes of sample from before and after the concentration treatment to determine the yield. To further avoid precipitation, load the sample directly onto the gel after heating to 95°C (do not spin down). Use an imaging system, such as Molecular Dynamics Typhoon, to quantify the two bands. Since the concentration of the histone before concentration is known, this can be used to determine the concentration of the histone afterward by comparing the fluorescence signals for the two samples. The gel can also be stained using SYPRO red as an independent way of quantifying the histone concentration (see Note 5).
10. At this stage, the Microcon-treated histones should be shielded from light and can be stored at room temperature overnight if needed.
11. Octamers can then be assembled by mixing appropriate amounts of the labeled histone in its guanidinium buffer with other histones dissolved in the unfolding buffer (see refs. 10, 11).

3.2. Preparation of Histone Octamer from Purified Core Histones

This method describes the assembly of histone octamer using a labeled histone subunit and other nonlabeled subunits (adapted from (10, 11)). The octamer is folded by dialysis against high salt and purified over an FPLC column. The H3/H4 tetramer is difficult to resolve from the octamer on the column; therefore, it is important to use a slight molar excess of H2A/H2B to make sure that octamer formation goes to completion.

1. Dissolve each lyophilized histone to a concentration of 2 mg/ml in unfolding buffer. Pipette up and down to dissolve. Do not vortex.
2. Allow unfolding to proceed for at least 30 min at room temperature and for not more than 3 h.
3. Mix the four histones so that the molar ratio of H2A:H2B: H3:H4 is 1.2:1.2:1:1 (see Note 6).
4. Adjust the final protein concentration of the mix to 1 mg/ml total protein using unfolding buffer.
5. For volumes up to ~8 ml of mixed protein, dialyze in 6–8 kDa cutoff dialysis bag three times against 1 l of refolding buffer in the cold room. At least one dialysis should be overnight. The rest should be at least 3 h each. For large-scale assemblies, scale up the dialysis buffer proportionately.
6. After the final dialysis, spin down sample to remove any precipitated material. Save the supernatant for loading on the column.
7. Begin washes and equilibration of Superdex 200 HR 10/30 column the day before. A wash step with 2-column volumes of water is required if the column has been stored in 20% ethanol or another similar storage buffer. Then, equilibrate with at least 2-column volumes of refolding buffer (24 ml × 2).
8. Concentrate the assembly to ~200 or 500 μl in a Microcon 10 or other concentration device. Save the flow through until it is verified that histones have been retained in the concentrating chamber.
9. Filter the concentrated sample using a 0.22-μm Millipore Ultrafree or equivalent centrifugal filter and load on the column (save ~5 μl for running on a gel). Run the buffer at 0.5 ml/min and collect 0.5-ml fractions (see Note 7).
10. Run the flow through, load and peak fractions on a gel, and pool bona fide octamer fractions that contain equimolar H2A, H2B, H3, and H4. Concentrate to greater than 1 mg/ml using a Microcon 10 or other concentration device.
11. Determine octamer concentration by SYPRO staining using BSA standards.
12. Flash freeze in liquid N_2 and store at −80°C. The octamer is stable at −80°C for at least 6 months.

3.3. Large-Scale PCR Preparation of Fluorescently Labeled DNA

The following protocol is for the amplification of a 202-bp DNA fragment on a 10-ml scale. The protocol uses the 601 template (8) and a Cy3-labeled forward primer. This protocol can be adapted to amplify DNA fragments from any template using any 5′-labeled primer. For amplification on a larger scale, simply adjust the PCR protocol proportionally. A general PCR protocol is used with the modifications outlined below. The amplified product is then gel purified and precipitated with EtOH. Keep exposure to light at a minimum by shielding samples whenever possible.

1. Set up PCR as follows:

10× buffer (with 15 mM Mg^{2+})	1× final
Fluorescent forward primer	0.5 μM final
Reverse primer	0.5 μM final
dNTP	0.2 mM final
601 template	0.1 ng/μl final
Taq polymerase	0.05 U/μl final
H_2O (Milli-Q)	Adjust to 10 ml
Total volume	10 ml

2. Mix all components on ice in a 15-ml falcon tube.
3. Add 100 μl PCR mix in each well of a 96-well plate using the 12-channel multi-pipettor and seal.
4. Place in thermal cycler and use 35 cycles (sample program):

195°C	4 min
95°C	45 s
58°C	30 s
72°C	1 min
Loop to step 2, repeat 35 times	
72°C	10 min
4°C	

5. Confirm amplification by running 5 μl of sample on a 5% 1× TBE mini-acrylamide gel next to a 100-bp ladder. Stain with EtBr. A good yield is ~50 ng/ μl.
6. Transfer the PCR product into a 30-ml Oak Ridge tube.
7. Add one-tenth volume 3 M sodium acetate, pH 5.0, and three volumes of 200 proof ethanol.
8. Incubate at –20°C over night (or for 3 h if time is limiting) to precipitate DNA.

9. Spin down DNA at 4°C at 25,000×g for 20 min.
10. Carefully remove supernatant and allow the DNA pellet to air dry for ~30 min. There is no need to wash the pellet at this stage (see Note 8).
11. Redissolve pellet in ~400 μl native DNA-loading buffer (no dye).
12. Earlier in the day, pour a 5% acrylamide (29:1 acrylamide:bis) and 1× TBE gel (16 cm long×40 cm wide). Use a comb with a large well (12 cm wide).
13. Load the DNA into the large well. Load xylene cyanol and bromophenol blue dyes in an adjacent smaller well for tracking. For a 200-bp fragment, run the gel at 100–150 V until the bromophenol blue has reached the bottom of the gel.
14. Open the gel plates such that the gel remains on one of the two plates. Cover with plastic film wrap and then aluminum foil to shield from light.
15. Take off the foil and view the DNA band under a short-range handheld UV lamp in the darkroom. Cut out the desired band using a clean razor blade. Exposure to UV should be kept to a minimum.
16. Plunge the gel piece through a 10-ml syringe (no needle) into 15-ml tube. This step crushes the gel matrix.
17. Add ~10 ml TE buffer, pH 8.0, wrap in aluminum foil, and rock overnight at room temperature. This step extracts the DNA from the gel.
18. Pour gel and TE buffer mixture over a 50-ml conical-top filter unit, attach to a vacuum, and collect the flow through. DNA is in the flow through.
19. Transfer flow through to an Oak Ridge tube and repeat steps 7–10, this time washing the pellet with cold 70% ethanol.
20. After it is dry, dissolve pellet into ~50–100 μl TE buffer, pH 8.0.
21. Quantify by measuring absorbance at 260 nm. The DNA can be stored at –20°C till further use.
22. The purified labeled DNA and labeled octamer can then be assembled into mononucleosomes as described in 10, 11 (see Chapters 15 and 21).
23. Nucleosomes are purified from free DNA over a 10–30% glycerol gradient containing 0.1% NP40 to stabilize the nucleosomes (12) (see Chapter 21, Subheading 3.1.4). A Microcon 100 is used for subsequent concentration of the nucleosomes as the size of the micelles formed by NP40 is ~90 kDa.
24. Nucleosomes can be quantified according to steps 23–25 below. These steps are also applicable to quantification of

nonlabeled nucleosomes. Incorporation of DNA into nucleosomes quenches ethidium bromide fluorescence of DNA by 2.5-fold. This value is used when quantifying nucleosome concentration. DNA is stained with EtBr and scanned on a Typhoon or similar scanning device.

25. Run a few different amounts of nucleosomes on 5% acrylamide gel. The nucleosomes off the gradient already contain ~20% glycerol and can be loaded directly onto the gel. Do not use any dye.
26. Load 3–4 DNA amounts between 10 and 100 ng to generate a standard curve. Use loading buffer without dye. A 100-bp ladder and bromophenol blue can be loaded in a separate well for tracking. Run gel until dye reaches one-fourth of the way from the bottom of the gel.
27. Stain gel with EtBr and scan using the Typhoon or similar scanning device in the EtBr channel. Use the DNA standard to calculate the amount of nucleosomal DNA. This value must be adjusted for the 2.5× quenching effect of EtBr when staining nucleosomal DNA. 1 mole of nucleosomes contains 1 mole of nucleosomal DNA.

3.4. Measurement of Remodeling End Points

This protocol allows measurement of the final remodeling-induced FRET changes (Fig. 20.1b). We describe here how to measure the effects of nucleosome remodeling by SNF2h, the ATPase subunit of the human ACF complex.

1. Keep exposure of fluorescently labeled nucleosomes to a minimum.
2. Use all appropriate protocols and precautions when using your particular fluorometer. We use an ISS K2 fluorometer.
3. Prepare two reactions on ice, each 80 μl, as outlined below. One reaction contains ATP (+ATP) and the other does not (–ATP). The specific reaction volume depends on the size of the cuvette:

   +ATP reaction

   (a) 16 μl 15 mM $MgCl_2$ (3 mM final)
   (b) 16 μl 10 mM ATP–$MgCl_2$ (2 mM final obtained by mixing equimolar ATP and $MgCl_2$)
   (c) 16 μl 50 nM nucleosomes in nucleosome buffer (10 nM final)
   (d) 32 μl 300 nM SNF2h in BC100 (180 nM final) (see Note 9)

   –ATP reaction

   (a) 16 μl 15 mM $MgCl_2$ (3 mM final)
   (b) 16 μl water

   (c) 16 μl 50 nM nucleosomes in nucleosome buffer (10 nM final)
   (d) 32 μl 300 nM SNF2h in BC100 (180 nM final) (see Note 9)

4. Mix each reaction well and incubate at 30°C for 30 min. Then, transfer sample to the cuvette and place inside fluorometer.
5. Excite sample at the excitation wavelength of the donor fluorescent label (Cy3) and take an emission scan that spans the emission wavelengths of the donor and acceptor (Fig. 1b) (see Note 10).

3.5. Measurement of Remodeling Kinetics

Either donor unquenching or acceptor quenching can be measured in real time. The data shown in Fig. 1c follows Cy3 unquenching of the nucleosome shown in Fig. 1a as a function of SNF2h remodeling. Reactions are initiated by addition of ATP, and data collected at 30°C for at least 10 min and sampled once per second.

1. Prepare the ATP mix (40 μl):
   (a) 16 μl BC100 with 1 mg/ml FLAG peptide (Sigma)
   (b) 8 μl Nucleosome buffer
   (c) 8 μl BC100 without glycerol
   (d) 6 μl 100 mM $MgCl_2$
   (e) 2 μl 120 mM Mg–ATP
2. Set up the reaction mix on ice, keep in the dark (80 μl):
   (a) 16 μl 15 mM $MgCl_2$ (3 mM final)
   (b) 16 μl BC100 buffer
   (c) 16 μl 37.5 nM nucleosomes in nucleosome buffer (5 nM final)
   (d) 32 μl 1.125 μM SNF2h in BC100 with 1 mg/ml FLAG (300 nM final)
3. Immediately after addition of the enzyme in step 2, separately incubate both the ATP stock and reaction mix at 30°C for 5 min in the dark.
4. Add the 80 μl reaction mix to a cuvette that has been prewarmed to 30°C in the fluorometer and perform an emission scan as described in Subheading 3.4.
5. Start collecting time-course data of the 80 μl reaction mix. After verifying that the fluorescence intensity is stable over at least 10 s, pause the data collection and manually mix in the 40 μl of prewarmed ATP mix. Mix well, but avoid introducing air into the sample. Then, resume data recording. The time elapsed between the initiation of mixing and resumption of data acquisition should be accurately recorded; as this time-lag is needed when fitting the data (usually, about 10–15 s).

6. After the Cy3 signal stops increasing with time, the reaction is complete. At this point, a final emission scan can be performed to ensure the proper change in donor and acceptor fluorescence intensities.

The final concentrations for the reaction are as follows: 12 mM HEPES, pH 7.9, 4 mM Tris, pH 7.5, 60 mM KCl, 3 mM $MgCl_2$, 0.32 mM EDTA, 12% glycerol, 0.02% NP-40, 0.4 mg/ml FLAG peptide, 2 mM Mg–ATP, 5 nM nucleosomes, 300 nM SNF2h.

3.6. Fitting Models to the Data and Interpretation

Data are best fit by single or multiple exponentials using a program, such as Kaleidagraph or MATLAB (Fig. 1c). Variables include rate constants as well as fluorescence values for the starting material, any intermediates, and the end product. The fluorescence intensity obtained from the fit at time = zero should be very close to the value obtained from dilution of the reaction mix with the ATP mix. In the above experiment, the fluorescence intensity at time = zero should equal ~2/3 the fluorescence intensity before addition of ATP (80 μl diluted to 120 μl final). The first data point that is collected may differ significantly from the initial value as the reaction may have proceeded partially between manual mixing and data acquisition. Thus, one must also correct for this lag at every time point.

The rate constants obtained from fitting the data only represent the change in fluorophore distances that is detectable by FRET, which is based on the Förster radius of the fluorophores. The fit of data to multiple exponentials may indicate the presence of intermediates in the remodeling reaction or the presence of two populations of nucleosomes, which react with different rates (Fig. 1c). While the absolute value of FRET cannot be used to determine the exact distances between dyes due to orientation effects, this technique can reliably detect changes in distance between dyes.

4. Notes

1. A good control is to perform the labeling in parallel with H2A lacking any cysteines to ensure that labeling is specific to the single cysteine.
2. If fluorescein 5-maleimide or tetramethylrhodamine (TMR) maleimide are being used instead, these two dyes should be dissolved in anhydrous DMSO (100 mM final).
3. Each concentration step takes approximately 30 min for 0.4 ml. The free dye passes through the Microcon filter, but the histone does not, so at the end the free dye concentration is reduced by ~27-fold.
4. The labeled histone may migrate more slowly than unlabeled histone if labeled with fluorescein or rhodamine. The Cy dyes do not alter the mobility significantly.

5. The SYPRO gel can also be used to estimate the concentration of the labeled histone by comparison with a relevant and previously quantified histone standard.
6. A slight excess of H2A and H2B is used to ensure complete octamer formation. This is because octamer and H2A/H2B dimer can be easily resolved on the Superdex200 column, but the H3/H4 tetramer cannot be easily resolved from the octamer (M.W.: H2A – 13,960 Da, H2B – 13,774 Da, H3 – 15,273 Da, H4 – 11,236 Da).
7. For this particular column, the octamer should elute at between ~12 and 13 ml and the dimer between ~15 and 16 ml.
8. Do not discard supernatant until you have confirmed DNA recovery. Unlabeled DNA has a clear pellet that is harder to visualize by eye. Fluorescent labels on DNA result in a colored pellet that is clearly visible for this scale PCR. If no pellet is visible, respin the supernatant and if pellet is still not visible repeat the precipitation.
9. The specific buffer conditions for the remodeling reaction can be altered as needed for other remodeling enzymes.
10. While the excitation maximum for Cy3 is 550 nm, we often excite at 515 nm, which is the smaller excitation peak of Cy3, so as to minimize direct excitation of Cy5.

References

1. Clapier CR, Cairns BR: The biology of chromatin remodeling complexes. *Annu Rev Biochem* 2009, 78:273–304.
2. Lakowicz J: *Principles of Fluoroscence Spectroscopy.*: Springer; 1999.
3. Stryer L: Intramolecular resonance transfer of energy in proteins. *Biochim Biophys Acta* 1959, 35:242–244.
4. Yang JG, Narlikar GJ: FRET-based methods to study ATP-dependent changes in chromatin structure. *Methods* 2007, 41:291–295.
5. Luger K, Mader AW, Richmond RK, Sargent DF, Richmond TJ: Crystal structure of the nucleosome core particle at 2.8 A resolution. *Nature* 1997, 389:251–260.
6. Davey CA, Sargent DF, Luger K, Maeder AW, Richmond TJ: Solvent mediated interactions in the structure of the nucleosome core particle at 1.9 a resolution. *J Mol Biol* 2002, 319: 1097–1113.
7. Park YJ, Dyer PN, Tremethick DJ, Luger K: A new fluorescence resonance energy transfer approach demonstrates that the histone variant H2AZ stabilizes the histone octamer within the nucleosome. *J Biol Chem* 2004, 279: 24274–24282.
8. Li G, Levitus M, Bustamante C, Widom J: Rapid spontaneous accessibility of nucleosomal DNA. *Nat Struct Mol Biol* 2005, 12:46–53.
9. Yang JG, Madrid TS, Sevastopoulos E, Narlikar GJ: The chromatin-remodeling enzyme ACF is an ATP-dependent DNA length sensor that regulates nucleosome spacing. *Nat Struct Mol Biol* 2006, 13:1078–1083.
10. Dyer PN, Edayathumangalam RS, White CL, Bao Y, Chakravarthy S, Muthurajan UM, Luger K: Reconstitution of nucleosome core particles from recombinant histones and DNA. *Methods Enzymol* 2004, 375:23–44.
11. Luger K, Rechsteiner TJ, Richmond TJ: Expression and purification of recombinant histones and nucleosome reconstitution. *Methods Mol Biol* 1999, 119:1–16.
12. Lee KM, Narlikar G: Assembly of nucleosomal templates by salt dialysis. *Curr Protoc Mol Biol* 2001, Chapter 21:Unit 21 26.

Chapter 21

Preparation of Nucleosomes Containing a Specific H2A–H2A Cross-Link Forming a DNA-Constraining Loop Structure

Ning Liu and Jeffrey J. Hayes

Abstract

ATP-dependent chromatin-remodeling complexes use the energy of ATP hydrolysis to alter nucleosome structure, increase the accessibility of trans-acting factors, and induce nucleosome movement on the nucleosomal DNA. Recent studies suggest that bulge propagation is a major component of the mechanism for SWI/SNF remodeling. We describe in detail a method to prepare a mononucleosomal substrate in which the two H2A N-terminal tails are cross-linked in an intranucleosomal fashion, forming a closed loop around the two superhelical winds of DNA. This substrate is useful for researchers who wish to test processes in which the DNA is transiently or permanently lifted off the histone surface, such as in the bulge propagation model. Our method allows assessment of the extent of cross-linking within the population of nucleosomes used in small-scale experiments, such as assays to test SWI/SNF-remodeling activities.

Key words: Intra-histone cross-linking, SWI/SNF, Nucleosome remodeling, Bulge propagation

1. Introduction

Packaging of genome DNA into nucleosomes and higher order chromatin structures in the eukaryotic cell nucleus greatly inhibits the availability of DNA for various nuclear processes, such as transcription, replication, and DNA repair, due to the decreased accessibility of the DNA to *trans*-acting factors (1). Eukaryotic cells have developed several mechanisms to modulate chromatin structures and increase the accessibility of DNA (2), one of which involves nucleosome-remodeling complexes that use energy derived from ATP hydrolysis to rearrange chromatin structure (3–5). Remodelers can recognize epigenetic marks and modulate chromatin structure through histone ejection, dimer replacement, dimer removal, and sliding along the DNA (4, 6).

Randall H. Morse (ed.), *Chromatin Remodeling: Methods and Protocols*, Methods in Molecular Biology, vol. 833,
DOI 10.1007/978-1-61779-477-3_21,

The multiple subunit SWI/SNF complex is one of the most extensively studied chromatin-remodeling complexes and was originally identified and purified from yeast as required for the activation of a variety of genes (7, 8). A related complex was also purified from human cells (9). The catalytic subunit of this complex contains a DNA-stimulated ATPase activity, which belongs to the SNF2 superfamily (10). The Cairns group showed that the ATPase domain exhibits a 3′-5′ DNA translocase activity while others showed that the complex can create torsional stress in DNA fragments DNA which is likely related to the nucleosome mobilization process (11–13). However, the precise mechanism by which SWI/SNF alters the accessibility of nucleosomal DNA and induces nucleosome mobilization is not well-understood. One mechanism, originally proposed by van Holde and Yager (14), is known as the twist diffusion model and is supported by nucleosome crystal structure studies (15, 16). The data suggest that a "twist defect" generated by gain or loss of as little as one base pair of DNA can be tolerated internally within the nucleosome without disrupting DNA–histone interactions. If this twist defect could propagate through the entire segment of nucleosomal DNA, the histone octamer would shift along the DNA by 1 bp. However, remodeling of nucleosomes containing DNA nicks, branched DNA, DNA hairpins, and DNAs attached to beads, has shown that nucleosome mobilization is not prevented by these modifications, suggesting that pure twist diffusion is unlikely to be the major mechanism of mobilization (17–20). On the other hand, recent single-molecule studies have provided evidence for remodeling intermediates in which a loop or a bulge is generated within the nucleosome, resulting in octamer sliding on the DNA (21, 22). However the detailed information, such as the size of the generated loops or the position of the loops compared to the nucleosome, has not been well-characterized. To further examine the bulge propagation model, we designed a nucleosome substrate with a modification that sterically hinders formation or passage of a DNA loop.

Previous DNA–histone cross-linking experiments have shown that the H2A N-terminal tails interact with nucleosomal DNA on the back of the nucleosome, opposite where the nucleosome dyad passes through the DNA helix (Fig. 1) (23). We describe an experimental system to prepare nucleosomes in which the two H2A N-terminal tails are cross-linked together via reaction with the bifunctional cross-linker, $BM[PEO]_4$. We show that the H2A tails in nucleosomes reconstituted with H2A G2C are appropriately positioned to be efficiently cross-linked in an intranucleosomal fashion with this reagent. Cross-linking forms a closed proteinaceous loop that encompasses the two superhelical winds of DNA around the nucleosome and is thus expected to inhibit the formation and/or propagation of a DNA bulge on the nucleosome surface and has potential use in the analysis of the mechanism of ATP-dependent nucleosome remodeling by SWI/SNF. Our method balances the

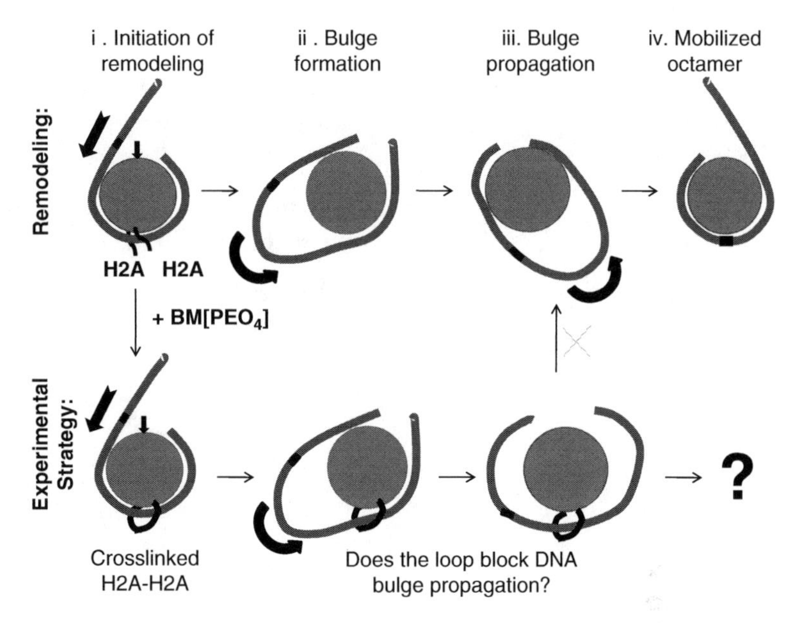

Fig. 1. Bulge propagation model and experimental strategy. In the bulge propagation model, the remodeling activity forms a transient DNA loop on the histone octamer surface. The loop can be propagated around the surface of the nucleosome, ultimately dissipating out the opposite side of the nucleosome. This advances (mobilizes) the histone octamer along the DNA by the amount of DNA in the loop. The two cysteine residues on the H2A N-terminal tails within the same nucleosome are cross-linked together by reaction with the bifunctional cross-linker BM[PEO]$_4$. The effect of cross-linking can be determined by restriction enzyme accessibility and mobility assays.

need to prepare sufficient quantities of material to allow determination of the extent of cross-linking by protein analyses within the actual fraction of nucleosomes to be remodeled with the fact that typical preparations of SWI/SNF enzyme are able to remodel only ng quantities of nucleosomes. Cross-linking is carried out using a nucleosome dilution and cross-linker concentration that maximize bidentate intranucleosome cross-linking over internucleosome reactions. We describe incorporation of radiolabel within the DNA either before or after nucleosome preparation, and a restriction enzyme assay and nucleosome mobilization assay to analyze remodeling activities.

2. Materials

2.1. Nucleosome Preparation with DNA Labeled at Both Ends

2.1.1. DNA Preparation

1. DNA template to prepare fragment for nucleosomal reconstitution.
2. 10× Taq polymerase reaction buffer: 200 mM Tris–HCl, pH 8.8, 100 mM $(NH_4)_2SO_4$, 100 mM KCl, 20 mM $MgSO_4$, 1% Triton X-100.
3. A mixture containing all four dNTPs at 10 mM for each.

4. Oligonucleotide primers.
5. Taq DNA polymerase: 5,000 U/ml.
6. Thermocycler and appropriate microtubes.
7. 3 M sodium acetate stock, pH 5.2.
8. 30 ml Corex tubes, previously siliconized and cleaned; appropriate super-speed rotor and centrifuge.
9. 95% ethanol, –20°C.
10. 70% ethanol, –20°C.
11. TE buffer: 10 mM Tris–HCl, pH 8.0, 1 mM EDTA (store at room temperature after filter sterilizing).
12. UV spectrophotometer.

2.1.2. Core Histone Protein Preparation

1. Expression plasmid for recombinant histone H2A and H2B proteins.
2. Lysis buffer: 10 mM EDTA, 0.5% Triton X-100, 0.2 mg/ml lysozyme, 0.25 mM PMSF.
3. HCl.
4. 10 M solution of NaOH .
5. Bio-Rex 70 Resin 50–100 (Bio-Rad).
6. β-mercaptoethanol.
7. Purified histones H3 and H4.

2.1.3. Large-Scale Nucleosome Reconstitution

1. 5 M NaCl.
2. Cold dialysis buffer: 2L TE solutions containing 1.2, 1.0, 0.8, and 0.6 M NaCl (store at 4°C after filter sterilizing).
3. Slide-A-Lyzer Dialysis Cassettes (Pierce Biotechnology): 10K MWCO, 0.5–3 ml.
4. Sterile 10-ml syringe and syringe needles (supplied with Dialysis Cassette).

2.1.4. Nucleosome Gradient Purification

1. Agarose gel: 0.7% agarose gel prepared with 0.5× TBE buffer (1× is 89 mM Trizma base, 89 mM boric acid, 2 mM EDTA).
2. Stock of glycerol solution for gradient: 5 and 30% glycerol, 10 mM Tris–HCl, 1 mM EDTA, 10 mM β-mercaptoethanol. Solution is stored at 4°C, shake well before use.
3. Beckmann Ultra-Clear centrifuge tube (14×89 mm).
4. BSA, 0.2 mg/ml.
5. Fraction collector.

2.1.5. H2A–H2A Intranucleosome Cross-Linking

1. Ultrafiltration microfuge spin columns, 10,000 MWCO.
2. TE buffer: 10 mM Tris–HCl, pH 8.0, 1 mM EDTA, freshly made and prechilled at 4°C.

3. Centriprep Centrifugal Filter Unit with Ultracel-10 membrane (Millipore).
4. TE solution containing 10% glycerol: Store at 4°C.
5. BM[PEO]$_4$: 50 mg dry compound is obtained from commercial supplier (Pierce) and stored at 4°C. Stock solution (10 mg/ml) is aliquoted into 500 μl/tube and kept frozen at −80°C. Make diluted solutions fresh just before use and keep on ice.
6. Western blotting apparatus.
7. Nonfat dry milk.
8. Anti-Histone H2A antibody, acidic patch (LP BIO): Prepare a 1:500 dilution with PBS buffer before use and keep on ice.
9. Anti-Rabbit IgG alkaline phosphatase conjugate: Use at 1:10,000 dilution.
10. ECF substrate for Western Blotting.
11. PVDF membrane for Western Blotting.

2.1.6. Nucleosome Labeling and Purification

1. Modified 10× T4 polynucleotide kinase buffer: 700 mM Tris–HCl, pH 7.6, 20 mM $MgCl_2$, 50 mM DTT.
2. T4 polynucleotide kinase: 10,000 U/ml.
3. [γ-^{32}P] ATP: 6,000 Ci/mmol.
4. Microfuge concentrator apparatus (Vivaspin 500, Sartorius Stedim Biotech recommended).
5. 40% (w/v:19:1) acrylamide:*N,N'*methylene-*bis*-acrylamide solution.

2.2. H2A–H2A Intra-cross-linked Nucleosome with a Single Unique Radioactive DNA End Label

2.2.1. DNA Fragment Preparation and Labeling

1. Bacterial plasmid preparation via commercial kit or standard protocol.
2. *Eco*RI (20,000 U/ml); *Hin*dIII (20,000 U/ml).
3. Alkaline phosphatase (10,000 U/ml); polynucleotide kinase (PNK) (20,000 U/ml).
4. Stock solution of 10% sodium dodecyl sulfate (SDS).
5. Phenol/chloroform/isoamyl alcohol (25:24:1): Stored at 4°C in dark bottle.
6. 3 M NaOAc, 95 and 70% ice-cold ethanol.
7. [γ-^{32}P] ATP: 6,000 Ci/mmol.
8. 40% (w/v:19:1) acrylamide:*N,N'*methylene-bis-acrylamide solution.
9. Phosphorescent markers: Kept in dark.

2.3. Nucleosome Remodeling

1. Purified remodeling complex (e.g., ref. 24).
2. *Hha*I or other appropriate restriction enzyme from commercial supplier (5,000–20,000 U/ml).

3. 10× remodeling buffer: 100 mM Tris–HCl, pH 8.0, 500 mM $MgCl_2$, 50 mM MgCl2. 10 mM DTT, and 1 μg/μl BSA are prepared separately from 10× remodeling buffer. Keep at –20°C.
4. 10 mM ATP kept at 4°C.
5. 6× SDS DNA loading buffer: 0.25% bromophenol blue; 0.25% xylene cyanol FF; 30% glycerol. Store at 4°C.
6. Stopping buffer: 50% glycerol, 150 ng/μl calf thymus DNA; store at 4°C.
7. 40% (w/v:35.36:1) acrylamide: *N*,*N'*methylene-bis-acrylamide solution.

3. Methods

The following protocols include the reconstitution of nucleosomes in vitro, detailed information regarding H2A–H2A cross-linking within the nucleosome, and nucleosome purification. A nucleosome preparation in which the vast majority of H2As are cross-linked in an intranucleosome fashion is critical for the analysis of the SWI/SNF-remodeling mechanism. Therefore, we optimized the conditions to bring cross-linking efficiency to near 95%. The protocol also describes a new method of radiolabeling DNA ends within a reconstituted nucleosome. We obtain a specific activity of about 3,000 cpm/ng radiolabeled nucleosomes, which is suitable for detection in the assays described.

SWI/SNF mobilizes nucleosomes and enhances DNA accessibility in an ATP-dependent manner. We also describe detection of nucleosome movement by a gel-based mobilization assay, and by monitoring accessibility of nucleosomal DNA by restriction enzyme accessibility (REA).

3.1. Nucleosome Preparation with DNA Labeled at Both Ends

3.1.1. DNA Preparation

Any DNA fragment can be used that is long enough to harbor one nucleosome (≥147 bp). For our experiments, we employed a 343-bp DNA fragment containing a centrally located 601 nucleosome-positioning sequence, derived from the plasmid CP1024 (generous gift of Craig Peterson, U Mass Worcester).

1. Design two primers to PCR yield a DNA fragment from the template DNA of length 200–350 bp with a centrally located nucleosome-positioning sequence. Longer fragments may be used, but reconstitution of a single nucleosome species becomes problematic.
2. In a PCR tube, add the following:
   (a) 50 ng template DNA
   (b) 10 μl 10× Taq polymerase reaction buffer

(c) 2 μl of dNTP mix (10 mM each)

(d) 100 pM of primers

(e) 2 μl of $MgCl_2$ (50 mM)

(f) 1 U Taq, added just before the start of the PCR

(g) ddH_2O to a final of 100 μl

3. Set up the thermocycler program:

(a) One round: 2 min at 95°C

(b) Thirty-five rounds: 30 s at 95°C, 30 s at 55°C, 30 s at 72°C

(c) One round: 5 min at 72°C

4. Monitor the product of the reaction by running 10 μl on an agarose gel. If a single strong band is observed, indicating that the PCR works, prepare enough mix for 200–300 PCRs and repeat the PCR. After the PCRs, combine all reactions together for a total volume of ~30 ml. Precipitate the DNA by adding 3 ml of 3 M NaOAc and distribute 5.5 ml into 6 30-ml Corex tubes (previously siliconized). Add 2.5 volumes (13.5 ml) of 95% ethanol, vortex to mix, and centrifuge immediately at 15,000 ×*g* for 30 min at 4°C. Carefully remove the supernatant and discard, wash the DNA pellet with ice-cold 70% ethanol, dry under vacuum (if available, otherwise, air dry), and resuspend the DNA in 500 μl of TE buffer. The concentration of DNA is measured by measuring absorbance at 260 nM.

3.1.2. Core Histone Protein Preparation

1. Wild-type *Xenopus* H2A and H2B proteins are expressed in bacteria cells and purified as described in 25. Briefly, after expression, resuspend the cells with 20 ml lysis buffer and incubate on ice for 30 min in a 50-ml conical tube. Nucleic acids are removed by adding HCl to the mixture to a final concentration of 0.4 M and incubating on ice for 30 min. Insoluble material is removed by centrifugation at 13,000×*g* (avg. RCF) for 15 min. The clear supernatant is carefully removed from the tube and neutralized with 10 M NaOH to make the final pH 8.0 (see Note 1). Adjust the NaCl concentration in the solution to 1 M with TE buffer. The proteins in this solution are stable for several months when stored at −80°C.

2. H2A/H2B dimers are purified by mixing 1:1 stoichiometric ratios of H2A and H2B as determined by quantitative 15% SDS-PAGE, and the complexes purified by chromatography with Bio-Rex resin beads (50–100) exactly as described in 26. The fractions containing the purified dimers are determined by running samples on 15% SDS-PAGE.

3. In order to generate nucleosomes in which two H2A N-terminal tails are cross-linked together, we generated the H2A mutant H2A G2C containing a glycine-to-cysteine substitution at position 2. The coding sequence for H2AG2C was generated using

the stratagene QuikChange Site-Directed mutagenesis kit. The mutant dimer was purified as described above. Note that in order to protect the sulfhydryl group from oxidation, a final concentration of 10 mM β-mercaptoethanol is included in all steps, except the wash and elution buffers. Purified dimer is frozen with dry ice immediately after elution from the column. The wild-type tetramer (H3/H4) is purified from chicken erythrocyte nuclei as described in 26. Alternatively, recombinant H3/H4 can be used (see ref. 7).

3.1.3. Large-Scale Nucleosome Reconstitution

We describe a method for the large-scale reconstitution of homogenous nucleosomes that can be used for cross-linking and labeling. This allows determination of the cross-linking efficiency within the actual sample used in the remodeling experiment. Classic salt dialysis is used for reconstitution of nucleosome containing either wild-type H2A/H2B dimer or H2AG2C/H2B (27).

1. Combine 500 μg of PCR-generated DNA fragments, 232 μg dimer, and 168 μg H3/H4 together with 800 μl 5 M NaCl and add the appropriate volume of TE containing 10 mM DTT to adjust the total volume to 2 ml (see Note 2).
2. After incubation at 37°C for 30 min, carefully transfer the solution to a prehydrated Slide-A-Lyzer Dialysis Cassette (10K MWCO, Pierce Biotechnology) with a syringe by carefully following the manufacturer's specifications. Note that when working with this cassette, it is important to use caution to avoid contacting the cassette membrane with the needle and not to allow the membrane to dry out during the process.
3. Nucleosomes are reconstituted by gradually decreasing the salt concentration in the dialysis buffer containing 10 mM β-mercaptoethanol through five rounds of dialysis at 4°C (26). Typically, 2 L of buffers containing TE/10 mM β-mercaptoethanol and 1.2, 1.0, 0.8, and 0.6 M NaCl are used, with a final dialysis into 2 L of TE/10 mM β-mercaptoethanol overnight.

3.1.4. Nucleosome Gradient Purification

1. The next day, carefully remove the reconstitution solution from the cassette with a 10-ml syringe and transfer to a 5-ml conical sterilized tube that is prechilled on ice. Measure the solution volume, which is typically around 2.2 ml, using a 5-ml pipet. Determine the extent of nucleosome reconstitution by adjusting ~5 μl of the reconstitution to 5% glycerol and loading onto a native agarose gel (0.7% agarose gel, 0.5× TBE) using a running time of ~2.5 h at 120 V at RT.
2. To remove free histones, free DNA and non-nucleosomal DNA–histone complexes from the reconstitution and purify the nucleosomes by glycerol gradient. Each batch of 100 μg nucleosomes (400 μl) is loaded onto 10 ml of 5–30% glycerol

gradients (10 mM Tris–HCl, 1 mM EDTA, and 10 mM β-mercaptoethanol) and centrifuged at 15,000 × *g* (34 krpm in the SW41 Ti rotor) for 15–18 h at 4°C.

3. Gradients are fractionated by collecting 18 fractions (~600 μl/each) into tubes pretreated with BSA (see Note 3). Fractions are analyzed by running a 0.7% native agarose gel (1/2× TBE) that is stained with ethidium bromide to visualize nucleosome DNA.
4. The nucleosome-containing fractions are combined together (total ~3 ml). Gradient-purified nucleosome fractions contain at least 10% glycerol and so can be safely stored at −80°C for several months.

3.1.5. H2A–H2A Intranucleosome Cross-Linking

In order to cross-link two cysteine residues on the H2A N-terminal tails together with the cross-linker, reducing agent has to be removed from the nucleosome reconstitution. We developed two ways to do this.

Dialysis Method

1. Condense the ~3 ml nucleosome solution to 500–700 μl by using a centrifugal filtration unit (Vivaspin 6, 10K MWCO) spun at 10,000 × *g* for about 20 min at 4°C (see Note 4).
2. Transfer the condensed solution to a new tube and keep on ice. To get optimal recovery, after removal of the retained nucleosome solution, thoroughly rinse the filter membrane with 100 μl TE buffer and combine with the previous solution.
3. Next, apply this solution into a new Slide-A-Lyzer Dialysis Cassette (10K MWCO) and dialyze for four rounds against fresh 1 L TE (pH 8.0) buffer for 2 h each at 4°C (see Note 5). The concentration of β-mercaptoethanol in the reconstitution should be diluted from 10 mM to ~10 fM, which does not affect the cross-linking experiment.
4. These samples can be used directly for cross-linking or immediately mixed with prechilled glycerol to 10% final concentration, frozen on dry ice, and stored at −80°C until use. In our experience, the cysteine residues on H2A maintain their activity for at least several weeks when kept frozen.

Centrifugation Method

Alternatively, the reducing agent can be removed by a centrifugation/filtration method.

1. Place the ~3 ml of nucleosome peak fractions into the chamber of an Amicon YM-10 concentrator, add fresh, cold 13 ml of TE containing 10% glycerol, and then carefully mix by pipeting. Replace the interior chamber into the unit and spin the concentrator in a large-bore rotor (unit does not have to fit snugly) at no more than 1,000 × *g* (2,500 rpm in Sorvall GSA type rotor) for 1 h at 4°C.

2. Remove the waste solution from the interior chamber by using a 5-ml pipet and repeat the centrifugation for an additional two times. Note that of the original 16 ml sample, which contains 10% glycerol, the waste solution to be removed after each of the three centrifugations is about 6, 5, and 3 ml, respectively. After this process, about 2 ml of reconstitution is left in the big chamber.
3. Add an additional 14 ml of TE buffer containing 10% glycerol and repeat the centrifugation steps as above. This centrifuge step is repeated eight more times.
4. Finally, spin the solution for additional 30 min and condense to a final volume of ~700 μl–1 ml. The concentration of β-mercaptoethanol after this procedure should be diluted to less than 10 pM. Each centrifugation process takes 3–4 h to obtain the 2 ml final sample, so the total procedure takes 2–3 days to finish (see Note 6).

H2A–H2A Intranucleosome Cross-Linking

The maleimide groups of the cross-linker, $BM[PEO]_4$, react specifically with the cysteine sulfhydryl group and the cross-linking reaction is not reversible. Our previous cross-linking studies (27) indicate that the cysteine residues on the two H2A N-terminal tails in nucleosomes reconstituted with H2A G2C are spaced approximately 20 Å apart and thus are appropriately positioned for cross-linking by $BM[PEO]_4$. This cross-linking should create a hindrance for the formation of a DNA bulge or loop on the surface of the nucleosome at the location of the tail domains.

1. Dissolve 50 mg of $BM[PEO]_4$ with 5 ml ddH_2O to make the final concentration of 10 mg/ml (28 mM). Add 3 ml of ddH_2O to the original bottle and warm up the whole bottle in 37°C water bath for about 10 min to increase the solubility. Mix by vortexing and transfer the fully resuspended solution to a 15-ml sterile conical tube and bring the volume up to 5 ml with ddH_2O. To avoid multiple freeze/thaw cycles, aliquot the stock 500 μl/tube and keep frozen at –80°C. Diluted solutions are made fresh just before use and kept on ice.
2. Thaw a sample of nucleosomes from which the β-mercaptoethanol has been depleted and keep on ice. Determine the concentration of nucleosomes by absorption spectroscopy (using a 1:200 μl dilution) at 260 nm, which should be ~400 ng/μl. The cross-linker is diluted with fresh ddH_2O to 2.28 μM just before use.
3. In theory, exactly one cross-linker/nucleosome should give the highest cross-linking efficiency. To empirically determine the best molar ratio of nucleosome to cross-linker, 2 μg (~5 μl) of nucleosomes are mixed with diluted cross-linker at increasing

ratios and carefully mixed by pipeting, and the cross-linking reactions are then incubated for 1 h at RT or 30 min at 37°C (Fig. 2b).

4. Stop the reaction by adding DTT to 10 mM final concentration and analyze the cross-linking on a 15% SDS-PAGE gel. We find that H2A N-terminal tails are cross-linked by $BM[PEO]_4$ in our reaction conditions. The cross-linked H2A–H2A species migrates to a higher position than monomeric H2A on the gel. Most efficient cross-linking (95%) occurs when the apparent nucleosome:cross-linker ratio is 1:2, so we performed subsequent cross-linking experiments at this ratio (Fig. 2c).
5. Next, a large-scale cross-linking reaction can be carried out. This reaction is performed in the same manner, except that in order to ensure bidentate intranucleosome cross-linking the free cross-linker is kept at concentration well below that of the unreacted nucleosome by titrating in the cross-linker in four separate additions (see Note 7). Mix 200 μg (400 ng/μl) of nucleosome with 7.5 μl of 1,000-fold diluted $BM[PEO]_4$ (28.3 μM) and incubate at 37°C for 30 min. At this point, the nucleosome:cross-linker ratio is 0.5:1.
6. Repeat this step three times. The final nucleosome:cross-linker ratio is 1:2. Stop the reaction as described above followed by analysis of a small sample on 15% SDS-PAGE and Coomassie staining (Fig. 2d). The cross-linking efficiency is about 95%, and we find that glycerol in the nucleosome solution does not affect the cross-linking efficiency.
7. Cross-linking can also be analyzed by Western blot using anti-H2A antibodies (Fig. 2e) according to the manufacturer's instructions. Separate 2 μg of uncross-linked and cross-linked nucleosomes on 15% SDS-PAGE. Proteins on the gel are transferred to PVDF membrane by electrophoresis at 100 V for 10 min, and then 60 V for 20 min. The membrane is blocked with 5% nonfat milk for 30 min and then blotted with 10 ml PBS buffer containing 500-fold diluted anti-H2A antibody overnight at 4°C. After washing with PBS for three times, the membrane is incubated with the secondary antibody for 1 h at 4°C. The result is detected by ECF. In our experience, the cross-linked H2A–H2A protein species typically exhibits a reduced Coomassie staining band intensity compared to normal H2A, but shows a relatively more prominent band compared to monomeric H2A when blotted by anti-H2A antibody. The cross-linked nucleosomes can be stored at 4°C for several weeks.

3.1.6. Nucleosome Labeling and Purification

We developed a method to label DNA ends on the reconstituted nucleosome with [γ-^{32}P] ATP. Due to the fact that the 5′ end of each strand of the PCR-generated DNA does not bear 5′-terminal phosphates, nucleosome DNA can be radiolabeled at both ends by kinase without prior phosphatase treatment.

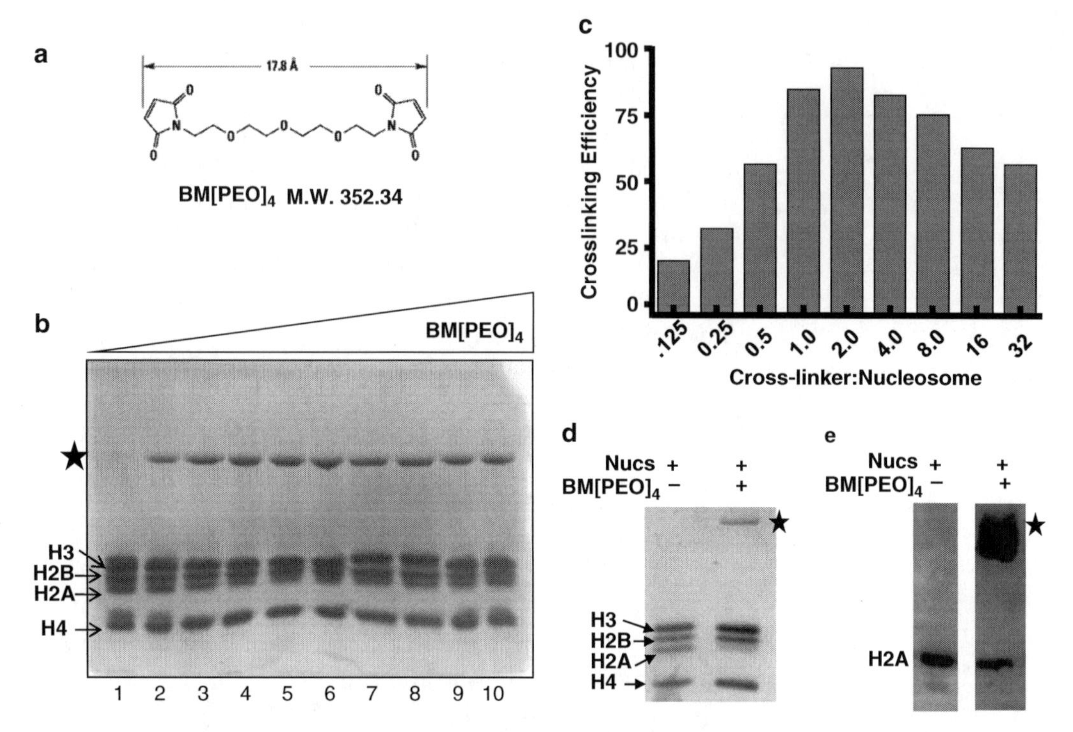

Fig. 2. H2A–H2A intranucleosome cross-linking. (**a**) Diagram of $BM[PEO]_4$. (**b**) Extent of cross-linking assayed by 15% SDS-PAGE. Gradient-purified 601 nucleosomes (~2 μg) were reacted with increasing amounts of $BM[PEO]_4$ at 37°C for 30 min, the reaction stopped by addition of DTT, and cross-linking analyzed by SDS-PAGE and Coomassie blue staining. Lane 1 is a no-cross-linking control, lanes 2–10, cross-linking was performed with nucleosome:cross-linker ratios of 1:0.125, 1:0.25, 1:0.5, 1:1, 1:2, 1:4, 1:8, 1:16, and 1:32, respectively, incubated at 37°C for 30 min. Reactions were stopped by 10 mM DTT final concentration. Samples were mixed with 2× SDS loading buffer and heated at 95°C for 10 min. After a quick spin, samples are loaded directly onto the 15% SDS-PAGE and electrophoresed at 120 V for 2.5 h at RT, and then the gel is stained with Coomassie blue. Star indicates the position of cross-linked H2A–H2A products. (**c**) Cross-linking efficiency was quantified from the gel in (**b**). (**d, e**) Efficient H2A–H2A cross-linking within 601 nucleosomes. Cross-linking was performed by adding four successive 7.5 μl volumes of 28 μM $BM[PEO]_4$ to 200 μg nucleosomes as described in the text. Five microliters of product was removed and cross-linking analyzed by 15% SDS-PAGE followed by Coomassie staining (**d**) or Western Blotting (**e**).

1. Determine concentrations of uncross-linked and cross-linked nucleosomes by absorption spectroscopy as described above. In a 20 μl reaction, incubate 500 ng of nucleosome with 2 μl of 1× PNK reaction buffer (see Note 8), 6 μl of [γ-^{32}P] ATP (6,000 Ci/mmol), and 10 U of PNK for 30 min at 37°C.
2. After labeling, remove free [γ-^{32}P] ATP by centrifugation and filtration. Add 480 μl of TE buffer to the reaction and then apply the mixture to a Vivaspin 500 Centrifugal Concentrator spun at 12,000 × *g* for about 4 min (see Note 9). Carefully remove the retained 100 μl of radiolabeled nucleosome solution.
3. Purify nucleosomes by glycerol gradient centrifugation at 15,000 × *g* (34 krpm in an SW41 Ti rotor) for 15–18 h at 4°C and identify nucleosome fractions by electrophoresis on a 0.7%

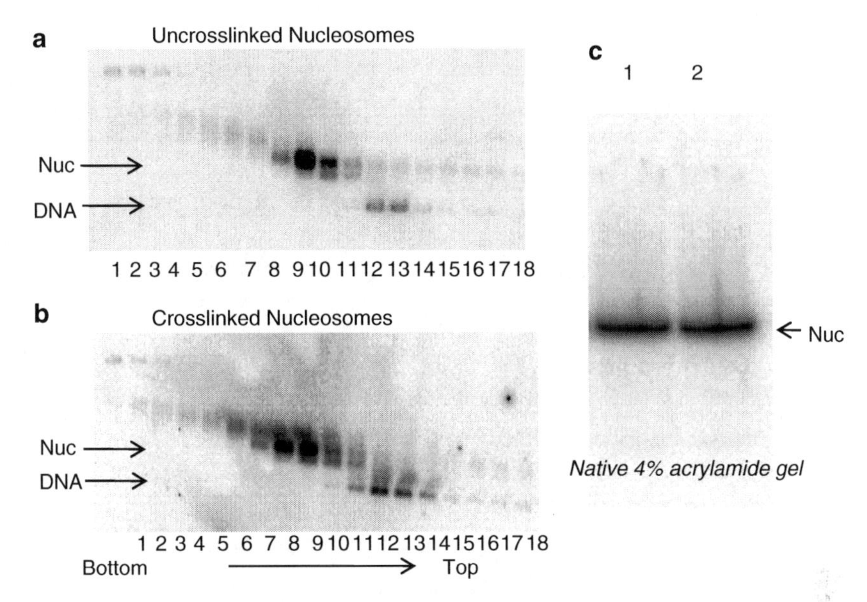

Fig. 3. Nucleosome purification. (**a**, **b**) Glycerol gradient purification. Radiolabeled uncross-linked and cross-linked nucleosomes were separated on 5–30% glycerol gradients and fractionated as described in the text. *Lanes 1–18* show consecutive fractions (600 μl/each) collected from bottom to top of the gradient. Ten microliters of each fraction was loaded to native a 0.7% native agarose (0.5× TBE) gel and electrophoresed at 120 V for 2 h at RT. Then, the gel was dried and analyzed by phosphoimagery. *Lane 9* in (**a**) and *lanes* 8 and 9 in (**b**) were the nucleosome-containing fractions. (**c**) Polyacrylamide gel diagnostic for nucleosome translational positions. *Lane 1*, gradient-purified 601 nucleosome. *Lane 2*, gradient-purified cross-linked nucleosomes. Specifically, 10 μl of fraction 9 in (**a**, **b**) were run on a native 4% acrylamide gel run at 120 V for 2 h at RT.

native agarose gel as described above. Fractions containing radiolabeled nucleosome are stored at −80°C until used for remodeling experiments. As shown in (Fig. 3a, b), we find that nucleosomes are successfully and efficiently radiolabeled in this way regardless of cross-linking. Note that purified cross-linked nucleosomes exhibit the same sedimentation pattern in gradients as uncross-linked nucleosomes, suggesting that cross-linking occurs intra- rather than internucleosomally. Also the sedimentation pattern indicates that cross-linking does not significantly alter the native conformation or stability of the nucleosome. Additional support for this conclusion comes from the fact that cross-linked and uncross-linked nucleosomes exhibit identical migration rates through native polyacrylamide gels (4%, 0.5× TBE ran at 120 V for 2 h) (Fig. 3c).

3.2. Preparing H2A–H2A Cross-Linked Nucleosomes with DNA Containing a Unique Radioactive End Label

Having both ends of the DNA fragment radiolabeled increases specific activity of the DNA fragment and thus improves detection of templates in restriction enzyme accessibility and nucleosome mobility assays. However, a single-DNA end-labeled nucleosome substrate is required to test exact nucleosome translocation position on the DNA by footprinting methods, such as the *Exo*III digestion assay.

3.2.1. DNA Fragment Preparation and Labeling

A DNA fragment is prepared for nucleosome reconstitution. If cross-linked nucleosomes are to be used to analyze ATP-dependent remodeling-induced changes in nucleosome translational position, the fragment needs to be long enough to allow alternative translational positions to be formed and distinguished via the gel assay but not so long so as to allow more than one nucleosomes to be formed during reconstitution. We have found that fragment sizes in the 220–350 bp range are ideal for this purpose. For remodelers, such as the SWI/SNF complex that move nucleosomes from the center to the ends of DNA fragments, PCR primers should be designed to place a nucleosome-positioning element in the center of the resulting fragment. In the current example, we generated a 343-bp fragment excised from the plasmid CP1024 using *Eco*RI and *Hin*dIII. The fragment is purified and uniquely end radiolabeled using a standard protocol.

1. Purify the plasmid DNA. We typically employ the Qiagen Plasmid Mega Kit. Digest 10 μg of plasmid with 20 U of the appropriate restriction enzyme according to the manufacturer's instructions. It is most convenient if a single-cutter enzyme is employed at this step since each cut results in two 5′ end-labeling events. Precipitate the DNA and rinse with 70% ethanol as described above.
2. Resuspend the digested plasmid DNA in phosphatase buffer and treated with 20 U of alkaline phosphatase for 45 min at 37°C. Stop the reaction by adding SDS to 1% final concentration and then adjust the reaction volume to 200 μl with TE buffer. Remove the phosphatase by phenol/chloroform extraction as follows. Add 200 μl of phenol/chloroform/isoamyl alcohol (25:24:1) and vortex vigorously, and then spin for 6 min at 4°C at 15,000 ×*g* (13 krpm) in the microfuge. Carefully remove ~90% of the upper aqueous layer and place into a new microfuge tube. Add 100 μl of TE back to the organic layer and repeat the extraction. Combine these two upper aqueous layers and add 200 μl of phenol/chloroform, and repeat the extraction once more to completely remove the phosphatase (see Note 10). Add 3 M NaOAc to bring the solution to a 0.3 M final concentration and 2.5 volumes of ice-cold 95% ethanol, vortex, and then spin at 15,000 ×*g* for 30 min in a microfuge. Remove the supernatant carefully with a transfer pipette, being careful to avoid disturbing the DNA pellet. Resuspend the pellet in 180 μl of TE, add 20 μl of 3 M NaOAc and 550 μl of ice-cold 95% ethanol, and repeat the precipitation step once more. Carefully rinse the DNA pellet with 70% ethanol without agitation and dry completely in a Speedvac. Resuspend the DNA in 15 μl TE.
3. The following steps require a Plexiglass shield to protect the researcher from radioactivity. To 5′ end label the DNA, add 3 μl 10× PNK buffer, 12 μl of [γ-^{32}P] ATP (6,000 Ci/mmol),

and 20 U (1 μl) of PNK to the sample and incubate the reaction at 37°C for 30 min. Terminate the reaction and precipitate the DNA by standard ethanol precipitation as described in step 2, being careful to discard the supernatants into the 32P waste. Resuspend the labeled DNA in 27 μl TE buffer and transfer into a new tube. Note that all contaminated materials should be properly discarded into the 32P waste.

4. Digest the end-labeled plasmid DNA with 20 U of the appropriate restriction enzyme (~1 μl enzyme along with 3 μl manufacturer's 10× digestion buffer) for 1 h in digestion buffer. The enzyme does not have to be a single cutter, but does have to cut at the appropriate distance from one of the end labels and nucleosome-positioning element so as to release the desired radiolabeled fragment and also cut an appropriate distance from the other labeled end so as to release a fragment that can be resolved from the desired fragment on the preparative gel. In the example here, a 343-bp DNA fragment containing a centrally located 601 nucleosome-positioning element is released from the plasmid by digestion with *Hin*dIII. Since both enzymes are single cutters, the other fragment is >2.5 kbp in length and thus is easily separated from the 343-bp fragment. Terminate the reaction by adding SDS loading dye and load the digest onto a preparative 5% polyacrylamide gel containing 0.5× TBE and 0.02% SDS. Run the gel at 120 V for 2 h at room temperature to separate the radiolabeled 343-bp fragment from the remainder of plasmid (see Note 11).
5. When the gel is finished running, the wet gel is wrapped with plastic wrap (e.g., Saran wrap) and exposed to X-ray film for 60 s. The position of 343-bp radiolabeled fragment is determined by comparison with radiolabeled DNA marker and attachment of a light-sensitive phosphorescent tape (e.g., from Stratagene Corp.) to the wrapped gel. Excise the gel slice containing the appropriate-sized DNA band with a clean razor blade and place in a fresh microfuge tube. Crush the gel slice completely with a small pestle or plastic rod and add 700 μl TE buffer containing 0.02% SDS and allow to soak overnight at room temperature to elute the radiolabeled DNA. The next day, precipitate the DNA twice using standard ethanol precipitation, rinse with ice-cold 70% ethanol, and resuspend in 100 μl TE buffer. Transfer the DNA solution to a new tube, store at –20°C until use, and determine the specific radioactivity by scintillation of a small portion of the sample (1 μl). In our experience, 10 ng labeled 343-bp fragment contains ~10,000 cpm counts. Store the DNA at –20°C until use.

3.2.2. Nucleosome Reconstitution

A portion of the radiolabeled fragment can now be used for nucleosome reconstitution. We mix 1,000 ng of radiolabeled DNA fragment which contains ~1,000 kcpm radioactivity, 500 μg of the 343-bp DNA fragment, 282 μg H2A/H2B, 218 μg H3/H4, 800 μl of 5 M NaCl, and 20 μl of 1 M DTT in a total volume of 2 ml. Nucleosomes are reconstituted by standard salt dialysis (as above) and the products from reconstitution are visualized by autoradiography of native agarose gels as described above.

3.2.3. H2A–H2A Intranucleosome Cross-Linking

Cross-linking is performed as above with nucleosomes reconstituted with H2A G2C. However, we find that the efficiency of cross-linking on single-end-radiolabeled nucleosomes prepared in this manner is only about 70–80%, which is not as efficient as in preparations in which nucleosomes are reconstituted on homogenous, unlabeled DNA fragments. It is possible that in the radiolabeled reconstituted nucleosome samples the presence of free dimer, hexamer, and other histone–DNA complexes that do not efficiently cross-link reduces the overall efficiency.

3.2.4. Nucleosome Purification

Separate radiolabeled nucleosomes from nucleosomes reconstituted on the cold plasmid DNA and other by-products of the reconstitution by purifying uncross-linked and cross-linked nucleosomes on glycerol gradients (see *above*). Fractions are analyzed by native 0.7% agarose gel as above.

3.3. Nucleosome Remodeling

3.3.1. Restriction Enzyme Assay

A powerful strategy to test the activity of ATP-dependent remodeling complexes exploits the ability of these complexes to use the energy derived from ATP hydrolysis to increase the accessibility of a restriction enzyme recognition site located internally within the nucleosome. The 601 nucleosome used in our example, from plasmid CP1024, harbors an *Hha*I restriction enzyme site near the nucleosome dyad, which is occluded by the histone octamer (Fig. 4a). We employed this site to determine whether histone–histone cross-linking affects nucleosome remodeling by SWI/SNF, and use this as an example in the protocol below. Specifics should be tailored to the remodeling complex and substrate employed.

1. Incubate approximately 1 ng of nucleosome in a 20 μl reaction mixture with 1 nM SWI/SNF and 10 U *Hha*I in 1× remodeling buffer (10 mM Tris, 50 mM NaCl, 5 mM $MgCl_2$, 1 mM DTT, and 100 ng/μl bovine serum albumin) in the presence or absence of 1 mM ATP. The reaction is initiated by careful mixing and incubated at 30°C. The reaction is terminated at the desired time by adding 4 μl 6× SDS DNA-loading dye containing 2 mM EDTA.
2. The accessibility of nucleosome DNA is analyzed by 5% polyacrylamide gel (containing 0.04% SDS final) at 120 V for 2.5 h at RT. As shown in Fig. 4a, lane 1, nucleosome DNA is almost 100% resistant to *Hha*I cleavage, indicating that nucleosomes are efficiently assembled in the expected location on the DNA fragment. The addition of SWI/SNF in the presence of ATP

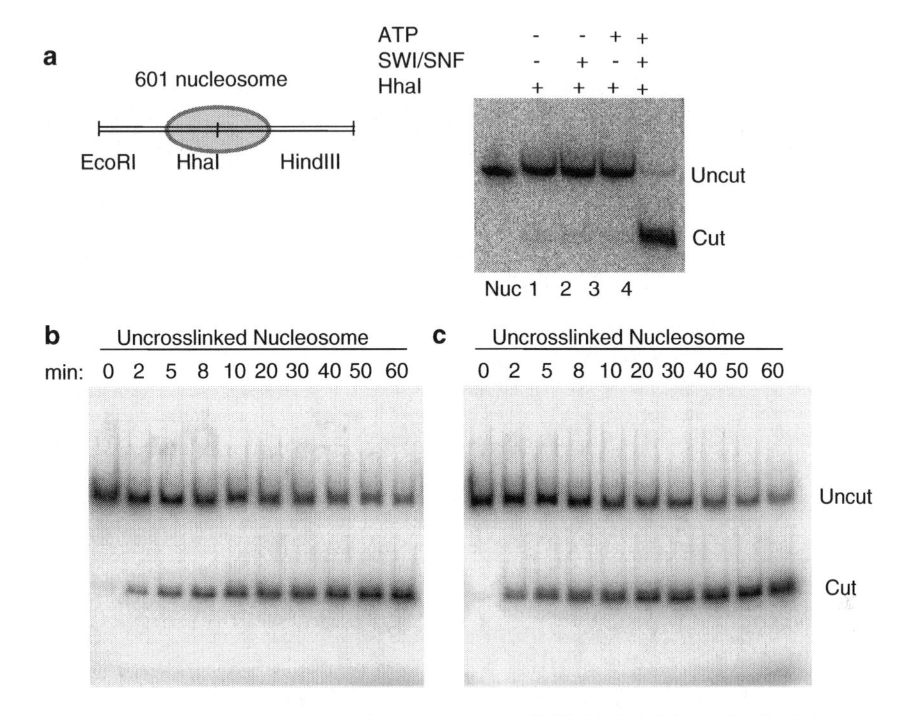

Fig. 4. Chromatin remodeling detection by restriction enzyme assay (REA). (**a**) *Left*:Schematic of 343 bp 601 nucleosome substrate. The *HhaI* site is close to the nucleosome dyad. *Right*:Nucleosomes were incubated in 1× remodeling buffer, with 10 U*HhaI*, with or without 1 mM ATP or1 nM SWI/SNF, as indicated above the gel. The reactions were incubated at 30°C for 60 min and then stopped by addition of SDS loading buffer and samples were analyzed on a 5% acrylamide–0.04% SDS gel, 0.5× TBE, electrophoresed at 120 V for 2.5 h at RT. Gel was dried and analyzed by phosphoimager. *Lane 1*, nucleosomes incubated with *HhaI* in remodeling buffer. *Lane2*, nucleosomes incubated with *HhaI* and SWI/SNF; *Lane3*, nucleosomes with *HhaI* and ATP; *Lane 4*, nucleosomes with *HhaI* and both SWI/SNF and ATP. (**b**, **c**) Restriction enzyme accessibility time course. Reaction conditions were the same as shown in (**a**) with reactions stopped at indicated times.

leads to increased cleavage of the nucleosome DNA such that more than 90% of nucleosome DNA is cleaved completely by *Hha*I (Fig. 4a, lane 4). We next attempted to test whether cross-linking decreases the access of *Hha*I to nucleosome DNA by performing a REA time course. As shown here (Fig. 4c), the cross-linking does not decrease the *Hha*I accessibility, and SWI/SNF shows similar remodeling activity on cross-linked and uncross-linked nucleosomes.

3.3.2. Nucleosome Mobility Assay

Next, we tested the remodeling activity of SWI/SNF by a nucleosome mobility assay that is diagnostic for the translational position of the nucleosome. The SWI/SNF complex uses the energy of ATP hydrolysis to move the histone octamer bidirectionally from central locations to the ends of the DNA fragment. The end-positioned nucleosomes migrate faster than centrally positioned nucleosomes on the gel, so the alteration of nucleosome positions can be determined by native polyacrylamide gels.

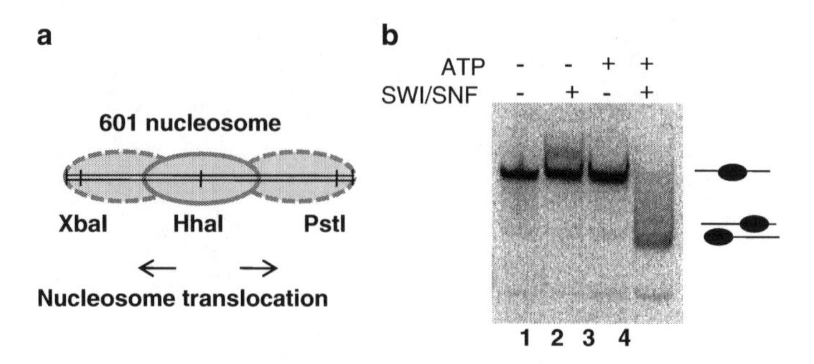

Fig. 5. Gel analysis of nucleosome mobilization by chromatin remodeling. (**a**) Diagram of nucleosome movement. (**b**) Nucleosome-remodeling reactions with 1 ng nucleosome in the presence or absence of SWI/SNF or ATP as indicated. Reactions were incubated at 30°C for 60 min and stopped by addition of 50% glycerol and excess of unlabeled CT DNA. After incubation for 5 min on ice, nucleosome positions were analyzed by 4% nondenaturing polyacrylamide gel run at 100 V for 3 h at RT.

1. Nucleosomes are mixed with remodeling buffer (as described above), 100 ng/μl BSA, and 3 nM SWI/SNF in the presence of 3 mM ATP. Mixture is incubated at 30°C. The reaction is stopped at indicated times by mixing with prechilled 2 μl stop buffer (50% glycerol, 300 ng calf thymus DNA) and incubated for 5 min on ice.
2. The remodeled nucleosomes are analyzed by 4% native polyacrylamide gels for 3 h at 100 V at RT. For the mobility gel assay, the prepared gel needs to be prewarmed by running at 100 V for 1.5 h at RT. As diagrammed in Fig. 5a, SWI/SNF induces nucleosome movement on the DNA template in the presence of ATP. Nucleosomes migrate according to translational position in the gel (Fig. 5b). Migration is not altered by incubation with SWI/SNF or ATP alone, but in the presence of both SWI/SNF and ATP nucleosomes migrate much more rapidly through the gel, signifying a change in translational position. Digestion of nucleosome products with *Xba*I or *Pst*I (Fig. 5a) indicates that nucleosomes are moved to both ends of the DNA fragment (result not shown).

4. Notes

1. After neutralizing the supernatant with NaOH, if the solution is not clear, centrifuge the supernatant for 15 min at 15,000 × *g* again. Carefully transfer the supernatant to a new tube.
2. To keep cysteines active, H2A G2C/H2B dimer is added last to the reconstitution solution. Keep the dimer stock frozen at −80°C when it is not in use. The conditions for reconstitution

have to be empirically optimized dependent on the different size and sequence of DNA templates. Normally, the weight ratio of DNA:histones is 1:1–0.6, and the H2A/H2B:H3/H4 is 1.3:1.

3. When collecting fractions from the glycerol gradients, use 0.6-ml Axygen maximum recovery tubes pretreated with BSA by rinsing the walls of the tubes with 600 μl of cold BSA solution (0.2 mg/ml) several times and placing 6 μl of BSA (10 mg/ml) at the bottom of the tubes. When fractions are collected into these tubes, each fraction is mixed immediately with BSA by inverting the tubes several times followed by a brief microfuge spin, yielding a final BSA concentration of 0.1 mg/ml.
4. When using centrifugal filtration units for condensing solutions, buffer changing, or component removal, make sure to prespin the filter with the buffers that are employed in the following samples. In addition, the flow rate of the solution could be estimated in this way. Note that flow rate varies depending on sample concentration, centrifugal temperature, and rotor type, so it is always a good idea to estimate the flow rate before working on the samples. This helps avoid the risk of filtering the sample to dryness and it is important to check the retained sample volume every few minutes.
5. There is no reducing agent in the dialysis TE buffer for reducing agent removal. The oxidation reaction rate is decreased at low temperature, so it is important to use cold TE buffer and dialyze at 4°C.
6. The centrifugation can be stopped and left in the centrifuge at 4°C overnight after the final 1-h spin of the day. The cap of the concentrator cannot be opened and the waste solution may remain in the unit overnight.
7. We have found that the volume of the reaction and ratio of cross-linker:nucleosome are important for the cross-linking efficiency. In our experience, dilution of the cross-linker such that the volume of cross-linker to be added to the reaction is less than ~10% of the volume of nucleosome provides maximal efficiency.
8. Manufacturer-supplied 10× PNK buffer typically contains 10 mM $MgCl_2$. In order to maximize nucleosome stability during labeling, we decreased the concentration of $MgCl_2$ to 4 and 2 mM. We found that the nucleosome is efficiently radiolabeled by PNK and the nucleosomes maintain stability when supplied with one of these buffers.
9. Nucleosomes from the gradient contain about 10–20% glycerol, so the centrifugation step is critical not only for removing the unincorporated [γ-^{32}P] ATP, but also to remove the glycerol from nucleosomes before application to the second glycerol gradient.

10. Phenol/chloroform extraction should be performed in the hood because this reagent is volatile and toxic.
11. After loading the radiolabeled samples to the polyacrylamide gel, wash the tube with another 20 μl TE buffer containing 1× SDS loading dye to maximize recovery.

References

1. Kornberg, R. D., and Lorch, Y. (1999) Twenty-five years of the nucleosome, fundamental particle of the eukaryote chromosome, *Cell 98*, 285–294.
2. Jenuwein, T., and Allis, C. D. (2001) Translating the histone code, *Science 293*, 1074–1080.
3. Kingston, R. E., and Narlikar, G. J. (1999) ATP-dependent remodeling and acetylation as regulators of chromatin fluidity, *Genes Dev 13*, 2339–2352.
4. Gangaraju, V. K., and Bartholomew, B. (2007) Mechanisms of ATP dependent chromatin remodeling, *Mutat Res 618*, 3–17.
5. Liu, N., Balliano, A., and Hayes, J. J. (2010) Mechanism(s) of SWI/SNF-Induced Nucleosome Mobilization, *Chembiochem*.
6. Cairns, B. R. (2007) Chromatin remodeling: insights and intrigue from single-molecule studies, *Nat Struct Mol Biol 14*, 989–996.
7. Côté, J., Quinn, J., Workman, J. L., and Peterson, C. L. (1994) Stimulation of GAL4 derivative binding to nucleosomal DNA by the yeast SWI/SNF complex, *Science 265*, 53–60.
8. Peterson, C. L., and Herskowitz, I. (1992) Characterization of the yeast SWI1, SWI2, and SWI3 genes, which encode a global activator of transcription, *Cell 68*, 573–583.
9. Schnitzler, G., Sif, S., and Kingston, R. E. (1998) Human SWI/SNF interconverts a nucleosome between its base state and a stable remodeled state, *Cell 94*, 17–27.
10. Flaus, A., Martin, D. M., Barton, G. J., and Owen-Hughes, T. (2006) Identification of multiple distinct Snf2 subfamilies with conserved structural motifs, *Nucleic Acids Res 34*, 2887–2905.
11. Havas, K., Flaus, A., Phelan, M., Kingston, R., Wade, P. A., Lilley, D. M., and Owen-Hughes, T. (2000) Generation of superhelical torsion by ATP-dependent chromatin remodeling activities, *Cell 103*, 1133–1142.
12. Saha, A., Wittmeyer, J., and Cairns, B. R. (2002) Chromatin remodeling by RSC involves ATP-dependent DNA translocation, *Genes Dev 16*, 2120–2134.
13. Gavin, I., Horn, P. J., and Peterson, C. L. (2001) SWI/SNF chromatin remodeling requires changes in DNA topology, *Mol Cell 7*, 97–104.
14. van Holde, K., and Yager, T. (2003) Models for chromatin remodeling: a critical comparison, *Biochem Cell Biol 81*, 169–172.
15. Richmond, T. J., and Davey, C. A. (2003) The structure of DNA in the nucleosome core, *Nature 423*, 145–150.
16. Luger, K., Mader, A. W., Richmond, R. K., Sargent, D. F., and Richmond, T. J. (1997) Crystal structure of the nucleosome core particle at 2.8 A resolution, *Nature 389*, 251–260.
17. Aoyagi, S., Wade, P. A., and Hayes, J. J. (2003) Nucleosome sliding induced by the xMi-2 complex does not occur exclusively via a simple twist-diffusion mechanism, *J Biol Chem 278*, 30562–30568.
18. Aoyagi, S., and Hayes, J. J. (2002) hSWI/SNF-catalyzed nucleosome sliding does not occur solely via a twist-diffusion mechanism, *Mol Cell Biol 22*, 7484–7490.
19. Langst, G., and Becker, P. B. (2001) Nucleosome mobilization and positioning by ISWI-containing chromatin-remodeling factors, *J Cell Sci 114*, 2561–2568.
20. Langst, G., and Becker, P. B. (2001) ISWI induces nucleosome sliding on nicked DNA, *Mol Cell 8*, 1085–1092.
21. Zhang, Y., Smith, C. L., Saha, A., Grill, S. W., Mihardja, S., Smith, S. B., Cairns, B. R., Peterson, C. L., and Bustamante, C. (2006) DNA translocation and loop formation mechanism of chromatin remodeling by SWI/SNF and RSC, *Mol Cell 24*, 559–568.
22. Zofall, M., Persinger, J., Kassabov, S. R., and Bartholomew, B. (2006) Chromatin remodeling by ISW2 and SWI/SNF requires DNA translocation inside the nucleosome, *Nat Struct Mol Biol 13*, 339–346.
23. Yang, Z., Zheng, C., and Hayes, J. J. (2007) The core histone tail domains contribute to sequence-dependent nucleosome positioning, *J Biol Chem 282*, 7930–7938.

24. Smith, C. L., Horowitz-Scherer, R., Flanagan, J. F., Woodcock, C. L., and Peterson, C. L. (2003) Structural analysis of the yeast SWI/SNF chromatin remodeling complex, *Nat Struct Biol* *10*, 141–145.
25. Wang, X., and Hayes, J. J. (2007) Site-specific binding affinities within the H2B tail domain indicate specific effects of lysine acetylation, *J Biol Chem* *282*, 32867–32876.
26. Hayes, J. J., and Lee, K. M. (1997) In vitro reconstitution and analysis of mononucleosomes containing defined DNAs and proteins, *Methods* *12*, 2–9.
27. Lee, K. M., and Hayes, J. J. (1997) The N-terminal tail of histone H2A binds to two distinct sites within the nucleosome core, *Proc Natl Acad Sci USA* *94*, 8959–8964.

Chapter 22

Sulfyhydryl-Reactive Site-Directed Cross-Linking as a Method for Probing the Tetrameric Structure of Histones H3 and H4

Andrew Bowman and Tom Owen-Hughes

Abstract

The structural characterisation of protein–protein interactions is often challenging. Where interactions are not amenable to high-resolution approaches, alternatives providing lower resolution information are often of value. One such approach is site-directed cross-linking. Here, through the introduction of cysteine residues at strategic locations in histone proteins, we use site-directed cross-linking to monitor the association of chromatin subunits. This approach is informative for the study of both recombinant and native chromatin complexes consisting either of histone subunits alone or in association with accessory proteins, in this case histone chaperones. The approaches described may be generally applicable for monitoring the interactions of a diverse range of multi-protein complexes.

Key words: Chromatin, Histone octamer, Tetramer, Nucleosome, Histone chaperone, Chromatin remodelling, Site-directed cross-linking

1. Introduction

Proteins and protein complexes are often not amenable to high-resolution structural analysis. Targeted cross-linking that utilizes the unique reactivity of cysteine side chains has provided valuable structural information regarding topology, conformation, and structural rearrangements of such proteins. The sulfhydryl group of cysteine has a valency of two and an ionization constant of ~8.2, conferring reactivity within a biologically applicable pH range. Cysteines are comparatively rare, accounting for only 0.6% of amino acids in *Saccharomyces cerevisiae* (1), which makes them ideal for site-directed cross-linking analysis. Protein domains of moderate size usually contain only a handful of cysteine residues

Randall H. Morse (ed.), *Chromatin Remodeling: Methods and Protocols*, Methods in Molecular Biology, vol. 833, DOI 10.1007/978-1-61779-477-3_22, © Springer Science+Business Media, LLC 2012

that, generally, can be mutated to a homologous amino acid, such as alanine or serine, without affecting the overall structure or function of the protein. Using recombinant technologies, site-directed mutagenesis can be employed to substitute cysteine residues at strategic locations within the cysteine null protein of interest. The proximity of two cysteine residues can then be analysed using targeted cross-linking of their sulfhydryl groups, either through direct disulphide bond formation or through the use of a homo-bifunctional cross-linking reagent. Cross-linking generally affects the migration of the target protein when analysed by denaturing sodium dodecyl sulphate-polyacrylamide gel electrophoresis (SDS-PAGE), even when cross-linking occurs between residues within the same polypeptide. Thus, SDS-PAGE is a convenient and cheap method for analysing the extent of cross-linking at any given location. Combining these approaches, proximity constraints within the tertiary structure can be placed upon the primary sequence, yielding valuable structural information.

The core histone octamer forms the scaffold for wrapping ~147 bp of DNA within the nucleosome core particle. The histone octamer is composed of two H2A–H2B dimers and an $(H3–H4)_2$ tetramer. Assembly of the nucleosome core particle follows an ordered series of events, H3 and H4 being deposited onto DNA first, proceeded by the deposition of two H2A–H2B dimers to form the nucleosome proper. In vivo assembly and disassembly involve accessory proteins known as histone chaperones. Histone chaperones have been shown to be important in moulding the thermodynamic landscape of histone–DNA interactions, thereby promoting correct nucleosome formation or dissolution (2, 3). In addition, these accessory proteins are often found as components of ATP-dependent, chromatin-remodelling enzymes and histone post-translational modifying complexes (4, 5).

One question that has arisen in the field is the conformation of soluble histones H3 and H4 prior to deposition while in complex with histone chaperones. Interestingly, although adopting a tetrameric conformation in the nucleosome, H3 and H4 exist as an obligate dimer while associated with the ubiquitous chaperone Asf1 (6–8). Recently, however, it has been shown that while in complex with another class of histone chaperone H3 and H4 can adopt their tetrameric conformation previously observed within the nucleosome (9). In this chapter, we describe a method which utilises a site-directed cross-linking strategy to probe the conformation of H3 and H4 (Fig. 1). This is also applicable to H3 and H4 while in complex with accessory proteins, such as histone chaperones, with the previously well-characterised interaction between Asf1 and H3–H4 used as an example of how the methodology can be implemented.

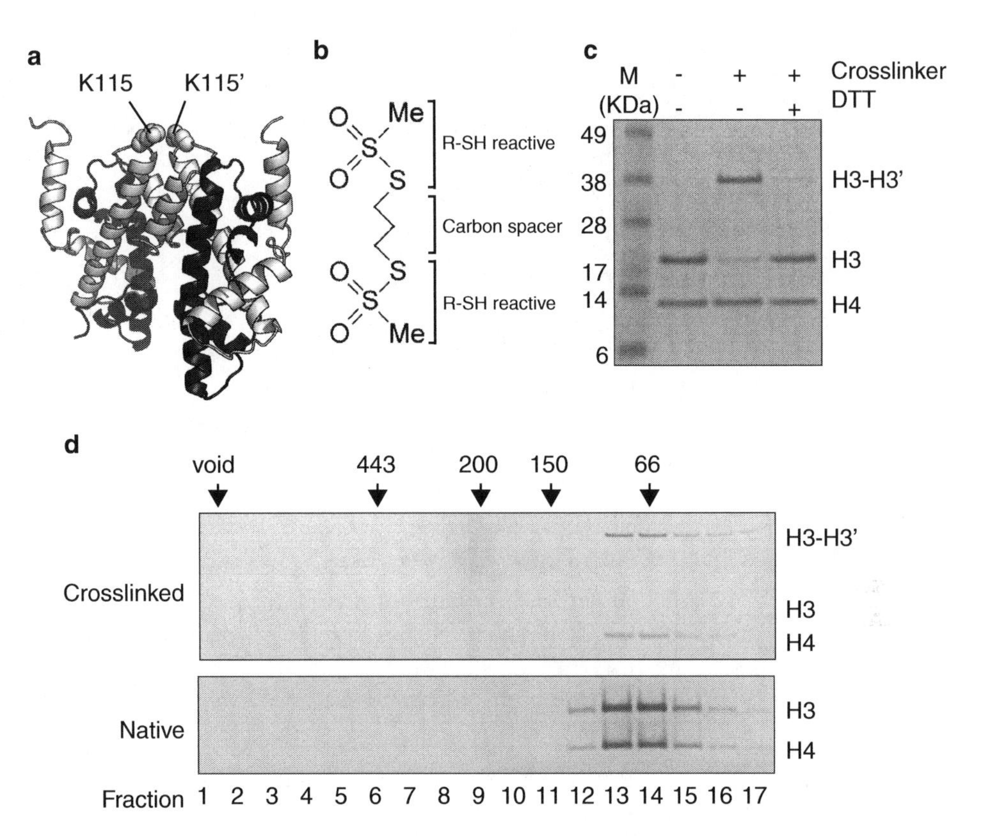

Fig. 1. Site-directed cross-linking as a probe of tetramer structure. (**a**) Histones H3 and H4 shown in their tetrameric structure observed within the nucleosome (coordinates taken from PDB: 1KX5). H3 is shown in *white*, whereas H4 is shown in *black*. The site-directed cross-linking positions, H3K115, are shown as *spheres*. (**b**) Chemical structure of the sulfhydryl-reactive cross-linker 1,3-propanediyl bismethanethiosulfonate (M3M). (**c**) Cross-linking of the $(H3\text{–}H4)_2$ tetramer at position H3 K115C. Cross-linking across the dyad interface of H3–H3′ results in a slower migrating species that can be resolved from non-cross-linked H3 by SDS-PAGE. Incubation with the reducing agent DTT reverses the cross-linking. (**d**) Gel filtration analysis shows that the cross-linked tetramer has the same size as non-cross-linked (native) tetramer, demonstrating the fidelity of cross-linking across the dyad interface of H3.

2. Materials

2.1. Preparation of Reduced Histone Tetramers

1. Recombinant histones H3 K115C and H4.
2. Unfolding buffer: 7 M guanidinium chloride, 20 mM HEPES, pH 7.5, 1 mM EDTA. Care should be taken when handling guanidinium chloride as it is toxic, and with EDTA as it is an irritant.
3. Refolding buffer: 1 M sodium chloride, 20 mM HEPES, pH 7.5, 1 mM EDTA.
4. β-mercaptoethanol (β-ME). Care should be taken when preparing β-ME as it is toxic.
5. Dialysis tubing (≤8,000 molecular weight cut-off, such as Spectra/Por® Dialysis Membrane MWCO: 6–8,000, Spectrum Laboratories Inc.).

6. Superdex™ S200 packed gel filtration column (such as GL 10/300, GE Healthcare) and chromatography system.
7. 4× SDS-PAGE loading buffer: 40% glycerol, 240 mM Tris–HCl, pH 6.8, 8% SDS, 0.04% bromophenol blue, 5% β-ME (or pre-made, e.g. by Invitrogen).
8. Dithiothrietol (DTT). Care should be taken when preparing DTT as it is harmful.
9. SDS-PAGE equipment for protein separation and analysis.
10. Centrifugal concentrators (≤10,000 molecular weight cut-off, such as Amicon® Ultra, Millipore™).
11. Liquid nitrogen. Personal protective equipment should be worn while handling liquid nitrogen.

2.2. Targeted Cross-Linking as a Probe of Tetramer Structure

1. Buffer A: 0.5 M sodium chloride, 20 mM HEPES–KOH, pH 7.5, and 1 mM EDTA.
2. Purified histone chaperone of interest in buffer conditions that allow binding to histones. The histone chaperone Asf1 is used as an example in this protocol. Buffer conditions that allow histone binding by Asf1 had previously been determined (buffer A).
3. 1,3-Propanediyl bis-methanethiosulfonate (M3M) (see Note 1).
4. A mono-reactive, small-molecule methanethiosulfonate (MTS) compound, such as propyl methanethiosulfonate (PMTS) (see Note 2).
5. SDS-PAGE equipment for protein separation and analysis.
6. Coomassie protein stain (for example, "Instant*Blue*", Expidion).
7. Dimethyl sulfoxide (DMSO). Care should be taken when handling DMSO as it is an irritant.

2.3. Targeted Cross-Linking as a Method of Tetramer Stabilisation

1. Superdex™ S200 PC 3.2/30 column (e.g. GE Healthcare) and micro-chromatography system (see Note 3).
2. Microcentrifugal filter units (such as Ultrafree® 0.22-μm Centrifugal Filters, Millipore™) (optional).

3. Methods

3.1. Preparation of Reduced H3 K115C Tetramers

For purification of individually expressed recombinant core histones in bacteria and their refolding in vitro, we refer the reader to previous publications (10–12). *Xenopus laevis* histones were used in this study. The single, non-conserved cysteine (C110) of H3 was mutated to alanine and the K115C mutation inserted by site-directed mutagenesis. H3 and H4 are insoluble on their own, but highly soluble after denaturation and refolding in equal stoichiometries. However, refolding is never 100% efficient, and requires a further chromatographic step to isolate correctly folded tetramers from misfolded aggregates and monomers. We have adapted this step to incorporate the removal of reducing agents from the H3 K115C mutant tetramer prior to cross-linking.

1. Dissolve purified, lyophilised histones H3 K115C and H4 at equal stoichiometries in unfolding buffer + 20 mM DTT at a final concentration of 20–100 μM (individual histone monomer).
2. After 1 h, transfer the unfolded histones to dialysis tubing with a molecular weight cut-off of ≤8,000, and dialyse overnight versus 2 L of refolding buffer + 5 mM β-ME at room temperature or 4°C.
3. Retrieve the refolded histones from the dialysis tubing and centrifuge at 10,000 × *g* to remove any precipitated material.
4. Transfer the soluble portion of the refolding reaction to a centrifugal concentrator (molecular weight cut-off of ≤10,000). The extent of concentration depends on the type of gel filtration column used during size-exclusion chromatography. Typically, resolution among soluble aggregates, histone monomers, and refolded tetramer is maintained up to a loading volume of 1/60th of the column's bed volume. We have found that using a Superdex™ 200 10/300 GL column with a 24-mL volume, a maximum load volume of 400 μL, works well.
5. After concentration, add 20 mM DTT and leave at room temperature for 1 h to ensure that cysteine residues are fully reduced.
6. Chromatography is carried out in refolding buffer without reducing agents. Thus, in addition to isolating the correctly folded tetramer, it serves as a desalting step to remove reducing agents from the sample prior to cross-linking. Equilibration of the gel filtration column can be carried out during the preparation of the refolded tetramer sample.
7. Collect fractions spanning the void to bed volumes of the column. We typically run a Superdex™ S200 10/300 GL column at 0.25 mL/min, collecting 0.5-mL fractions from 8 to 24 mL, using an ÄKTA™ purifier liquid chromatography system (GE Healthcare).
8. After gel filtration, it is important to work as quickly as possible to minimise the oxidising effect of dissolved atmospheric oxygen on the reduced cysteines. An indication of the fractions containing tetramer can be obtained by monitoring the absorbance at 276 nm. This can be confirmed by SDS-PAGE and Coomassie staining. Typically, the tetramer elutes with an

apparent molecular weight of 80–100 kDa, and with experience the appropriate fractions can be selected for step 9 before the SDS gel has been run.

9. Pool tetramer containing fractions and concentrate using a centrifugal concentrator (molecular weight cut-off of <10,000). Monitor the concentration of the tetramer during centrifugation by its absorbance at 280 nm (see Note 5). Once it has reached ~200 μM, distribute it into 10-μL aliquots, flash freeze in liquid nitrogen, and store at −80°C. We have found that the tetramer remains in its reduced state for at least 6 months under these conditions.

3.2. Targeted Cross-Linking as a Probe of Tetramer Structure

In this chapter, two complementary approaches for using site-directed cross-linking as a probe of H3–H4 conformation while in complex with histone chaperones are described. In the first approach detailed in this section, the histone tetramer containing the H3 K115C mutation is incubated with increasing concentrations of the chaperone of interest. After binding has equilibrated, cross-linking is carried out using a homo-bifunctional sulfhydryl reactive cross-linker. Free histone tetramers are covalently linked across the dyad interface at the introduced H3 K115C cross-linking site. As two covalently linked H3 molecules (H3–H3′) have twice the molecular weight of a single H3, these two species are easily separated by denaturing SDS-PAGE (Fig. 1c). A histone chaperone which disrupts the H3–H3′ interface, such as Asf1, hinders cross-linking. Thus, increasing the concentration of such a chaperone, prior to cross-linking, results in reduction of the H3–H3′ species and an increase in the H3 species. Alternatively, a chaperone which interacts with H3–H4 in their tetrameric configuration has minimal effect on the H3–H3′ species. Before starting, it is necessary to determine interaction conditions between H3–H4 and your chaperone of interest. It is also beneficial to know the stoichiometry of the interaction so that relevant titration points can be made.

The quantities of chaperone required are dependent on the method of detection. In the approaches described below, histones were detected using Coomassie staining at a concentration of 20 pmoles (tetramer) per lane. Thus, assuming a 1:1 stoichiometry of binding and allowing for repetition, quantities in the range of ~0.5–1.0 nmoles of chaperone are required. This is typically well within the range of recombinant expression in bacteria and also within the range of TAP tag-purified yeast proteins of moderate abundance. However, if material is scarce, one can resort to immunoblotting for detection, thus lowering the required amount of chaperone by 10–100-fold.

As this methodology is based on exploiting the unique reactivity of the sulfhydryl containing side chain of cysteine, it is important that all buffers are free from reducing agents, such as DTT and β-ME. Furthermore, the presence of cysteine residues within the chaperones can potentially complicate the interpretation of cross-linking. When using recombinant chaperones, such problems may be circumvented through mutation of native cysteines (to either alanine or serine) by site-directed mutagenesis. If this is not possible or if the chaperone is purified from an endogenous source, the chaperone can be incubated with PMTS to cap any reduced histones, followed by desalting or dialysis to remove the small molecule, before cross-linking. Cross-linking in this protocol details the use of MTS groups which form reducible disulphide bonds with cysteines separated by a carbon spacer. However, other methods of sulphydryl cross-linking are also possible (see Note 4).

In the following protocol, the binding of H3–H4 by Asf1 (*S. cerevisiae*) shall be used as a model histone–chaperone interaction that disrupts the H3–H3′ interface of the histone tetramer. The affinity of Asf1 for H3–H4 has previously been shown to be within the low nanomolar range (13), and insensitive to high ionic conditions (6, 8), of up to 1 M sodium chloride (9), with a binding stoichiometry of two Asf1 per $(H3–H4)_2$ tetramer (7).

1. Retrieve an aliquot of histone chaperone (in this case, Asf1 purified as described previously (9) and stored in aliquots dialysed into buffer A at −80°C). After thawing, make four 30 μL dilutions of 1, 2, 3, and 4 μM in buffer A and place on ice.
2. Retrieve an aliquot of refolded H3 K115C tetramer from −80°C storage, thaw, and place on ice. As the cross-linking reaction is sensitive to the stoichiometry of tetramer to cross-linker (Fig. 2b), it is worthwhile centrifuging the thawed aliquot, removing the supernatant and redetermining the concentration of tetramer by spectrophotometric analysis in case any precipitation has occurred during freeze thawing (see Note 5).
3. Dilute the tetramer to 2 μM (4 μM H3–H4 dimer) with buffer A. Note: Due to the large dilution factor, the salt content of the refolding buffer, in which the tetramer is dissolved, is negligible.
4. Aliquot 25 μL of the tetramer solution into six 1.5-mL microcentrifuge tubes labelled “0”, “0.5”, “1.0”, “1.5”, “2.0”, and “Control” with the suffix “T” (tetramer) and keep on ice. Add 25 μL of the 1, 2, 3, and 4 μM Asf1 dilutions to samples “0.5T”, “1.0T”, “1.5T”, and “2.0T”, respectively, to achieve the corresponding concentrations. To the remanding “0T” and “Control T” samples, add 25 μL of buffer A. Mix well by pipetting. Incubate at room temperature for 10 min to allow binding to equilibrate. Return to ice.
5. You should now have six sample tubes, all with a final tetramer concentration of 1 μM, four of which have Asf1 titrations of 0.5, 1.0, 1.5, and 2 μM. Remember that the binding stoichiometry

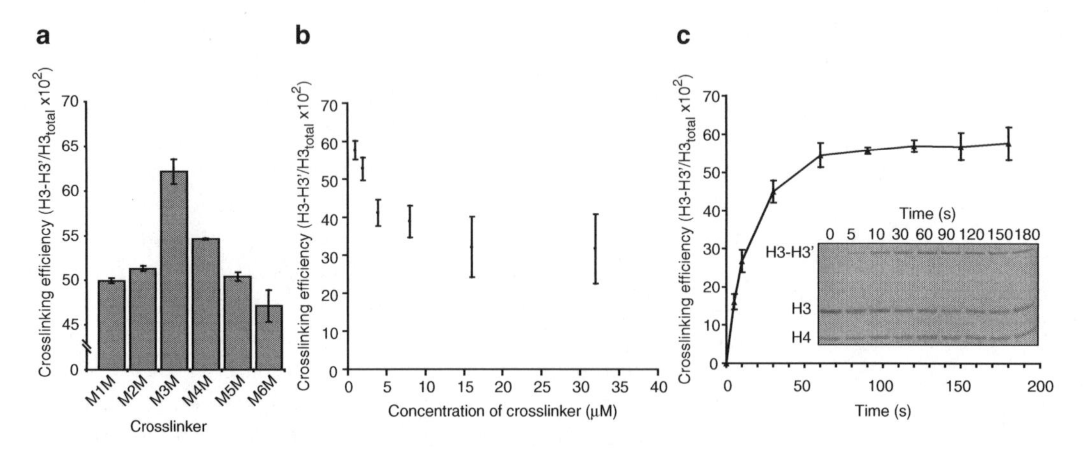

Fig. 2. Parameters affecting cross-linking across the dyad interface of histone H3. (**a**) Cross-linking efficiency with respect to cross-linker length. Methanethiosulfonate cross-linkers containing 1–6 carbon spacers (M1–6M, respectively) were tested for their cross-linking efficiency against 1 μM tetramer at equal stoichiometry by quantification of Coomassie stained bands. A 3-carbon spacer (M3M) was determined to be the most efficient. (**b**) Cross-linker efficiency with respect to the stoichiometry of cross-linker to cross-linking sites. Histone tetramer at a concentration of 1 μM was incubated with increasing amounts of M3M. Cross-linking was found to be the most efficient at a stoichiometry of one cross-linker per tetramer or one cross-linker per two cysteine residues. (**c**) The cross-linking reaction monitored over time. 1 μM tetramer was incubated with 1 μM M3M and cross-linking allowed to proceed for the displayed times before quenching. Cross-linking occurs rapidly and is all but complete after 60 s. The insert is a gel from one of the time courses. In all three experiments, cross-linking reactions were carried out in triplicate, with error bars representing the standard deviation.

of Asf1 to the histone tetramer has been previously determined as 2:1; therefore, the final titration represents an equal stoichiometry in binding.

6. In the meantime, retrieve a 10-nmole aliquot of dried cross-linking reagent M3M (see Note 6) and dissolve in 250 μL DMSO to achieve a 40 μM solution. Retrieve a 2-μmole aliquot of quenching reagent PMTS and dissolve in 10 μL of DMSO to achieve a 200 mM solution.
7. Aliquot out 1 μL of the cross-linking reagent to five micro-centrifuge tubes labelled "0", "0.5", "1.0", "1.5", and "2.0" (the final concentration of Asf1) with the suffix "XL" (cross-linker), 1 μL of the quenching reagent to a sixth labelled "control Q". Additionally, aliquot out 1 μL of quenching reagent to five micro-centrifuge tubes, again, labelled "0", "0.5", "1.0", "1.5", and "2.0", but with the suffix "Q" (quencher), and 1 μL of cross-linking reagent to a sixth (labelled "control XL"). Do not put these tubes on ice as the solvent, DMSO, freezes and prevents rapid mixing. It is helpful to organise these tubes into two corresponding rows in a micro-centrifuge tube rack: "0 XL", "0.5 XL", "1.0 XL", "1.5 XL", "2.0 XL", and "Control Q" (top row) and "0 Q", "0.5 Q", "1.0 Q", "1.5 Q", "2.0 Q", and "Control XL" (bottom row).
8. 40 μL of each of the Asf1 tetramer samples ("0.5T", "1.0T", "1.5T", and "2.0T") and the tetramer-alone control ("0T") is,

in turn, rapidly mixed with the cross-linking reagent ("0 XL", "0.5 XL", "1.0 XL", "1.5 XL", and "2.0 XL"), left for 60 s at room temperature (see Note 7) and then rapidly mixed with the quenching reagent. The second tetramer control ("Control T") is mixed first with the quenching reagent ("Control Q"), left for 60 s and then mixed with the cross-linking reagent ("Control XL"). This serves as a control to monitor the effectiveness of the quenching reagent. As MTS groups are highly reactive (see Note 8), it is important that mixing of the protein and cross-linker is as rapid as possible. We find that the following procedure works well. (a) Remove 40 μL of the protein sample and keep it in the pipette tip. (b) Start a timer at 70 s. (c) Open the corresponding micro-centrifuge tube which contains the cross-linking reagent and ready the tip at the bottom of the tube just above the aliquot of cross-linker. (d) As the timer counts down to 60 s, pipette rapidly up and down ten times. As the protein concentration is low and no detergents are used in the buffer, bubbles are not usually a problem. (e) Recover the sample in the tip of the pipette and discard the tube. (f) Open the corresponding micro-centrifuge tube containing the quenching reagent and ready the tip just above the aliquot, which should be sitting at the bottom of the tube. (g) As the timer hits 0 s, pipette rapidly up and down ten times, as before. We have found that this procedure works more effectively, and is more reproducible than other methods of mixing, such as using a vortex. With regards to the quenching control, the same procedure is carried out, just reversed: first, the tetramer is mixed with quenching reagent, and then cross-linking reagent.

9. Allow quenching to continue for a minimum of 5 min. Remove 20 μL of each sample and mix with SDS-PAGE loading buffer. Heat the sample at 65°C to ensure complete denaturation before separating the polypeptides by SDS-PAGE (see Note 9). After separation, stain the gel with Coomassie to visualise the extent of the cross-linking. Non-cross-linked H3 migrates slightly slower than its theoretical molecular weight, ~18 kDa, whereas H3–H3′ migrates at ~38 kDa (Figs. 1c and 3b). Asf1 is a chaperone that binds to the dyad interface of H3–H4, disrupting the H3–H3′ interaction. Thus, as would be expected, as the concentration of Asf1 is increased, the H3–H3′ at ~38 kDa gradually disappears (Fig. 3b). If quantitation of cross-linking is required, at 20 pmoles, tetramer histones H3 and H4 are well within the linear range for densitometric analysis (see Note 10).

3.3. Targeted Cross-Linking as a Method of Tetramer Stabilisation

The second approach capitalises on the observation that covalent linkage across the dyad interface of H3 and H3′ effectively stabilises the conformation of the histone tetramer (9). Using pre-cross-linked tetramer as a substrate, histone chaperones that

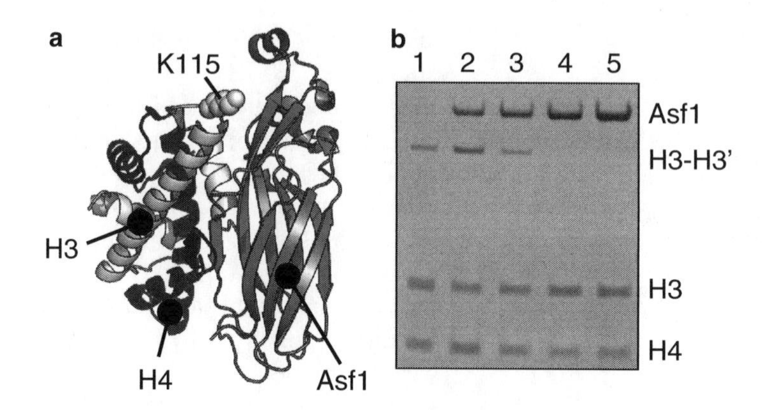

Fig. 3. Probing the structure of histones H3 and H4 while in complex with the histone chaperone Asf1. (**a**) Co-crystal structure of the histone-binding domain of *S. cerevisiae* Asf1 in complex with *X. laevis* histones (PDB code: 2HUE). (**b**) Pre-incubation with increasing amounts of Asf1 results in the inhibition of H3–H3′ formation. *Lane 1*, no Asf1; *lane 2*, 0.5 μM Asf1; *lane 3*, 1.0 μM Asf1; *lane 4*, 1.5 μM Asf1; *lane 5*, 2.0 μM Asf1; all lanes containing 1.0 μM $(H3–H4)_2$ tetramer.

specifically interact with a dimer of H3–H4 (such as Asf1) display reduced binding compared to that of non-cross-linked histones. Following the same line of reasoning, histone chaperones whose binding is not affected by the presence of the covalent linkage do not require access to the H3–H3′ interaction interface for binding (9). Binding or, more correctly, the loss of binding could be monitored thermodynamically using quantitative methods, such as isothermal titration calorimetry, surface plasmon resonance, or Forster resonance energy transfer. However, we detail here a simpler approach utilising gel filtration chromatography to monitor the association of H3–H4 with the histone-binding domain of Asf1 (Asf1g).

Before binding and separation of cross-linked tetramer and chaperone can be carried out, a number of controls must first be implemented: one must determine the elution volumes of the chaperone alone, H3–H4 alone, and the H3–H4–chaperone complex (without cross-linking). This protocol is set out with the assumption that the researcher is competent in using gel filtration chromatography to separate protein complexes of different sizes. In this protocol, a Superdex™ S200 PC 3.2/30 column attached to a SMART system (Amersham Pharmacia) micro-chromatography unit is used (see Note 3).

1. Equilibrate the gel filtration column with buffer A. Retrieve an aliquot of histone chaperone (Asf1g) from −80°C storage. To avoid precipitates that may have formed during freeze thawing being loaded on the micro-gel filtration column, either centrifuge the thawed sample in a micro-centrifuge and remove the supernatant to a fresh tube or apply the sample to a centrifugal filter unit. Asf1g was previously purified and stored at 400 μM

in buffer A. Make a 20 μM dilution of Asf1g in 40 μL in buffer A. Load 20 μL of this into the pre-washed injection loop of the micro-chromatography system using a 100-μL Hamilton syringe. Handling such small volumes can be tricky. To precisely inject 20 μL without any air bubbles, the full 40 μL is taken up into the syringe. 10 μL is then expelled to remove any air bubbles from the needle and insure drop-to-drop contact with the injection loop. The needle is inserted into the loop and the remaining sample volume noted (~30 μL). 20 μL can then be precisely injected leaving ~10 μL of residual sample in the syringe so that no air bubbles follow the sample into the loop.

2. Chromatography is then carried out at a flow rate of 20 μL/min. 50-μL fractions are collected spanning the void to bed volume and collected in 0.5-mL tubes: using a Superdex™ S200 PC 3.2/30 column, this amounts to 30 fractions spanning an elution volume of 0.8–2.4 mL. Once all fractions have been collected, add 17 μL of 4× SDS-PAGE loading buffer to each, mix, and heat at 65°C for 10 min before storing at -20°C.
3. Carry out the same procedure for H3–H4 alone (carrying the H3 K115C mutation, but not cross-linked). The 10-μL aliquots stored at −80°C in refolding buffer can be used for gel filtration analysis. Again, centrifuge or filter the sample before chromatography to remove any precipitation that may have occurred during freeze thawing. Dilute the tetramer to 10 μM in 40 μL of buffer A and apply to the injection loop. Carry out chromatography as detailed above. Again, to each of the 30 fractions, add 17 μL of SDS-PAGE loading buffer, heat at 65°C for 10 min, and store at −20°C.
4. Next, the elution volume for the H3–H4–chaperone complex is determined. Retrieve an aliquot of H3 K115C tetramer and an aliquot of chaperone from −80°C storage. Centrifuge or filter the samples before chromatography to remove any precipitation that may have occurred during freeze thawing. Dilute the tetramer to 10 μM in buffer A and transfer 40 μL to a fresh tube. To this, add 2 μL of Asf1g and leave to equilibrate for 10 min at room temperature (Asf1g binding to H3–H4 is rapid). Carry out gel filtration analysis as before, collecting 50-μL fractions and storing in 1× SDS-PAGE loading buffer at −20°C.
5. Now, the effect of the covalent cross link on histone chaperone binding can be analysed. Retrieve an aliquot of H3 K115C tetramer and an aliquot of chaperone from −80°C storage. Centrifuge or filter the samples before chromatography to remove any precipitation that may have occurred during freeze thawing. Redetermine the concentration of histone tetramer by spectrometric absorption, dilute to 10 μM in buffer A, and transfer 50 μL to a fresh tube.

6. Now, to cross-link the histone tetramer across the dyad interface: Dissolve a 10-nmole aliquot of cross-linking reagent (M3M) in 25 μL DMSO to make a 500 μM stock solution, and dissolve a 2-μmole aliquot of quenching reagent (PMTS) in 10 μL to make a 200 mM stock solution. Add 1 μL of the cross-linker and 1 μL of quenching reagent to two labelled micro-centrifuge tubes. To the tube containing 1 μL of cross-linking reagent, add 40 μL of the diluted histone tetramer and mix rapidly by pipetting up and down. After 60 s, transfer the reaction to the tube containing 1 μL of quenching reagent and mix rapidly by pipetting. In the previous experiment, cross-linking was carried out at 1 μM tetramer, which resulted in ~60% cross-linking (Fig. 2). At 10 μM tetramer and keeping the tetramer–cross-linker ratio stoichiometric, the cross-linking efficiency is increased to ~80%, which aids in the analysis of the dimer–tetramer preference of the chaperone.
7. After quenching has proceeded for 5 min, add 2 μL of Asf1g to the cross-linked tetramer and leave for 10 min at room temperature to allow binding to equilibrate. Addition of cross-linker and quenching reagent slightly dilutes the sample. However, this has negligible effect on the elution profile and, as the stoichiometry is retained, minimal effect on the cross-linking efficiency. Carry out chromatography as before. Once all fractions are collected, again, mix with 17 μL 4× SDS-PAGE loading buffer and heat for 10 min at 65°C.
8. Comparison of the cross-linked and control chromatograms: The four gel filtration runs can be compared in two ways. Firstly, if the chromatography system is equipped with a UV flow cell, peak profiles from each experiment can be compared directly. Secondly, fractions can be analysed for their content by separation using SDS-PAGE. As the peak profile does not directly report on the component of each peak and as multiple complexes can be formed, we suggest that SDS-PAGE analysis of each chromatographic profile is carried out. The UV trace can aid in determining which fractions to be analysed by SDS-PAGE.
9. Interpretation of the results: The three control runs show the elution profiles of (a) the histone chaperone alone, (b) H3–H4 alone, and (c) H3–H4–chaperone complex. If stabilisation of the tetramer structure (through the covalent linkage across the dyad interface introduced by cross-linking) inhibits the binding of the histone chaperone, an increase in the free histone chaperone and free histone tetramer species should be observed. Whereas, if cross-linking does not affect the interaction between histone tetramer and histone chaperone, the elution profile should be identical to that seen for the non-cross-linked tetramer in complex with the histone chaperone (9).

4. Notes

1. MTS-based cross-linking reagents lose their reactivity when in an aqueous environment for extended periods. Typically, minute quantities of this compound are needed per reaction. Therefore, weighing out milligrams of dry material each time, the experiment is to be carried out is not cost-effective. In the interests of economy and accuracy, stock material received from the supplier can be dissolved in trichloroethane or dimethylformamide, distributed into 10-nmole aliquots in 1.5-mL micro-centrifuge tubes, and vacuum dried. The dried aliquots are then stored at –20°C.
2. As for the cross-linking reagent, this compound is moisture sensitive; thus, we tend to store dried-down aliquots of 2 μmole at –20°C (see Note 1, above).
3. The micro-chromatography system we use is a discontinued SMART system produced by Amersham Pharmacia, which has subsequently been taken over by GE Healthcare. GE Healthcare has superseded the SMART system with the ÄKTAmicro™. In addition, a number of other manufacturers have produced chromatography system amenable to micro-gel filtration. We have also used the same Superdex™ 200 PC 3.2/30 column (GE Healthcare) attached to a Dionex manufactured Ultimate® 3000 Quaternary Micro LC System, with similar results. If material is not limiting and a micro-chromatography system is not available, everything can be scaled up tenfold for use with a 10/300 GL column (GE Healthcare), or similar, attached to a standard FPLC system.
4. For some applications, reducing conditions may be required after cross-linking is carried out. For this, we recommend using a maleimide-based cross-linker whose thioether linkage is non-cleavable with reducing agents. We have found bis-maleimidoethane (BMOE) to work just as efficiently and with similar characteristics to M3M (9). Additionally, a disulphide bond can be catalysed between the two K115C residues using copper chelated with 1,10-phenanthroline. We refer the reader to the following sources for more information on using this approach (9, 14). Oxidised and reduced glutathione have also been used in disulphide bond formation between cysteine residues introduced to histones (15).
5. Absorption coefficients for calculating protein concentration were determined from the primary sequences of *X. laevis* H3 and H4 using the tool ProtParam (http://expasy.org/tools/protparam.html) as 4,470 and 5,960 M^{-1} cm^{-1} for H3 and H4, respectively.

6. We have investigated the effect of the length of the cross-linker with respect to cross-linking efficiency using the H3 K115C tetramer as a substrate. We tested cross-linkers with 1–6 carbon atoms in the spacer region and found that cross-linking efficiencies peaked at the 3 carbon spacer mark (Fig. 2a). Although there were significant improvements in cross-linking between cross-linkers, the overall increase in cross-linking efficiencies were not drastic, representing ~15% increase from most efficient to least efficient (Fig. 2a). This is most likely due to the conformational flexibility of both the cross-linker (16) and the L2 loop region in which K115C resides (Fig. 1a).
7. We have found that cross-linking at 1 μM tetramer and 1 μM M3M is complete by 60 s (Fig. 2c). Shorter time periods for the reaction may be less accurate as a few seconds' difference in quenching times may significantly alter the cross-linking efficiency.
8. We have found that cross-linking efficiency is significantly affected by the ratio of cross-linker to cross-linking sites. There is a tendency to err towards a higher concentration of cross-linker in order to achieve a more efficient cross-linking ratio. However, upon titrating cross-linker against the H3 K115C tetramer, we found that cross-linking efficiency was maximal at a ratio of one cross-linker per cross-linking site (or one cross-linker per two cysteine residues) (Fig. 2b). This can be explained by saturation of the cysteine residues with mono-reacted cross-linker or the so-called hanging cross-linkers, highlighting the reactivity of MTS groups.
9. For SDS-PAGE analysis, we typically use the NuPAGE® Novex® precast bis–tris 1.0 mm mini gels (Invitrogen) with an MES buffer system (Invitrogen). However, the more traditional Laemmli tris–glycine systems also work fine. As the molecular weight difference between H3 and H3–H3′ is so large, they can be efficiently separated over a wide range of polyacrylamide concentrations; however, we routinely use 12%.
10. For densitometric analysis, we use the software AIDA (Raytest). Separation on a single-percentage acrylamide gel, as opposed to a gradient, helps in defining the background baseline.

References

1. Martini, A. E., Miller, M. W., and Martini, A. (1979) Amino acid composition of whole cells of different yeasts, *J Agric Food Chem 27*, 982–984.
2. Andrews, A. J., Chen, X., Zevin, A., Stargell, L. A., and Luger, K. (2010) The Histone Chaperone Nap1 Promotes Nucleosome Assembly by Eliminating Nonnucleosomal Histone DNA Interactions, *Molecular Cell 37*, 834–842.
3. Laskey, R. A., Honda, B. M., Mills, A. D., and Finch, J. T. (1978) Nucleosomes are assembled by an acidic protein which binds histones and transfers them to DNA, *Nature 275*, 416–420.

4. De Koning, L., Corpet, A., Haber, J. E., and Almouzni, G. (2007) Histone chaperones: An escort network regulating histone traffic, *Nature Structural and Molecular Biology 14*, 997–1007.
5. Park, Y. J., and Luger, K. (2008) Histone chaperones in nucleosome eviction and histone exchange, *Current Opinion in Structural Biology 18*, 282–289.
6. English, C. M., Adkins, M. W., Carson, J. J., Churchill, M. E. A., and Tyler, J. K. (2006) Structural Basis for the Histone Chaperone Activity of Asf1, *Cell 127*, 495–508.
7. English, C. M., Maluf, N. K., Tripet, B., Churchill, M. E. A., and Tyler, J. K. (2005) ASF1 binds to a heterodimer of histones H3 and H4: A two-step mechanism for the assembly of the H3-H4 heterotetramer on DNA, *Biochemistry 44*, 13673–13682.
8. Natsume, R., Eitoku, M., Akai, Y., Sano, N., Horikoshi, M., and Senda, T. (2007) Structure and function of the histone chaperone CIA/ASF1 complexed with histones H3 and H4, *Nature 446*, 338–341.
9. Bowman, A., Ward, R., Wiechens, N., Singh, V., El-Mkami, H., Norman, D. G., and Owen-Hughes, T. (2011) The histone chaperones Nap1 and Vps75 bind histones H3 and H4 in a tetrameric conformation, *Molecular Cell doi: 10.1016/j.molcel.2011.01.025.*
10. Luger, K., Rechsteiner, T. J., Flaus, A. J., Waye, M. M. Y., and Richmond, T. J. (1997) Characterization of nucleosome core particles containing histone proteins made in bacteria, *Journal of Molecular Biology 272*, 301–311.
11. Tanaka, Y., Tawaramoto-Sasanuma, M., Kawaguchi, S., Ohta, T., Yoda, K., Kurumizaka, H., and Yokoyama, S. (2004) Expression and purification of recombinant human histones, *Methods 33*, 3–11.
12. Anderson, M., Huh, J. H., Ngo, T., Lee, A., Hernandez, G., Pang, J., Perkins, J., and Dutnall, R. N. (2010) Co-expression as a convenient method for the production and purification of core histones in bacteria, *Protein Expr Purif 72*, 194–204.
13. Park, Y. J., Sudhoff, K. B., Andrews, A. J., Stargell, L. A., and Luger, K. (2008) Histone chaperone specificity in Rtt109 activation, *Nature Structural and Molecular Biology 15*, 957–964.
14. Kobashi, K. (1968) Catalytic oxidation of sulfhydryl groups by o-phenanthroline copper complex, *Biochim Biophys Acta 158*, 239–245.
15. Dorigo, B., Schalch, T., Kulangara, A., Duda, S., Schroeder, R. R., and Richmond, T. J. (2004) Nucleosome arrays reveal the two-start organization of the chromatin fiber, *Science 306*, 1571–1573.
16. Green, N. S., Reisler, E., and Houk, K. N. (2001) Quantitative evaluation of the lengths of homobifunctional protein cross-linking reagents used as molecular rulers, *Protein Sci 10*, 1293–1304.

Chapter 23

Genomic Approaches for Determining Nucleosome Occupancy in Yeast*

Kyle Tsui, Tanja Durbic, Marinella Gebbia, and Corey Nislow

Abstract

The basic unit of chromatin is double-stranded DNA wrapped around nucleosome core particles, the classic "beads-on-a-string" described by Kornberg and colleagues. The history of chromatin studies has experienced many peaks, from the earliest studies by Miescher to the biochemical studies of the 1960s and 1970s, the appreciation for the influence of histone modifications in controlling gene expression in the 1990s to the genome-wide studies that began in 2006 and show no signs of abating with the introduction of next generation sequencing technologies. Genome-wide studies not only have provided a base line to understand relationships between chromatin structure and gene function but also have begun to provide new insights into chromatin remodelling. Here, we describe the use of genome-wide approaches to determining nucleosome occupancy in yeast.

Key words: Nucleosome occupancy, Tiling array, High-throughput sequencing, Illumina sequencing

1. Introduction

The current wave of interest in chromatin, in particular the role of nucleosome occupancy, positioning and modification with regard to effects on gene expression began with a genome-scale study of nucleosome occupancy of yeast chromosome 3 by Oliver Rando and colleagues (1), and soon thereafter by the publication of the first whole-genome high-resolution occupancy map for any organism (2). These studies used high-resolution (20 and 4 bp, respectively) tiling microarrays. This array data agrees quite well with subsequent studies that used next generation sequencing (in particular 454 pyrosequencing and Illumina's reversible dye-terminator chemistry)

*Kyle Tsui and Tanja Durbic have equally contributed to this chapter.

Randall H. Morse (ed.), *Chromatin Remodeling: Methods and Protocols*, Methods in Molecular Biology, vol. 833, DOI 10.1007/978-1-61779-477-3_23, © Springer Science+Business Media, LLC 2012

using either nucleosomal DNA or chromatin-IP using anti-histone antibodies. As interest in chromatin genomics grows, experiments based on these genome-wide datasets will permit a systems biology approach to chromatin, as evidenced by other chapters in this volume. Accordingly there is a compelling need to provide validated, step-by-step protocols to allow integration of datasets across platforms and laboratories. In this chapter, we present our protocol for the isolation of yeast mononucleosomal DNA (principally but not exclusively *Saccharomyces* spp.) for both tiling microarray hybridization and next-generation sequencing. We provide the basics on how to map these data to genomic coordinates; for details on how these data can be analyzed for biological discovery, please see Chapters 26 and 27.

2. Materials

2.1. Nucleosome Isolation

1. 1 l Yeast-extract peptone dextrose (YPD) medium (20 g of agar, 10 g yeast extract, 20 g bacto-peptone in 860 mL de-ionized distilled H_2O). Autoclave for 40 min, then add 100 mL 20% sterile glucose.
2. Formaldehyde (see Note 1).
3. A 2.5 M Glycine stock is prepared by adding 19 g of glycine and dissolving into 100 mL of ddH_2O.
4. Zymolyase buffer: 1 M Sorbitol, 50 mM Tris–HCl, pH 7.4 with freshly added 10 mM β-mercaptoethanol.
5. Zymolyase at a stock of 12.5 mg/mL (dissolved in the Zymolyase buffer above).
6. Phosphate buffered saline (PBS): 137 mM NaCl, 2.7 mM KCl, 10 mM Na_2HPO_4 (dibasic, anhydrous), 2 mM KH_2PO_4 (monobasic, anhydrous) (pH 7.4).
7. 1× MNase buffer: 1 M Sorbitol, 50 mM NaCl, 10 mM Tris–HCl, pH 7.4, 5 mM $MgCl_2$, 1 mM $CaCl_2$ and 0.075% Igepal. Add β-mercaptoethanol to a final concentration of 1 mM from a 2.5-M stock and Spermidine from a 0.25-M stock to a final concentration of 500 μM to the MNase buffer prior to enzymatic digestion.
8. MNase dissolved in ddH_2O at a stock concentration of 10 U/μL. Store at −20°C.
9. Stop buffer: 5%SDS, 50 mM EDTA and 5 mg/mL of proteinase K.
10. Phenol equilibrated to pH 8.
11. Phenol:chloroform:isoamyl alcohol, 24:24:1, at pH 8.
12. 3 M NaAc, pH 5.2.

13. 100% Ethanol.
14. RNaseA.
15. 1× TBE buffer: 89 mM Trizma base, 89 mM boric acid, 2 mM EDTA.
16. Molecular weight standard, such as Fermentas 6× MassRuler™ (any vendor is OK).
17. 6× DNA Loading Dye: 0.25% xylene cyanol, 0.25% bromophenol blue, 30% glycerol.
18. Gel stain, such as SYBR Safe DNA Gel Stain (Invitrogen) (any vendor is OK).
19. Kit for DNA extraction, such as QIAquick Gel Extraction Kit.
20. If using the Agilent bioanalyzer, the Agilent DNA 1000 Kit 1 is required.

2.2. Microarray Hybridization

1. Deoxyribonuclease I (DNase I) (>10,000 U/mg), such as Invitrogen Amplification Grade DNase I.
2. 10× DNase I buffer: 10 mM Tris–acetate, pH 7.5, 10 mM magnesium acetate and 50 mM potassium acetate.
3. Biotin-N6-ddATP (1 nmole/μL) for biotin end labelling.
4. Terminal deoxynucleotidyl transferase (TdT).
5. 5× TdT buffer: 1 M potassium cacodylate, 125 mM Tris–HCl, 0.05% (v/v) Triton X-100, 5 mM $CoCl_2$ (pH 7.2 at 25°C) (can also be purchased from Invitrogen).

2.3. Parallel Sequencing

2.3.1. Quality Control #1

1. Microplate reader with 485(20)/535(25) fluorescence filter [e.g. Tecan Infinite 200 PRO multiplate reader (Tecan Group)].
2. Quant-iT dsDNA HS Assay Kit (high-sensitivity range), 100 assays "0.2–100 ng" (Invitrogen).
3. Quant-iT dsDNA BR Assay Kit (broad-range), 100 assays "2–1,000 ng" (Invitrogen).
4. Fluorometer [e.g. Qubit 1.0 or 2.0 (Invitrogen)].
5. Agilent 2100 Bioanalyzer (Agilent Technologies).
6. Agilent 2100 Bioanalyzer High Sensitivity DNA LabChips (Agilent Technologies).

2.3.2. Library Preparation

1. Molecular biology grade water, any vendor.
2. 10 mM Tris–Cl, pH 8.5 [e.g. EB buffer (Qiagen)].
3. Nucleotide removal kit [e.g. QIAquick Nucleotide Removal Kit (Qiagen)].
4. Spin column-based gel extraction kit [e.g. QIAquick Gel Extraction Kit (Qiagen)].
5. NEBNext DNA Sample Prep Reagent Set 1 (NEB).

6. T4 DNA Ligase (NEB).
7. Any high-fidelity PCR mix [e.g. Hifi Hot Start Readymix (Kapa Biosystems)].
8. Any low-range DNA ladder [e.g. GeneRuler low-range DNA ladder 25–700 bp (Fermentas)].
9. Any loading dye [e.g. 6× MassRuler Loading Dye Solution (Fermentas)].
10. Agarose powder, e.g.:
    (a) High-resolution agarose (Sigma-Aldrich).
    (b) Certified Molecular Biology agarose (Bio-Rad).
    (c) Certified Low-Melt agarose (Bio-Rad).
11. LabChip XT System (Caliper Life Sciences).
12. E-Gel iBase and E-Gel Safe Imager Combo Kit (Invitrogen).
13. E-Gel SizeSelect 2% Agarose (Invitrogen).
14. Any qPCR system, e.g. CFX96™ Real-Time PCR Detection System (Bio-Rad).
    (a) Please note that Kapa Library Quantification Kit/Illumina GA/Universal (Kapa Biosystems, KK4824) was optimized for the above system. Please check compatibility of this kit with the on-site qPCR instrument.
15. Plate roller for sealing multi-well plates, any vendor (e.g. VWR, part # 60941–118).
16. Kapa Library Quantification Kit/Illumina GA/ Universal (Kapa Biosystems, KK4824).
17. Indexed, amp-free adapter (100 μM): /5Phos/GAT CGG AAG AGC GGT TCA GCA GGA ATG CCG AGA CCG NNN NNN NNA TCT CGT ATG CCG TCT TCT GCT TG, IDT, HPLC purified.
18. Common, amp-free adapter (100 μM): 5′AATGATACGGCG ACCACCGAGATCTACACTCTTTCCCTACACGACGCT CTTCCGATC*T, IDT, HPLC purified, 3′ C contains a phosphothioate.
19. Ligation-mediated primer stocks (100 μM).
20. truncPCR primer 1: 5′ AAT GAT ACG GCG ACC ACC GAG A, IDT, desalted.
21. truncPCR primer 2: 5′ CAA GCA GAA GAC GGC ATA CGA G, IDT, desalted.
22. Read 1 SEQ primer (100 μM): 5′ AC ACT CTT TCC CTA CAC GAC GCT CTT CCG ATC T, IDT, desalted.
23. INDEX Read SEQ primer (100 μM): 5′ GGT TCA GCA GGA ATG CCG AGA CCG, IDT, desalted.

24. Read 2 SEQ primer (100 μM): 5′ CGG TCT CGG CAT TCC TGC TGA ACC GCT CTT CCG ATC T, IDT, desalted.

2.4. Sequencing: Cluster Generation

1. 10 N NaOH, any vendor.
2. TruSeq PE Cluster Kit v2.5 – cBot – HS (Illumina, part# PE-401-2510).
3. TruSeq SR Cluster Kit v2.5 – cBot – HS (Illumina, part# GD-401-2510).
4. PhiX Control Kit v3 (Illumina, part# FC-110-3001).

2.5. Sequencing

1. Optical Conical Tubes w/Plug Seal Caps, 175 mL.
2. TruSeq SBS Kit v5 – GA (36-cycle) Illumina, part# FC-104-5001.
3. TruSeq SBS Kit v5 – GA (20 pack/36-cycle) Illumina, part# FC-104-5020.
4. TruSeq SBS Kit – HS (200-cycle) Illumina, part# FC-401-1001.
5. TruSeq SBS Kit – HS (50-cycle) Illumina, part# FC-401-1002.

3. Methods

3.1. Growth of Cultures and Nucleosome Isolation

Timeline: the entire protocol, from starting the first culture to preparing samples for hybridization or sequencing takes 3 or 4 days, respectively.

1. (Day 0: afternoon) Inoculate a 100 mL overnight culture at 30°C or the appropriate temperature for other yeast strains (see Note 2).
2. (Day 1) In the early morning inoculate pre-culture into pre-warmed media to achieve an OD_{600} of 0.4 and shake in a baffled flask containing 400 mL of media in a 1-L flask in a water bath shaker at 225 rpm. Water bath shakers are preferable for maintaining temperature consistency and are essential for experiments with temperature-sensitive strains (see Notes 3 and 4).
3. Add 44 mL 10% formaldehyde for a final volume of 1% (see Note 5).
4. Cross-link with formaldehyde for 30 min while shaking at 30°C.
5. Quench the free formaldehyde by adding 2.5 M glycine (in water) for a final glycine concentration of 125 mM. Shake well immediately, then place in shaker for an additional 5 min (see Note 6).

6. Spin down the fixed, quenched cells in 250–500 mL centrifuge bottles at 2,000 × *g* at 4°C for 10 min. Discard supernatant into formaldehyde waste container. If all cultures are ready, proceed with spheroplasting, if not, drop freeze samples in liquid nitrogen and hold in LN2 or at −80°C until all samples have been collected. If freezing is required to schedule samples for the next step of processing, the final samples collected should also be frozen, so that all samples are thawed at the same time prior to spheroplasting (see Note 7).
7. Take fresh or thawed samples, resuspend cell pellet in 10-mL Zymolyase buffer and transfer to a 15- or 50-mL screw cap centrifuge tube. Add zymolyase from stock to a final concentration of 2 mg/mL. Place tubes on a Nutator or similar rocker shaker at 37°C for 30 min.
8. Centrifuge cells at 3,000 × *g* for 10–15 min at 4°C. Carefully pour off supernatant and resuspend cell pellet in 40 mL ddH_2O in a 50-mL conical. Pellet cells again at 2,000 × *g*, resuspend in 1 mL 1× PBS and transfer to a microfuge tube(s) so you can store your samples in a freezer box rather than having them loose in the freezer. Keep pellets to be further processed on ice and drop freeze the remainder in LN2 and store at −80°C (see Note 8).
9. Micrococcal nuclease digestion (see Note 9): Add 8 mL of MNase buffer. Mix samples gently and divide into six equal aliquots (1.2 mL per tube). Spheroplasted samples are digested with six concentrations of MNase ranging from 0 to 500 U for 45 min at 37° (the appropriate enzyme concentration depends on cell concentration and zymolyase activity and thus needs to be determined empirically). Inactivate MNase by adding 150 μL of stop buffer.
10. Reverse cross-links by incubating overnight at 65°C in water bath (see Note 10).
11. Samples are then extracted with an equal volume of phenol by vortexing for 1 min, followed by centrifugation for 5 min at 10,000 × *g*. The upper aqueous phase is further extracted with an equal volume of phenol:chloroform:isoamyl alcohol.
12. Samples are precipitated by addition of 1/10 volumes 3 M NaAc, pH 5.2, 2.5 volumes of 100% ethanol and centrifuged for 10 min at 13,000 × *g*. Pellets are washed with 70% ethanol, air-dried inverted at RT or for 5 min at low heat in a Speedvac centrifugal evaporator.
13. Dissolve DNA in 80 μL of molecular grade water. Add RNaseA to the DNA at a final concentration of 100 μg/mL for 1 h at 37°C.
14. MNased and RNased samples are then separated on a 2% agarose gel for 1 h at 100 V (1× TBE buffer) and stained with SYBR

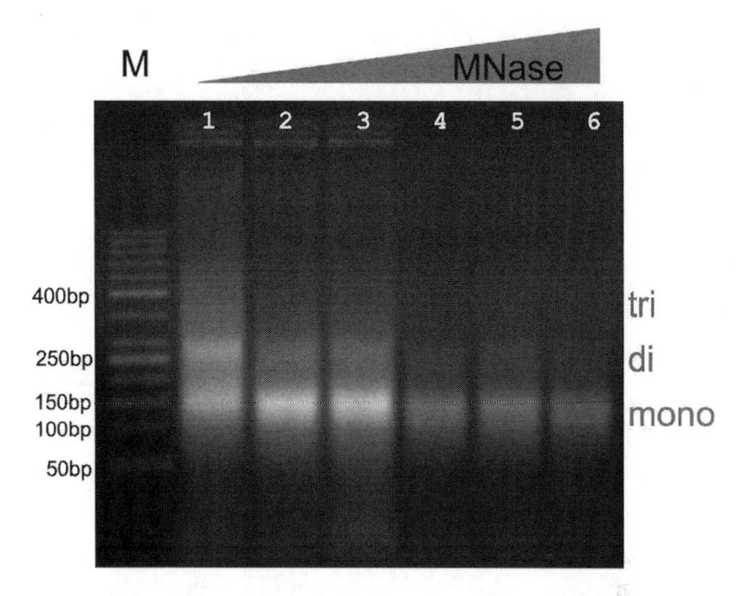

Fig. 1. Micrococcal nuclease digestion. Increasing concentrations of MNase were used to digest cross-linked DNA followed by gel electrophoresis. 1 μg of DNA was loaded in a 2% TAE gel at a running voltage of 100 V for 1 h. *Lanes 4* and *5* contain primarily mononucleosomes and would be selected for band isolation. *Lane 6*, in which only mononucleosomes are detectable, would be considered overdigested. M = DNA size marker.

Safe DNA Gel Stain. The MNase digestion that results in primarily (>70%) mononucleosomal DNA is selected and the gel band corresponding to 146 bp (Fig. 1) is excised and purified using a gel extraction kit (e.g. QIAquick Gel Extraction Kit) and eluted in 80 μL of molecular grade water.

15. 1 μL of the gel purified material is analyzed using an Agilent bioanalyzer (or equivalent) using the Agilent DNA 1000 Kit. A range of 100–200 bp is expected with an average size of 130–150 (Fig. 2). These samples can then be used for either array hybridization following DNase I digestion (5 μg material required) or Illumina sequencing library preparation (1 μg material required).

3.2. Microarray Hybridization (see Notes 11 and 12)

3.2.1. Sample Labelling

1. 5 μg of purified mononucleosomal DNA is digested with DNase I for 1 min at 37°C using 0.1 U of DNase I in a total volume of 100 μL of 1× DNase I buffer. The reaction is stopped by heat inactivation at 65°C for 10 min. To simplify the workflow, these reactions should be performed in an open PCR machine. Initially, several digestion attempts may be necessary.
2. Have 2% agarose gel ready to load and analyze samples immediately after DNase I digestion. Load 2 μL of sample and, after electrophoresing, visualize the gel on a UV light box. The ideal DNase I digestion should appear as a smear centered at ~75 bp (see Fig. 3). If the DNA appears as a large smear or if its distribution

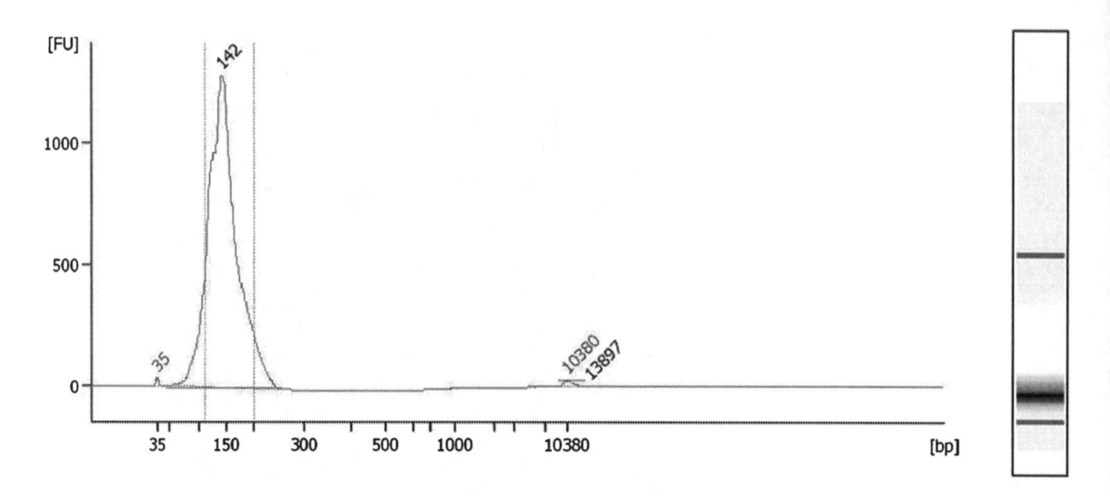

Fig. 2. High-sensitivity kit bioanalyzer output. The electropherogram shows a yeast nucleosome fragment distribution following MNase digestion with a peak at 140–150 bp.

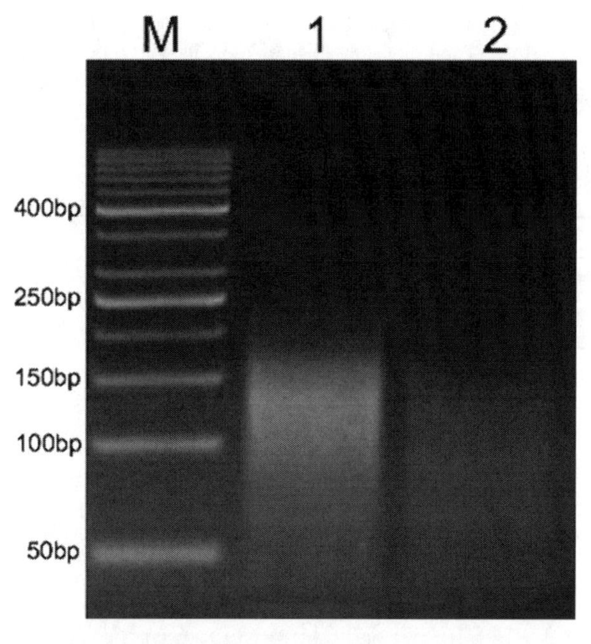

Fig. 3. Digestion of isolated mononucleosomes with DNase I. Approximately 5 μg of gel purified mononucleosomes was digested with 0.1 U of DNase I for 1 min at 37°C and then immediately heat inactivated at 65°C for 10 min. *Lane 1* contains mononucleosomes after gel purification (average size 150 bp) and *lane 2* is a similar sample following digestion with DNase I, with an average size of 75 bp). 250 ng of DNA was loaded in a 2% TAE gel at a running voltage of 100 V for 1 h. M = DNA size markers.

is not centered at 75 bp add an additional 1.0 μL of 0.1 U of DNase I to the reaction and incubate for an additional 1 min at 37°C. Additional rounds of digestion may be required with shorter incubation times until the ideal digestion size is obtained.

3. End label the DNase I-digested DNA fragments with 1 μL of Biotin-N6-ddATP, 1.54 μL of 20 U/μL TdT and 1× TdT buffer in a total reaction volume of 120 μL by adjusting with ddH_2O and incubate 2 h at 37°C.
4. Labelled DNA can be stored at 4°C for up to 1 week.
5. Process samples according to manufacturer's protocol for the array being utilized (see Notes 11 and 12).

3.2.2. Data Analysis

Data analysis of raw data (.CEL files) can be done using Affymetrix Tiling Analysis Software (http://www.affymetrix.com/support/developer/downloads/TilingArrayTools/index.affx) and visualized with Affymetrix Integrated Genome Browser (http://www.affymetrix.com/support/developer/tools/download_igb.affx). Set up a tiling analysis group (.TAG file) for a two-sample analysis containing the nucleosomal experiments as the "treatment" and the whole-genome samples as the "control". The data is normalized together with the built-in quantile normalization and probe-level analysis with both perfect match and mismatch (PM/MM) probes and run with a bandwidth of 40. This .TAG file is then subjected to "Analyze Intensities", resulting in a .bar file (Binary Analysis Results), a binary file that needs to be converted to a .txt file. This can be achieved by using a program that is part of the Affymetrix DevNet tools (http://www.affymetrix.com/partners_programs/programs/developer/tools/devnettools.affx) by using the program bar2txt. This new .txt file contains chromosome # on column 1, probe center on column 2 and log2 intensity values on column 3. To define nucleosome occupancy genome annotations are downloaded from Saccharomyces Genome Database (ftp://ftp.yeastgenome.org/yeast). Probe positions can be aligned according to the genomic feature and the intensity values are plotted to represent the nucleosome occupancy at the genomic features.

3.3. Parallel Sequencing Protocol

3.3.1. Sequencing: Quality Control #1 (Prior to Library Preparation)

1. Samples are stored at −20°C with freeze–thaw cycles minimized. Oligonucleotides are resuspended to 100 μM in molecular biology grade water and aliquoted into 1.5-mL low-retention tubes and stored at −20°C.
2. Quality control prior to library preparation consists of two essential steps (1) determining sample concentration and (2) determining size distribution of the digested nucleosomal DNA.
   (a) Depending on throughput, concentration and size can be determined by measuring fluorescence in a single step using one of two systems:
      - For high-throughput applications we recommend running the HT High Sensitivity DNA assay on the Caliper GX system.

- For low-throughput applications we recommend running 1 μL of sample on the Agilent 2100 Bioanalyzer High Sensitivity LabChip (Agilent Technologies) (Fig. 2). The Bioanalyzer High Sensitivity kit has a quantitative range of 0.005–0.5 ng/μL. For single-broad peak samples, concentration should be in the 0.05–1 ng/μL range for best results. Increased sample concentrations improve chances for detection of low-levels of contamination. If the sample is too dilute, we recommend concentrating the sample by vacuum concentration (not by column) and resuspending in 10 mM Tris–Cl, pH 8.5 (Qiagen) or molecular grade water. For resuspension of samples, we *do not recommend* using buffers that contain EDTA, e.g. TE buffer.
- Alternatively, for size determination, we recommend running a 2% or higher high-resolution agarose gel (Sigma) in 1× TBE buffer, alongside a marker lane containing appropriate size standards such as GeneRuler low-range DNA ladder 25–700 bp (Fermentas).

(b) For high-throughput applications, concentration (not integrity) can be determined by fluorescence in a suitable multiplate reader. We use Quant-iT dsDNA Assay Kits (Invitrogen) on a Tecan Infinite 200 PRO microplate reader with excitation/emission maxima 485(20)/535(25) nm. Similarly, the Qubit 1.0 or 2.0 fluorimeter (Invitrogen) can be used for low-throughput applications.

3.3.2. Library Preparation

Here, we report the standard Illumina ChIP-Seq protocol, with minor modifications, including: use of NEB libary construction chemistry an overnight ligation step, modifications to the cleanup protocol between steps, and the substitution of standard Illumina adapters with custom full-length, amplification-free, indexed adapters. Illumina-specific, amplification-free adapter sequences were first published by Kozarewa et al. (3). We have modified those adapters to contain indexes for multiplexing purposes.

1. Determine the appropriate library preparation protocol based on the input amount of the digested nucleosomal DNA (Fig. 4):

   (a) For 1–5 μg follow the standard Illumina PE protocol, with minor protocol modification (above). These libraries can be constructed and sequenced without ligation-mediated amplification step.

   (b) For 250 ng–1 μg follow the standard Illumina PE protocol, with minor protocol modification (as in a). Post-ligation these libraries need 8–12 cycles of PCR.

   (c) For 100 ng–250 μg follow Illumina Ch1p-seq protocol with 14 cycles of PCR.

   (d) For 15 ng–100 μg as (c) with 16–18 cycles of PCR.

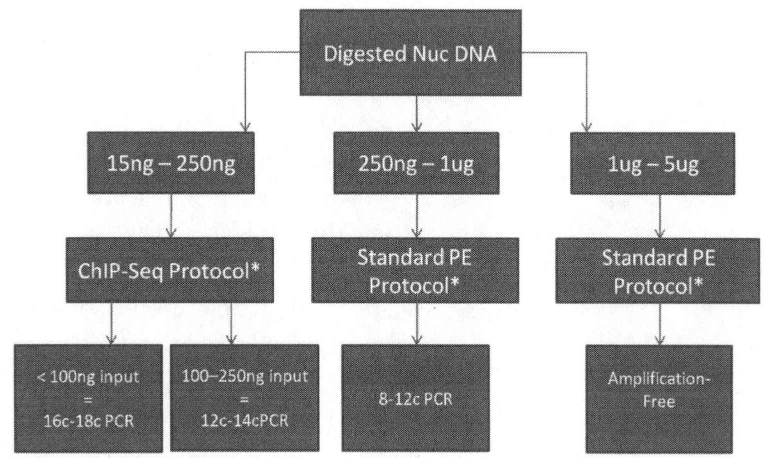

Fig. 4. Model for optimized library construction based on varying concentrations of input material.

Table 1
End-repair reaction mix

Reagent	Volume (μL)
Digested, nucleosomal dsDNA	40
10× Phosphorylation reaction buffer	5
(100 mM) Deoxynucleotide solution mix	2
T4 DNA polymerase	1
(1:5) DNA polymerase I, Large (Klenow) Fragment	1
T4 Polynucleotide kinase	1
Total volume (μL)	50

2. All library construction reagents listed below are components from the NEBNext DNA Sample Prep Reagent Set 1 kit (NEB, part# E6000L). Individual components can be purchased separately from various vendors; however, enzyme and buffer concentrations will vary and optimization may be needed.
3. Construct libraries by following the modified ChIP-Seq protocol (<250 ng input):
   (a) End-Repair
   - Dilute Klenow DNA polymerase 1:5 in nuc-free water for a final enzyme concentration of 1 U/μL.
   - Prepare the reaction mix shown in Table 1:
   - Incubate for 30 min at 20°C.

- Remove nucleotides from end-repaired nucleosomal DNA using a spin column system, such as QIAquick Nucleotide Removal Kit.
- We use the following modified protocol for oligonucleotide (17–40 mers) and DNA (40 bp to 10 kb) cleanup from enzymatic reactions, using the QIAquick Nucleotide Removal Kit:
  - Add 10 volumes of Buffer PN to 1 volume of the reaction sample and mix.
  - Place a MinElute spin column in the provided 2-mL collection tube.
  - To bind DNA, apply the sample to the MinElute spin column, incubate at room temperature for 2 min and centrifuge for 1 min at 5,000 × *g*. Repeat by re-loading flow-through onto same column, incubating at room temperature for 5 min and centrifuging for 1 min at 5,000 × *g*.
  - To wash the MinElute spin column, add 750 μL of Buffer PE and centrifuge for 1 min at 5,000 × *g*.
  - Discard the flow-through and place the MinElute spin column back in the same tube, which should be empty. Centrifuge for additional 2 min at 13,000 rpm (17,900 × *g*).
  - Place the MinElute spin column in a clean, labelled 1.5-mL microcentrifuge tube. To dry the column, leave the column open at room temperature for 10 min.
  - To elute DNA, add half the desired elution volume of EB buffer (10 mM Tris–Cl, pH 8.5) or water (pH 7.0–8.5) to the center of the MinElute spin column, incubate at room temperature for 2 min, and centrifuge the column for 1 min at 13,000 rpm (17,900 × *g*). Repeat by adding the other half of desired elution volume to the column and incubating at room temperature for 5 min, centrifuge for 3 min at 13,000 rpm (17,900 × *g*). This is a safe stopping point. Store the construct at –20°C.

(b) A-tailing:

- Prepare the reaction mix shown in Table 2:
- Incubate for 30 min at 37°C (pre-heated water bath is recommended).
- Remove nucleotides from nucleosomal DNA using spin column, such as QIAquick Nucleotide Removal

Table 2
A-tailing reaction mix

Reagent	Volume (μL)
Digested, end-repaired, nuc dsDNA	34
10× NE Buffer 2 for Klenow Fragment (3′→5′ exo–)	5
1 mM 2′-deoxyadenosine-5′-triphosphate (dATP)	10
Klenow exo– (3′→5′ exo minus)	1
Total volume (μL)	50

Table 3
Adapter-ligation reaction mix

Reagent	Volume (μL)
A-tailed, nuc dsDNA	23.5
10× T4 DNA Ligase buffer	3
(1:20) Diluted adapter oligo mix (0.75 μM→25 nM)	1
(400 U/μL) T4 DNA Ligase	2.5
Total volume (μL)	30

Kit (Qiagen). See last step in End Repair section immediately preceding. This is *not* a safe stopping point. Please proceed to adapter ligation.

(c) Adapter-ligation:

- Adjust adapter concentration for the smaller quantity of input DNA. Use water to dilute 15 μM annealed adapter oligo mix 1:20 to a final concentration of 0.75 μM. Label 1:20 dilution tube with Oligo ID and date of dilution. Store at –20°C after use.
- Prepare the reaction mix shown in Table 3, adding enzyme last:
- Incubate overnight at 16°C.
- Purify nucleosomal DNA from adapter oligos *using standard PCR purification protocol (e.g. MinElute protocol)*, eluting in 25 μL manufacturer's recommended elution buffer or water. This is a safe stopping point. Proceed or store the construct at –20°C.

(d) Ligation-mediated PCR:

- Dilute a small aliquot of ligation-mediated primer stocks (100 μM) to 25 μM in water.

Table 4
Ligation-mediated PCR reaction mix

Reagent	Volume (μL)
Adapter-ligated, nuc dsDNA	23
(25 μM) truncPCR primer 1	1
(25 μM) truncPCR primer 2	1
2× KAPA HiFi HotStart ReadyMix (contains 2.5 mM Mg^{2+} at 1×)	25
Total volume (μL)	50

Label dilution tubes with Oligo ID and date of dilution. Store at −20°C after use.

- For 1–5 μg follow the standard Illumina PE protocol, with minor protocol modification. These libraries can be constructed and sequenced without ligation-mediated amplification step.
- For 250 ng–1 μg follow the standard Illumina PE protocol, with minor protocol modification. Post-ligation these libraries need 8–12 cycles of PCR.
- For 100 ng–250 μg as (b) with 14 cycles of PCR.
- For 15 ng–100ug as (b) with 16–18 cycles of PCR.

• Prepare the reaction mix shown in Table 4 (add enzyme mix last):

• Amplify using the following PCR protocol:
  - 2 min at 98°C
  - 16–18 cycles of steps 2–4:
  - 30 s at 98°C
  - 30 s at 56°C
  - 30 s at 72°C
  - 5 min at 72°C
  - Hold at 4°C

• Purify nucleosomal DNA from adapter oligos using *standard PCR purification protocol* (*e.g. MinElute protocol*), eluting in 35 μL manufacturer's recommended elution buffer or water. This is a safe stopping point. Proceed or store the construct at −20°C.

(e) The size-selection step can be performed using several systems, including but not limited to E-gel SizeSelect gels (Invitrogen), Agencourt AMPure XP bead-based system (Beckman-Coulter Genomics), Caliper LabChip XT fractionation system and manual size selection. We describe here the protocol for manual size selection.

Manual Size Selection and Cleanup (up to 8 h Hands-on Time) Protocol

1. Using a clean, 14 + cm gel tray prepare a 100 mL, 3% agarose gel in 1× TBE buffer containing ethidium bromide or Sybr safe or other appropriate DNA staining agent. Use broad-well comb (or combine wells using tape).
2. For non-indexed samples, run one sample per gel to avoid cross-contamination between samples. For indexed samples, you can run several samples per gel with 2–3 wells space between each sample. Indexed samples run on the same gel contain distinct indexes.
3. Add 7 μL of 6× loading buffer to 25 μL of the DNA from the purified ligation reaction.
4. Load 500 ng of a marker containing appropriate size standards, such as GeneRuler low-range DNA ladder 25–700 bp (Fermentas). Load the entire sample (41 μL) in another lane of the gel, leaving at least one empty lane between ladder and sample. Do not overload wells.
5. For best separation, run gel at 120 V for 5 min, followed by:
   (a) Low-melt gel (Bio-Rad): 6 h at 60 V to preserve integrity of the gel.
   (b) High-resolution (Sigma) and molecular biology grade gels (Bio-Rad): 3–4 h at 80 V.
6. Monitor the run frequently.
7. View the gel using a Dark Reader (Clare Research) transilluminator to avoid UV light damage.
8. Image the gel before and after the slice is excised.
9. Excise region of interest with a clean scalpel. Use a new/clean scalpel for each sample. The gel slice should contain material in the 180 ± 40-bp range if original digestion was in the 60-bp range, and 270 ± 40 bp if the original digestion was in the 150-bp range.
10. Follow standard gel extraction protocol, e.g. QIAquick Gel Extraction protocol, with following exceptions:
    (a) Use MinElute columns if possible.
    (b) Adjust incubation times as outlined above in the modified protocol for oligonucleotide (17–40 mers) and DNA

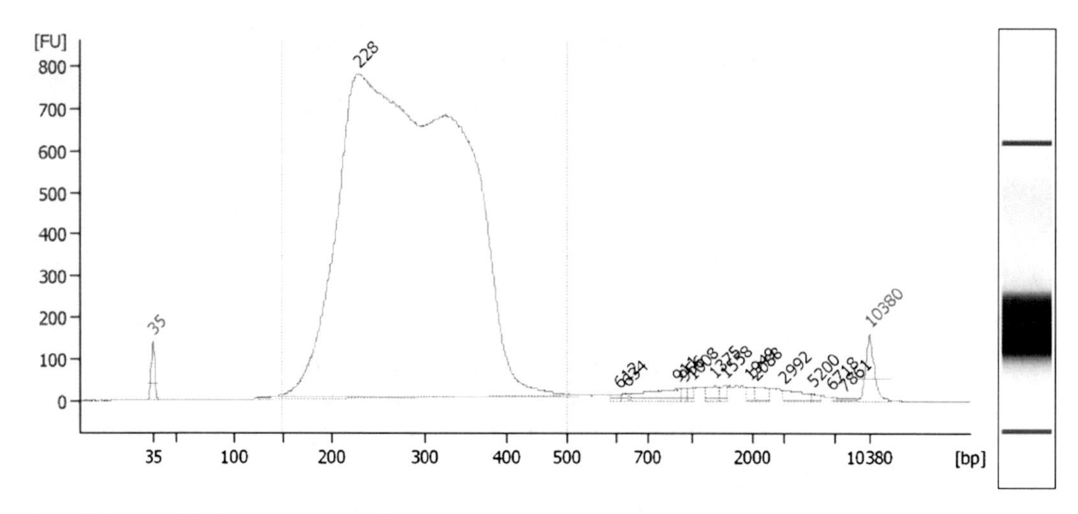

Fig. 5. High-sensitivity bioanalyzer trace of a nucleosomal DNA sample suitable for sequencing.

(40 bp to 10 kb) cleanup from enzymatic reactions, using the QIAquick Nucleotide Removal Kit

11. Perform two 20-μL elutions per column in EB buffer (10 mM Tris–Cl, pH 8.5) or water (pH 7.0–8.5).

3.3.3. Sequencing: Quality Control #2 (Prior to Sequencing)

1. If continuing from ChIP-Seq library prep protocol, concentrate the sample by speed-vacuum and resuspend in 10 μL elution buffer (EB) or molecular grade water. Not necessary if continuing from a standard Illumina PE protocol.
2. Run 1 μL of sample on a high-sensitivity DNA bioanalyzer chip or at least 1 ng on the Caliper GX HT High Sensitivity DNA assay (Fig. 5).

3.3.4. Sequencing: Quality Control #3 (Library Quantification with Quantitative PCR)

This section describes absolute quantification of templates that have adapter sequences ligated to both ends of target DNAs using Sybr Green assay on a qPCR instrument.

Quantitative PCR reagents and standards listed below are components from the Kapa SYBR Library Quantification (Illumina-compatible) Universal Kit (Kapa Biosystems, KK4824).

1. Based on the fluorescence measurements, dilute library stock (original template) to 30 nM.
2. Dilute 25 μL of Tween 20 in 50-mL molecular biology water to a final concentration of 0.05% Tween to be used in subsequent dilution reactions
3. Prepare serial dilutions of templates. Dilute each template as follows:
   (a) Dilution A: Dilute original template 1:1,000 in 0.05% Tween.

Table 5
PCR master mix

Library quantification master mix (D-Mark BioSciences)	1× (μL)
Molecular biology water (minus Tween 20)	4
2× Kapa SYBR FAST Universal qPCR Master Mix	12
Total volume	16

   (b) Dilution B: Dilute *Dilution A* 1:2 in 0.1% Tween (e.g. 200 μL Dil.1 in 200 μL 0.1% Tween).

4. Prepare master mix (in a DNA Template-free hood):

   (a) Prepare the master mix as shown in Table 5.

   (b) Vortex for 10 s and place on ice.

   (c) Plate setup (see Table 6) – work quickly to reduce exposure to direct light and evaporation:

   - Load each well with 16 μL chilled Kapa library quantification master mix using 5–100-μL repeater pipette (in template-free hood).
   - Load 6 standards supplied in the Kapa SYBR Library Quantification (Illumina-compatible) Universal Kit (Kapa Biosystems, KK4824) using 0.5–10-μL repeater pipette (DNA hood):
   - Well A1-3: 4 μL undiluted standard 1 into each well A1-3.
   - Well A4-6: 4 μL undiluted standard 2 into each well. Continue until you have loaded all 6 standards in triplicates. Standards are as follows: Std1 20pM, Std2 2pM, Std3 0.2pM, Std4 0.02pM, Std5 0.002pM, Std6 0.0002pM.
   - Load diluted templates using 0.5–10-μL repeater pipette as follows:
   - Well C1-3: 4 μL of 1:1,000 diluted library (unknown) 1 into each well.
   - Well C4-6: repeat above step.
   - Well D1-3: 4 μL of (1:1,000) diluted PhiX control library.
   - Well D4-6: 4 μL of (1:2,000) diluted PhiX control library.

   (d) Keep plate on ice.

Table 6
Plate setup

Well ID	1	2	3	4	5	6	7	8	9	10	11	12
A	20pM	20pM	20pM	2pM	2pM	2pM	0.2pM	0.2pM	0.2pM	0.02pM	0.02pM	0.02pM
B	0.002pM	0.002pM	0.002pM									
C	Unkn#1 (1:1,000)	Unkn#1 (1:1,000)	Unkn#1 (1:1,000)	Unkn#1 (1:2,000)	Unkn#1 (1:2,000)	Unkn#1 (1:2,000)	Unkn#2 (1:1,000)	Unkn#2 (1:1,000)	Unkn#2 (1:1,000)	Unkn#2 (1:2,000)	Unkn#2 (1:2,000)	Unkn#2 (1:2,000)
D	PhiX (1:1,000)	PhiX (1:1,000)	PhiX (1:1,000)	PhiX (1:2,000)	PhiX (1:2,000)	PhiX (1:2,000)						

Table 7
PCR protocol

Procedure	Temperature (°C)	Time
Denaturation	95	5 min
40 cycles	95 60	30 s 1 min

(e) Load 16 μL master mix per well and apply plastic seal gently.

(f) Centrifuge plate for 1 min at 1,000 rpm (100 × *g*).

(g) Remove plastic seal, add 4 μL diluted template per well, apply plastic seal tightly using plate roller.

(h) Centrifuge plate for 1 min at 1,000 rpm (100 × *g*).

(i) Run the protocol shown in Table 7 on a Bio-Rad CFX96 qPCR instrument (any calibrated instrument should work but will require validation).

For each dilution series, average out Mean Starting Quantity (SQ) values for the three technical replicates. Use average size in base pairs for each template to calculate nM concentration values, using the following formula:

For 1:1,000 dilution, nM = Average of mean SQ × (452 bp/ave template size in bp)

For 1:2,000 dilution, nM = (Average of mean SQ × (452 bp/ave template size in bp)) × 2.452 bp *is the template size in bp for the six standards.*

Proceed with 25–30 nM of material into template denaturation and cluster generation steps on Illumina's cBot platform based on protocol outlined in cBot guide, part # 15006165 Rev D. October 2010.

Analysis of sequencing data. The analysis of sequencing data can be more daunting than that of the tiling array data, primarily because the number of available sequencing parameters is large and the available analysis tools are diverse and growing. Analysis will depend on length of sequences and single-end vs. paired-end reads. Sequence reads in qseq or fastq format can be mapped to the yeast genome using several programs (e.g. Maq, Soap, Novocraft, Bowtie and Casava). Typically we use unique reads and then define the chromosome location, i.e. the start and end of each read on the genome. For paired end reads, mapping is straightforward because the read start and stop represent both ends, and therefore the entire mono-nucleosome fragment. In contrast, for single end reads, the start and stop are equivalent to the read length and an additional normalization step is needed to define the nucleosomal DNA ends. One method is to extend the reads to 147 bp from the start while another method is to define the reads that map to the forward strand and those that map consecutively to the reverse strand then defining the center. After the nucleosomal fragments are mapped using either SE or PE reads, the number of reads for each base is used to calculate nucleosome occupancy across the genome.

4. Notes

1. We use Polysciences 10% methanol-free formaldehyde. Other formaldehyde preparations often contain methanol or other stabilizers that can interfere with downstream processing. Two months after opening the formaldehyde, do not use for nucleosome preps, although it can be repurposed for immunocytochemistry, etc.
2. For species other than *Saccharomyces cerevisiae* or mutants, verify strain identity by PCR with strain-specific primers and grow cultures at their optimum temperature.
3. Temperature sensitive allele strains will vary in their response to different temperatures so it is important to find the appropriate temperature and duration of the restrictive temperature for the particular strain (4). Once this is determined, the temperature must be reached rapidly by mixing the culture with an equal volume of media at an elevated temperature. For example, if the initial temperature is 25°C and the desired temperature is 37°C add equal amounts of media at 25 and 50°C then mix immediately before returning the flask to the water bath. After 3 h check the OD_{600}. A final OD_{600} as close to 1.0 is the goal (see Note 4).
4. Some yeast mutants will change their morphology at restrictive temperatures, which can result in an increase in OD_{600} without an increase in cell number; therefore, the first time you work with a strain, check it microscopically.
5. All formaldehyde-containing solutions must be disposed of properly.
6. Formaldehyde cross-links DNA-protein interactions forming covalent linkages with any exposed amino and imino groups (e.g. those in lysine and arginine side chains). This forms a Schiff's base that can participate in a second linkage, creating methylene bridges between amino acids that are in close proximity) (5). The chemistry that results in quenching of the cross-linking reaction is not completely understood.

Glycine can act to quench the reaction by reacting with remaining formaldehyde molecules as well as with the formaldehyde-modified residues (6).

7. Frozen samples can be stored for at least 12 months at –80.
8. The spheroplasting step is inherently variable, depending on strain, growth media, growth phase of the culture, etc. Check the degree of spheroplasting for each preparation microscopically. Place a 3-μL drop of sample on 1 × 3″ slide and cover with a 25 mm^2 coverslip. This will flatten the cells, such that while you are observing them in phase contrast, a small amount of added pressure applied with a pen or pipette tip will cause the cells to burst and form transparent ghosts, while the undigested cells will remain refractile. We aim for 95% spheroplasts/sample (Fig. 6).
9. Micrococcal nuclease has been the enzyme of choice for isolating nucleosomal fragments for several decades, both on the single gene level and for genome-wide studies, based on its

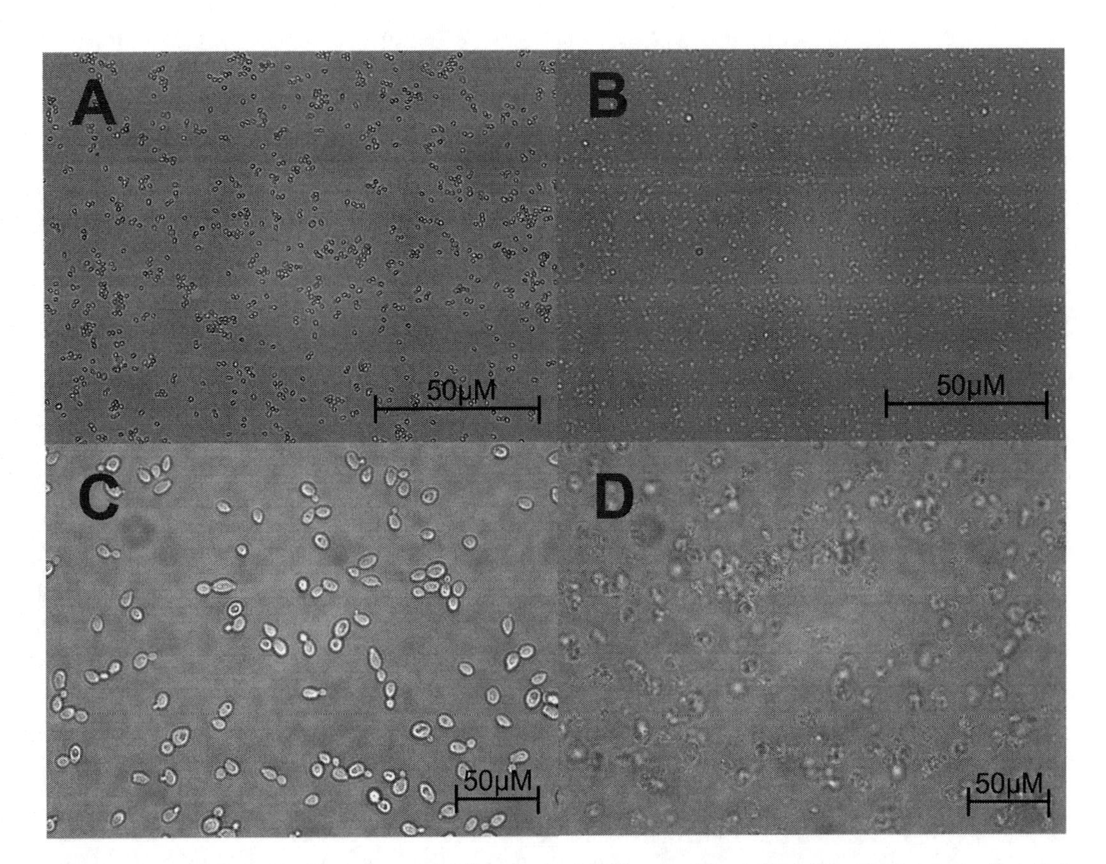

Fig. 6. Zymolyase treatment to digest the yeast cell wall prior to MNase treatment. (**a**, **c**) Show fixed, intact cells prior to digestion at 200× and 630× respectively. (**b**, **d**) Show the same cells following zymolase treatment. The sharply defined, phase-dense cells are converted into phase-opaque "ghosts" that lack sharply defined edges after treatment.

preference for linker DNA vs. nucleosome-wrapped DNA. It has long been appreciated that the activity of this enzyme depends upon several factors, including enzyme source and lot, and sample composition. Therefore, MNase studies typically involve performing a titration to empirically determine the "best" sample for further study. A careful analysis of what defines such an optimal sample has been carried out by Rando's group (7). Specifically, they under-digested and overdigested with MNase and used next generation sequencing to characterize the samples. They found that maps from the three titration steps broadly agree, and the primary (though not exclusive) differences were changes in occupancy rather than positions of nucleosomes (8). The sensitivity of DNA to MNase digestion underscores the need for careful technique, meticulous record keeping and implementing quality checks throughout the protocol. This vigilance is important regardless of the final readout, but particularly for next generation sequencing. Because the selectivity of MNase cleavage of DNA within linkers vs. nucleosomes is limited to a factor of ~25, the effect of potential biases introduced by MNase have attracted renewed attention, e.g. (7, 9). It is clearly important to confirm nucleosome mapping studies with other methodologies. To date, some promising alternatives include; DNase I and restriction enzyme digestion and the use of modified histones (for review see ref. 10). These alternative approaches will become increasingly important for analyses that interrogate specific subsets of nucleosomes and their dynamic behaviour.

10. Cross-link reversal at high temperature presumably disrupts any novel hydrogen bonds formed by formaldehyde cross-linking (11).

11. This protocol is for specifically for Affymetrix arrays. Please see Rando (12) for methods for other array types, e.g. Agilent, Nimblegen and home spotted arrays.

12. Microarray Design (PN 520055) (see ref. 13) contains 25-mer probes spaced every 8 bp covering one strand of the entire *S. cerevisiae* genome sequence and a second set of probes offset 4 bp from the first set corresponding to the other strand. Therefore, when combined, these probes provide whole genome coverage with 4-bp resolution for double-stranded hybridization samples. The array contains 6.5 million 5-μM oligonucleotide features and is compatible with commercially available Affymetrix scanners. Other Affymetrix yeast tiling arrays are also commercially available, such as the *S. cerevisiae* Tiling 1.0R Array, which also contains 25-mer probes tiled every 5 bp for one strand of the genomic DNA. For these arrays, the protocol for nucleosome preparation, labelling, staining, washing and scanning are exactly the same.

References

1. Yuan, G.C., et al., *Genome-scale identification of nucleosome positions in S. cerevisiae.* Science, 2005. **309**(5734): p. 626–30.
2. Lee, W., et al., *A high-resolution atlas of nucleosome occupancy in yeast.* Nat Genet, 2007. **39**(10): p. 1235–44.
3. Kozarewa, I., et al., *Amplification-free Illumina sequencing-library preparation facilitates improved mapping and assembly of (G+C)-biased genomes.* Nat Methods, 2009. **6**(4): p. 291–5.
4. Badis, G., et al., *A library of yeast transcription factor motifs reveals a widespread function for Rsc3 in targeting nucleosome exclusion at promoters.* Mol Cell, 2008. **32**(6): p. 878–87.
5. Schmiedeberg, L., et al., *A temporal threshold for formaldehyde crosslinking and fixation.* PLoS One, 2009. **4**(2): p. e4636.
6. Sutherland, B.W., J. Toews, and J. Kast, *Utility of formaldehyde cross-linking and mass spectrometry in the study of protein-protein interactions.* J Mass Spectrom, 2008. **43**(6): p. 699–715.
7. Weiner, A., et al., *High-resolution nucleosome mapping reveals transcription-dependent promoter packaging.* Genome Res, 2010. **20**(1): p. 90–100.
8. Pugh, B.F., *A preoccupied position on nucleosomes.* Nat Struct Mol Biol, 2010. **17**(8): p. 923.
9. Fan, X., et al., *Nucleosome depletion at yeast terminators is not intrinsic and can occur by a transcriptional mechanism linked to 3′-end formation.* Proc Natl Acad Sci USA, 2010. **107**(42): p. 17945–50.
10. Chatterjee, C. and T.W. Muir, *Chemical approaches for studying histone modifications.* J Biol Chem, 2010. **285**(15): p. 11045–50.
11. Rait, V.K., et al., *Modeling formalin fixation and antigen retrieval with bovine pancreatic RNase A II. Interrelationship of cross-linking, immunoreactivity, and heat treatment.* Lab Invest, 2004. **84**(3): p. 300–6.
12. Rando, O.J., *Genome-wide mapping of nucleosomes in yeast.* Methods Enzymol, 2010. **470**: p. 105–18.
13. David, L., et al., *A high-resolution map of transcription in the yeast genome.* Proc Natl Acad Sci USA, 2006. **103**(14): p. 5320–5.

Chapter 24

Genome-Wide Approaches to Determining Nucleosome Occupancy in Metazoans Using MNase-Seq

Kairong Cui and Keji Zhao

Abstract

The precise location of nucleosomes in functional regulatory regions in chromatin is critical to the regulation of transcription. The nucleosome structure protects DNA from micrococcal nuclease (MNase) digestion and leaves a footprint on DNA that indicates the position of nucleosomes. Short sequence reads (25–36 bp) from ends of mononucleosome-sized DNA generated from MNase digestion of chromatin can be determined using next-generation sequencing techniques. Mapping of these short reads to the genome provides a powerful genome-wide approach to precisely define the nucleosome positions in any genome with known genomic sequence. This chapter outlines the reagents and experimental procedures of MNase-Seq for mapping nucleosome positions in the human genome.

Key words: Nucleosome mapping, MNase-Seq, Chromatin structure

1. Introduction

Nucleosomes, consisting of approximately 146 bp of DNA wrapped around a histone octamer, are the fundamental structural units of chromatin in metazoans (1, 2). The translational positioning of nucleosomes along DNA is implicated in profoundly influencing gene expression (3–6). Thus, defining the nucleosome positioning and occupancy is critical to understand the mechanisms of regulation of transcription by chromatin.

Nucleosome structure is resistant to micrococcal nuclease (MNase) digestion, leaving a footprint of about 150 bp that reflects the position of a nucleosome (7). Therefore, determining the boundaries of these footprints indicates the positions of nucleosomes in the genome. Since the genomic sequences of most model organisms

Randall H. Morse (ed.), *Chromatin Remodeling: Methods and Protocols*, Methods in Molecular Biology, vol. 833,
DOI 10.1007/978-1-61779-477-3_24, © Springer Science+Business Media, LLC 2012

are already available, sequencing a short tag from DNA at each end of the nucleosome is sufficient to determine its position in the genome. Thus, the next-generation sequencing techniques are perfectly suited for this purpose (8). We have generated genome-wide maps of nucleosome positions in both resting and activated human CD4+ T cells by direct sequencing of nucleosome ends using the Illumina Genome Analyzer Platform (MNase-Seq) (9). As the next-generation sequencing techniques improve, the capacity and cost of sequencing become lower. For example, one sequencing run on the Illumina Genome Analyzer II can produce 100–200 million sequencing reads, which is sufficient to reach 10× coverage for all nucleosomes in the human genome.

We describe two different methods to prepare nucleosome templates used for sequencing. One is digestion of native chromatin and the other is digestion of formaldehyde-cross-linked chromatin by MNase. The native nucleosome protocol works well to reveal stable nucleosome structure and avoid cross-linking of non-histone proteins; the cross-linking protocol may stabilize "unstable" nucleosomes, but may also stabilize non-nucleosome structures that are resistant to MNase digestion.

2. Materials

2.1. Preparation of Nucleosome DNA Templates

1. 1× PBS: 137 mM NaCl, 2.7 mM KCl, 10 mM Na_2HPO_4, 2 mM KH_2PO_4, without calcium and magnesium, pH7.4.
2. Lysis buffer: 10 mM Tris–HCl, pH7.5, 10 mM NaCl, 3 mM MgCl2, 0.5% NP-40, 0.15 mM spermine, 0.5 mM spermidine.
3. Formaldehyde (37%).
4. 1× PBS + 0.5% Triton X-100 solution.
5. MNase digestion buffer: 10 mM Tris–HCl, pH7.4, 15 mM NaCl, 60 mM KCl, 0.15 mM spermine, 0.5 mM spermidine.
6. $CaCl_2$, 1 M solution.
7. MNase.
8. Stop buffer: 20 mM EDTA, 20 mM EGTA, 0.4% SDS, and 0.5 mg/ml proteinase K.
9. Phenol/chloroform.
10. Glycogen, 2 mg/ml stock solution.
11. Ethanol, 100%.
12. Sodium acetate (NaAc), 3 M, pH 5.3.
13. 70% ethanol.
14. 1× TE buffer: 10 mM Tris–HCl, 1 mM EDTA, pH7.4.
15. E-Gel EX 2% agarose (Invitrogen).

16. E-Gel iBase Power System (Invitrogen).
17. E-Gel Safe Imager Real-time Transilluminator (Invitrogen).
18. Gel knife or sterile razor blade.
19. QIAquick Gel Extraction Kit (Qiagen).

2.2. Preparation of DNA for Sequencing

1. Epicentre DNA END-Repair kit (Epicentre Biotechnologies).
2. 10× end repair buffer: 330 mM Tris–acetate, pH7.8, 660 mM potassium acetate, 100 mM magnesium acetate, 5 mM DTT.
3. 2.5 mM dNTPs, PCR grade.
4. 10 mM ATP.
5. End-Repair Enzyme mix (from END-Repair kit): (T4 DNA Pol + T4 polynucleotide kinase).
6. QIAquick PCR Purification Kit (Qiagen).
7. 10× Taq buffer: 200 mM Tris–HCl, 100 mM $(NH_4)_2SO_4$, 100 mM KCl, 20 mM $MgSO_4$, 1% Triton X-100, pH 8.8.
8. dATP, 1 mM solution.
9. Taq DNA Polymerase with ThermoPol Buffer (NEB).
10. 10× T4 DNA ligase buffer: 500 mM Tris–HCl, 100 mM $MgCl_2$, 100 mM DTT, 10 mM ATP, pH7.5.
11. PE Adapter Oligo Mix, PCR Primer PE 1.0 and 2.0 (Illumina).
12. T4 DNA ligase.
13. 2× Phusion HF Master Mix (NE Biolabs).
14. Thermal cycler.
15. 5 M NaCl.
16. Sodium dodecyl sulphate (SDS, 10% [w/w] solution).
17. 100 mM dATP Solution PCR Grade (Invitrogen, Cat#: 10216–018).
18. Proteinase K, recombinant PCR Grade (Roche, Cat#: 03115828001).

2.3. Sequencing and Data Analysis

1. Access to Genome Analyzer IIX (Illumina).

3. Methods

3.1. Preparation of Nucleosome DNA Templates

1. Native nucleosomes: Harvest cells (10–20 million). Wash the cells 2× with ice-cold 1× PBS (5–10 ml). Spin down the cells at 350 × *g* for 5 min at 4°C. Lyse the cells in 1 ml of ice-cold lysis buffer, and incubate for 5 min on ice.

   Cross-linked nucleosomes: Cross-link 10–20 million cells by adding formadehyde to 1% and incubating for 10 min at

37°C. Wash 2× with 10–20 ml 1× PBS. Spin down the cells at 350×*g* for 5 min at 4°C. Lyse cells in 1 ml 1× PBS+0.5% Triton X-100 for 3 min on ice.

2. Pellet the nuclei by spinning at 350×*g* for 5 min at 4°C. Wash the nuclei with 1 ml MNase digestion buffer, spin down at 350×*g* for 5 min at 4°C, and resuspend the pellet in 800 μl of the same buffer (at a concentration of 10–20 million nuclei per ml). Adjust the final Ca^{2+} concentration to 1 mM with 1 M $CaCl_2$.
3. Aliquot the nuclei suspension into eight tubes (100 μl each), to which 0, 0.01, 0.03, 0.05, 0.1, 0.3, 0.5, and 1 U of MNase are added, respectively. Incubate the reaction mixture at 37°C for 5 min (see Note 1), and then stop the reaction by adding 150 μl of stop buffer.
4. Incubate the mixture at 65°C for 6 h or overnight.
5. Extract the mixture using an equal volume of phenol/chloroform.
6. Add 20 μg of glycogen from a stock solution to the aqueous phase, precipitate the DNA with 750 μl of ethanol and 25 μl of 3 M NaAc, pH5.3, wash the DNA pellet once with 750 μl of ice-cold 70% ethanol, and dissolve the pellet in 30 μl of 1× TE buffer.
7. Load 10 μl DNA from each reaction onto a 2% E-Gel EX-(Invitrogen). Run for 10 min at EX Gel condition (see Note 2).
8. Using E-Gel Safe Imager Real-time Transilluminator (Invitrogen), identify and excise the mononucleosome bands (see Note 3).
9. Purify the DNA using Qiagen gel extraction kit (see Note 4), and elute the DNA in 30 μl EB from Qiagen gel extraction kit (see Note 5).

3.2. Preparation of DNA for Sequencing

1. Repair DNA ends using the Epicentre DNA END-Repair kit This step generates blunt-ended DNA. Mix ingredients as follows:

   1–34 μl DNA (0.1–0.5 μg)

   5 μl 10× end repair buffer

   5 μl 2.5 mM each dNTPs

   5 μl 10 mM ATP

   × μl H_2O to adjust the reaction volume to 49 μl

   1 μl End-Repair Enzyme mix (T4 DNA Pol + T4 PNK)

   Incubate the reaction mixture at room temperature for 45 min. Purify the DNA using QIAquick PCR Purification Kit. Elute DNA in 30 μl of EB (see Note 5)

2. Add "A" to 3′ ends of DNA templates. Mix ingredients as follows:

   30 μl DNA from above

2 μl H_2O

5 μl 10× Taq buffer

10 μl 1 mM dATP

3 μl 5 U/ml Taq DNA polymerase

Incubate the reaction mixture at 70°C for 30 min. Purify the DNA as before. Elute DNA in 20 μl of EB (see Note 5).

3. Adaptor ligation. Mix ingredients as follows:

   20 μl DNA (300 ng)

   3.9 μl H2O

   3 μl 10× T4 DNA ligase buffer

   0.1 μl Adaptor oligo mix (mixture of two adaptors from Illumina)

   3 μl T4 DNA ligase (400 U/ml)

   Incubate the reaction mixture at room temperature for 30 min.

4. Size selection using 2% E-Gel:
   Load 30 μl adaptor-ligated DNA onto a 2% E-Gel EX (Invitrogen). Run for 10 min at EX Gel condition. Using E-Gel Safe Imager Real-time Transilluminator (Invitrogen), cut the gel around 200–400-bp region (DNA may not be visible). Extract the DNA using QIAquick Gel Extraction Kit (see Note 4). Elute in ~23 μl EB (see Notes 5 and 6).

5. Amplify the DNA using Illumina PCR primers and Phusion HF Master Mix. Mix ingredients as follows:

   23 μl of DNA

   25 μl of master mix

   1 μl of PCR primer 1 (2× diluted, Illumina)

   1 μl of PCR primer 2 (2× diluted, Illumina)

   Total volume: 50 μl

   Denature at 98°C for 30 s

   98°C, 10″; 65°C, 30″; 72°C, 30″

   Try 18 cycles first, and check 2.5 μl of product on 1.8% gel. If the band is not clearly visible, do three more cycles. Check again.

6. Purify the amplified products on 2.5% agarose gel. Excise the band near 220 bp and purify the DNA using Qiagen gel extraction kit. Measure the DNA concentration using Qubit fluorometer (see Note 4).

3.3. Sequencing and Data Analysis

1. Purified DNA is used directly for cluster generation and sequencing analysis on Illumina IIX Genome Analyzer following manufacturer's protocols.

2. Data Analysis: Sequenced reads of mostly 25 bp are obtained using the Illumina Analysis Pipeline. All reads are mapped to the human genome (hg18) or other reference genomes and all uniquely matching reads are retained. Nucleosome profiles are obtained by applying a scoring function to the sequenced reads. A sliding window of 10 bp is applied across all chromosomes, and at each window all reads mapping to the sense-strand 80 bp upstream of the window and reads mapping to the antisense-strand 80 bp downstream of the window contribute equally to the score of the window.

4. Notes

1. Incubate the reaction for 8–10 min at 37°C for formadehyde-cross-linked nuclei.
2. The E-Gel® agarose gel electrophoresis system is a complete bufferless system for agarose gel electrophoresis of DNA samples. It provides fast, safe, consistent, and high-resolution electrophoresis and minimizes sample contamination. E-Gel® EX pre-cast agarose gels are generally used gels which contain a proprietary fluorescent nucleic acid stain with high sensitivity, allowing (1) detection of down to 1 ng/band of DNA, (2) compatibility with blue light transillumination to dramatically reduce DNA damage, and (3) easy opening of cassette with gel knife. If this system is not available, traditional gel purification methods can be used, but cross-contamination may result.
3. Unlike UV transilluminators, the E-Gel® Safe ImagerTM Real-time Transilluminator does not produce UV light and does not require UV-protective equipment during use. Blue light transillumination also results in dramatically increased cloning efficiencies compared to UV transillumination.
4. Dissolve gel slices at room temperature with frequent mixing, but not at elevated temperature. This helps to preserve AT-rich DNA that can be easily denatured at higher temperature and could then be lost at the column binding step.
5. Warm up EB at 65°C. Ensure that the EB is dispensed directly onto the QlAquick membrane for complete elution of bound DNA.
6. NEVER run PCR-amplified samples of Step 7 with the linker-ligated products of Step 5 together on the same gel because the latter can be contaminated.

References

1. Kornberg, R.D., and Lorch, Y. (1999). Twenty-five years of the nucleosome, fundamental particle of the eukaryote chromosome. Cell. 98, 285–294.
2. Kornberg, R.D., and Lorch, Y. (2002). Chromatin and transcription: where do we go from here. Curr. Opin. Genet. Dev. 12, 249–251.
3. Henikoff, S., Furuyama, T., and Ahmad, K. (2004). Histone variants, nucleosome assembly and epigenetic inheritance. Trends Genet. 20, 320–326.
4. Kingston, R.E., and Narlikar, G.J. (1999). ATP-dependent remodeling and acetylation as regulators of chromatin fluidity. Genes Dev. 13, 2339–2352.
5. Kouzarides, T. (2007). Chromatin modifications and their function. Cell. 128, 693–705.
6. Li, B., Carey, M., and Workman, J.L. (2007). The role of chromatin during transcription. Cell. 128, 707–719.
7. Nobile, C., Nickol, J., and Martin, R. G. (1986). Nucleosome phasing on a DNA fragment from the replication origin of simian virus 40 and rephrasing upon cruciform formation of the DNA. Mol Call Biol. 6(8), 2916–2922.
8. Barski, A., Cuddapah, S., Cui. K., Roh. T., Schones, D. E., Wang, Z., Wei, G., Chepelev, I., and Zhao, K. (2007), High-Resolution Profiling of Histone Methylations in the Human Genome. Cell. 129, 823–837.
9. Schones D.E., Cui, K., Cuddapah, S., Roh, T., Barski, A., Wang, Z., Wei, G., and Zhao, K. (2008). Dynamic regulation of nucleosome positioning in the human genome. Cell. 132, 887–898.

Chapter 25

Salt Fractionation of Nucleosomes for Genome-Wide Profiling

Sheila S. Teves and Steven Henikoff

Abstract

Salt fractionation of nucleosomes, a classical method for defining "active" chromatin based on nucleosome solubility, has recently been adapted for genome-scale profiling. This method has several advantages for profiling chromatin dynamics, including general applicability to cell lines and tissues, quantitative recovery of chromatin, base-pair resolution of nucleosomes, and overall simplicity both in concept and execution. This chapter provides detailed protocols for nuclear isolation, chromatin fragmentation by micrococcal nuclease digestion, successive solubilization of chromatin fractions by addition of increasing concentrations of salt, and genome-wide analyses through microarray hybridization and next-generation sequencing.

Key words: Salt extraction, Nucleosome solubility, Chromatin organization

1. Introduction

Dynamic chromatin organization maintains DNA compaction while allowing for accessibility during active processes, such as transcription (1). These active processes are regulated through the action of nucleosome remodeling, histone modifications and variants, and chromatin-associated proteins. A variety of methods have been developed to study chromatin dynamics. Traditional methods utilize DNA cleavage systems coupled with chromatin probing, such as DNase I hypersensitivity (2–4) and micrococcal nuclease (MNase) (5, 6) mapping assays, which, respectively, measure chromatin accessibility and nucleosome occupancy using nuclease digestion. Other methods rely on chromatin solubility or partitioning differences that occur after formaldehyde cross-linking (FAIRE, Sono-Seq) (7, 8). In addition, there are methods that measure chromatin dynamics directly by mea-

Randall H. Morse (ed.), *Chromatin Remodeling: Methods and Protocols*, Methods in Molecular Biology, vol. 833, DOI 10.1007/978-1-61779-477-3_25, © Springer Science+Business Media, LLC 2012

suring nucleosome turnover, either using protein-encoded tags (9) or metabolic labeling of histones (10). Protein-related information can be obtained using either chromatin immunoprecipitation (ChIP), which relies on affinity capture of the protein of interest (11–13), or DamID, which relies on DNA methylation by tethered Dam methyltransferase (14). All of these methods for chromatin characterization have been adapted for genome-wide profiling, taking advantage of the extraordinary improvements in microarray and sequencing technologies that have occurred over the past several years.

Another traditional method for assaying chromatin is salt fractionation (15). Chromatin digested with an enzyme, such as MNase, can be separated into soluble and insoluble fractions in the presence of physiological Mg^{2+} and low Na^{+} concentrations (16). Subsequently, increasing Na^{+} concentrations allow for separation of the insoluble fraction into high-salt soluble and insoluble fractions (15). Low-salt concentrations solubilize about 5–10% of chromatin composed primarily of mononucleosomes, whereas high salt solubilizes the majority of the nucleosomes (Fig. 1). Genome-wide profiling of low-salt soluble, high-salt soluble, and high-salt insoluble fractions versus total MNase-treated nuclei reveals that salt fractionation can differentially extract chromatin based on distinct physical properties (16). Highly accessible chromatin is enriched in the low-salt soluble fraction while

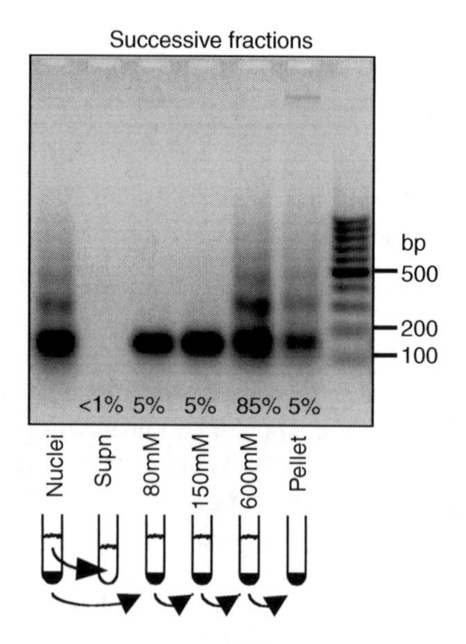

Fig. 1. Size distribution of salt fractions by agarose gel electrophoresis. Drosophila S2 cells were subjected to salt fractionation as described (16). DNA was extracted from each fraction and electrophoresed on a 1.5% agarose gel. *Lane 1* (Nuclei) corresponds to MNase-treated total chromatin. *Lane 2* (Supn) corresponds to the supernatant after the MNase-treated nuclei are pelleted. *Lane 3, 4, 5,* and *6* are the 80, 150, 600 mM, and insoluble pellet fractions, respectively. The percentage shown indicates the amount of chromatin solubilized in each fraction. The diagram below depicts the process of successive solubilization of chromatin with increasing amounts of salt (Reproduced from (16), with permission).

high salt solubilizes the majority of condensed chromatin, revealing insights into chromatin structure. The insoluble fraction is enriched in transcriptionally active chromatin, rendered insoluble presumably due to its association with large multiprotein complexes. Salt fractionation can, therefore, be used to map differences in physical properties and organization of chromatin. Furthermore, affinity capture of histones from each fraction can reveal differences in composition and modification of nucleosomes in their respective fraction. Although originally developed for studying nucleosomes, salt fractionation has recently been used to map paused RNA polymerase (17).

Salt fractionation has several advantages over other methods for characterizing chromatin dynamics. No antibodies, transgenes, or special treatments are needed so that salt fractionation can be applied to essentially any eukaryotic cell type, whether from cell lines or tissues. By assaying the low-salt-soluble (active) fraction, the high-salt fraction, and the insoluble ("nuclear matrix") pellet, essentially 100% of native chromatin is characterized. An important advantage of salt fractionation over other methods such as X-ChIP, FAIRE, Sono-Seq, and DamID, is that mononucleosome resolution is achieved, which allows for mapping of active chromatin at single base-pair level using massively parallel sequencing (Fig. 2). Furthermore, the simplicity of the salt fractionation process makes it an attractive method for characterizing epigenomes.

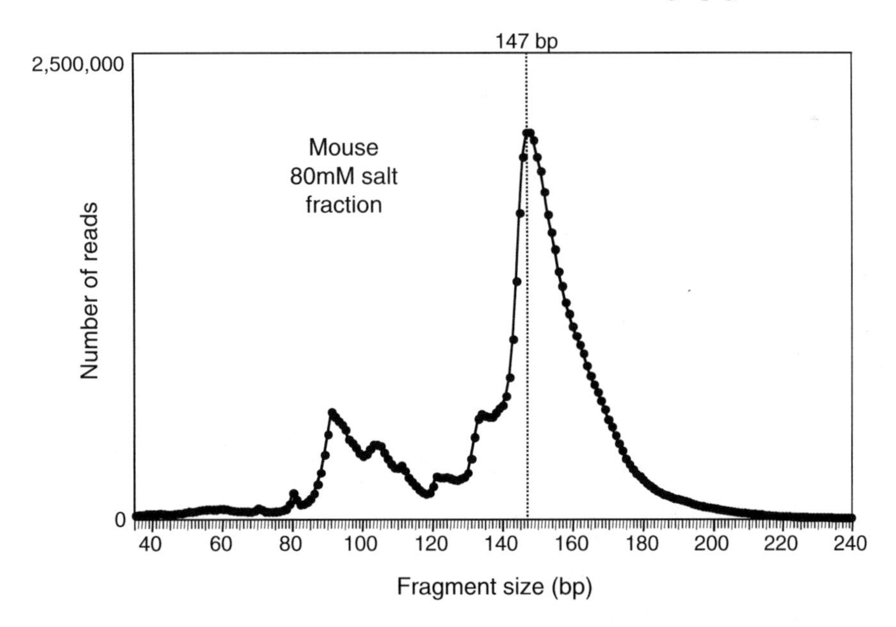

Fig. 2. Size distribution of a low-salt fraction by paired-end Solexa sequencing. Immortalized mouse pre/pro B cells PD31A were subjected to salt fractionation as described here. A paired-end Solexa library was generated from the 80-mM fraction following Subheading 3.3.2. Paired-end sequencing was performed in a single lane of an Illumina Hi-Seq 2000 instrument following the Illumina protocol, and reads were mapped to the mouse genome using the Bowtie alignment program (http://www.bowtie-bio.sourceforge.net/index.shtml). The length distribution of all 84 million mapped paired-end reads is shown at base pair resolution. Note that the dominant peak of nucleosomal DNA is centered around 147 bp, with small-sized fragments indicative of MNase cleavages within the nucleosome.

We divide the salt fractionation procedure into three stages: (1) preparation of nuclei and MNase digestion, (2) chromatin extraction and DNA isolation, and (3) preparation for genome-scale assays. The use of EGTA, instead of EDTA, retains free Mg^{2+} ions, which are critical for nuclear and chromatin integrity (15), allowing for ease of nuclear isolation with mild nonionic detergents. Protein analysis of each fraction can be performed using SDS-polyacrylamide gel electrophoresis and immunoblotting. Affinity capture and subsequent DNA isolation can be used to identify changes in nucleosome composition within each fraction. Finally, DNA isolated from each fraction can be used to generate genome-wide profiles of chromatin structure and physical properties using microarray hybridization or next-generation sequencing.

2. Materials

2.1. Preparation of Nuclei and MNase Digestion

1. Mid-log-phase cultured cells.
2. 14-mL polypropylene tubes.
3. Phosphate-buffered saline (PBS): 11.9 mM phosphates, 137 mM NaCl, 2.7 mM KCl.
4. Protease-inhibitor tablets, EDTA-free (Roche).
5. *TM2 buffer: 10 mM Tris–HCl, pH 7.4, 2 mM $MgCl_2$, 0.5 mM phenylmethylsufonyl fluoride (PMSF).
6. 10% Nonidet P-40.
7. 0.2 M $CaCl_2$.
8. MNase 200 U resuspended to 0.2 U/μL.
9. 0.2 M EGTA.

2.2. Chromatin Extraction and DNA Isolation

1. *80 mM Triton buffer: 70 mM NaCl, 10 mM Tris–HCl, pH 7.4, 2 mM $MgCl_2$, 2 mM EGTA, 0.1% Triton X-100, 0.5 mM PMSF.
2. *150 mM Triton buffer: 140 mM NaCl, 10 mM Tris–HCl, pH 7.4, 2 mM $MgCl_2$, 2 mM EGTA, 0.1% Triton X-100, 0.5 mM PMSF.
3. *600 mM Triton buffer: 585 mM NaCl, 10 mM Tris–HCl, pH 7.4, 2 mM $MgCl_2$, 2 mM EGTA, 0.1% Triton X-100, 0.5 mM PMSF.
4. *TNE buffer: 10 mM Tris–HCl, pH 7.4, 200 mM NaCl, 1 mM EDTA.
5. 5 M NaCl.
6. 0.5 M EDTA.
7. RNase, DNase-free 500 μg/mL solution.

8. Proteinase K 20 mg/mL RNA grade.
9. Phenol–chloroform–isoamyl alcohol (25:24:1 v/v).
10. 200 proof ethanol.
11. TE, pH 8: 10 mM Tris–HCl, 1 mM EDTA.
12. Phase-lock gel tubes (Heavy 1.5 mL – 200 tubes) (5 Prime).
13. Glycogen, 20 mg/mL.

2.3. Genome-Scale Profiling

1. 5′ Cy labeled NimbleGen Validated Random 7-mer (Tri-Link).
2. 40 mM dNTPs (10 mM each).
3. Nuclease-free H_2O.
4. Klenow fragment 3′→5′ exo- (50 U/μL).
5. Isopropanol.
6. Paired-end Sample Preparation Kit (Illumina).
7. Qiagen gel purification kit or similar substitute.
8. Qiagen MinElute PCR purification kit or similar substitute.

2.4. General Equipment

1. Refrigerated tabletop centrifuge.
2. Non-refrigerated centrifuge for 15-mL conical tubes (e.g., IEC Centra CL2).
3. End over end Eppendorf tube rotator (e.g., Labquake Shaker – Thermo Scientific).
4. PCR thermocycler.
5. Vortexer.
6. Water bath at 37°C.
7. Nanodrop spectrometer.

* Buffers are supplemented with 1× Protease inhibitor (Roche).

3. Methods

Briefly, cells are harvested and lysed with mild NP-40 to release nuclei. Washed nuclei are subjected to limited MNase digestion to fractionate the chromatin. Successive incubation with buffers containing increasing salt concentrations differentially solubilizes the chromatin. DNA can then be purified from each fraction for genome-wide analysis using microarray hybridization or massively parallel sequencing, such as the Illumina platform.

3.1. Preparation of Nuclei and MNase Digestion

1. Grow Drosophila Schneider 2 (S2) cell line cells in preferred media to exponential growth phase and 90% confluency (~2×10^6 cells per cm^2; see Note 1).

2. Scrape cells off one 75-cm^2 flask and collect them in 14-mL polypropylene, round-bottom tubes. Pellet the cells from the media in a room-temperature tabletop centrifuge (IEC Centra CL2) for 3 min at 1500 rcf.
3. Discard the media and wash cells in cold 1× PBS and pellet cells as above.
4. Resuspend cells in 1 mL TM2 buffer and cool on ice for 3 min.
5. To lyse the cells while keeping the nuclei intact, slowly add 60 μL 10% NP-40 while gently vortexing the tube. Incubate the cells on ice for 3 min with 5-s gentle vortexing every minute (see Note 2).
6. Separate the nuclei from cellular debris by gentle centrifugation (100 × *g*) for 10 min in a refrigerated tabletop centrifuge. The nuclear pellet appears as a white, loose pellet that is easily disrupted. Carefully remove the supernatant to prevent disrupting the nuclear pellet. Wash the nuclei with 1 mL of TM2 buffer by gently pipetting the buffer into the tube. The pipetting action easily disrupts most of the nuclei pellet and the rest can be fully resuspended by gentle flicking of the tube (see Note 3). Pellet the nuclei for 10 min at 100 × *g* and remove the supernatant as before.
7. Resuspend nuclei in 400 μL TM2 and warm to 37°C in a water bath for 5 min.
8. To fractionate the chromatin, add 2 μL of 0.2 M $CaCl_2$ to final concentration of 1 mM and 2.5 μL of MNase (final concentration of 1.25 U/mL). Return to 37°C in a water bath for 10 min with intermittent mixing to prevent aggregation of nuclei at the bottom of the tube (see Note 4).
9. Addition of 4 μL of 0.2 M EGTA to final concentration of 2 mM stops the MNase reaction. Remove 40 μL (10% of reaction) and label as "Nuc" for total MNase-treated chromatin, and an additional 40 μL for protein analysis. To remove the MNase, pellet the nuclei for 10 min at 100 × *g* and carefully remove the supernatant, and save as "Supn" fraction. Remove 30 μL for protein analysis of the Supn fraction. Wash the nuclei carefully with 1 mL of TM2 buffer, pellet, and remove the supernatant as above. Proceed with the nuclei pellet to the salt fractionation step (Subheading 3.2).

3.2. Salt Fractionation and DNA Isolation

1. From step 9 of Subheading 3.1, resuspend the nuclear pellet in 700 μL of 80 mM Triton buffer (see Note 5) and incubate in constant agitation by placing the tube in a rotator (e.g., Labquake Shaker) at 4°C for 2 h. This releases nucleosomes soluble in 80 mM salt concentration into the supernatant.

2. To extract the low-salt-soluble nucleosomes, pellet the nuclei at 100 × *g* for 10 min at 4°C and save the supernatant labeled as "80 mM" fraction. The loose nuclear pellet is often slightly disrupted during this process, causing some of the nuclei to be aspirated with the supernatant. To clear the 80-mM fraction, respin the supernatant for 2 min at maximum centrifugation speed and transfer the cleared 80-mM fraction to a new tube. Remove 30 μL for protein analysis.
3. Optional: Resuspend the nuclei from the 80 mM salt extraction with 700 μL 150 mM Triton buffer and incubate with constant agitation at 4°C for 2 h. Extract and clear the supernatant as in Subheading 3.2, step 2, to release 150 mM soluble nucleosomes and remove 30 μL for protein analysis. The 150-mM fraction consists of primarily mononucleosomes (Fig. 1) and is enriched at the 5′ ends of active genes (16). This fraction is very similar to the 80-mM fraction, but with lower resolution.
4. The low-salt buffers solubilize ~5–10% of total chromatin. To solublize the majority of nucleosomes, resuspend the nuclei in 600 mM Triton buffer and incubate at 4°C in constant agitation in a rotator for 2 h to overnight. Extract and clear the supernatant labeled as "600 mM" fraction as in step 2 of Subheading 3.2. Remove 30 μL for protein analysis.
5. The remaining pellet fraction corresponds to ~5–10% of chromatin. Resuspend the pellet in 700 μL of TNE buffer and label as "Pel" fraction. Hold all fractions on ice prior to DNA extraction.
6. Aliquots of each fraction can be electrophoresed on an SDS gel to visualize histones and probe for the presence of specific proteins.
7. ChIP assays can be performed on each salt fraction using standard native ChIP protocols. Save an aliquot of the salt fraction for "input" DNA and use the remainder for affinity purification.
8. For each fraction, add 1/50th volume of 0.5 M EDTA (14 μL). For the nuclei, Supn, 80 mM and the optional 150-mM fractions, add 1/50th volume (14 μL) of 5 M NaCl. The 600 mM and pellet fractions contain sufficient amounts of NaCl for DNA precipitation purposes.
9. Prepare 1:10 dilution of RNase enzyme in H_2O and add 5 μL of diluted RNase to each fraction. Allow for RNA digestion to proceed for 10 min in a 37°C water bath.
10. To remove proteins, add 1/16th volume of 10% SDS (44 μL) for a final concentration of 0.63%. Then, add 2.5 μL of proteinase K and incubate at 75°C for 10 min.
11. Extract DNA by adding one volume of phenol–chloroform–isoamyl alcohol. Transfer the samples to phase-lock gel tubes and vortex for 2 min. Centrifuge the samples at maximum

speed in a refrigerated microfuge for 10 min. Transfer the supernatant into a new phase-lock tube and repeat the extraction one more time. Transfer the aqueous solution into Eppendorf tubes. Alternatively, a standard phenol–chloroform–isoamyl alcohol extraction may be performed, provided the aqueous solution is carefully extracted from the interphase and organic phase.

12. To precipitate the DNA, add 2 μL of glycogen and 2.5 volumes of ice-cold 100% ethanol. Incubate on ice for 20 min and centrifuge at maximum speed for 15 min in a refrigerated microfuge. Remove the supernatant and wash the pellet with 1 mL of ice-cold 80% ethanol. Centrifuge the samples for 5 min at maximum speed, remove the supernatant, and allow the pellet to dry.
13. Once fully dried, resuspend the DNA with 0.1× TE, pH 8. Determine the concentration of DNA using Nanodrop. Electrophorese an aliquot in a 1.5% agarose gel with ethidium bromide.

3.3. Genome-Scale Profiling

3.3.1. Microarray Analysis

1. The following protocol is a modified version of Nimblegen labeling methods specifically adapted to Drosophila S2 cells. Bring 0.2–1 μg of DNA from each fraction to 20 μL with H_2O and add 20 μL of Cy5 dye in 0.6-μL thin-walled PCR tubes. Use the same amount of DNA for Nuc and add 20 μL of Cy3 dye (see Notes 6 and 7).
2. Incubate each sample at 95°C for 10 min on a thermocycler. Immediately place the samples on ice water for 2 min (see Note 8).
3. After instant chill, add 5 μL of 50 mM dNTPs, 4 μL of nuclease-free H_2O, and 1 μL of Klenow fragment. Allow the labeling reaction to proceed for 4.5 h at 37°C in a thermocycler (see Note 9).
4. To stop the labeling reaction, add 5 μL of 0.5 M EDTA. Precipitate the labeled DNA by adding 5.75 μL of 5 M NaCl and 55 μL of isopropanol. Incubate at room temperature for 10 min and pellet the DNA at maximum speed for 10 min in a room-temperature microfuge. Wash the colored pellet with ice-cold 80% ethanol and recentrifuge for 2 min at maximum speed at room temperature. Discard the supernatant carefully and speed-vac dry the samples for at most 5 min to prevent overdrying.
5. Resuspend the pellet in 20 μL of nuclease-free H_2O. To quantify the labeled DNA, dilute 0.5 μL of sample in 4.5 μL H_2O and use 1.5 μL for nanodrop measurement.
6. Combine 34 μg of Cy5-labeled salt fraction with 34 μg of Cy3-labeled Nuc and concentrate the volume to 12.5 μL using

a speed-vac. Proceed with Nimblegen hybridization protocol with high-density Drosophila microarrays.

3.3.2. Paired-End Solexa Library Preparation

1. Library preparation for sequencing of salt fractions follows closely the Illumina protocol provided with its Paired-End Sample Preparation Kit. Use 500 ng of DNA and follow the Illumina protocol for end repair, 3′ adenylation, and adapter ligation.
2. After adapter ligation, samples must be size selected and purified from free adapters. Electrophorese the samples on a 2% agarose gel and excise DNA in the 100- to 600-bp range (see Note 10). This isolates double-stranded DNA derived from mono- to trinucleosomes. Extract DNA from the agarose gel following the Qiagen gel extraction kit protocol using MinElute columns and elute the DNA with 30 μL of EB buffer.
3. To amplify the library, follow Illumina's protocol on PCR amplification and clean up. For salt fractionation, it is not necessary to perform a secondary size selection process after amplification. Quality control analysis varies depending on the sequencing facility's specification, but may include PicoGreen quantification, Bioanalyzer (Agilent) analysis, and qPCR quantification. The resulting paired-end library can be sequenced using the Illumina Genome Analyzer platform.

4. Notes

1. This protocol is specifically designed for S2 cells. However, it can be adapted to any cultured cells from any species, provided that the cells are undergoing exponential growth. Changes in the growth phase lead to changes in the transcriptional program, which may also lead to changes in chromatin structure. As such, salt fractionation methods are most reproducible and reliable for cells in the same log-phase growth. For adherent cells, follow established trypsin conditions for cell harvest and proceed to step 3 of Subheading 3.1.
2. The conditions for cell lysis must be empirically determined for each cell type and for each species. This can be done by altering the final concentration of NP-40 from 0.08 to 0.8% in TM2 solution. Using the lysis protocol described in step 5 of Subheading 3.1, check for complete lysis of the cellular membrane while maintaining nuclei integrity by removing an aliquot and examining the nuclei under a microscope in comparison to intact cells. Alternatively, one can use Trypan-blue exclusion.

Remove an aliquot of the lysis, add an equal volume of Trypan-blue solution, and visualize the nuclei under a microscope.

3. The nuclear membrane is sensitive to mechanical disruption, which can lead to lysis and subsequent release of chromatin into the solution. This results in a nuclear pellet that is difficult to resuspend in solution. Formation of clumps in resuspended nuclei is a telltale sign of nuclear lysis. Discard samples and repeat nuclei preparation using less NP-40 in the lysis buffer or gentler handling of nuclei.

4. MNase conditions must be determined empirically for consistent digestion. The optimal conditions yield mostly mononucleosomes with decreasing amounts of di- and trinucleosomes (Fig. 1). To determine MNase conditions, prepare nuclei from 150×10^6 cells as described in Subheading 3.1 and divide the nuclei into 5–10 aliquots, depending on the number of MNase conditions to be tested. Add increasing amounts of MNase starting with 0.5 U per reaction and incubate each sample in a 37°C water bath for 10 min. Isolate DNA as described in Subheading 3.3 and electrophorese an aliquot in a 1.5% agarose gel. Increasing amounts of MNase should yield increasing intensity of the mononucleosome band at 150 bp. Determine which amount of MNase yields the distinct ladder of mono- to trinucleosomes and repeat the conditions to ensure replicability. Furthermore, intermittent mixing of the MNase reaction is important as nuclei can pellet in the span of the 10-min digestion, which can lead to unequal and incomplete digestion.

5. The concentration of nuclei in the salt buffers is about 2×10^5 nuclei per μL of buffer. This ratio is critical for the maintenance of nuclear integrity through the interaction of Mg^{2+} ions with the chromatin and nuclear complex. When adapting this protocol for using less starting number of cells, different cell types, or different species, this ratio must be maintained for proper nuclear integrity.

6. For control purposes, dye swaps may be necessary to determine dye-labeling biases. In these cases, the Nuc fraction can be labeled with Cy5 and salt fractions with Cy3. Alternatively, the Nuc fraction can be labeled with Cy5 and Cy3 so that hybridization should result in zero enrichment and depletion in the microarray profile. Labeling biases can be identified by lack of inverse correlation with the dye swap or nonzero profiles in the nuclei hybridization.

7. The labeling reactions described are half of the total volume of the Nimblegen dye-labeling protocol. This is sufficient for Drosophila samples because of their smaller genome size, but for mammalian systems, such as mouse, a full reaction is optimal with 1 μg of starting material. In this case, follow the Nimblegen dye-labeling protocol closely.

8. It is important that samples are chilled in ice water for faster and more uniform cooling. Otherwise, efficiency of the labeling reaction is decreased.
9. The Nimblegen protocol calls for 2-h labeling reactions. However, a single labeling reaction does not usually produce enough material needed for hybridization. For the smaller Drosophila genome, a longer labeling reaction produces sufficient material for hybridization without introducing labeling biases. However, for larger and more complex mammalian genomes, labeling biases become more pertinent. Therefore, set up 2 or more full-volume reactions (Nimblegen protocol) and limit the length to 2 h.
10. The standard Illumina protocol for size selection of libraries calls for extraction of a relatively small range of sizes for sequencing. However, one advantage of paired-end sequencing is the ability to measure sizes of the sequenced population. This allows for mapping of nucleosomes at base-pair resolution. It should be noted that breaks caused by MNase result in primarily ~147-bp fragments from gel-purified mono-, di-, and trinucleosomes (17), with smaller species indicative of internal cleavages (Fig. 2).

References

1. Henikoff S (2008) Nucleosome destabilization in the epigenetic regulation of gene expression. Nat Rev Genet 9:15–26
2. Weintraub H, Groudine M (1976) Chromosomal subunits in active genes have an altered conformation. Science 193:848–56
3. Crawford GE, Holt IE, Mullikin JC, Tai D, Blakesley R, Bouffard G, Young A, Masiello C, Green ED, Wolfsberg TG, Collins FS (2004) Identifying gene regulatory elements by genome-wide recovery of DNase hypersensitive sites. Proc Natl Acad Sci USA 101:992–7
4. Hesselberth JR, Chen X, Zhang Z, Sabo PJ, Sandstrom R, Reynolds AP, Thurman RE, Neph S, Kuehn MS, Noble WS, Fields S, Stamatoyannopoulos JA (2009) Global mapping of protein-DNA interactions in vivo by digital genomic footprinting. Nat Methods 6:283–9
5. Lee W, Tillo D, Bray N, Morse RH, Davis RW, Hughes TR, Nislow C (2007) A high-resolution atlas of nucleosome occupancy in yeast. Nat Genet 39:1235–44
6. Tsankov AM, Thompson DA, Socha A, Regev A, Rando OJ (2010) The role of nucleosome positioning in the evolution of gene regulation. PLoS Biol 8:e1000414
7. Giresi PG, Kim J, McDaniell RM, Iyer VR, Lieb JD (2007) FAIRE (Formaldehyde-Assisted Isolation of Regulatory Elements) isolates active regulatory elements from human chromatin. Genome Res 17:877–85
8. Auerbach RK, Euskirchen G, Rozowsky J, Lamarre-Vincent N, Moqtaderi Z, Lefrancois P, Struhl K, Gerstein M, Snyder M (2009) Mapping accessible chromatin regions using Sono-Seq. Proc Natl Acad Sci USA 106: 14926–31
9. Dion MF, Kaplan T, Kim M, Buratowski S, Friedman N, Rando OJ (2007) Dynamics of replication-independent histone turnover in budding yeast. Science 315:1405–8
10. Deal RB, Henikoff JG, Henikoff S (2010) Genome-wide kinetics of nucleosome turnover determined by metabolic labeling of histones. Science 328:1161–64
11. Sikes ML, Bradshaw JM, Ivory WT, Lunsford JL, McMillan RE, Morrison CR (2009) A streamlined method for rapid and sensitive chromatin immunoprecipitation. J Immunol Methods 344:58–63
12. Bonner J, Dahmus ME, Fambrough D, Huang RC, Marushige K, Tuan DY (1968) The Biology of Isolated Chromatin: Chromosomes,

biologically active in the test tube, provide a powerful tool for the study of gene action. Science 159:47–56

13. Gilchrist DA, Fargo DC, Adelman K (2009) Using ChIP-chip and ChIP-seq to study the regulation of gene expression: genome-wide localization studies reveal widespread regulation of transcription elongation. Methods 48:398–408
14. van Steensel B, Delrow J, Henikoff S (2001) Chromatin profiling using targeted DNA adenine methyltransferase. Nat Genet 27:304–8
15. Sanders MM (1978) Fractionation of nucleosomes by salt elution from micrococcal nuclease-digested nuclei. J Cell Biol 79:97–109
16. Henikoff S, Henikoff JG, Sakai A, Loeb GB, Ahmad K (2009) Genome-wide profiling of salt fractions maps physical properties of chromatin. Genome Res 19:460–9
17. Weber CM, Henikoff JG, Henikoff S (2010) H2A.Z nucleosomes enriched over active genes are homotypic. Nat Struct Mol Biol 17:1500–7

Chapter 26

Quantitative Analysis of Genome-Wide Chromatin Remodeling

Songjoon Baek, Myong-Hee Sung, and Gordon L. Hager

Abstract

Recent high-throughput sequencing technologies have opened the door for genome-wide characterization of chromatin features at an unprecedented resolution. Chromatin accessibility is an important property that regulates protein binding and other nuclear processes. Here, we describe computational methods to analyze chromatin accessibility using DNaseI hypersensitivity by sequencing (DNaseI-seq). Although there are numerous bioinformatic tools to analyze ChIP-seq data, our statistical algorithm was developed specifically to identify significantly accessible genomic regions by handling features of DNaseI hypersensitivity. Without prior knowledge of relevant protein factors, one can discover genome-wide chromatin remodeling events associated with specific conditions or differentiation stages from quantitative analysis of DNaseI hypersensitivity. By performing appropriate subsequent computational analyses on a select subset of remodeled sites, it is also possible to extract information about putative factors that may bind to specific DNA elements within DNaseI hypersensitive sites. These approaches enabled by DNaseI-seq represent a powerful new methodology that reveals mechanisms of transcriptional regulation.

Key words: Chromatin, Chromatin remodeling, DNaseI hypersensitivity, Global DHS-seq analysis, Global ChIP-seq analysis, Genome-wide, High-throughput deep sequencing, Computational methods, Footprinting

1. Introduction

Chromatin remodeling is the mechanism by which tightly packaged chromatin in the nucleus of a eukaryotic cell is modified such that transcription factors or other nuclear proteins can access their cognate sites on the DNA (1). The chromatin exerts a significant regulatory effect upon the expression of genes, for example. Transcription factor binding sites can be remarkably clustered and access to these elements is tightly regulated in a cell- and condition-specific manner (2–4).

Randall H. Morse (ed.), *Chromatin Remodeling: Methods and Protocols*, Methods in Molecular Biology, vol. 833,
DOI 10.1007/978-1-61779-477-3_26, © Springer Science+Business Media, LLC 2012

The accessibility of the chromatin template can be measured by the susceptibility of DNA to cutting by DNaseI (3). Chromosomal regions that have an open chromatin structure are hypersensitive to DNaseI cutting and are consequently termed DNaseI hypersensitive sites (DHSs). DHSs often coincide with regulatory elements, such as promoters, enhancers, insulators, silencers, and locus control regions (1, 5–7). DNaseI hypersensitivity coupled with chip (DNaseI-chip) and more recently with sequencing (DNaseI-seq) has provided genome-wide identification of functional regulatory elements (Fig. 1) (3, 8, 9). This approach makes it possible to achieve a continuous high-resolution profile of DNaseI sensitivity along the genome that represents a map of quantitative chromatin accessibility (2, 10).

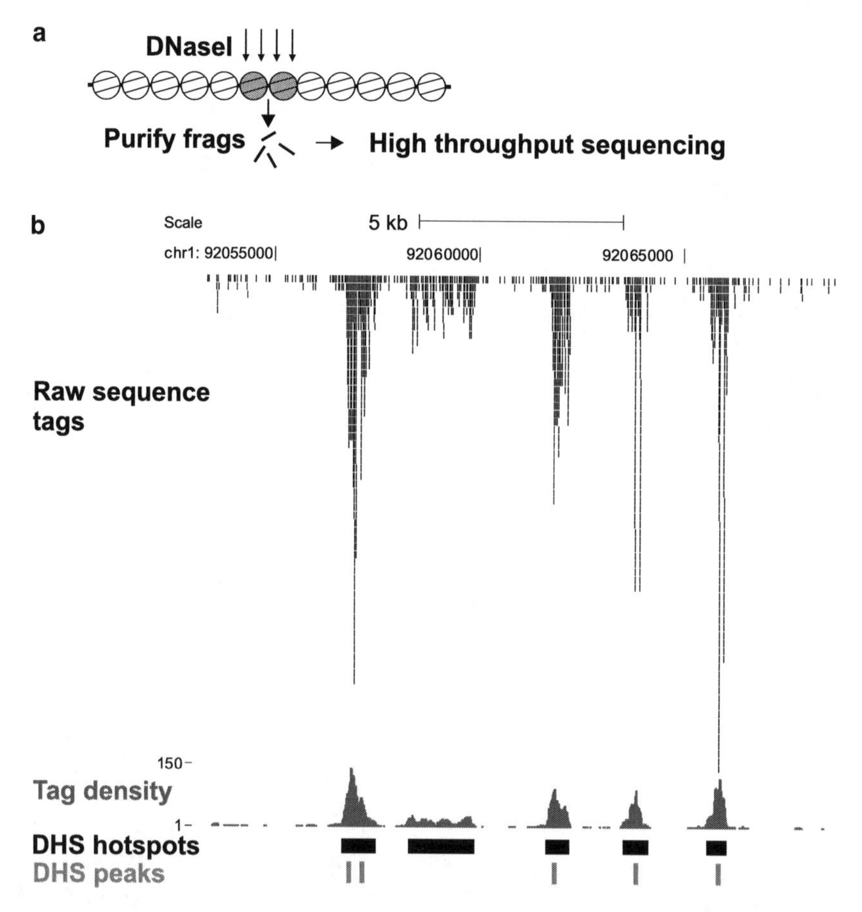

Fig. 1. Example sequence tag profiles of DHS elements. (**a**) DNaseI hypersensitive sites (DHSs) are measured by deep sequencing of small, size-selected DNA fragments released by DNaseI treatment from regions of locally reorganized chromatin (2). (**b**) The DNaseI-seq data are presented as tag density profiles, summarizing the number of sequence tags that are aligned to the region. Density values can be normalized by the total number of aligned sequence tags to aid comparison of multiple samples. A DNaseI-seq profile obtained from a mammary carcinoma cell line is shown for a region in chromosome 1 (genome build: mm8). Regions of increased DNaseI sensitivity are detected by the computational algorithms as "hotspots" (DHS hotspots track). Within these hotspots are found localized peaks of hypersensitivity (DHS peaks track).

If the DNA fragments cut and released by DNaseI are sequenced deeply to yield very large numbers of reads, narrow regions of protection against DNaseI cleavage, caused by transcription factor binding, can be identified as "footprints" (7, 10). Although the cost of sequencing becomes an issue in practice, sufficient tag coverage allows pinpointing of specific binding sites at nucleotide resolution.

This chapter provides a technical description of the computational procedures that can be employed to analyze DNaseI-seq data. These algorithms are categorized into different types of computational analyses: (1) generation of tag density maps for visual exploration; (2) hotspot detection based on a background probability model and calculation of statistical measure of enrichment; (3) artifact filtering; (4) footprint detection for extra-deep sequencing data; (5) annotation of the identified regions with respect to other genomic features or related data; and (6) downstream analyses that the individual studies call for.

2. Data and Sources of Bias

Current high-throughput sequencing of a DNaseI-seq sample routinely produces tens of millions of short sequence reads of 25–80 bp in length. In single-end sequencing, these short reads or tags belong to one end of the DNA fragments that were doubly cut by DNaseI and size selected for 100 bp–1 kb range. Sequence reads are then aligned to the reference genome and regions with densely accumulated tags are putative DHSs (2, 3).

Although DNaseI-seq data are highly reproducible and visually convincing (Fig. 1), there are systematic biases that are readily observed and should be corrected. For example, all the reads are aligned to the reference genome with the assumption that the genomic sequence of the cells used for the experiment is identical to the reference sequence, whereas many commonly used cell lines harbor genomic aberrations, such as polyploidy, translocations, or other mutations. Even karyotypically normal cells may have variations from the reference genome. Amplified regions would contribute more input to the DNA sample and deleted regions would not support any sequence reads.

Another source of bias arises from the highly variable tag "mappability" together with the fact that the initial processing of sequencing data relies on alignment of short reads for assigning their genomic coordinates. A given n-mer may occur at a unique location or at multiple genomic positions. Most alignment algorithms produce genomic coordinates only for uniquely mappable reads, which creates "dark spots" across the genome that directly affect the background probability of observing tags at any given position.

Identification of the genomic regions, where tags are truly enriched over the background, must take into account these and other sources of bias and artifacts in the sequencing data. The objective of an algorithm for DHS detection is to find all of the truly enriched sites while minimizing the false-positive rate. Although various software packages have been developed for calling peaks or enriched sites from ChIP-seq data, most publicly available software are not optimally designed for handling DNaseI-seq data. An ad hoc solution has been to use a ChIP-seq algorithm that does not require input DNA control (11). In the following section, we describe an algorithm that identifies regions of significant tag enrichment [termed hotspots (2, 10)] from DNaseI-seq data.

3. Methods

3.1. Building a Tag Density Profile for Data Visualization in a Browser

A tag density profile is a direct summary of the sequencing data showing the distribution of tag counts over the genome. This basic information serves as a first-pass assessment of the data, since some data quality problems can be found at the early stage. Because the size-selected DNA fragments (100–1,000 bp) are longer than the sequenced reads (25–80 bp) from the 5′ ends, positional adjustments are necessary to estimate the real distribution of fragments in the DNaseI-treated sample. If the average fragment length is known, each sequence tag can be extended in the 3′ direction up to that length. When paired-end sequencing is employed, the fragments are obtained directly and can be used to generate the tag (or fragment) density profile. After the adjustment, the tag density profile is calculated simply by counting the number of tags overlapping a given nucleotide (or a bin if a lower resolution is acceptable).

Simultaneous visualization of the tag density profile and identified enriched sites (Fig. 1, see Subheading 3.2) is essential to confirm and understand the data and analysis output. There are several publicly available browser software tools that accept genomic data files in predefined formats and display them in the context of annotation tracks, such as known genes, ESTs, and comparative genomic features.

The University of California Santa Cruz (UCSC) Genome Browser is a popular Web-based browser (http://genome.ucsc.edu). It provides a rich database of useful information, including known genes and ncRNA, repeat elements, and publicly available data tracks from the ENCODE project, which can be used to explore the data with additional correlative features. Currently, the UCSC Genome Browser supports various data formats, such as BED, BedGraph, GFF, WIG, and BAM. The UCSC Web site also provides the Table Browser from which one can download data tracks seen in the browser (http://genome.ucsc.edu/cgi-bin/hgTables).

Integrated Genome Browser (IGB) is also a powerful browser for high-throughput genomic data (http://www.bioviz.org/igb/). IGB supports many reference genomes through the DAS/2 protocol and a custom genome can be created and loaded by the user. An attractive feature of IGB over the UCSC Genome Browser is the speed of browsing: zooming or updating the data tracks is very fast because it runs as a local application and the accompanying annotation tracks are minimal. Moreover, IGB can call upon the same genomic region in the UCSC Genome Browser with a single click, making it easy for the user to identify regions of interest quickly by browsing with IGB and then accessing the vast annotation database of UCSC Genome Browser for more detailed information on the relevant regions.

3.2. Hotspot Detection Algorithm

Our algorithm identifies regions of local enrichment, termed hotspots, of short-read tags mapped to the reference genome as follows (see also Note 1). Enrichment of tags in a 250-bp target window relative to a 200-kb surrounding window (local background) is gauged by the model based on the binomial distribution. Each tag is extended to be 150 bp in length, or the average size from selection, into its strand. Then, the 250-bp target window centered at the extended tag is assigned a *z* score as explained below using the background window centered at the tag. An unthresholded hotspot is defined as a contiguous cluster of 250-bp windows whose *z* scores are nominally significant (greater than 2). Once a hotspot is called, the *z* score of the hotspot is taken to be the highest *z* score from the constituent tags.

3.2.1. Mappability-Adjusted Z Scores

If there are n observed tags and N tags that overlap the target window and the local background window, respectively, then the probability p of a tag in the background window overlapping the target window is given by the ratio [# of uniquely mappable tags for the 250-bp window]/[# of uniquely mappable tags for the 200-kb window]. Because not all 36-mers (a common read length from sequencers) in a window can be aligned uniquely to the reference genome, p differs greatly from genomic region to region. The expected number of tags overlapping the target window is then $\mu = Np$ after this important step of adjusting for the differential mappability of short reads. The standard deviation of the expectation is: $\sigma = \sqrt{Np(1-p)}$. The z score for the observed number of tags in the target window is defined by: $z = \dfrac{n-\mu}{\sigma}$.

3.2.2. Peak Detection

Hotspots vary in size because wide regions of DNaseI hypersensitivity can span several kb. We employ a peak finding procedure to detect narrow 150-bp regions of tag enrichment within hotspots. For each hotspot, we compute a new *z* score "peak *z*" similarly as for hotspot detection, but instead using a 150-bp window within the hotspot and the hotspot as the target window and local background window, respectively. The peak z score is calculated by scanning through the hotspot with bp resolution for all possible 150-bp windows. A putative peak is defined as a 150-bp window whose peak z score is above 30. If nearby putative peaks overlap within a hotspot, the 150-bp window with the highest peak z score is selected to be the peak among the cluster of overlapping putative peaks. Therefore, a hotspot may not have any peak or may have multiple peaks (Fig. 1b).

3.2.3. Repeat Sequence Filtering

In sequencing reads, artifacts are inevitably found and typically concentrated on relatively small regions. Before calling hotspots and peaks, these artifacts are eliminated by filtering out sequence reads which overlap satellites, long interspersed repetitive elements (LINE), and short tandem repeats (STR) after extending tags into 150 bp. Repeat masking files can be downloaded from the UCSC Web site http://hgdownload.cse.ucsc.edu/downloads.html.

3.2.4. False Discovery Rate Calculations

The final z-score threshold is imposed on candidate hotspots for each data set based on false discovery rate (FDR). To calculate FDR, random sampling of uniquely mappable tags is performed to generate a random data set with the same number of tags as for the observed data set. Hotspots are called for the random data and their z scores are calculated. Then, the FDR for the observed hotspots at a given z score threshold z_0 is estimated as [# of hotspots with $z > z_0$ in the random data]/[# of hotspots with $z > z_0$ in the observed data]. Hotspots are selected by setting a z-score threshold that corresponds to the desired FDR.

3.2.5. Replicate Concordance

When two biological replicates of the data are available, a replicate concordant set of hotspots is defined as follows (see Note 2). First, for each replicate, minimally thresholded (z score > 2) hotspots are called and nearby hotspots are merged according to a distance threshold (within 150 bp, for example). The initial replicate concordant regions are defined as the intersection of these hotspots between the replicates. Then, the tags from both replicates are combined and the detection algorithm calls FDR-thresholded hotspots from the pooled tags. Finally, FDR-thresholded replicate concordant set of hotspots is defined as the intersection of the FDR-thresholded set of hotspots from the combined data and the set of replicate concordant regions.

3.3. Digital Footprinting Analysis

Unlike ChIP which identifies binding sites for a given protein, digital footprinting can detect sites bound by any factors from the same sample with a nucleotide precision (7, 10). To find footprints, one looks for narrow regions (8–30 bp) on which DNaseI cleavage is significantly reduced in comparison with that for the immediately surrounding regions. The analysis requires extremely deep sequencing to achieve reasonable coverage of cleavage events. John Stamatoyannopoulos and coworkers used a detection algorithm to identify footprints in DNaseI-seq data from *Saccharomyces cerevisiae*. The software (available at http://www.nature.com/nmeth/journal/v6/n4/extref/nmeth.1313-S2.zip) does not scale very well with larger mammalian genomes. We are currently developing computational methods that can be applied to data sets for mammalian genomes.

3.4. Annotation and Integrative/Downstream Analyses

Once the hotspots and the peaks are obtained according to the above procedures, it is often desirable to associate the sites with additional correlative information, such as the closest genes or regions found from other relevant experiments. For instance, one can examine the proportions of accessible sites located at promoters, introns, or intergenic regions. When there are several experimental conditions or treatments, the maximum tag density of hotspots can be used to perform cluster analysis or supervised classification to reveal chromatin accessibility changes across the samples (4).

Motif discovery is a downstream analysis, where one looks for putative protein-binding events within specified DHSs. There are two types of such sequence analyses. One method is simply scanning the DNA sequences for the presence of motifs for known factors. It requires prior knowledge of binding motifs, but the computation is straightforward and focused (FIMO available at http://meme.sdsc.edu/meme/fimo-intro.html; MAST at http://meme.nbcr.net/meme4_5_0/mast-intro.html) (12). The other analysis discovers novel motifs enriched in the target DNA sequences, which is computationally very intensive due to the large number of DHSs that are often taken as search input (see Note 3). For preparing input, the UCSC Table Browser can be used to extract the DNA sequences of specified genomic regions in the FASTA format, although there is a limit on the number of sites. The widely used de novo motif discovery tool, MEME (13, 14), uses an expectation maximization (EM) algorithm and reports ranked potential motifs (see Note 4). Among the motifs from the MEME output, those matching known binding motifs can be identified by motif database query tools, like TomTom (http://meme.nbcr.net/meme4_5_0/cgi-bin/tomtom.cgi) (15).

4. Software

The software package implementing the hotspot detection algorithm is available upon request (see Note 5). The software was written in C++ and compiled by GNU C++ compiler. As output, the program reports the genomic location, maximum tag density, average tag density, and statistics for each hotspot. Accompanying R programs perform some downstream analyses, such as generation of an annotation table for hotspots and rendering Venn diagrams from comparison of multiple data sets.

5. Notes

1. Our algorithms for detecting hotspots and replicate concordance are similar to the two-pass hotspot detection algorithm developed by Robert Thurman for the ENCODE project (2). The two-pass algorithm addresses the issue, where significantly enriched regions are sometimes shadowed by a neighboring extreme hotspot. After obtaining hotspots, it detects a second set of hotspots using the data from which all the tags in the first set of hotspots are deleted. Therefore, the algorithm generally detects more hotspots than our approach while comparison of the significance z scores of different-pass hotspots becomes less intuitive.
2. Analogous definitions of replicate concordance can be formulated for three or more replicates.
3. It is often worthwhile to group sites from DNaseI-seq data into more biologically defined subsets. This can be guided by additional data or available information, such as gene expression, proximity to promoters, or colocalization with factor binding or motifs. The input for MEME can be further reduced by taking the top proportion of sites and/or trimming the width of individual sites.
4. Interpretation of an MEME report requires care and proper understanding of the discovery algorithm to sort out the motifs that are likely to be meaningful. Often, the objective is to discover candidate protein factors associated with chromatin remodeling at the DHSs. However, strong signals from sequence-intrinsic motifs that disfavor nucleosome occupancy (thus, abundant in the DNaseI-treated sample) or other repetitive sequence patterns can dominate the top-ranking motifs and mask more subtle but specific motifs that arise from protein binding.
5. The software requires the mappability file of the matching genome. The mappability file for k-mers is a binary file with the value (0–255) at the nth byte representing the number of occurrences of the k-mer sequence starting from the genomic coordinate n. Our software uses the same mappability file format as in PeakSeq, and the user can generate a new file using the mappability generation program of PeakSeq (http://archive.gerstei.02nlab.org/proj/PeakSeq/Mappability_Map/Code/Mappability_Map.tar.gz).

References

1. Wiench, M., Miranda, T. B., and Hager, G. L. (2011) Control of nuclear receptor function by local chromatin structure. *FEBS J.*, **278**(13): 2211–30.
2. John, S., Sabo, P. J., Thurman, R. E., Sung, M. H, Biddie, S. C., Johnson, T. A., Hager, G. L., and Stamatoyannopoulos, J. A. (2011) Chromatin accessibility pre-determines glucocorticoid receptor binding patterns. *Nat. Genet.*, **43**, 264–268.
3. Sabo, P. J., Kuehn, M. S., Thurman, R., Johnson, B. E., Johnson, E., M, Cao, H., Yu, M., Rosenzweig, E., Goldy, J., Haydock, A., Weaver, M., Shafer, A., Lee, K., Neri, F., Humbert, R., Singer, M. A., Richmond, T. A., Dorschner, M. O., McArthur, M., Hawrylycz, M., Green, R. D., Navas, P. A., Noble, W. S., and Stamatoyannopoulos, J. A. (2006) Genome-scale mapping of DNase I sensitivity in vivo using tiling DNA microarrays. *Nat. Methods*, **3**, 511–518.
4. Siersbaek, R., Nielsen, R., John, S., Sung, M. H., Baek, S., Loft, A., Hager, G. L., and Mandrup, S. (2011) Adipogenic development is associated with extensive early remodeling of the chromatin landscape and establishment of transcription factor 'hotspots'. *EMBO J.*, **30**, 1–14.
5. Hager, G. L., Nagaich, A. K., Johnson, T. A., Walker, D. A., and John, S. (2004) Dynamics of nuclear receptor movement and transcription. *Biochim. Biophys. Acta*, **1677**, 46–51.
6. Hager, G. L., Elbi, C., Johnson, T. A., Voss, T. C., Nagaich, A. K., Schiltz, R. L., Qiu, Y., and John, S. (2006) Chromatin dynamics and the evolution of alternate promoter states. *Chromosome. Res.*, **14**, 107–116.
7. Hager, G. L. (2009) Footprints by deep sequencing. *Nat. Methods*, **6**, 254–255.
8. Crawford, G. E., Davis, S., Scacheri, P. C., Renaud, G., Halawi, M. J., Erdos, M. R., Green, R., Meltzer, P. S., Wolfsberg, T. G., and Collins, F. S. (2006) DNase-chip: a high-resolution method to identify DNase I hypersensitive sites using tiled microarrays. *Nat. Methods*, **3**, 503–509.
9. Boyle, A. P., Davis, S., Shulha, H. P., Meltzer, P., Margulies, E. H., Weng, Z., Furey, T. S., and Crawford, G. E. (2008) High-resolution mapping and characterization of open chromatin across the genome. *Cell*, **132**, 311–322.
10. Hesselberth, J. R., Zhang, Z.,, Sabo, P. J., Chen, X., Sandstrom, R., Reynolds, A. P., Thurman, R. E., Neph, S., Kuehn, M. S., Noble,W.S.,Fields,S.,andStamatoyannopoulos, J. A. (2009) Global mapping of protein-DNA interactions in vivo by digital genomic footprinting. *Nat. Methods*, **6**, 283–289.
11. Ling, G., Sugathan, A., Mazor, T., Fraenkel, E., and Waxman, D. J. (2010) Unbiased, genome-wide in vivo mapping of transcriptional regulatory elements reveals sex differences in chromatin structure associated with sex-specific liver gene expression. *Mol Cell Biol*, **30**, 5531–5544.
12. Bailey, T. L. and Gribskov, M. (1998) Combining evidence using p-values: application to sequence homology searches. *Bioinformatics.*, **14**, 48–54.
13. Bailey, T. L., Williams, N., Misleh, C., and Li, W. W. (2006) MEME: discovering and analyzing DNA and protein sequence motifs. *Nucleic Acids Res.*, **34**, W369–W373.
14. Bailey, T. L. and Elkan, C. (1994) Fitting a mixture model by expectation maximization to discover motifs in biopolymers. *Proc. Int. Conf. Intell. Syst. Mol Biol*, **2**, 28–36.
15. Gupta, S., Stamatoyannopoulos, J. A., Bailey, T. L., and Noble, W. S. (2007) Quantifying similarity between motifs. *Genome Biol*, **8**, R24.

Chapter 27

Computational Analysis of Nucleosome Positioning

Itay Tirosh

Abstract

Genome-wide patterns of nucleosome occupancy and positioning have greatly impacted on studies of chromatin structure, yet these studies require extensive computational analysis which is crucial for the quality of the resulting datasets and inferred conclusions. This chapter describes the computational steps required in order to estimate genome-wide patterns of nucleosome occupancy and positioning from raw data obtained from high-throughput sequencing of mononucleosome DNA fragments. Potential pitfalls that may be encountered in such analysis and computational quality controls are further discussed in Subheading 3.

Key words: Nucleosome positioning, Bioinformatics, High-throughput sequencing

1. Introduction

The development of microarray technology and, more recently, advances in high-throughput sequencing have revolutionized the study of chromatin structure. As described in other chapters, it is now relatively straightforward to measure the genome-wide distribution of MNase-digested DNA in practically any sequenced organism, and recent studies employing this approach have provided tremendous insight into the organization of nucleosomes, their determinants, and functional roles (1–8).

However, as for any large-scale experiment, these datasets must be correctly processed, normalized, and visualized in order to obtain a meaningful description of nucleosome positioning. These crucial steps are the subject of this chapter, which is aimed as a guide for analysis of MNase-digestion genomic data. The rapid advances and decreasing cost of high-throughput sequencing suggest that this becomes the method of choice for most future

Randall H. Morse (ed.), *Chromatin Remodeling: Methods and Protocols*, Methods in Molecular Biology, vol. 833,
DOI 10.1007/978-1-61779-477-3_27,

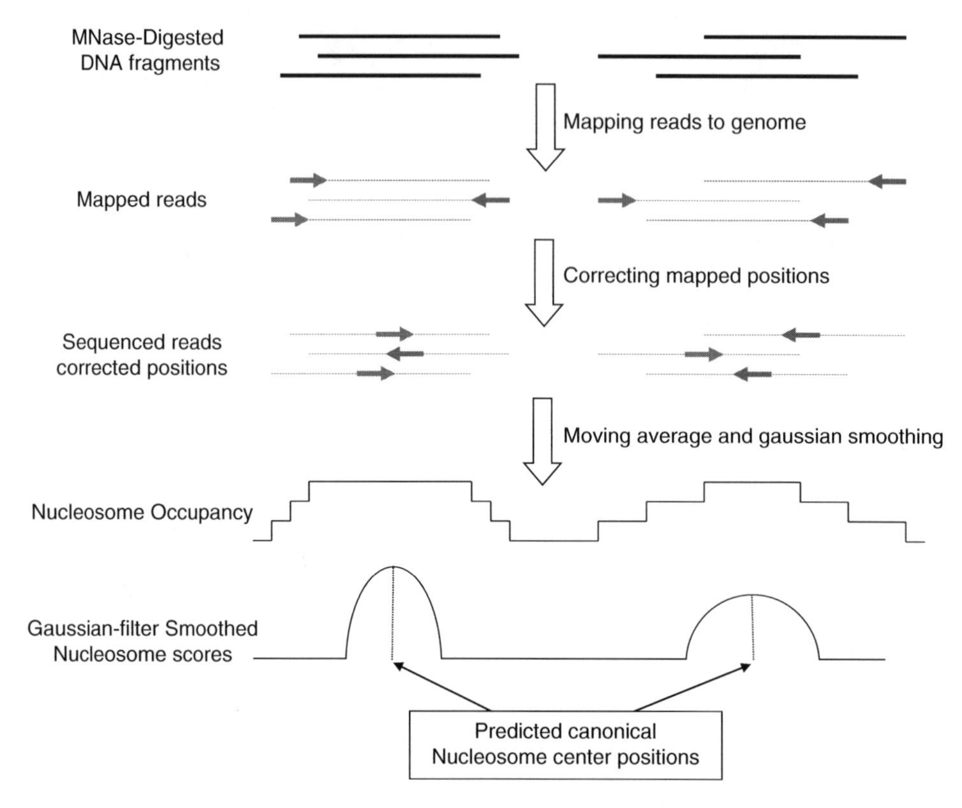

Fig. 1. Schematic illustration of analysis of high-throughput sequencing data. MNase-digested DNA fragments are subjected to short-read, high-throughput sequencing, generating reads from either end of the DNA fragment. These reads are first mapped to the relevant genome. Read positions are then corrected for the shift between the forward and backward strands by adding or subtracting half the characteristic length of a digested DNA fragment. The corrected read positions are than averaged with a sliding window of 147 bp to produce nucleosome occupancies or smoothed with Gaussian filter to produce scores whose peaks represent predicted centers of canonical nucleosome positions. The figure shows two nucleosomes, with high (*right*) and low (*left*) fuzziness.

genomic studies of nucleosome positioning. This chapter is, therefore, focused on analysis of high-throughput sequencing data, with some notes on how these methods should be modified for microarray analysis (see Note 1).

Subheading 2 describes the basic processing steps from raw sequencing data to genome-wide estimates of nucleosome occupancy and positions. The main steps include mapping the sequence reads to the genome, correcting mapped positions according to their strand, summing over the read counts to obtain nucleosome occupancy, and Gaussian smoothing of the read counts to predict nucleosome center positions (see Fig. 1). Subheading 3 includes comments on how this analysis can be modified and on how to assure the quality of the data and avoid potential pitfalls.

2. Methods

2.1. Mapping Reads to the Genome

This is the first step in any analysis of high-throughput sequencing data. Many commercial as well as publicly available programs exist for this purpose and each of them can be used to map the reads into the genome of interest. These programs include MAQ (9), RMAP (10), Bowtie (11), ELAND, and many others. The output of these mapping programs should consist of the number of reads that map to each base pair in the genome, for both DNA strands. Note that it is important to retain the strand information of each read and avoid summing the read counts of the two DNA strands together.

2.2. Estimating Characteristic DNA Fragment Length

MNase-digested DNA fragments, which reflect mononucleosomes, are typically of length 110–160 bp, but the reads that are obtained from high-throughput sequencing (e.g., with Illumina platform) are typically much shorter and reflect the ends of the DNA fragments. Thus, in the typical case of short-read sequencing platforms without paired-end sequencing, the genomic position of the entire mononucleosome DNA fragment (rather than only the sequenced reads) needs to be inferred. To this end, the standard approach is to approximate a characteristic length for all DNA fragments within a single sample and use it to extend the observed reads up to the length of a characteristic DNA fragment. This approximation, as described below, is performed by calculating a score for each possible fragment length (e.g., between 100 and 180 bp), defined as the correlation between the data of the two strands after shifting the forward strand by the estimated fragment length. The result is an estimated fragment length that obtained the highest score.

If the dataset is not paired end and includes reads that are shorter than the MNase-digested DNA fragments, then the following procedure should be repeated for each chromosome.

1. For each DNA strand, smooth the read counts with a moving average of 147 base pairs.
2. For $i = 100–180$:
   (a) Shift the smoothed read counts of the forward strand by i base pairs: $S_f'(x) = S_f(x - i)$, where S_f, S_f' are the smoothed read counts of the forward strand before and after the shift, and x are all chromosomal positions larger than i.
   (b) Calculate the correlation between the smoothed read counts of the backward strand and the (shifted) smoothed read count of the forward strand, S_f'. Note that this correlation might be skewed by few genomic positions with extremely high values, and thus it is preferable to either use Spearman Rank correlation or to identify and exclude such genomic regions (see Note 2).

3. The estimated fragment length for this chromosome is the one for which the correlation in step 2(b) was the highest.

The estimated fragment length should be highly similar for different chromosomes and the average value should be used in further analysis. The estimated fragment length is a good estimator of the degree of digestion, and may thus indicate whether the DNA was over- or underdigested, which in turn can affect patterns of nucleosome occupancy and positioning (5). At this step, it is also easy to estimate the consistency of the data among the two DNA strands, which is a highly important quality control (see Note 3).

2.3. Calculating Nucleosome Occupancy

Nucleosome occupancy is the most straightforward result of MNase-digestion datasets, and typically the subject of most analysis. The measure of nucleosome occupancy should reflect the number of reads that cover each base pair, taking into account that each read represents the 5′-end of a mononucleosome DNA fragment.

1. In order to convert the 5′-end read positions into the estimated center positions of the mononucleosome fragments, shift the mapped positions by L/2, where L is the estimated fragment length (see previous section). Positions of reads from the forward strand are, thus, increased by L/2 and positions of reads from the backward strand are decreased by L/2 (see Fig. 1). This step is required only for short-read sequencing that is not paired end.
2. Sum the converted positions over the forward and backward strands, as they both correspond to the centers of mononucleosome fragments.
3. Smooth the read-counts data with a moving average of 147 bp.

This measure of nucleosome occupancy could be modified in three main ways. First, it is often convenient to convert this absolute measure of occupancy (how many mononucleosome fragments cover each base pair) to a relative measure of nucleosome occupancy which would be more suitable for comparison among different samples. This can be done by taking the $\log_2$ of the ratio between nucleosome occupancy at each position and the genomic average of nucleosome occupancy such that zero reflects average occupancy, and plus or minus one reflect twice or half the average occupancy, respectively. Second, it is important to note that these estimates of nucleosome occupancy are confounded by MNase bias, and it may be possible to estimate or partially correct such biases (see Note 4). Finally, in most analyses, we are ultimately interested in the nucleosome patterns of genes and thus it might be convenient to convert the data from genomic coordinates into relative gene positions (gene-centric database). Notably, this conversion enables an analysis of the average gene profile, which can serve as an important initial quality control and may also highlight general trends in the data (see Note 5).

2.4. Estimating Nucleosome Center Positions

In addition to nucleosome occupancy, we are often interested in estimating the exact positions of nucleosomes. Nucleosome positions vary among different cells, which are averaged in the MNase-digested data, and thus any estimation of nucleosome positioning actually reflects the average or canonical positions. This variability, often referred to as fuzziness or positioning, is another measure that might be of interest (see Note 6).

To estimate the canonical center positions of nucleosomes, we start from the read-counts data, after these were converted from 5′-end to center positions and were summed over the two strands (steps 1 and 2 in previous section). In principle, nucleosome center positions should be those base pairs with the highest read counts. However, read counts of single bases are typically of limited accuracy due to low coverage, MNase bias, sequencing bias, and other technical effects. We, thus, calculate a score for each base that represents the likelihood that this base is the center of the canonical nucleosome position. This score is calculated by smoothing the read counts with a Gaussian filter, which incorporates the data from adjacent nucleosomes but uses nonuniform weights such that the score at each base pair is most influenced by its own read count, followed by directly adjacent bases and with decreasing contribution of more distant bases. Such scores typically display very sharp local peaks which reflect the estimated canonical positions (see Fig. 1).

1. Smooth the read-counts data with a Gaussian filter. This filter should be the probability density function of a Gaussian distribution with mean zero and standard deviation of 25. For example, this step can be implemented in Matlab (MathWorks) by Gaus = normpdf(−50:50,0,25); Smoothed = filter(Gaus,1/sum(Gaus),ReadCounts).
2. Identify local peaks in the smoothed data and sort them by their values.
3. In order of decreasing value, add each local peak to the list of estimated nucleosome center positions if its minimal distance from local peaks that are already in the list is larger than 120 bp.

As for nucleosome occupancy, this analysis is also affected by MNase bias which could be reduced (see Note 4). It is also possible to slightly modify this procedure in order to search for overlapping nucleosome positions. For example, in cases where a certain nucleosome can occupy two alternative positions that are spaced by 10–20 bp, it might be possible to identify these positions by decreasing the standard deviation of the Gaussian filter and remove the constraint on distance between nucleosomes (1).

3. Notes

1. An alternative to high-throughput sequencing is to use tiling microarrays. Microarray hybridization intensities from hybridization of MNase-digested DNA should directly reflect nucleosome occupancies and thus require only conventional normalization procedures and avoid most of steps 1–3 in Subheading 2. Prediction of nucleosome positions and fuzziness is more difficult with microarrays than with high-throughput sequencing and has frequently relied on the use of hidden Markov models (HMMs) (2, 7).
2. A technical issue that is often encountered with high-throughput sequencing data is that specific genomic positions are highly enriched with mapped reads in a manner that may bias certain analysis (e.g., correlations that are calculated over entire chromosomes). It may, thus, be beneficial to search for particular base pairs or short genomic regions with read counts considerably higher than the rest of the genome. In the context of nucleosome occupancy, such high values cannot be biologically meaningful and should thus be excluded from further analysis.
3. MNase-digested DNA is double stranded, but high-throughput sequencing generates single-stranded reads. It is, thus, expected that read counts of the two strands (after correcting for the fragment-length shift) will be highly correlated. However, we and others have noted that this assumption is often violated and that between-strand correlations over entire chromosomes may be very low and even approach zero. These correlations should naturally decrease in samples with smaller coverage, but other unknown factors appear to also decrease it, resulting in some samples having high coverage but very low correlations. Such samples produce low-quality datasets of nucleosome patterns and should thus be excluded from analysis. Thus, a simple analysis of the correlation between data of the two strands [which is calculated in step 2(b) of Subheading 2.2] provides an important quality control.
4. Although MNase preferentially digests linker DNA, it also has a sequence-specific bias that is negatively correlated with nucleosome occupancy and may thus influence estimation of nucleosome patterns (12, 13). Although this bias cannot be eliminated or precisely controlled, Albert et al. have previously proposed a simple approach that may reduce the effect of this bias on estimated nucleosome patterns (1).
5. Analysis of multiple organisms have shown that promoters are typically depleted of nucleosomes, compared with coding regions, and that nucleosomes display a similar periodicity

within coding regions of different genes. These patterns are typically evident in the average pattern of nucleosome occupancy across all genes, which are also often used to display global differences between samples of different species, conditions, or mutants (3–8). Thus, visualizing this average pattern may serve as an initial quality control and can help to identify differences between multiple samples.

6. Nucleosome fuzziness indicates that, within a certain region, nucleosomes can adopt multiple different configurations and that the observed pattern of occupancy reflects their combined profile. A simple measure for the degree of fuzziness for a given nucleosome can be obtained by the standard deviation of read locations for all reads within a small region (e.g., 100 bp) around the nucleosome's predicted center coordinate (1). For example, the right nucleosome in Fig. 1 is fuzzier than the left nucleosome.

References

1. Albert, I., Mavrich, T.N., Tomsho, L.P., Qi, J., Zanton, S.J., Schuster, S.C. and Pugh, B.F. (2007) Translational and rotational settings of H2A.Z nucleosomes across the Saccharomyces cerevisiae genome. *Nature*, **446**, 572–576.
2. Lee, W., Tillo, D., Bray, N., Morse, R.H., Davis, R.W., Hughes, T.R. and Nislow, C. (2007) A high-resolution atlas of nucleosome occupancy in yeast. *Nat Genet*, **39**, 1235–1244.
3. Shivaswamy, S., Bhinge, A., Zhao, Y., Jones, S., Hirst, M. and Iyer, V.R. (2008) Dynamic remodeling of individual nucleosomes across a eukaryotic genome in response to transcriptional perturbation. *PLoS Biol*, **6**, e65.
4. Tirosh, I., Sigal, N. and Barkai, N. (2010) Divergence of nucleosome positioning between two closely related yeast species: genetic basis and functional consequences. *Mol Syst Biol*, **6**, 365.
5. Weiner, A., Hughes, A., Yassour, M., Rando, O.J. and Friedman, N. (2010) High-resolution nucleosome mapping reveals transcription-dependent promoter packaging. *Genome Res*, **20**, 90–100.
6. Whitehouse, I., Rando, O.J., Delrow, J. and Tsukiyama, T. (2007) Chromatin remodelling at promoters suppresses antisense transcription. *Nature*, **450**, 1031–1035.
7. Yuan, G.C., Liu, Y.J., Dion, M.F., Slack, M.D., Wu, L.F., Altschuler, S.J. and Rando, O.J. (2005) Genome-scale identification of nucleosome positions in S. cerevisiae. *Science*, **309**, 626–630.
8. Tirosh, I., Sigal, N. and Barkai, N. (2010) Widespread remodeling of mid-coding sequence nucleosomes by Isw1. *Genome Biol*, **11**, R49.
9. Li, H., Ruan, J. and Durbin, R. (2008) Mapping short DNA sequencing reads and calling variants using mapping quality scores. *Genome Res*, **18**, 1851–1858.
10. Smith, A.D., Chung, W.Y., Hodges, E., Kendall, J., Hannon, G., Hicks, J., Xuan, Z. and Zhang, M.Q. (2009) Updates to the RMAP short-read mapping software. *Bioinformatics*, **25**, 2841–2842.
11. Langmead, B., Trapnell, C., Pop, M. and Salzberg, S.L. (2009) Ultrafast and memory-efficient alignment of short DNA sequences to the human genome. *Genome Biol*, **10**, R25.
12. Fan, X., Moqtaderi, Z., Jin, Y., Zhang, Y., Liu, X.S. and Struhl, K. (2010) Nucleosome depletion at yeast terminators is not intrinsic and can occur by a transcriptional mechanism linked to 3′-end formation. *Proc Natl Acad Sci USA*, **107**, 17945–17950.
13. Horz, W. and Altenburger, W. (1981) Sequence specific cleavage of DNA by micrococcal nuclease. *Nucleic Acids Res*, **9**, 2643–2658.

INDEX

Randall H. Morse (ed.), *Chromatin Remodeling: Methods and Protocols*, Methods in Molecular Biology, vol. 833, DOI 10.1007/978-1-61779-477-3, © Springer Science+Business Media, LLC 2012

T

V

X

Y

Z